Protein Design and the Development of New Therapeutics and Vaccines

NEW HORIZONS IN THERAPEUTICS
Smith Kline & French Laboratories Research Symposia Series

Series Editors: Jerry B. Hook and George Poste
SmithKline Beecham Pharmaceuticals, Philadelphia, Pennsylvania

---

Protein Design and the Development of New Therapeutics and Vaccines

Edited by

JERRY B. HOOK *and*
GEORGE POSTE

SmithKline Beecham Pharmaceuticals
Philadelphia, Pennsylvania

Technical Editor
JUDY SCHATZ

PLENUM PRESS • NEW YORK AND LONDON

Library of Congress Cataloging-in-Publication Data

Protein design and the development of new therapeutics and vaccines /
   edited by Jerry B. Hook and George Poste.
      p.   cm. -- (New horizons in therapeutics)
   Based on a symposium held in Nov. 1988 in King of Prussia, Penn.,
sponsored by Smith, Kline, & French.
   Includes bibliographical references.
   Includes index.
   ISBN-13: 978-1-4684-5741-4   e-ISBN-13: 978-1-4684-5739-1
   DOI: 10.1007/978-1-4684-5739-1
   1. Protein engineering--Congresses.   2. Drugs--Design--Congresses.
3. Vaccines--Congresses.   I. Hook, Jerry B.   II. Poste, George.
III. Smith, Kline, & French.   IV. Series.
   [DNLM: 1. Protein Engineering--congresses.   2. Proteins--analysis-
-congresses.   3. Proteins--therapeutic use--congresses.   QU 55
P966233]
RS431.P75P76   1990
615'.19--dc20
DNLM/DLC
for Library of Congress   90-7444
                                                               CIP

© 1990 Plenum Press, New York

Softcover reprint of the hardcover 1st edition 1990

A Division of Plenum Publishing Corporation
233 Spring Street, New York, N.Y. 10013

Contributors

Steven C. Almo, Department of Chemistry, Massachusetts Institute of Technology, Cambridge, Massachusetts 02139

Robert L. Baldwin, Department of Biochemistry, Beckman Center, Stanford University Medical Center, Stanford, California 94305

Stephen J. Benkovic, Department of Chemistry, Pennsylvania State University, University Park, Pennsylvania 16802

Klaus Biemann, Department of Chemistry, Massachusetts Institute of Technology, Cambridge, Massachusetts 02139

Lutz Birnbaumer, Departments of Cell Biology and Molecular Physiology and Biophysics, Baylor College of Medicine, Houston, Texas 77030

Tom L. Blundell, Laboratory of Molecular Biology, Department of Crystallography, Birkbeck College, London University, London WC1E 7HX, England

Arthur M. Brown, Department of Molecular Physiology and Biophysics, Baylor College of Medicine, Houston, Texas 77030

Frank Brown, SmithKline Beecham Pharmaceuticals, King of Prussia, Pennsylvania 19406

Tauseef R. Butt, SmithKline Beecham Pharmaceuticals, King of Prussia, Pennsylvania 19406

Brian F. C. Clark, Division of Biostructural Chemistry, Department of Chemistry, Aarhus University, 8000 Aarhus C, Denmark

Juan Codina, Department of Cell Biology, Baylor College of Medicine, Houston, Texas 77030

M. Courtney, Transgene S.A., 67082 Strasbourg Cedex, France;   *present address:* Delta Biotechnology, Nottingham NG7 AFD, United Kingdom

Stanley T. Crooke, Isis Pharmaceuticals, Carlsbad Research Center, Carlsbad, California 92008

E. Degryse, Transgene S.A., 67082 Strasbourg Cedex, France

Scott Dixon, SmithKline Beecham Pharmaceuticals, King of Prussia, Pennsylvania 19406

David J. Ecker, Isis Pharmaceuticals, Carlsbad Research Center, Carlsbad, California 92008

Ben Fane, Department of Biology, Massachusetts Institute of Technology, Cambridge, Massachusetts 02139

Gregory K. Farber, Department of Chemistry, Massachusetts Institute of Technology, Cambridge, Massachusetts 02139

Kenji Fujimori, Theoretical Immunology Section, Laboratory of Mathematical Biology, DCBD, National Cancer Institute, National Institutes of Health, Bethesda, Maryland 20892

Kathleen L. Gould, Molecular Biology and Virology Laboratory, The Salk Institute, San Diego, California 92138

Cameron Haase-Pettingell, Department of Biology, Massachusetts Institute of Technology, Cambridge, Massachusetts 02139

Janos Hajdu, Laboratory of Molecular Biophysics, University of Oxford, Oxford OX1 3QU, England

John D. Hildebrandt, Worcester Foundation for Experimental Biology, Shrewsburg, Massachusetts 01545

P. Lynne Howell, Department of Chemistry, Massachusetts Institute of Technology, Cambridge, Massachusetts 02139

Tony Hunter, Molecular Biology and Virology Laboratory, The Salk Institute, San Diego, California 92138

Ravi Iyengar, Department of Pharmacology, Mount Sinai School of Medicine, New York, New York 10029

S. Jallat, Transgene S.A., 67082 Strasbourg Cedex, France

Michael Jensen, Division of Biostructural Chemistry, Department of Chemistry, Aarhus University, 8000 Aarhus C, Denmark

Mark S. Johnson, Laboratory of Molecular Biology, Department of Crystallography, Birkbeck College, London University, London WC1E 7HX, England

Sobhanaditya Jonnalagadda, SmithKline Beecham Pharmaceuticals, King of Prussia, Pennsylvania 19406

M. P. Kieny, Transgene S.A., 67082 Strasbourg Cedex, France

Jonathan King, Department of Biology, Massachusetts Institute of Technology, Cambridge, Massachusetts 02139

Morten Kjeldgaard, Division of Biostructural Chemistry, Department of Chemistry, Aarhus University, 8000 Aarhus C, Denmark

Peter Kollman, Department of Pharmaceutical Chemistry, School of Pharmacy, University of California at San Francisco, San Francisco, California 94143

Lee F. Kuyper, Wellcome Research Laboratories, Research Triangle Park, North Carolina 27709

J. P. Lecocq, Transgene S.A., 67082 Strasbourg Cedex, France

Richard A. Lindberg, Molecular Biology and Virology Laboratory, The Salk Institute, San Diego, California 92138; and Department of Biology, University of California at San Diego, La Jolla, California 92093

Jan Markussen, Novo Research Institute, Novo Allé, DK-2880 Bagsvaerd, Denmark

Jill Meisenhelder, Molecular Biology and Virology Laboratory, The Salk Institute, San Diego, California 92138

P. Meulien, Transgenc S.A., 67082 Strasbourg Cedex, France

David S. Middlemas, Molecular Biology and Virology Laboratory, The Salk Institute, San Diego, California 92138

Christopher K. Mirabelli, Isis Pharmaceuticals, Carlsbad Research Center, Carlsbad, California 92008

Anna Mitraki, Department of Biology, Massachusetts Institute of Technology, Cambridge, Massachusetts 02139

Brett P. Monia, Department of Pharmacology, University of Pennsylvania, Philadelphia, Pennsylvania 19104; and Isis Pharmaceuticals, Carlsbad Research Center, Carlsbad, California 92008

Luciano Mueller, SmithKline Beecham Pharmaceuticals, King of Prussia, Pennsylvania 19406

Michael G. Mulkerrin, Biomolecular Chemistry Department, Genentech, Inc., South San Francisco, California 94080

Eva J. Neer, Department of Medicine, Brigham and Women's Hospital and Harvard Medical School, Boston, Massachusetts 02115

Robin E. Offord, Department of Medical Biochemistry, University of Geneva, Geneva 4, Switzerland

John P. Overington, Laboratory of Molecular Biology, Department of Crystallography, Birkbeck College, London University, London WC1E 7HX, England

A. Pavirani, Transgene S.A., 67082 Strasbourg Cedex, France

L. Jeanne Perry, Biomolecular Chemistry Department, Genentech, Inc., South San Francisco, California 94080; *present address:* Molecular Biology Institute, University of California at Los Angeles, Los Angeles, California 90024

Gregory A. Petsko, Department of Chemistry, Massachusetts Institute of Technology, Cambridge, Massachusetts 02139

Scott J. Pollack, Department of Chemistry, University of California, Berkeley, California 94720

L. Michael Randall, Biomolecular Chemistry Department, Genentech, Inc., South San Francisco, California 94080

Dagmar Ringe, Department of Chemistry, Massachusetts Institute of Technology, Cambridge, Massachusetts 02139

Andrej Šali, J. Stefan Institute, Ljublyana, Yugoslavia;   *present address:* Laboratory of Molecular Biology, Department of Crystallography, Birkbeck College, London University, London WC1E 7HX, England

Peter G. Schultz, Department of Chemistry, University of California, Berkeley, California 94720

Barry Stoddard, Department of Chemistry, Massachusetts Institute of Technology, Cambridge, Massachusetts 02139

Søren Thirup, Division of Biostructural Chemistry, Department of Chemistry, Aarhus University, 8000 Aarhus C, Denmark

David P. Thompson, Molecular Biology and Virology Laboratory, The Salk Institute, San Diego, California 92138; and Department of Biology, University of California at San Diego, La Jolla, California 92093

E. Tomlinson, Advanced Drug Delivery Research, Ciba-Geigy Pharmaceuticals, Horsham, West Sussex RH12 4AB, England;   *present address:* Somatix Corporation, Cambridge, Massachusetts 02139

Robert Villafane, Department of Biology, Massachusetts Institute of Technology, Cambridge, Massachusetts 02139

Carston R. Wagner, Department of Chemistry, Pennsylvania State University, University Park, Pennsylvania 16802

Paul Weber, SmithKline Beecham Pharmaceuticals, King of Prussia, Pennsylvania 19406

John N. Weinstein, Theoretical Immunology Section, Laboratory of Mathematical Biology, DCBD, National Cancer Institute, National Institutes of Health, Bethesda, Maryland 20892

Ronald Wetzel, Biomolecular Chemistry Department, Genentech, Inc., South

San Francisco, California 94080;   *present address:* Macromolecular Sciences
Department, SmithKline Beecham Pharmaceuticals, King of Prussia, Pennsylvania 19406

Don C. Wiley, Department of Biochemistry and Molecular Biology and Howard
Hughes Medical Institute, Harvard University, Cambridge, Massachusetts 02138

Atsuko Yatani, Department of Molecular Physiology and Biophysics, Baylor
College of Medicine, Houston, Texas 77030

Preface to the Series

The unprecedented scope and pace of discovery in modern biology and clinical medicine present remarkable opportunities for the development of new therapeutic modalities, many of which would have been unimaginable even a few years ago. This situation reflects the unprecedented progress being made not only in disciplines such as pharmacology, physiology, organic chemistry, and biochemistry that have traditionally made important contributions to drug discovery, but also in new disciplines such as molecular genetics, cell biology, and immunology that are now of sufficient maturity to contribute significantly to our understanding of the pathogenesis of disease and to the development of novel therapies. Contemporary biomedical research, embracing the entire spectrum of biological organization from the molecular level to whole body function, is on the threshold of an era in which biological processes, including disease, can be analyzed in increasingly precise and mechanistic terms. The transformation of biology from a largely descriptive, phenomenological discipline to one in which the regulatory principles underlying biological organization can be understood and manipulated with ever-increasing predictability brings an entirely new dimension to the study of disease and the search for effective therapeutic modalities. In undergoing this transformation into an increasingly mechanistic discipline, biology and medicine are following the course already charted by the sister disciplines of chemistry and physics, albeit still far behind.

The consequences of these changes for biomedical research are profound: new concepts; new and increasingly powerful analytical techniques; new advances generated at a seemingly ever-rapid pace; an almost unmanageable glut of information dispersed in an increasing number of books and journals; and the task of integrating this information into a realistic experimental framework. Nowhere is the challenge more pronounced than in the pharmaceutical industry. Drug discovery and development have always required the successful coordination of multiple scientific disciplines. The need to assimilate more and more disciplines within the drug discovery process, the extraordinary pace of discovery in all disciplines, and the growing scientific and organizational complexity of coordinating increasingly ultraspecialized and resource-intensive scientific skills in an ever-enlarging framework of collaborative research activities represent formidable challenges for the pharmaceutical industry. These demands are balanced, however, by the excitement and the scale of the potential opportunities for achieving dramatic improvements in health care and the quality of human life over the next 20 years via the development of novel therapeutic modalities for effective treatment of major human and animal diseases.

It is against this background of change and opportunity that the present symposium series, *New Horizons in Therapeutics,* was conceived as a forum for providing critical and up-to-date surveys of important topics in biomedical research in which significant advances were occurring and which offer new approaches to the therapy of disease. Each volume will contain authoritative and topical articles written by investigators who have contributed significantly to their respective research fields. While individual articles will discuss specialized topics, all papers in a single volume will be related to a common theme. The level will be advanced, directed primarily to the needs of the active research investigator and graduate students.

Editorial policy will be to impose as few restrictions as possible on contributors. This is appropriate since each volume is limited to the papers presented at the symposium and no attempt will be made to create a definitive monograph dealing with all aspects of the selected subject. Although each symposium volume will provide a survey of recent research accomplishments, emphasis will also be given to the examination of controversial and conflicting issues, to the presentation of new ideas and hypotheses, to the identification of important unsolved questions, and to future directions and possible approaches by which such questions might be answered.

The range of topics for future volumes in the symposium series will be broad and will embrace the full repertoire of scientific disciplines that contribute to modern drug discovery and development. We thus look forward to the publication of what we hope will be viewed as a worthy series of volumes that reflect the excitement and challenge of contemporary biomedical research in defining new horizons in therapeutics.

Jerry B. Hook
George Poste

Philadelphia

Contents

II. PROTEIN FOLDING AND STABILITY

III. PRINCIPLES OF RECEPTOR DESIGN AND REGULATION

Chapter 6

Protein–Tyrosine Kinases and Their Substrates:
Old Friends and New Faces 119

Tony Hunter, Kathleen L. Gould, Richard A. Lindberg,
Jill Meisenhelder, David S. Middlemas, and David P. Thompson

IV. THE GUANINE NUCLEOTIDE BINDING PROTEIN FAMILY

Chapter 7

G Proteins: A Family of Signal-Transducing Molecules 143

Eva J. Neer

VI. PROTEIN ENGINEERING AND ENZYME DESIGN

Chapter 13

Robin E. Offord

Chapter 14

*J. P. Lecocq, M. Courtney, E. Degryse, S. Jallat, M. P. Kieny,
P. Meulien, and A. Pavirani*

Chapter 17

John N. Weinstein and Kenji Fujimori

Chapter 18

Scott J. Pollack and Peter G. Schultz

IX. DESIGN OF NOVEL ANTIVIRAL AGENTS AND VACCINES

Chapter 20

Studies on the Structure and Function of Ubiquitin ....

Stanley T. Crooke, Christopher K. Mirabelli, David J. Ecker,
Tauseef R. Butt, Sobhanaditya Jonnalagadda, Scott Dixon,
Luciano Mueller, Frank Brown, Paul Weber, and Brett P. Monia

Chapter 21

Recognition at Membrane Surfaces: Influenza HA and Human HLA ..

Don C. Wiley

I

ADVANCES IN THE STRUCTURAL ANALYSIS OF PROTEINS

Dynamic Processes in Proteins
by X-Ray Diffraction

DAGMAR RINGE, STEVEN C. ALMO, GREGORY K. FARBER,
JANOS HAJDU, P. LYNNE HOWELL, GREGORY A. PETSKO,
and BARRY STODDARD

1. Dynamic Processes in Proteins

Proteins are dynamic systems. Their dynamic properties can be divided into three broad classes: individual atomic fluctuations, collective motions of bonded and nonbonded neighbouring atoms, and ligand-induced conformational changes (Ringe and Petsko, 1985). The first two classes represent small-amplitude excursions around the equilibrium conformation of a protein; triggered conformational changes lead to the formation of a new average structure. Although the time-scale of individual and collective fluctuations is relatively short (10^{-13} to 10^{-9} sec), they can be studied by a variety of spectroscopic techniques and can be simulated computationally by molecular dynamics calculations (Karplus and McCammon, 1983). It has even proved possible to map the spatial distributions of these motions by X-ray crystallography, because they produce a spreading of the electron density around each atom, which may be modeled by various distribution functions (Petsko and Ringe, 1984; Ringe and Petsko, 1985). Triggered conformational changes have proven much more difficult to study in detail. Their time-scales are too long (10^{-6} to 10^1 sec) for simulation by simple molecular dynamics techniques. Moreover, since they produce a change in the equilibrium conformation of the protein, they involve crossing relatively large potential ener-

DAGMAR RINGE, STEVEN C. ALMO, GREGORY K. FARBER, P. LYNNE HOWELL, GREGORY A. PETSKO, and BARRY STODDARD • Department of Chemistry, Massachusetts Institute of Technology, Cambridge, Massachusetts 02139.   JANOS HAJDU • Laboratory of Molecular Biophysics, University of Oxford, Oxford OX1 3QU, England.

gy barriers. Theoretical methods for simulating barrier crossings are only just being developed, and they require detailed knowledge of both the initial and final states of the molecule, as well as of any intermediate structures that have a lifetime longer than that of a single atomic vibration (10^{-15} sec). Such information has been very hard to obtain. Spectroscopic methods can demonstrate the existence of a ligand-induced conformational change, but the information content of most of these tools is low (the exception is nuclear magnetic resonance).

X-ray crystallography would seem to be the ideal method for examining triggered conformational changes: its resolving power is at the level of individual atomic positions, and it can provide a complete description of the protein structure in the absence and presence of the ligand. Indeed, in some cases, notably allosteric proteins, X-ray crystallography has proven to be very powerful for dissecting the details of the process (Perutz, 1970; Kantrowitz and Lipscomb, 1988). However, such cases are special: allosteric proteins switch between two conformational states, T and R, in response to the binding of an effector ligand that is not altered chemically after it binds. The two states are stable indefinitely. *Substrate*-induced conformational changes in enzymes are metastable; they cannot last longer than the time required for one complete catalytic event and may have even shorter lifetimes since the ligand is being transformed chemically into something else.

Ultimately, the mechanism of an enzyme-catalyzed reaction must be understood in terms of conformational changes. The mechanism is often understood well in chemical terms, because the nonenzymatic reaction can be studied in solution. Mechanistic enzymologists often describe the enzyme-catalyzed reaction in the same way as the reaction in solution, as though the participating groups were free in solution, and only bound to the protein for convenience. However, an enzyme not only provides the necessary reacting groups, but in most cases positions them properly for optimum reactivity. In this way, the entropic component of a transformation becomes favorable. For many proteins this requires conformational changes when a substrate binds or during the course of a reaction in response to changing geometry in the substrate. Thus, defining the machinery of catalysis is as important as defining the chemistry of a reaction. In fact, it may well be that the absence of such machinery is the reason that enzyme mimics having the correct chemistry do not work as effectively as the real thing. So the question arises: how can conformational changes in an enzyme be studied while the enzyme is reacting?

2. Enzyme Substrate Complexes by X-Ray Diffraction

It is possible to imagine looking at enzyme-substrate complexes by X-ray crystallography because most enzymes are active in the crystalline state (Makinen and Fink, 1977) and because protein crystals contain large, solvent-filled channels into which substrates can be diffused (Fig. 1). Protein crystals are typically 50–70% solvent by weight, which compares favorably with the 80% by weight of water found in living cells. Therefore, one could diffuse substrate into an enzyme crystal and, if the active site faced a solvent-filled channel, expect it to bind to the crystalline enzyme, induce any required conformational changes, and then be converted to product, all in a matter of minutes or less. However, that time-scale presents a huge problem.

Most enzymes have turnover numbers of 1–1000/sec. Consequently, it is necessary to have a structural method that can give information on a millisecond-to-second time-scale. However, the structures observed by protein crystallography are a double average: over all molecules in the crystal and over the time required for collection of diffraction data. Even with high-intensity sources and area detectors to measure many reflections simultaneously, data collection for protein crystals takes hours or days. Averaged over such long periods, the occupancy of any enzyme-substrate complex will be negligible. The time-average nature of the diffraction experiment prohibits determination of structures that have lifetimes less than the data collection time.

A further complication is introduced by spatial averaging over all molecules

Figure 1. The crystal lattice of tetragonal phosphorylase b crystals viewed down the c axis. The protein is shown as an α-carbon trace. One unit cell is marked in the center of the figure. Glucose-1-phosphate (Glc-1-P), a substrate of the enzyme, is drawn in one of the solvent-filled channels which run the whole length of the crystal. Adapted from a drawing by P.J. McLaughlin.

Glc-1-P

in the lattice. It would take several minutes for substrate diffusion and reaction, even though the catalytic event might only take a second or less. The difference between the total time and the reaction time is the diffusion time. Substrate diffusion is hindered by drag effects due to the smaller diameter of the protein-lined solvent channels (Fink and Petsko, 1981). Consequently, the diffusion time is usually long compared with the turnover time for the enzyme. As substrate diffuses into the crystal, a wave of product will follow it. At any given instant until diffusion is complete, the crystal will be inhomogeneous and the average structure will be a mess. Once diffusion is complete, so is the reaction, and it is too late.

For these reasons, protein crystallographers have, in general, been unable to look at productive enzyme–substrate (E–S) complexes. They have had to resort to the diffusion of inhibitors into the crystals on the assumption that the resulting enzyme–inhibitor (E–I) complex will be a good model for the unobservable Michaelis complex with substrate.

A considerable risk is associated with this assumption. An inhibitor by definition cannot display all of the interactions exhibited by the true substrate. If a particular interaction present with substrate but absent with inhibitor is respon-sible for a conformational change, that change may only be observed with the true substrate. Conformational changes that occur during the reaction as various intermediates are formed respond to—and may induce—chemical changes in the ligand and will not be observed if one studies the binding of an inert molecule.

Historically, there have been two methods for examining productive E–S complexes crystallographically. One approach involves cooling the enzyme crystal to subzero temperatures in a cryoprotective mother liquor that remains fluid even when cooled (Fink and Petsko, 1981; Douzou and Petsko, 1984). Substrate is then diffused into the crystal at low temperature. Most enzyme reactions have activation enthalpies between $+10$ and $+20$ kcal/mole, so a reduction in temperature from 20°C to -100°C will reduce the reaction rate by over a millionfold. A reaction that takes 1 sec at room temperature may thus take at least a week at a sufficiently low temperature, which is ample time for a complete X-ray data set to be recorded. The cryoprotectant must have a low viscosity so that diffusion will be fast compared with the days required for catalysis. The Michaelis complex will accumulate in the crystal and remain stable long enough for direct observation. Although this method has had some success (Alber et al., 1976), it is not a general solution to the problem. Low-temperature X-ray equipment is expensive to operate and must be tended care-fully. Suitable cryoprotectant mother liquors are hard to find for many protein crystals, especially those grown from high salts.

Enzymes that catalyze single-substrate/single-product equilibration (iso-merases, mutases, epimerases, racemases) offer a simple solution to the problem of E–S complex structure determination. One merely ignores the diffusion and reaction times and flows substrate into the crystal at a concentration in excess of

K_m. Substrate will bind to the active site and be converted to product, but product is just the substrate for the back reaction. So the system settles to equilibrium, and as long as a high substrate concentration is maintained, the thermodynamically favored species (E–S, E–P, or some intermediate, whichever has the lowest free energy) will predominate in the crystal. This structure can be observed by ordinary X-ray methods at room temperature. If the internal equilibrium constant (i.e., the equilibrium constant on the enzyme) is near 1, the structure will be an average of E–S and E–P and may be difficult to deconvolute, but even a K_{eq} of 3 or 0.3 will be far enough from unity to yield a single structure, since the noise level in a typical electron density map is on the order of 25%. Any species with a time-averaged occupancy of ¼ or less will not be observable in the X-ray structure.

Triosephosphate isomerase catalyses the interconversion of D-glyceralde-hyde-3-phosphate (D-GAP) and dihydroxyacetone phosphate (DHAP). Albery, Knowles, and their co-workers have determined the free-energy profile for this reaction (Albery and Knowles, 1976). The enzyme–DHAP complex predominates at equilibrium. Alber *et al.* (1988) determined the crystal structure of this complex by flowing a high concentration of DHAP into native enzyme crystals. Substrate binding induces a large conformational change in the enzyme: a loop of 10 amino acids moves over 8 Å to fold down onto the substrate and seal off the active site from contact with bulk solvent. A number of phosphate- and sulfate-derived molecules are known to inhibit triose phosphate isomerase. The crystal structures of many of them complexed to the enzyme have been determined; some induce the conformational change, but others do not. This study points up the importance of using the physiological substrate if one wishes to observe productive, ligand-induced protein dynamics.

In sum, the conventional crystallographic approach of looking at dead-end complexes is really not sufficient to characterize conformational changes. By definition, such complexes lack some of the essential features that may trigger the movement of the protein. Even methods that can look at stable intermediates or E–S or E–P complexes are not sufficient, since the important movements in a protein may occur in transit between the initial binding event and the formation of the final complex. What is needed is a way of viewing the catalytic process as it happens: this requires a crystallographic method that allows data collection at a rate faster than the turnover rate of a reaction (Ringe, 1987; Hajdu *et al.,* 1987b, 1988). Such a method is now available: it is called Laue diffraction (Moffat *et al.,* 1984).

3. Laue Diffraction

Laue diffraction is the use of a beam of white radiation to record the X-ray scattering from a crystal. The method is 76 years old: the first X-ray diffraction pattern ever observed, by Friedrich, Knipping, and von Laue in 1912, was obtained with polychromatic radiation. Yet, Laue diffraction was not used to collect protein crystal data, or even data from crystals of small molecules, for over 50 years. Monochromatic X-rays have long been the preferred incident radiation in crystallography, because diffraction geometry is simple when the primary beam has only a single wavelength, and because derivation of the structure factors from the measured reflection intensities is straightforward.

However, as stated earlier, conventional data collection with monochromatic sources is quite time-consuming. One reason is that, for crystals of protein and nucleic acids, there are a lot of data to collect. The number of reflections that must be measured is a function of the molecular weight of the macromolecule and of the resolution desired for the structure. The resolution dependence is particularly severe, since it is a cubic function. There are eight times as many measurements for a 1 Å resolution structure as are required to see the molecule at only 2 Å resolution. A protein of mol. wt. 50,000, for example, may need 17,500 measurements to 2.5 Å resolution, 35,000 measurements to 2 Å resolution, and 83,000 measurements to 1.5 Å (or atomic) resolution. If these were collected one at a time, on a single-counter diffractometer, no more than 4000 reflections a day could be measured, leading to a data collection time of 3 weeks in the 1.5 Å case. Using X-ray film as a detector can reduce this time considerably, but not enough. The reason is that only a limited number of reflections can be observed at one time for any given orientation of the crystal. To see why that is so, we need to consider the conditions for diffraction in more detail.

When monochromatic X-rays of wavelength λ fall onto a single crystal, they will be scattered in particular directions. Only those scattered waves that obey Bragg's law

$$n\lambda = 2d \sin \Theta$$

will be produced for any particular orientation of the crystal with respect to the incident beam. Bragg's law states that the crystal can be considered to consist of an array of planes of atoms, and reflections will only be produced at an angle Θ with respect to the source planes with spacing d that also make an angle Θ with respect to the direction of the source.

There is an easier way to illustrate this principle geometrically. The set of all possible reflections lies on a regular three-dimensional lattice (Fig. 2). A sphere of radius $1/\lambda$ drawn tangent to the incident beam direction at the origin of the lattice will only intersect a few of the reflection points. These are the reflections that are observed at that particular orientation of the crystal. Bragg's law can thus be restated: a reflection is produced whenever a reflection lattice point lies on the surface of the so-called *sphere of reflection* (also called *Ewald sphere*, after the inventor of this construction).

Immediately it can be seen that only a few reflections satisfy the reflecting

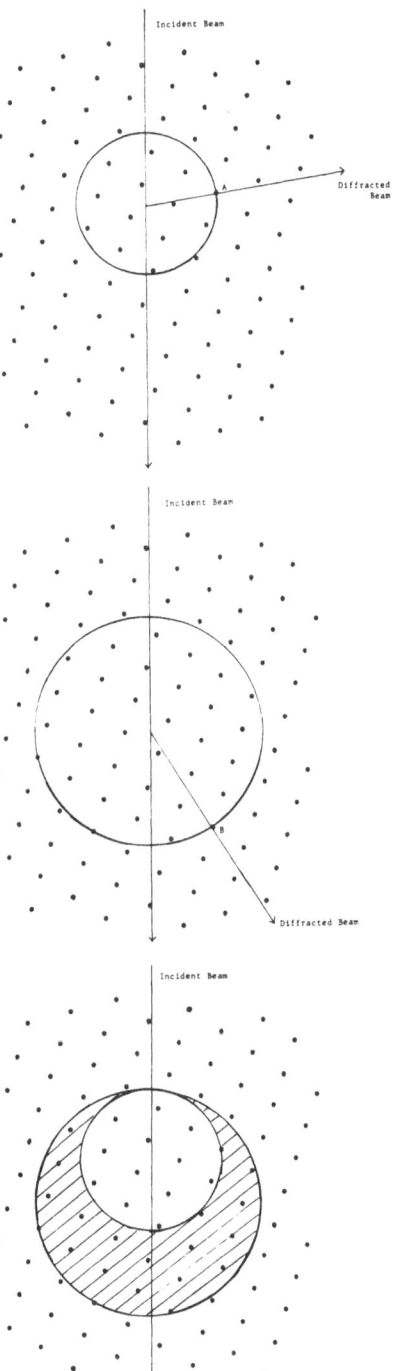

Figure 2. Condition for diffraction in terms of the Ewald construction (in two dimensions for simplicity). For a particular orientation of the crystal, a reflection is observed if it lies on the circle defined by the wavelength of the incident beam. Thus, reflection A lies on the circle defined by $1/\lambda_{max}$ and is observed if the incident beam has wavelength λ_{max}. Reflection B lies on the circle defined by $1/\lambda_{min}$ and is observed if the incident beam has wavelength λ_{min}. With a polychromatic incident beam, there is such a circle for every $\lambda_{min} < \lambda < \lambda_{max}$ and all reflections between the limits defined by $1/\lambda_{min}$ and $1/\lambda_{max}$ are observed with one orientation of the crystal (hatched area). The area between these limits becomes larger if λ_{min} and λ_{max} are further apart. Therefore, if the wavelength spread is large enough, greater than 90% of the possible reflections can be observed at once.

condition. More can be observed if the crystal is rotated or oscillated during data collection, since that will cause the lattice to move with respect to the sphere, but the requirement that a reflection remain on the surface of the sphere long enough for a statistically significant measurement to be made of its full integrated intensity necessitates that rates and ranges of oscillation be small. This is why conventional data recording on X-ray film, and data collection by area detectors (which are just electronic analogs of a sheet of X-ray film), is so time-consuming. The crystal is rotated or oscillated slowly through a small angular range (say, 1 or 2°) and the intensities of those reflections that cut the Ewald sphere are measured

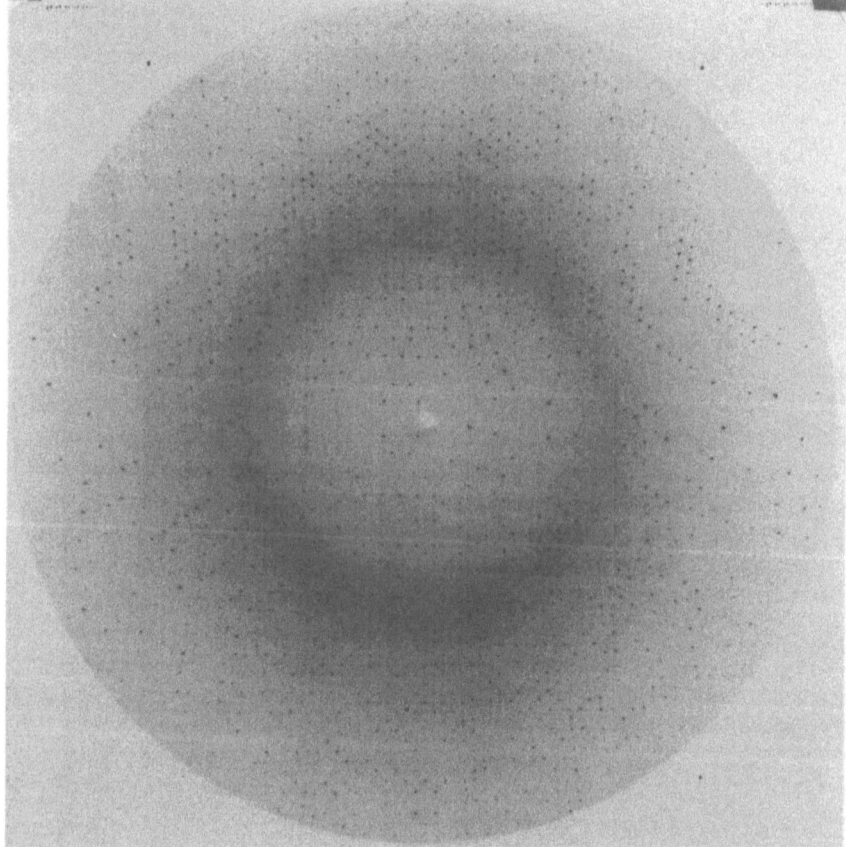

Figure 3. Oscillation photograph (oscillation angle 2°) of a single crystal of xylose isomerase. The photograph is the fourth film in a four-film pack, and is photograph number 4 in a set of 47. The spots on the edge of the film are at about 2 Å resolution. The photograph was recorded on the synchrotron at Brookhaven, using monochromatic X rays of 1 Å wavelength, and is part of a complete set of high-resolution native data that were collected in 8 hr. Less than 3% of the total unique data are obtained on this film.

(Fig. 3). The orientation of the crystal is then changed by a rotation comparable to the rocking range, and another oscillation is performed, allowing a new set of reflections to intersect the sphere and a new film or frame of data to be recorded. Sixty or more different orientations might be required to allow all reflections to be observed. If each position requires 2 hr of exposure for good measurement statistics, it will take 5 days to collect the data.

The entire process can be speeded up considerably by the use of an ultra-high-intensity X-ray source, synchrotron radiation. Particle accelerators produce a broad spectrum of intense radiation, called synchrotron radiation, when they accelerate elementary particles such as electrons to high energies. The X-rays output from a circular accelerator such as the Daresbury Synchrotron at Daresbury, England, or the National Synchrotron Light Source at Brookhaven, New York, are several orders of magnitude more intense than the beam from the most powerful laboratory generator. It is easy to monochromatize a synchrotron beam, so this very intense source can be used for photographic data collection by the rotation/oscillation method. Synchrotron radiation reduces the time required for a photograph to about 30 sec. Unfortunately, it still requires several dozen photographs for a complete data set, so the total collection time is still at least 20 min. Although this method was useful for following the reaction of phosphate with heptenitol to make heptulose-2-phosphate on the enzyme phosphorylase (because heptenitol is a slow substrate for this enzyme, taking hours to react while a complete set of data could be collected in less than 1 hr on the synchrotron), there are not many crystalline enzymes where such slow reactions exist (Hajdu *et al.*, 1987a). What is needed is a method for collecting data on a subminute time-scale.

The Ewald diagram in Fig. 2 shows how this can be done. Suppose a much different wavelength had been used. A different sphere would have to be drawn, one of larger radius if the new wavelength were smaller than the first, or of smaller radius if the new incident beam were of longer wavelength. This, of course, does not solve the problem. Again, only those reflections that lie on the surface of the sphere will be observed. They will be a different set of reflections, but still a small number. However, what if, instead of using one wavelength or the other, one used a continuum of wavelengths, a beam of white radiation between the lower and the higher value? Then *all* reflections between the two spheres (all the reflections in the shaded area in Fig. 2) will be observed at once. Each reflection is produced by a particular wavelength, and if the inner and outer sphere cutoffs are far enough apart, nearly all of the data can be observed in a single, stationary crystal orientation. The use of nonmonochromatic radiation to excite a larger number of reflections simultaneously is called *Laue diffraction.*

Figure 4 shows a Laue diffraction photograph of the enzyme xylose isomerase. It was taken for 1 sec using a beam of white radiation (wavelength range 0.2–2.1 Å) on the Daresbury Synchrotron. A Laue photograph looks completely different from the monochromatic rotation photograph shown in Fig. 3. There are

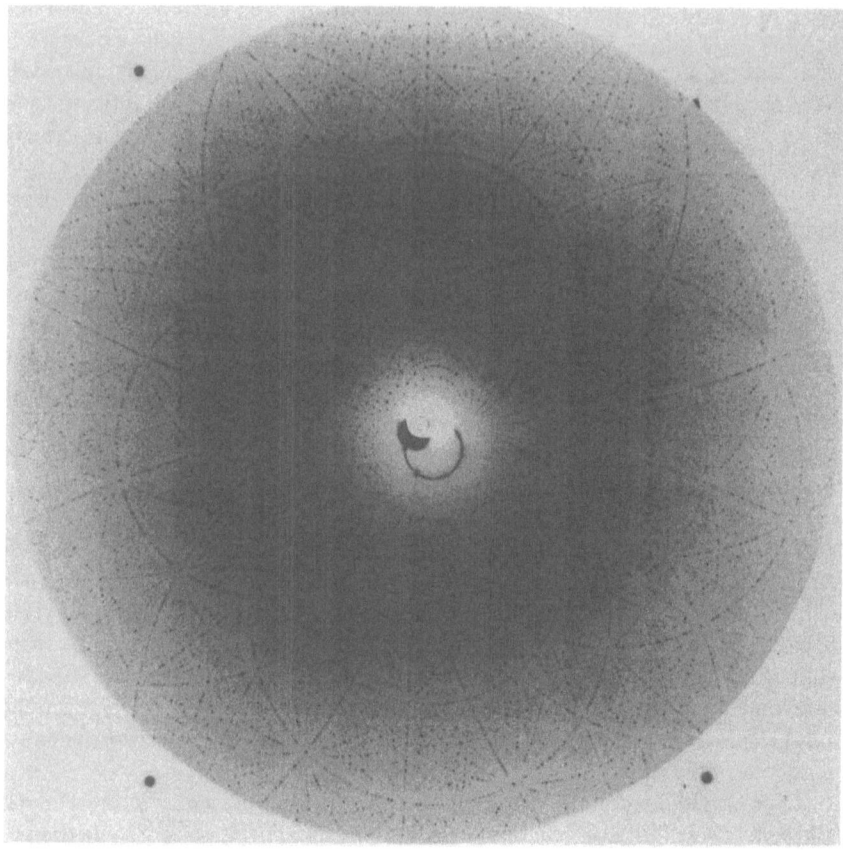

Figure 4. Laue diffraction photograph of xylose isomerase. Exposure time was for one second using a beam of white radiation, wavelength range from 0.2 to 2.1 Å. The photograph was recorded at the synchrotron at Daresbury, England. Twenty-five percent of the unique data to 2 Å resolution are obtained on this film.

a much larger number of reflections on the Laue photograph—thousands more. In fact, the single 1-sec Laue photograph in Fig. 4 contains 25% of the unique data to 2 Å resolution. The Laue photograph is also much more complex. The reflections do not lie in short rows within lines of spots. They lie on a series of intersecting ellipses.

Why wasn't the Laue method in routine use for data collection, if so many more reflections can be recorded by this technique? The complexity of the photographs was one obstacle, but modern computers and computer graphics have made it easy to assign the appropriate indices to each of the thousands of reflections on a Laue photograph and to integrate their intensities. The main barrier to the use of the Laue technique was the assumption, held by almost all crystallographers for over 50 years, that most of the reflections in a Laue photograph would be overlapped. It seemed a logical assumption: with so many reflections occurring simultaneously, those harmonically related to each other [e.g., the (2,0,2), (4,0,4), (6,0,6), etc.] would lie along a common diffraction direction, and so would fall on top of each other on the photograph. That reasoning was correct, but the assumption that such overlaps would represent the majority of reflections on the photograph was not correct. Cruickshank, Helliwell, and Moffat (1987) showed theoretically that even in the worst cases only 17% of the spots on a Laue photograph would be harmonic multiples, and in most cases the percentage would be even smaller. With so many spots recorded, the harmonic overlaps can be discarded with impunity.

Geometric factors influence what percentage of the complete data set can be recorded from a single Laue photograph corresponding to one orientation of the crystal. The two most important factors are the particular orientation of the crystal with respect to the incident beam and the symmetry of the crystal lattice. High-symmetry lattices have their unique data more tightly bunched in reciprocal space than crystals from low-symmetry space groups, and so yield a higher degree of completeness from a single shot. In favorable cases (cubic, tetragonal, or hexagonal crystals), more than three-quarters of the data set may be obtained on one Laue photograph.

Processing of Laue data is more complicated than processing of monochromatic data. First, selected nodal reflections are used to determine an approximate orientation matrix. Nodal reflections are those that lie at the intersections of several ellipses; they tend to have a clear area around them (Fig. 4). Nodal reflections are usually harmonic overlaps with simple index values and thus serve to guide an automatic orientation-fixing algorithm. The success of this procedure is determined by matching the observed Laue pattern to one calculated from the assumed orientation. Orientation parameters are then refined with the aid of high-resolution, nonnodal reflections. Once the observed and calculated patterns agree well, the individual reflections are integrated, and harmonic and spatial overlaps are discarded.

Since each reflection has been generated by X-rays of a different wavelength, it is necessary to normalize the intensities to produce a data set that appears to have come from a monochromatic source (one reason this is necessary is that X-ray film has a different absorption coefficient for different-wavelength X-rays). There are two ways to do this wavelength correction. The first method involves scaling several different Laue data sets to a reference Laue or monochromatic data set. Because only the differences between two data sets are needed to calculate a difference Fourier map, this relative scaling method will yield useful data (Hajdu *et al.*, 1987b). We prefer, however, to correct each data set internally. This procedure, which is called wavelength normalization, makes use of the fact that symmetry-related reflections will be produced by X-rays of different wavelengths. A curve can be constructed to normalize the data if it is assumed that, after correction, symmetry-related reflections should have the same intensities. Wavelength normalization produces very precise data, with replicated measurements agreeing to 5% on intensities.

We have conducted a number of tests to evaluate the accuracy of protein crystal data obtained by the Laue method. The first test was to see if the data were sufficiently accurate to locate a small molecule bound to a protein. The protein chosen was xylose isomerase, a large (mol. wt. = 160,000) tetrameric enzyme that requires a magnesium atom for catalytic activity. Xylose isomerase crystallizes in a pseudo-body-centered space group with all three unit cell dimensions near 100 Å (Farber *et al.*, 1987). Therefore, it is a good example of an average crystal of a large biomolecule. Europium inhibits xylose isomerase by competing with magnesium, and europium is a strong scatterer of X-rays. We decided to locate the magnesium site in xylose isomerase by finding where europium binds to the enzyme.

As a control, we collected complete sets of data to 3 Å resolution, by conventional, single-counter diffractometry using monochromatic radiation, on two crystals of xylose isomerase: one, a crystal of the native enzyme; the other, a crystal of the native enzyme soaked in europium chloride. One week was required for each data set, a total of 14 days of data collection time. Then we collected the same sets of data by Laue photography with white radiation on the Daresbury Synchrotron. To increase the completeness of the data sets, we took pictures at three different orientations for each crystal. Nevertheless, the total data collection time required for each data set, native and europium-substituted, was less than 5 sec (Farber *et al.*, 1988).

Figure 5 shows the difference Fourier map obtained with coefficients $(F_{europium} - F_{native})$ using phases from the solution of the crystal structure of the enzyme (Farber *et al.*, 1987). Europium binds in a double site with 4 Å separating the two lanthanide positions. Both the monochromatic data and the Laue data give clean, easily interpreted maps, but the Laue data were obtained in $1/100,000^{th}$ of the time.

For the second test, we decided to see if Laue data could be used to solve the structure of a protein by the molecular replacement method (Rossman, 1972). In this method, the known structure of a homologous protein is used as a model whose orientation in the unit cell of the unknown structure is determined by comparison of the model and unknown Patterson functions. The system chosen was turkey egg-white lysozyme, a small enzyme that crystallizes in a hexagonal unit cell. Turkey egg-white lysozyme is highly homologous with the lysozyme from hen egg white, which was the first enzyme whose three-dimensional structure was determined by X-ray crystallography (Phillips, 1966). Thus, an excellent model structure was available. A set of Laue data was collected from a single turkey egg-white lysozyme crystal; less than 3 sec was required to measure 55% of the data to 2.5 Å resolution.

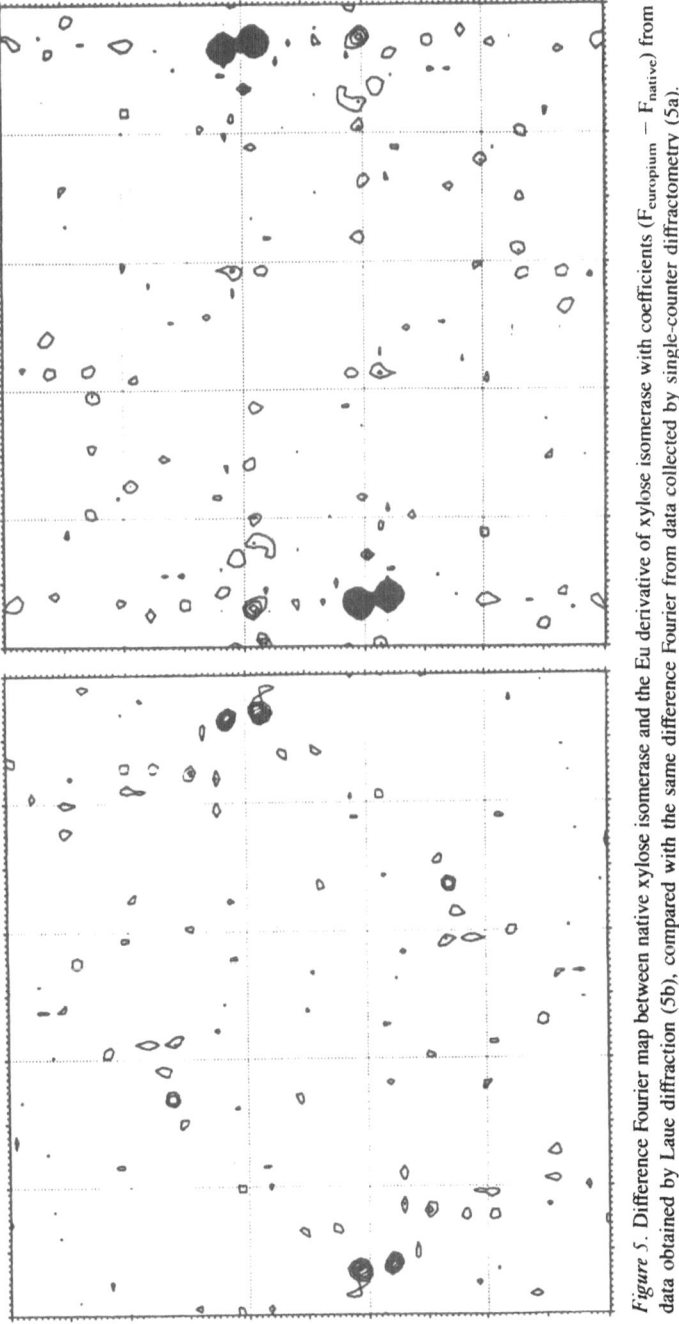

Figure 5. Difference Fourier map between native xylose isomerase and the Eu derivative of xylose isomerase with coefficients ($F_{europium} - F_{native}$) from data obtained by Laue diffraction (5b), compared with the same difference Fourier from data collected by single-counter diffractometry (5a).

Successful application of molecular replacement requires determination of both the rotational and translational orientation of the model structure. The rotation function for the Laue data from turkey egg-white lysozyme, using a partially refined ($R = 37\%$) hen egg-white lysozyme structure as a probe, was clean and unambiguous. The largest peak proved to give the correct rotational parameters for the model structure. The translation function was also easy to interpret and provided a unique, self-consistent position for the hen lysozyme model. Structure factors calculated from the rotated, translated model had an R-factor of 40% vs. the turkey lysozyme Laue data to 2.5 Å resolution. Refinement of this starting structure by molecular dynamics (Brünger *et al.*, 1987) reduced the R-factor to 19% in one pass, without any manual rebuilding. Electron density maps calculated at this step confirmed that the structure was solved.

4. Reaction Synchronization

Now that we have demonstrated that Laue data are of high enough quality to solve protein structures, and to locate small molecules bound to the active sites of enzymes, it would seem that we are ready to look at triggered conformational changes that occur on a time-scale of millisecond to second by taking a series of Laue photographs as the reaction proceeds. Unfortunately, although we have beaten the time-averaging problem in X-ray crystallography, the space-averaging problem remains.

Recall that the structure observed in any diffraction experiment (Laue or monochromatic) is an average over all molecules in the crystal, as well as over the time of data collection. If the reaction of interest is initiated by diffusion of substrate into the crystal, a process that typically takes minutes to complete, a Laue photograph taken for, say, 300 msec during diffusion will not represent a homogeneous population of states in the crystals. Protein molecules near the surface of the crystal will have bound product, while those in the center will not yet have seen the substrate. If the data are not collected until after diffusion is completed, most of the protein molecules will have already finished reacting and it will be impossible to follow the reaction. Either diffusion must be made fast compared to the process of interest, or some method must be found to initiate the reaction simultaneously throughout the crystal.

Bishop and Richards (1968) have measured diffusion times for small molecules into protein crystals. The numbers vary depending on the size of the small molecule and the diameter of the channels, ranging from a few seconds to many minutes, but organic molecules generally take a minute or so to reach the interior of a protein crystal. Thus, time-resolved X-ray diffraction studies of enzymatic reactions by Laue diffraction must reconcile three seemingly incompatible time-scales: that of milliseconds for fast enzyme turnover, that of seconds for data collection, and that of minutes for substrate diffusion.

There are two general solutions to this problem, both of which should yield homogeneous crystalline species and clean electron density maps. Neither type of solution is suitable for all systems, but it should be possible to find one for any given case.

The two methods are kinetic resolution of a slow reaction and triggering of the reaction following diffusion. Slow reactions can be achieved by the use of poor substrates, by low temperature, or by the use of mutant enzymes with poor turnover rates. Reaction triggers include metal ions, temperature and pH jump, and photochemical activation.

The first method is to choose a reaction that is slow compared with the diffusion time. If the catalytic process is not slow enough under normal conditions, it may be possible to select an abnormal substrate for which turnover takes many minutes. Many enzymes have "slow substrates" that are processed much less rapidly than the physiological substrate. Many also show sharp reductions in rate away from their pH optimum. If conditions can be found where turnover in the crystal is slow enough, diffusion will not be rate-limiting and the observed species in the crystal will depend on the time of observation and the kinetic mechanism of the enzyme. For example, consider an E–S reaction with a turnover time of 25 min, in which a single intermediate is formed after 5 min (i.e., breakdown of E–I is rate-limiting). Suppose that diffusion of the substrate into crystals of the enzyme takes 2 min to complete. Then, if one takes a single 1-sec Laue data set 1 min after the onset of diffusion, one will see the initial Michaelis complex at about 50% occupancy, with perhaps 10% contamination by the E–I complex. If one then takes a second 1-sec data set after 8 min, the crystal will contain E–I at about 90% occupancy, with 10% contamination from E–P. A third data set at 18 min may show the E–P complex if its lifetime on the enzyme is long enough. Laue diffraction makes it possible to look at reactions that are complete in minutes; with conventional data collection one could only study processes that required days to occur.

If the combined use of poor substrates and nonoptimum pH values cannot produce a turnover time slower than the diffusion time, it is possible to slow the reaction down further by cooling the enzyme. Unfortunately, large rate reductions require large temperature reductions, and these are difficult to achieve for many protein crystals. An alternative is to cripple the enzyme by site-directed mutagenesis. Protein engineering of improved enzymes has proven a difficult goal to achieve, but most molecular biologists have had no trouble producing mutant enzymes with reduced catalytic activity! In many cases, this reduction is substantial—three or four orders of magnitude in rate. If turnover takes 0.1 sec for the wild-type enzyme, but 600 sec for an active-site mutant, then crystals of the mutant are suitable for time-resolved study of the reaction by Laue diffraction. The danger with this method is that even a single point mutation in the active site may change the catalytic mechanism (Nickbarg *et al.*, 1988). Therefore, mutagenesis should only be employed in conjunction with thorough mechanistic characterization. But when it is clear that the mechanism of the mutant is the same as that of the wild type, mutagenesis provides a route to the production of some interesting, slow enzyme reactions, as in the case of aspartate aminotransferase (Smith *et al.*, 1988), which are amenable to structural study by Laue methods.

Triggering of enzymatic reactions is the other way to beat diffusion. If catalysis or binding requires a specific, rapid, controllable event, then it does not matter how long diffusion takes. One simply waits until the crystal has become filled with substrate, and then the reaction is initiated simultaneous with the start of data collection. The first Laue data set thus gives a picture of the initial E–S complex, and subsequent snapshots show the evolution to product.

How is the triggering to be accomplished? One way is to use pH or temperature jump, the classical tool of the pre-steady-state kineticist. The crystal is mounted in a flow cell at a pH temperature (or both) at which the reaction rate is much slower than the substrate diffusion time. One then rapidly changes either pH or temperature to initiate the reaction. pH jumps can be achieved by diffusing a new buffer into the crystal, since hydronium ions will travel throughout the lattice in less than a second, but a more elegant method might be to use volatile acids and bases to induce a rapid, controlled change (P. Marfey, personal communication). Photoactivatable acid/base compounds would be ideal. Temperature jumps of 20°/sec are possible given a suitable cryostat around the crystal and adequate control of cold gas flow. An elegant device has been designed and constructed by Dr. Bjarne Rasmussen for use in diffractometry or Laue diffraction.

Both pH- and temperature-jump triggering require that the crystal survive the change in environment. This requirement is unlikely to be met in all—or maybe even most!—cases. Metal atom triggering represents a good alternative when a metalloenzyme is being studied. Many metalloenzymes require a loosely held metal (Mg is common) for catalysis, but still bind their substrates tightly. In these cases, the crystal of the apoprotein can be loaded with substrate and then mother liquor containing the catalytic metal can be flowed in to initiate the reaction. Metal ions diffuse through protein crystals in seconds and make very good chemical triggers. Glucose isomerase (Farber *et al.*, 1987) and mandelate racemase (Neidhart *et al.*, 1988) both require Mg^{2+} for activity; in the latter case, the activation of the inert apoenzyme–mandelic acid complex takes several minutes with Mg^{2+} is added. Such a system is ideal for Laue diffraction study.

The problem with all of these triggering mechanisms is that they are not instantaneous. As long as their effects are rapid compared to turnover, they can be used, but millisecond X-ray diffraction really needs a submillisecond trigger if it is to be exploited to the fullest. Photochemical activation is the obvious solution. Indeed, "caged" substrate and cofactor molecules such as caged ATP have been synthesized, where the chemical "cage" is removable by photolysis.

Unfortunately, the decomposition of the caged substrate usually produces highly reactive products, such as nitroso ketones, that can react with nucleophilic groups on the protein, and active sites are usually rich in nucleophilic groups. We believe that a better method would be to "cage" the enzyme. A photocleavable inhibitor that covalently blocks the active site would allow an infinite time for substrate to be diffused into the crystal interior; a pulse of light would then cause the inhibitor to fall off the enzyme, and substrate, now surrounding the protein, would then bind immediately.

Ned Porter and associates (Turner *et al.*, 1987), at Duke University, have designed a series of photoreversible inhibitors for the serine proteases, especially thrombin. These inhibitors are based on the structure of the cinnamate group, and all exhibit isomerization about the double bond when irradiated at the appropriate frequency. Incubation of serine proteases with these compounds results in rapid, irreversible inactivation of the enzymes. Crystallographic study of the structure of the complex of one such inhibitor with gamma chymotrypsin, determined at high resolution in our laboratory, shows that the inhibitor is covalently attached to the active site serine 195 as a stable acyl enzyme. The stability of this complex seems to result from a failure of the carbonyl oxygen of the inhibitor to bind in the "oxyanion hole" on the enzyme. Instead, the carbonyl group is oriented out toward the solvent. Consequently, there is no stabilization of the oxyanion transition state for hydrolysis of the acyl enzyme, and the inhibitor does not deacylate at any appreciable rate until it is illuminated. Irradiation at the appropriate wavelength (320 nm for the compound shown here) results in a slow loss of inhibitor and restoration of full catalytic activity.

How does irradiation produce deacylation? Model acyl derivatives of the inhibitor hydrolyze by intramolecular attack of the aromatic hydroxyl after photoisomerization. Such a rearrangement is possible on the enzyme, but protein–inhibitor interactions would be expected to reduce its rate. The observed deacylation rate is slow, as expected, but that fact has an interesting consequence: the deacylation of the inhibitor itself can be followed, at atomic resolution, by Laue diffraction. These experiments are in progress.

Recent work by Moffat and associates at the Cornell Synchrotron have shown that Laue photographs can be taken on a picosecond-to-nanosecond time-scale by using the very intense, though narrow-band-width, radiation from an undulator magnet (Pool, 1988). These are important experiments, but the narrow band-pass means that only a small number of reflections are observed with one photograph. We have concentrated on broad-band radiation (λ range 0.2–2.1 Å), which permits complete data acquisition in the millisecond-to-second time range when the crystal has a high degree of symmetry. One such system is turkey egg-white lysozyme, which crystallizes in the hexagonal space-group $P6_122$. The crystal structure of this enzyme has now been solved at 2.5 Å resolution by molecular replacement using the well-known, related structure of hen egg-white lysozyme as a probe. The starting structure then refined smoothly and automatically from an R-factor of 40% to an R-factor of 19% by molecular dynamics refinement. All of the diffraction data used to solve this structure were collected on a single Laue diffraction photograph; the total data collection time was 1 sec.

This system has now been used to study the catalytic reaction of lysozyme. At pH 8, the hydrolysis of the substrate hexasaccharide $(NAG)_6$, where NAG = N-acetyl glucosamine, by lysozyme takes over 60 min in the crystal. Turkey lysozyme crystals mounted in flow cells (Petsko, 1985) were exposed to high concentrations of $(NAG)_6$, and a series of complete 1-sec data sets were collected by Laue diffraction approximately every 10 min for a period of over 1 hr. This set of snapshots is still being analyzed, but electron density difference maps between the Laue data set at $t = 55$ min and the data set at time zero show an elongated, beaded density in the active site cleft. Preliminary model building indicates that this density can be fit by a hexasaccharide. It would seem from these results that Laue diffraction is capable of providing an atomic resolution view of the dynamics of enzyme catalysis. Those systems that involve large-scale, triggered conformational changes may present serious problems when one attempts to study them with this technique, since big motions may cause crystal cracking or disintegration. But for crystals where such processes are tolerated in the lattice, Laue diffraction offers the promise of atomic-resolution, time-lapse pictures of protein machinery in action.

References

Alber, T., Petsko, G. A., and Tsernoglou, D., 1976, Crystal structure of an elastase-substrate complex at −55°C, *Nature* 263:297–300.

Alber, T. C., Davenport, R. C., Jr., Giammona, D. A., Lolis, E., Petsko, G. A., and Ringe, D., 1988, Crystallography and site-directed mutagenesis of yeast triose phosphate isomerase. What can we learn about catalysis from a 'simple' enzyme? *Cold Spring Harbor Symp. Quant. Biol.* LII:603–613.

Albery, W. J., and Knowles, J. R., 1976, Free-energy profile for the reaction catalyzed by triose phosphate isomerase, *Biochemistry* 15:5627–5631.

Bishop, W. H., and Richards, F. M., 1968, Properties of ligands in small pores, *J. Mol. Biol.* 38:315–328.

Brünger, A. T., Kuriyan, J., and Karplus, M., 1987, Crystallographic R factor refinement by molecular dynamics, *Science* 235:458–460.

Cruickshank, D. W. J., Helliwell, J. R., and Moffat, K., 1987, Multiplicity distribution of reflections in Laue diffraction, *Acta Cryst.* A43:656–674.

Douzou, P., and Petsko, G. A., 1984, Proteins at work: "Stop-action" pictures at sub-zero temperatures, *Adv. Prot. Chem.* 36:246–361.

Farber, G. K., Petsko, G. A., and Ringe D., 1987, The 3.0 Å crystal structure of xylose isomerase from *Streptomyces olivochromogenes*, *Protein Engineering* 1:459–466.

Farber, G. K., Machin, P., Almo, S. C., Petsko, G. A., and Hajdu, J., 1988, X-ray Laue diffraction from crystals of xylose isomerase, *Proc. Natl. Acad. Sci. USA* 85:112–115.

Fink, A. L., and Petsko, G. A., 1981, X-ray cryoenzymology, *Adv. Enzymol.* **52**:177–246.

Friedrich, W., Knipping, P., and von Laue, M., 1912, *Sitzungsberichte der Math. Phys. Klasse (Kgl.), Bayerische Akademie der Wissenschaften,* München, pp. 303–322.

Hajdu, J., Acharya, K. R., Stuart, D. I., McLaughlin, P. J., Barford, D., Klein, H. W., Oikonomakos, N. G., and Johnson, L. N., 1987a, Catalysis in the crystal: Synchrotron radiation studies with glycogen phosphorylase *b*, *EMBO J.* **6**:539–546.

Hajdu, J., Machin, P. A., Campbell, J. W., Greenough, T. J., Clifton, I. J., Zurek, S., Gover, S., Johnson, L. N., and Elder, M., 1987b, Millisecond X-ray diffraction and the first electron density map from Laue photographs of a protein crystal, *Nature* **329**:178–181.

Hajdu, J., Acharya, K. R., Stuart, D. I., Barford, D., and Johnson, L. N., 1988, Catalysis in enzyme crystals, *Trends Biochem. Sci.* **13**:104–109.

Kantrowitz, E. R., and Lipscomb, W. N., 1988, *E. coli* aspartate transcarbamylase: The relation between structure and function, *Science* **241**:669–674.

Karplus, M., and McCammon, J. A., 1983, Dynamics of proteins: Elements and functions, *Annu. Rev. Biochem.* **52**:263–300.

Makinen, M., and Fink, A. L., 1977, Reactivity and cryoenzymology of enzymes in the crystalline state, *Annu. Rev. Biophys. Bioeng.* **6**:301–342.

Moffat, K., Szebenyi, D. M. E., and Bilderback, D. H., 1984, X-ray Laue diffraction from protein crystals, *Science* **223**:1423–1425.

Neidhart, D. J., Powers, V. M., Kenyon, G. L., Tsou, A. Y., Ransom, S. C., Gerlt, J. A., and Petsko, G. A., 1988, Preliminary X-ray data on crystals of mandelate racemase, *J. Biol. Chem.* **263**:9268–9270.

Nickbarg, E. B., Davenport, Jr., R. C., Petsko, G. A., and Knowles, J. R., 1988, Triosephosphate isomerase from yeast: Removal of a putatively electrophilic histidine results in a subtle change in catalytic mechanism, *Biochemistry* **27**:5948–5960.

Perutz, M. F., 1970, Stereochemistry of cooperative effects in haemoglobin, *Nature* **228**:726–734.

Petsko, G. A., 1985, Flow cell construction and use, *Methods Enzymol.* **114**:141–146.

Petsko, G. A., and Ringe, D., 1984, Fluctuations in protein structure from X-ray diffraction, *Annu. Rev. Biophys. Bioeng.* **13**:331–371.

Phillips, D. C., 1966, The three-dimensional structure of an enzyme molecule, *Sci. Am.* **215**:78–90.

Pool, R., 1988, Molecular photography with an X-ray flash, *Science* **241**:295.

Ringe, D., 1987, Protein crystallography: Catching up with fast changes, *Nature* **329**:102.

Ringe, D., and Petsko, G. A., 1985, Mapping protein dynamics by X-ray diffraction, *Prog. Biophys. Mol. Biol.* **45**:197–235.

Rossman, M. G., 1972, *The Molecular Replacement Method,* Gordon Breach, New York.

Smith, D., Almo, S. C., Ringe, D., and Toney, M. D., 1989, 2.8 Å resolution crystal structure of an active site mutant of aspartate aminotransferase from *E. coli, Biochemistry* **28**:8161–8167.

Turner, A. D., Pizzo, S. V., Rozakis, G. W., and Porter, N. A., 1987, Photochemical activation of acylated α-thrombin, *J. Am. Chem. Soc.* **109**:1274–1275.

Mass Spectrometric Methods for Determination of the Structure of Peptides and Proteins

KLAUS BIEMANN

1. Introduction

Mass spectrometry (MS) has long been known as a sensitive method of characterizing compounds that are small, thermally stable, and sufficiently volatile to be vaporized into the vacuum system of the spectrometer. It was not, however, until the early 1980s that the method became applicable to large, polar molecules, with the development by Barber *et al.* (1981) of fast-atom-bombardment (FAB) ionization. This ionization method produces protonated molecules, $(M + H)^+$, by bombarding a solution of the compound of interest $(M)^+$, in a suitable liquid of very low vapor pressure, such as glycerol, with neutral atoms (or ions) of kilovolt kinetic energy.

Although mass spectral information on peptides of $M_r > 1000$ had previously been obtained by their ionization with field desorption (FD) (Shimonishi *et al.*, 1980) or plasma desorption (PDMS or ^{252}Cf-MS) (Hakansson *et al.*, 1982), FAB turned out to be a much simpler, more reliable, and more widely accessible technique for ionizing large peptides and soon even small proteins. Fortunately, the sensitivity of FAB-MS is high enough to be practically useful: 0.01–2 nmole (with increasing molecular size) is required for a measurement. Most important is the ability to analyze mixtures of peptides without prior separation, because the mass spectrometer can easily resolve components differing by 1 Da even at mass 5000, and sometimes beyond. Fortunately, peptides occurring

KLAUS BIEMANN • Department of Chemistry, Massachusetts Institute of Technology, Cambridge, Massachusetts 02139.

as a natural mixture (such as a proteolytic digest) generally differ much more in mass, with the exception of a pair that is produced by the partial hydrolysis of a -$CONH_2$ group to -COOH. Even such a pair differs by 1 Da and can thus be resolved.

Since the early days of FAB-MS, insulin has become a favored example. Figure 1 shows the molecular ion region of the FAB mass spectrum of porcine and human insulin, which differ in molecular weight by 30 Da, the mass difference between the C-terminal alanine and threonine of the B-chain of the porcine and human protein, respectively. The multiplicity of the peaks in the spectrum is due chiefly to the contributions of ^{13}C, which represents about 1.1% of the abundance of ^{12}C. Therefore, the likelihood that all of the 256 or 257 carbon atoms are ^{12}C is low. The peaks for the $(M + H)^+$ ions containing only ^{12}C (at m/z 5774.6 and 5804.6, respectively) are, therefore, much smaller than those 3 Da higher because, at that larger number of carbons, the species with three ^{13}C-atoms is statistically the most abundant.

It is not really necessary, however, to resolve isotopic multiplets; it suffices instead to determine their isotope weighted mass average, which requires much less resolution and thus increases sensitivity. This is demonstrated in Figure 2 for the molecular weight determination of a small protein (a serine protease inhibitor) which was found to be 6512.5 by centroiding the peak envelope, as shown.

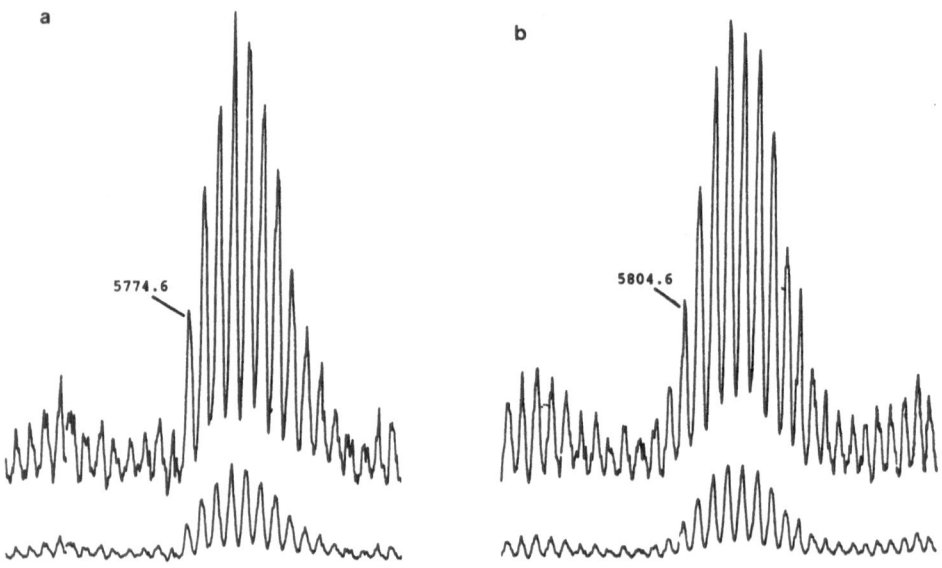

Figure 1. Resolved molecular ion, $(M+H)^+$, region of the FAB mass spectrum of porcine ($C_{256}H_{381}N_{65}O_{76}S_6$) (a) and human ($C_{257}H_{383}N_{65}O_{77}S_6$) (b) insulin.

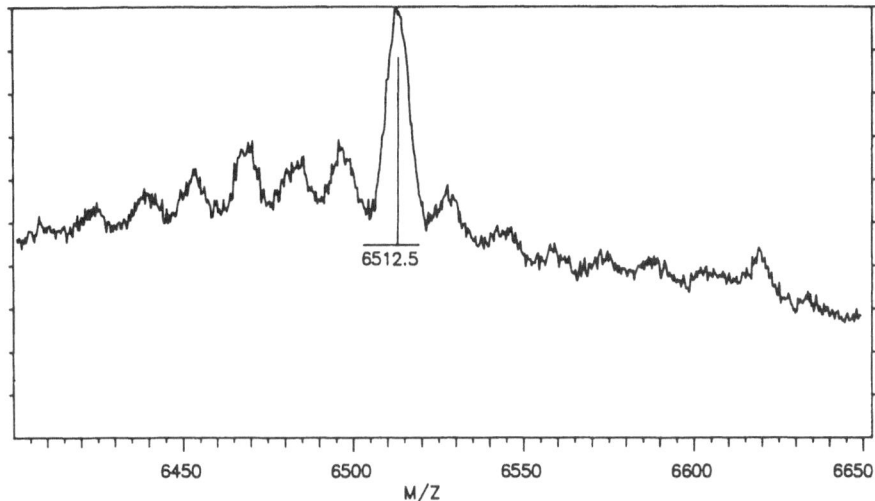

Figure 2. Molecular ion region of the FAB mass spectrum of a small protein (unresolved).

For calculation of the mass of the resolved $(M + H)^+$ ions (Fig. 1) the "physical" atomic weights ($^{12}C = 12.000000$) of all atoms present must be summed, while for the average mass (Fig. 2) the "chemical" atomic weights (abundance weighted sums of isotopic masses, i.e., $C = 12.011$) must be used.

FAB is a so-called "soft" ionization method which generates protonated molecules, $(M + H)^+$, or otherwise cationated molecules, such as $(M + Na)^+$, $(M + K)^+$, and so forth, if alkali salts are contaminating the sample. These ions have little excess energy and thus little tendency to fragment. As a result, the "mass spectrum" consists mainly of the molecular ions like those shown in Figures 1 and 2. Much smaller peptides do exhibit some limited fragmentation, but only if a large sample is used.

2. High-Mass Mass Spectrometry

There are many situations in which a problem of peptide or protein structure can be solved simply by molecular weight information. For example, a hypothetical or proposed primary structure of a protein can easily be confirmed and/or corrected by determining the molecular weights of the peptides generated by specific enzymatic or chemical cleavage of the protein. A comparison of the experimental data with the molecular weights predicted on the basis of the position in the hypothetical structure of the amino acids susceptible to specific cleavage by the enzyme or chemical reagent used reveals differences between the predicted and actual sequence. An early example is the verification and correction of amino acid sequences of proteins derived from the base sequence of the gene coding for a protein (Gibson and Biemann, 1984). Errors in the base sequence due to deletion/insertion errors or misidentifications of bases leading to premature stop codons or incorrect amino acids can easily be detected, located, and corrected. Since the mass spectrometer can reliably reveal the molecular weights of a dozen or so peptides in a mixture, this approach is suitable even for large proteins producing more than 100 tryptic peptides. Crude fractionation by high-performance liquid chromatography (HPLC) of such a digest, collecting 15–30 fractions and determining the FAB mass spectrum of each fraction, provides all the data required.

A specific example of such a situation is the determination of the point of beginning of the transcription of the DNA sequence coding for valyl-tRNA synthetase from *Saccharomyces cerevisiae* (Biemann and Scoble, 1987). This was difficult to ascertain with conventional methods because there were two ATG codons for methionine near the putative N-terminus in the DNA sequence of the gene encoding this protein, which appeared to be blocked at the N-terminus and precluded its identification by the Edman method (Chatton *et al.*, 1988). The two potential sequences were 1058 and 1104 amino acids long (Fig. 3), and the presumed positions of arginine and lysine would predict the generation of about 130 peptides upon digestion with trypsin. A tryptic digest of 2 nmole of cytoplasmic valyl-tRNA synthetase was prepared, partially separated by HPLC (Fig. 4), and collected in 27 fractions. Each of these fractions was subjected to FAB-MS, which permitted the determination of the molecular weights of 80 peptides. Of these, 79 matched those predicted from the DNA-derived sequence, covering

Met	Asn	Lys	Trp	Leu	Asn	Thr	Leu	Ser	Lys	Thr	Phe	Thr	Phe	Arg	15
ATG	AAT	AAG	TGG	TTA	AAC	ACA	TTA	TCT	AAG	ACA	TTC	ACT	TTT	CGG	45
Leu	Leu	Asn	Cys	His	Tyr	Arg	Arg	Ser	Leu	Pro	Leu	Cys	Gln	Asn	30
CTT	TTG	AAC	TGT	CAT	TAT	AGG	CGA	TCA	TTA	CCA	CTT	TGT	CAA	AAC	90
Phe	Ser	Leu	Lys	Lys	Ser	Leu	Thr	His	Asn	Gln	Val	Arg	Phe	Phe	45
TTT	TCT	CTG	AAG	AAG	TCG	TTA	ACT	CAT	AAT	CAA	GTC	AGG	TTC	TTT	135
Lys	Met	Ser	Asp	Leu	Asp	Asn	Leu	Pro	Pro	Val	Asp	Pro	Lys	Thr	60
AAA	ATG	AGC	GAT	CTT	GAT	AAT	TTG	CCT	CCA	GTT	GAC	CCA	AAG	ACT	180
...	...	...	...	...	...	...	...	...	...	...	...	...	...	...	....
...	...	...	...	...	...	...	...	...	...	...	...	...	...	...	....
Ile	Glu	Asn	Leu	Lys	Arg	Leu	Lys	Leu	1104						
AAT	GAA	AAC	TTG	AAG	CGT	TTG	AAA	TTG	3312						

Figure 3. Partial sequence of the gene that encodes valyl-tRNA synthetase from *Saccharomyces cerevisiae*, as determined by Chatton *et al.*, 1988. The two putative initiation translation sites are underlined.

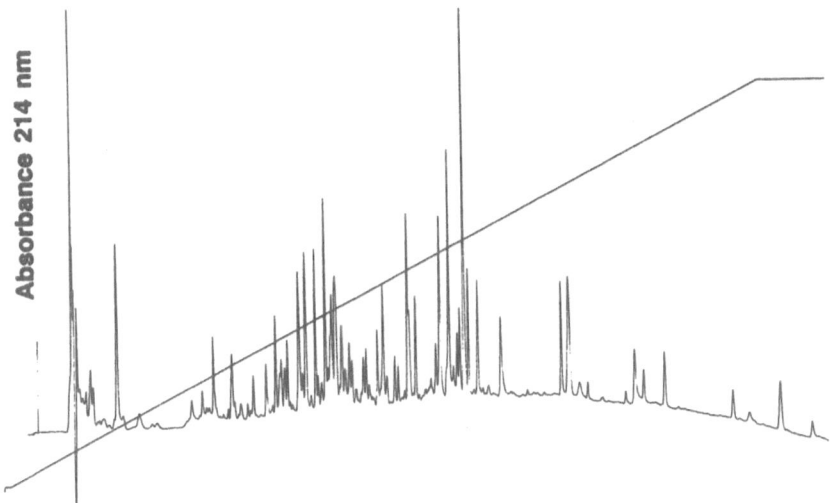

Figure 4. HPLC separation of the tryptic digest of 2 nmoles valyl-tRNA synthetase from *Saccharomyces cerevisiae*. The ascending line indicates the solvent gradient.

about 80% of it. This observation indicated that the DNA sequence was probably correct and definitely precluded any deletion and/or insertion errors that would have led to a frameshift. Furthermore, none of the tryptic peptides predicted for the region between Met-1 and Met-47 were observed, suggesting that the latter is probably the point of initiating transcription of the gene for the cytoplasmic enzyme.

There was, however, one peak (No. 7 in Fig. 5) at m/z 1351.7, indicating the presence of a peptide of mol.w. 1350.7. This did not match any of the predicted values, but was in the range expected for the tryptic peptide Met-47 through Lys-59, which would produce a $(M + H)^+$ ion of m/z 1440.7. If this peptide indeed represents the N-terminus of the cytoplasmic enzyme, post-translational processing must have involved the removal of the methionine, which decreases the molecular weight by 131 Da, followed by acetylation of the newly formed N-terminal serine, adding 42 Da (replacement of H by CH_3 CO). This interpretation was further proven to be correct by collision-induced decomposition (CID) of the ion of m/z 1351.7 in this mixture. Analysis of the fragment ions in a tandem MS (see Section 3) revealed the structure of this peptide to be CH_3CO-Ser-Asp-Leu-Asp-Asn-Leu-Pro-Pro-Val-Asp-Pro-Lys, unambiguously confirming the posttranslational processing sequence suggested earlier.

This example demonstrates the large amount of information that can be gleaned quickly by determining the molecular weights (accurate to a few tenths of a dalton) of peptides, even if they are present in relatively complex mixtures.

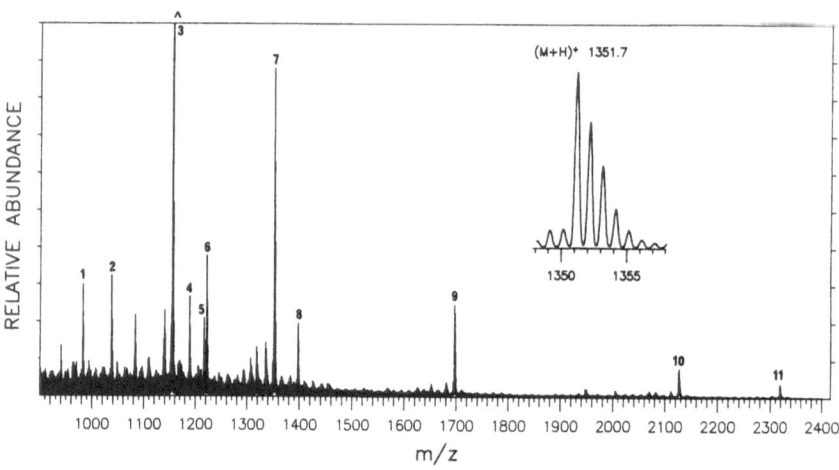

Figure 5. FAB mass spectrum of fraction 10 of the HPLC separation shown in Figure 4.

Fraction 10 of the tryptic digest of the valyl-tRNA synthetase (Fig. 5) contained at least 11 peptides.

Of interest is the fact that this structural gene was found by mutation experiment to actually encode both the cytoplasmic and the mitochondrial valyl-tRNA synthetase, the latter also including the stretch between Met-1 and Met-47. The N-terminal extension seems to target the cytoplasmically synthesized mito-chondrial synthetase to the mitochondrial outer membrane (Chatton *et al.*, 1988).

The same strategy can be applied to many problems related to native pro-teins, natural or biosynthetic mutants, or proteins modified by recombinant tech-nology, photoaffinity labeling, or other chemical or biological processes. Most of these involve modification or deletion of one amino acid or shortening of the protein from either the N-terminus or the C-terminus. Such shortening by ex-opeptidases often leads to "ragged ends" (i.e., a mixture of products differing in the number of successive amino acids removed). Here again, mass spectrometry is a simple, effective method that reveals in a proteolytic digest the presence of a series of peptides differing from each other by one amino acid. Such a ragged C-terminus is particularly difficult to recognize by the conventional Edman pro-cedure, which would require isolation and purification of each of these peptides and sequencing each one to the end, which is not always possible with a high degree of confidence.

A wide range of such modifications, often unplanned or unexpected, is encountered with proteins produced in relatively large quantities by recombinant DNA processes, which are frequently carried out under conditions of concentra-tion, growth rate, host organisms, and so forth, that are very different from their "natural" genesis. MS has a great potential for quickly and reliably demonstrating whether all the desired posttranslational modifications have occurred, and that by-products, due to the action of exopeptidases, or chemical side reactions, such as oxidation of methionine to the corresponding sulfoxide or sulfone, have not taken place.

Determining the sites of glycosylation is another important aspect of protein chemistry. This can be accomplished by treating the glycoprotein with endo-β-*N*-acetylglucosamidase ("Endo H"), which removes all carbohydrate moieties but leaves *N*-acetylglucosamine attached to asparagine. Digestion of the resulting modified protein with trypsin produces peptides where the original site of glycosylation is represented by a modified asparagine that is 203 Da heavier than unmodified asparagine. Again, simply comparing the expected molecular weights (based on the position of Arg and Lys in the primary structure) with those experimentally determined by FAB-MS of the tryptic digest permits pinpointing of the asparagines bearing carbohydrate units. These assignments are further facilitated by the known fact that the only asparagines that can be glycosylated are followed by *X-Y*, where *X* can be any amino acid but *Y* must be serine or threonine. This strategy was used to pinpoint the sites of glycosylation in external invertase from yeast (Reddy *et al.*, 1988).

Another approach involves complete removal of the carbohydrate moiety using the enzyme peptide: *N*-glycosidase F, which leaves an aspartic acid in place of the glycosylated asparagine. Since peptides carrying a large carbohydrate moiety are less easily ionized by FAB than the unglycosylated analog, the FAB spectrum of a tryptic digest obtained after treatment of the glycoprotein with the peptide: *N*-glycosidase F will exhibit peaks absent in the spectrum of the tryptic digest of the native protein. These new peaks are due to $(M + H)^+$ ions of peptides that were originally glycosylated at asparagine, which is now aspartic acid (i.e., they appear 1 Da heavier than predicted from the known amino acid sequence) (Carr and Roberts, 1986).

3. Tandem Mass Spectrometry

As mentioned earlier, the internal energy of the $(M + H)^+$ ions generated by FAB is not sufficient to cause fragmentation of this chemically very stable ion. This feature is an advantage in determining the molecular weights of the components of mixtures, because all the significant signals are probably due to such individual molecular ions, rather than to fragments. However, the lack of fragmentation is tantamount to a lack of structural information.

Fortunately, these two levels of information can be achieved almost simultaneously by the use of two MSs linked together: the first separates the molecular ions of the components of the mixture and determines their molecular weights, and the second analyzes the fragments formed from each component separately, by a process that leads to the specific cleavage of bonds in the molecular ion. The most common method generating the necessary vibrational energy involves collision of the molecular ion with a neutral gas in the region between the first and second MS.

The principle of such a "tandem mass spectrometer" is shown in Fig. 6. A mixture of peptides, P_1-P_5, is ionized by FAB in the ion source of the first MS (MS-1) and separated by scanning across the mass range of interest. The signals recorded correspond to the $(M + H)^+$ ions of the individual peptides and thus define their molecular weights. MS-1 is then reset to focus on the ^{12}C-component of the isotope cluster of one of the components, which then passes into the collision cell (CC) filled with $\sim 10^{-3}$ torr of helium. Upon collision with a neutral atom, a small fraction of the translational energy of the $(M + H)^+$ ion is converted to vibrational energy that suffices to cleave a bond (CID), thus leading to a fragment that carries the positive charge and a neutral fragment. The former is mass analyzed by scanning the second mass spectrometer (MS-2), resulting in a "product ion" spectrum that exhibits signals from remaining precursor ions and the fragment ions (F_1-F_6 in Fig. 6). This process is then repeated for the ^{12}C-component of the $(M + H)^+$ ion of the next peptide, and so forth.

The MSs presently used in tandem systems are either double-focusing magnetic instruments, quadrupole instruments, or some combination of the two ("hybrid" instruments). The major difference between the magnetic and quadrupole spectrometers is the kinetic energy of the precursor ion, which is typically 10 kV in the former and <100 V in the latter. Furthermore, in the mass range of protein-related peptides, only the magnetic instruments have sufficient resolving power and sufficient transmission to permit selection of the ^{12}C-only species of $(M + H)^+$ ions for CID, which thus produces simple product ion spectra unencumbered by isotopic multiplets (i.e., monoisotopic product ion spectra). This aspect is important for the unambiguous mass assignment of the product ions and the computer-aided interpretation of the CID spectra (see below).

The work and strategies discussed in this chapter used a tandem MS consist-

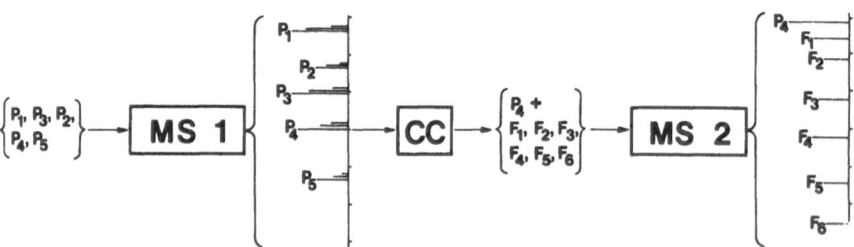

Figure 6. Principle of a tandem MS.

ing of a pair of double-focusing MSs with a mass range of 1–14,500 Da at full accelerating voltage (10 kV), which affords good sensitivity and mass accuracy (± 0.3 Da or better) in MS-2 (Sato *et al.*, 1987). Present technology limits efficient fragmentation of $(M + H)^+$ ions to peptides of molecular weights of less than 2500–3000 daltons (i.e., not exceeding 25 amino acids). Fortunately, this is the upper limit of the range for 90–95% of tryptic peptides from proteins of average lysine/arginine content.

From the very large number of CID spectra of peptides of known or initially unknown structure, we were able to elucidate the fragmentation processes that $(M + H)^+$ ions of peptides undergo upon high-energy collisions. The resulting fragments are depicted in Scheme 1. Those denoted a_n, b_n, c_n, x_n, y_n, and z_n are due to the simple cleavage of any one bond along the peptide chain and charge retention at the N-terminal (a_n, b_n, c_n) or at the C-terminal fragment (x_n, y_n, z_n), without or with (c_n, y_n) the rearrangement of a hydrogen atom. The mass difference (Table I) between fragments of the same ion type defines the corresponding amino acid, with the exception of the isomeric pair leucine/isoleucine ($\Delta m = 113$) and the isobaric pair glutamine/lysine ($\Delta m = 128$). The mass of ions a_n, b_n, c_n, x_n, y_n, and z_n can be calculated by the terms indicated in Scheme 1, where Σ_n is the sum of the residual masses (Table I) of the amino acids present in the fragment.

Figure 7 shows a somewhat atypical (because of its simplicity) CID spectrum of a peptide (valyl gramicidin A). It consists almost exclusively of a series of b_n and y_n ions, demonstrating the principle and advantage of only two overlapping extended ion series, one retaining the N-terminus and the other the C-terminus. This peptide does not contain any basic site. The N-terminal amino

Table I. Residue Masses of Common Amino Acids

Amino acid			Amino acid		
Three letter	Single letter	Residue mass[a]	Three letter	Single letter	Residue mass[a]
Gly	G	57.02	Asp	D	115.03
Ala	A	71.04	Lys	K	128.09
Ser	S	87.03	Gln	Q	128.06
Pro	P	97.05	Glu	E	129.04
Val	V	99.07	Met	M	131.04
Thr	T	101.05	His	H	137.07
Cys	C	103.01	Phe	F	147.07
Ile	I	113.08	Arg	R	156.10
Leu	L	113.08	Tyr	Y	163.06
Asn	N	114.04	Trp	W	186.08

[a]Mass of -NH-CHR-CO- rounded to the nearest 0.01 Da.

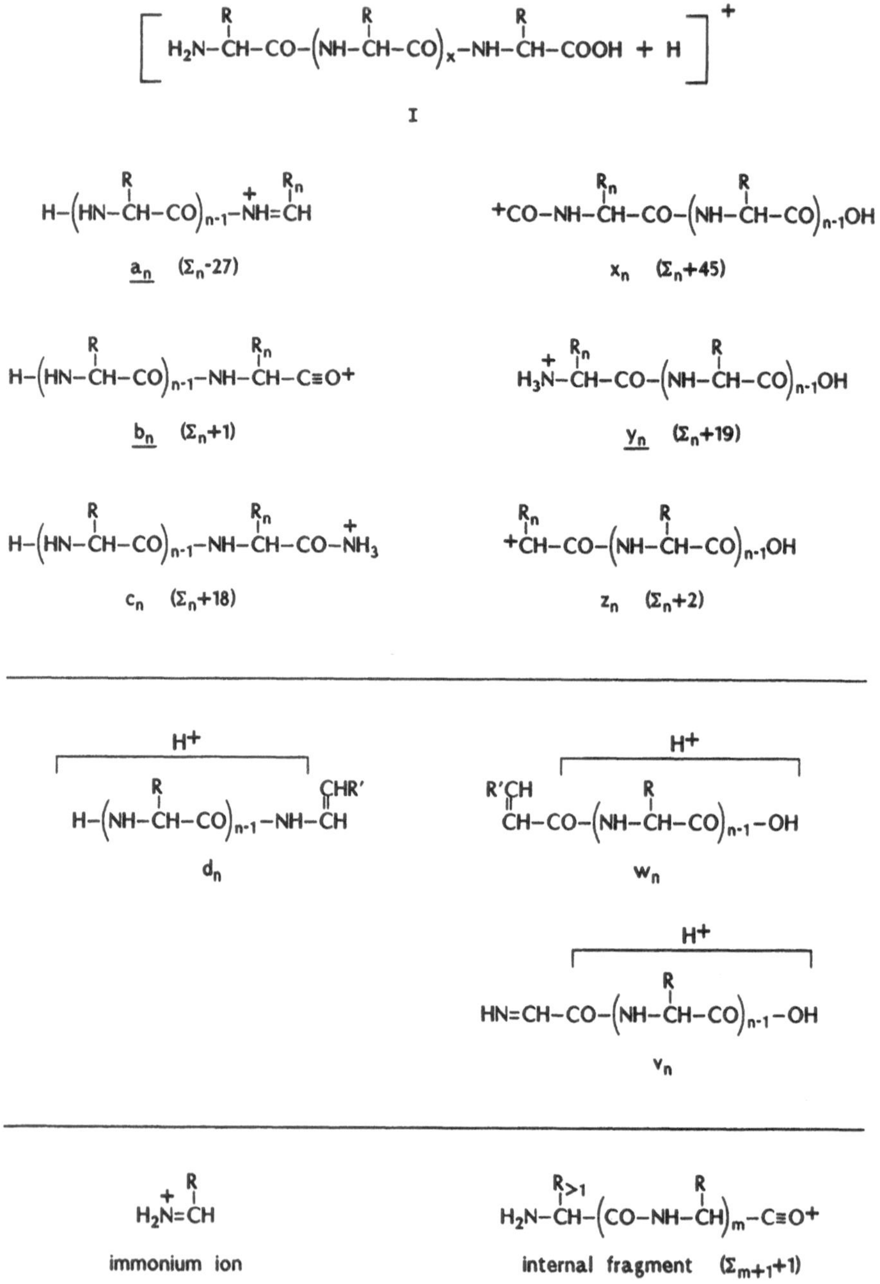

$$\left[H_2N-\overset{R}{\underset{|}{CH}}-CO-\left(NH-\overset{R}{\underset{|}{CH}}-CO\right)_x-NH-\overset{R}{\underset{|}{CH}}-COOH \; + \; H \right]^+$$

I

$$H-\left(HN-\overset{R}{\underset{|}{CH}}-CO\right)_{n-1}-\overset{+}{N}H=\overset{R_n}{CH}$$

$\underline{a_n}$ $(\Sigma_n\text{-}27)$

$$^+CO-NH-\overset{R_n}{\underset{|}{CH}}-CO-\left(NH-\overset{R}{\underset{|}{CH}}-CO\right)_{n-1}OH$$

x_n (Σ_n+45)

$$H-\left(HN-\overset{R}{\underset{|}{CH}}-CO\right)_{n-1}-NH-\overset{R_n}{\underset{|}{CH}}-C\equiv O^+$$

$\underline{b_n}$ (Σ_n+1)

$$H_3\overset{+}{N}-\overset{R_n}{\underset{|}{CH}}-CO-\left(NH-\overset{R}{\underset{|}{CH}}-CO\right)_{n-1}OH$$

$\underline{y_n}$ (Σ_n+19)

$$H-\left(HN-\overset{R}{\underset{|}{CH}}-CO\right)_{n-1}-NH-\overset{R_n}{\underset{|}{CH}}-CO-\overset{+}{N}H_3$$

c_n (Σ_n+18)

$$^+\overset{R_n}{\underset{|}{CH}}-CO-\left(NH-\overset{R}{\underset{|}{CH}}-CO\right)_{n-1}OH$$

z_n (Σ_n+2)

$$\overset{\displaystyle\overset{H^+}{\overbrace{}}}{H-\left(NH-\overset{R}{\underset{|}{CH}}-CO\right)_{n-1}-NH-\overset{CHR'}{\underset{|}{CH}}}$$

d_n

$$\overset{\displaystyle\overset{H^+}{\overbrace{}}}{\overset{R'CH}{\underset{\|}{CH}}-CO-\left(NH-\overset{R}{\underset{|}{CH}}-CO\right)_{n-1}-OH}$$

w_n

$$\overset{\displaystyle\overset{H^+}{\overbrace{}}}{HN=CH-CO-\left(NH-\overset{R}{\underset{|}{CH}}-CO\right)_{n-1}-OH}$$

v_n

$$H_2\overset{+}{N}=\overset{R}{\underset{|}{CH}}$$

immonium ion

$$H_2N-\overset{R>1}{\underset{|}{CH}}-\left(CO-NH-\overset{R}{\underset{|}{CH}}\right)_m-C\equiv O^+$$

internal fragment $(\Sigma_{m+1}+1)$

Scheme 1. Fragments produced from protonated linear peptides (1).

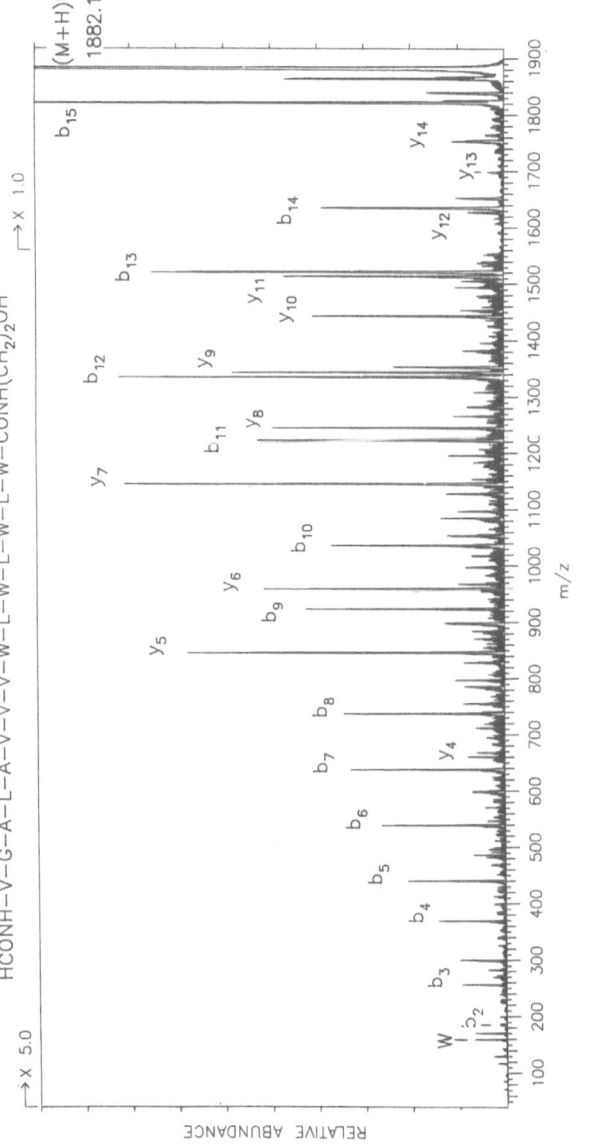

Figure 7. CID spectrum of valyl¹-gramicidin A.

group is formylated, rendering it nonbasic, and no basic amino acids are present. Therefore, protonation of this molecule in the FAB process leading to the $(M + H)^+$ ion that is subjected to collision takes place randomly on any of the amide nitrogens of the peptide bonds that trigger the cleavage of the resulting -CO-NH_2^+- bond. This leads to a b_n if the positive charge is retained at the carbonyl group and a neutral amine is formed, or a hydrogen is transferred from the N-terminal portion (Mueller *et al.*, 1988) to eliminate an ammonium ion, y_n. CID spectra of peptides generated by enzymatic cleavage of a protein usually contain at least one more basic site and are thus more complex (see Figs. 8, 10, and 12), because they also exhibit other types of fragments (Scheme 1). However, some of them are important in providing additional structural information, such as the d_n and w_n ions permitting the differentiation of leucine and isoleucine.

The differentiation of Leu/Ile is possible on the basis of the secondary fragmentation of the $\beta-\gamma$ bond of the side chain leading to C-terminal w_n ions, as shown in Scheme 1 (Johnson *et al.*, 1987), and N-terminal d_n ions (R S. Johnson *et al.*, 1988b). Mass spectrometric differentiation of Lys/Gln is more difficult and requires chemical conversion of Lys to a derivative, taking advantage of the basic ϵ-NH_2-group of this amino acid [i.e., acetylation or reaction with phenylisothiocyanate (PITC), which adds 42 or 135 Da, respectively]. For tryptic digests, the distinction can be made simply on the basis of the specificity of trypsin, which cleaves at lysine, unless it is followed by proline or flanked by acidic amino acids.

The major use of MS peptide sequencing is in cases where the conventional Edman method is ambiguous or not applicable. Foremost in this situation are peptides and proteins that are N-acylated and therefore do not react with PITC. The case of valyl-tRNA synthetase mentioned earlier is one example; other examples include peptides containing modified amino acids that either are converted back to the unmodified one (i.e., loss of carbohydrate from serine or threonine and decarboxylation of γ-carboxyglutamic acid to glutamic acid) or whose phenylthiohydantoin derivatives are too hydrophilic to be extracted into the nonpolar phase of the Edman cycle (i.e., glycosylated asparagine or phosphorylated hydroxy amino acids).

As an example, the first case of encountering *N*-acetyl-valine turned out to be the blocked N-terminus in the fatty acid binding protein from human heart (Offner *et al.*, 1988). Its N-terminal tryptic peptide resulted in the CID spectrum shown in Figure 8, which exhibits a b_1 ion at m/z 142 only compatible with an N-terminal acetyl-valine. The almost complete b_n series reveals the sequence but does not differentiate the last two amino acids, which must be Trp-Lys or Lys-Trp, based on the mass difference between the b_7 and the $(M + H)^+$ ion. The latter possibility can be eliminated because one is dealing with a tryptic peptide.

Another example is the detection and localization of phosphate groups attached to hydroxy amino acids. Such a group increases the mass of the amino

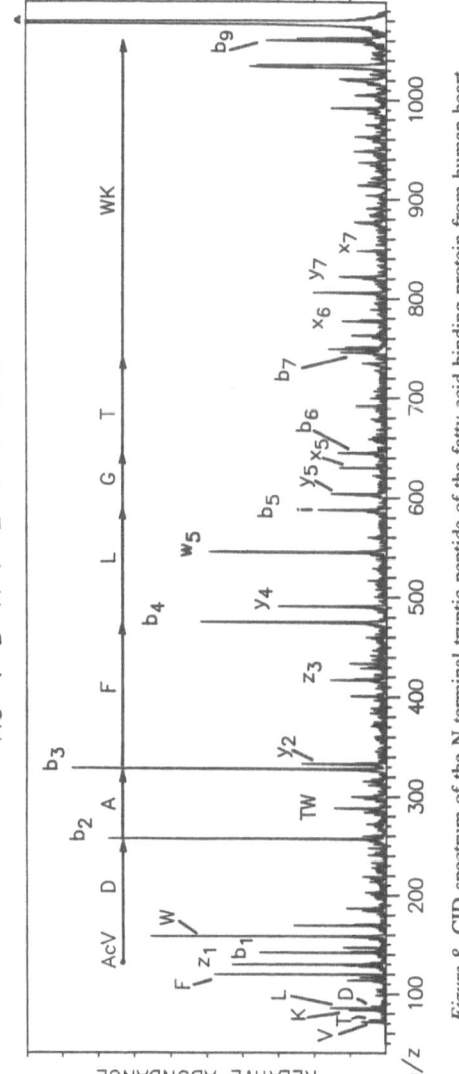

Figure 8. CID spectrum of the N-terminal tryptic peptide of the fatty acid binding protein from human heart.

acid by 80 Da. For phosphorylated serine and threonine, the phosphate ester group also can eliminate as H_3PO_4 upon collision, leading to ions of mass $(M - 98 + H)^+$ and mass differences between ions of the same series corresponding to dehydroalanine and dehydroaminobutyric acid, respectively. For example, the CID spectrum of the N-terminal peptide from troponin T from rabbit skeletal muscle revealed that the N-terminal amino acid is serine, which is not only N-acetylated (Ac) but is also phosphorylated at the hydroxyl group (Biemann and Scoble, 1987). The remainder of the spectrum indicated the sequence:

$$PO_3H_2$$
$$|$$
$$Ac\text{-}Ser\text{-}Asp\text{-}Glu\text{-}Glu\text{-}Val\text{-}Glu\text{-}His\text{-}Val\text{-}Glu\text{-}Glu$$

Another demonstration of the advantage of tandem MS over conventional techniques, reported in the same paper, involved a peptide synthesized by the automated Merrifield method. Two products were formed which could be separated by HPLC, but each gave the same amino acid analysis and Edman data for the expected sequence. The mass spectrum revealed that the earlier eluting component had the expected molecular weight, while that of the other product was 18 Da lower. The CID spectrum of the latter lacked the a_9 and b_9 ions which were abundant in the spectrum of the peptide with the correct molecular weight. Since the ninth amino acid is aspartic acid, it was obvious that cyclization by elimination of H_2O had taken place, which also prevented fragmentation at the aspartic acid position.

3.1. Sequencing of Proteins

We have developed a strategy for determining the primary structure of proteins by tandem MS. It is centered around the fact that mixtures of peptides produced by chemical or enzymatic cleavage of the protein do not have to be separated into individual components and purified before sequencing (Johnson and Biemann, 1987). This obviously saves considerable time and material. Another characteristic is the ability to determine the mass of peptides to better than 1 Da, providing information that permits the proper alignment of the peptides generated by one specific cleavage, generally by trypsin. Scheme 2 depicts this strategy. As a practical matter, it is advisable to sequence not only all the tryptic peptides, but also the majority of the secondary digest (generally using *Staphylococcus aureus* protease V8) to provide some redundancy of the sequence data. As mentioned earlier, the presence and position of basic amino acids in the peptide greatly influence the fragmentation of $(M + H)^+$ ions upon CID. Therefore, tryptic peptides (with a basic amino acid at the C-terminus) and *S. aureus* V8 peptides (that end in glutamic acid and may have a basic amino acid elsewhere) that originate from the same region will fragment differently, increasing

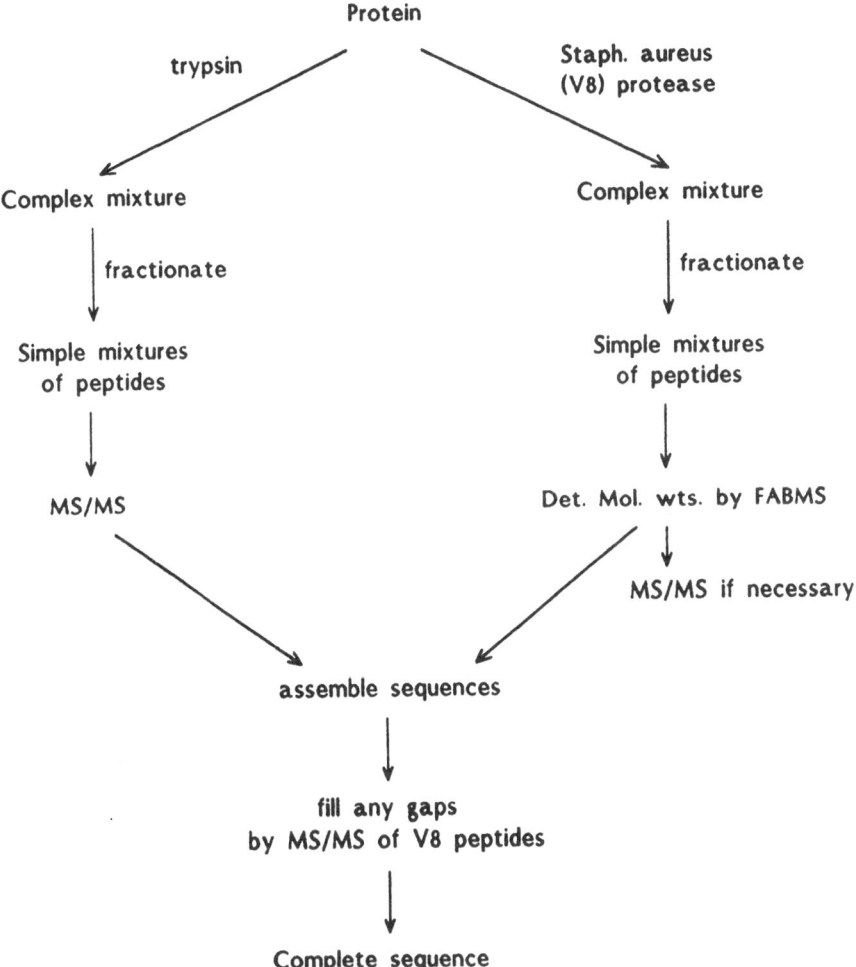

Scheme 2. Strategy for protein sequencing by tandem mass spectrometry.

the chances that all amino acids can be unambiguously placed and most, if not all, leucines and isoleucines are correctly identified.

Using this strategy, we have determined the primary structure of a number of thioredoxins, ubiquitous proteins about 105–110 amino acids long. Figure 9 shows a sequence isolated from *Rhodospirillum rubrum* and includes an indication of the sets of digest used (T. C. Johnson *et al.*, 1988). The heavy underlining denotes the sequence information derived from the CID spectra of the respective peptides. This composite figure not only demonstrates the identification of all

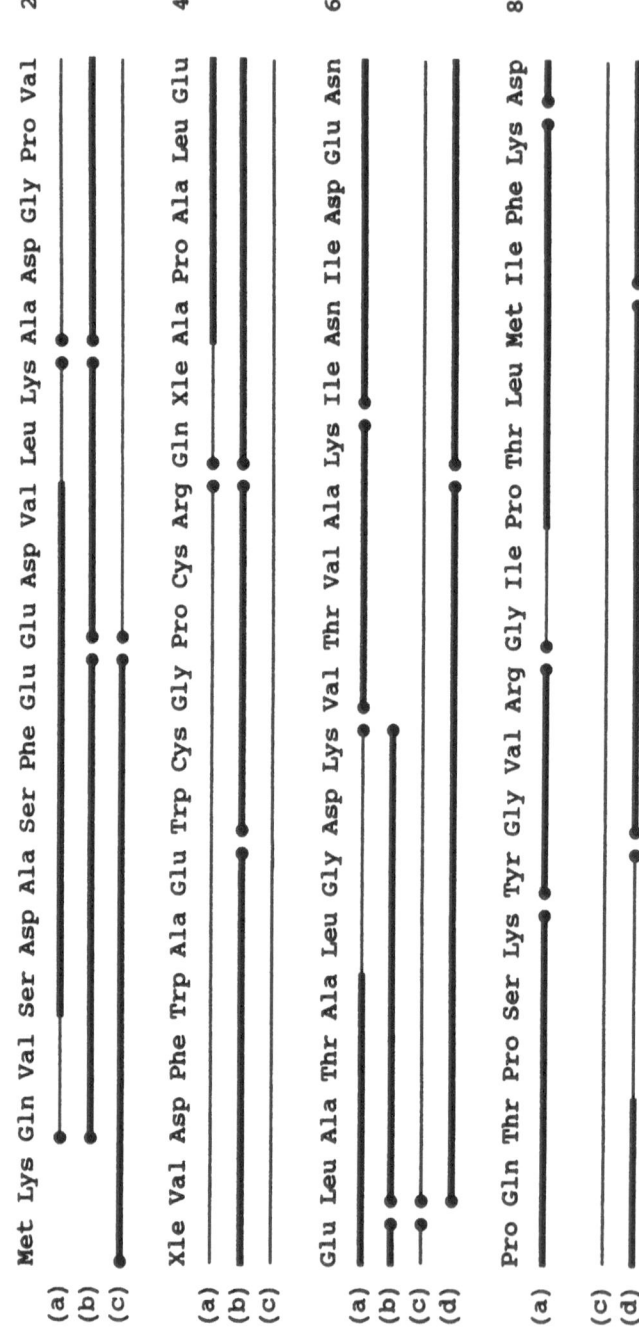

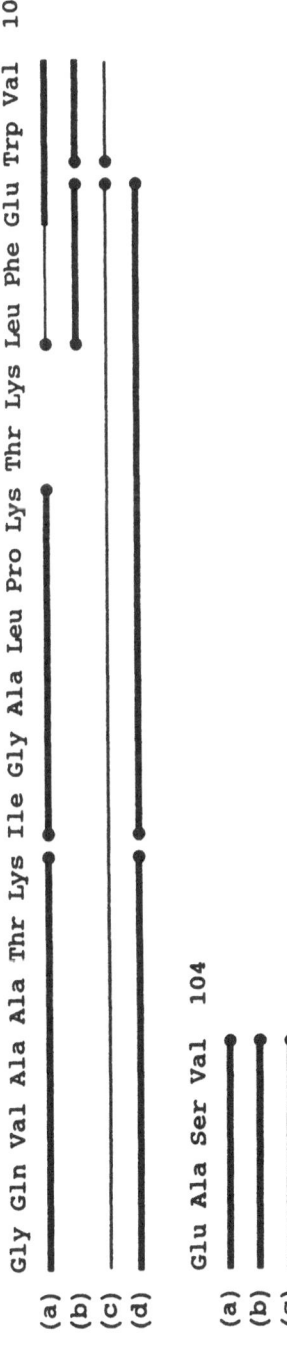

Figure 9. Amino acid sequence of *Rhodospirillum rubrum* thioredoxin: (a) tryptic peptides; (b) *Staphylococcus aureus* protease digestion of selected tryptic peptides; (c) *S. aureus* peptides; (d) further digestion of selected *S. aureus* proteolytic peptides with chymotrypsin. Heavy underlining: sequence derived from CID spectra. Solid lines without heavy markings indicate regions that match observed molecular weights of unsequenced peptides. Xle denotes positions where leucine could not be distinguished from isoleucine.

104 amino acids of this protein, but also shows that there is considerable redundance of the overlapping data. As an example, Figure 10 shows the CID spectrum of the tryptic peptide covering positions 71–79. It is characterized by extended series of complementary b_n and y_n ions but, as will be noted, it does not permit the exact definition of the first two amino acids because it begins with b_2 and ends with y_7. These were determined to be glycine followed by isoleucine (not leucine) by the CID spectrum of the overlapping peptide Gly(68)–Met(76) from a chymotryptic digest. Figure 11 shows the structure of four thioredoxins determined by tandem MS, including the first from a mammalian system (R. S. Johnson et al., 1988a). It will be noted that 80% of the leucines and isoleucines have been assigned on the basis of d_n or w_n ions in the CID spectra, but 20% remained undifferentiated. We have now developed a microderivatization technique that places a fixed positive charge at the N-terminus of a peptide, thus forcing the formation of d_n ions (Vath, 1990). This approach was used successfully in the course of the determination of the primary structure of the glutaredoxin isolated from rabbit bone marrow (Hopper et al., 1989), in which case all 19 leucines and isoleucines were unambiguously identified.

3.2. Increasing Sensitivity by Using an Array Detector

Obviously, in order to make tandem MS a useful complementary alternative to the Edman degradation, one must decrease the sample requirement for sequencing to the low picomole level. We have recently achieved this by designing a detector that increases the sensitivity of MS-2 at least 50-fold, bringing it into the required range. This was accomplished by using a focal plane detector that simultaneously collects 2048 data points over 6.6% of mass in a static mode, rather than collecting one point after another by sweeping the ion beam across a very narrow slit as a conventional detector does (Hill et al., 1988, 1989a). Consecutive segments of 6.6% each in mass are combined to the full spectrum by a computer and displayed. Figure 12 shows the CID spectrum obtained with 10 pmole of the tetradecapeptide renin substrate. This is well within, or perhaps below, the sample requirements of an automated gas-phase Edman sequencer, but is vastly superior in speed because the sequence of the peptide is defined by the spectrum, which takes only a few minutes to acquire. This is in sharp contrast to the 30–60 min required for each amino acid (i.e., 7–14 hr for a tetradecapeptide) by the Edman procedure.

This system has been further improved by the incorporation of a pair of quadrupole lenses placed between the exit of the magnet of MS-2 and its conventional detector. This arrangement greatly increases (to as much as 30%) the portion of the mass spectrum that can be focused simultaneously on the array detector, thus further increasing the sensitivity by a factor of five or decreasing the time required to record the spectrum by one-fifth (Hill et al., 1989b).

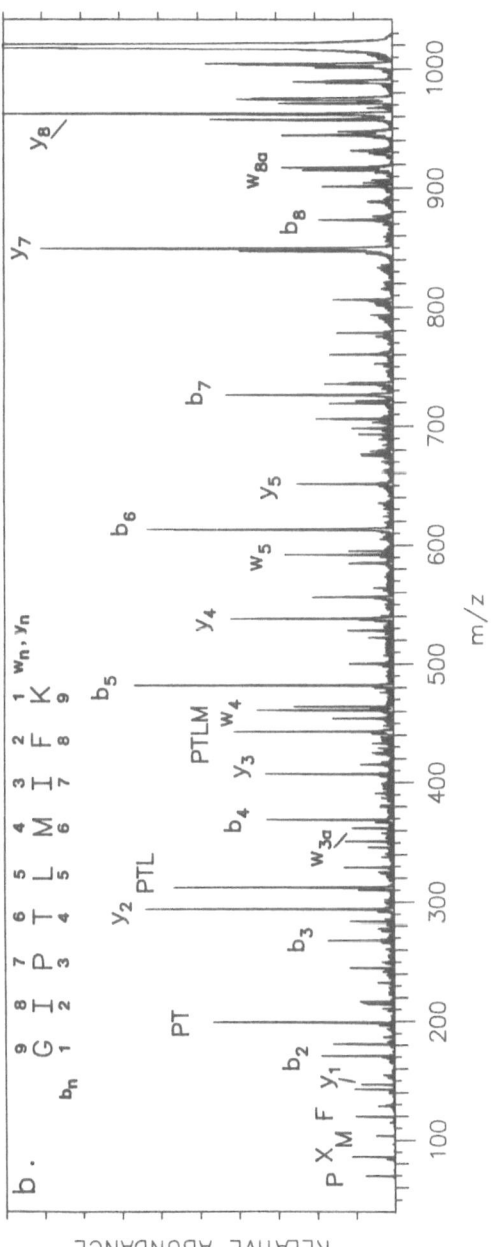

Figure 10. CID mass spectrum of the tryptic peptide representing amino acids 71–79 of the thioredoxin from *Rhodospirillum rubrum*.

```
Chlorobium    Ala Gly Lys Tyr Phe Glu Ala Thr Asp Lys Asn Phe Gln Thr Glu  15
Chromatium    Ser Asp Ser Ile Val His Val Thr Asp Asp Ser Phe Glu Glu Glu
R. rubrum             Met Lys Gln Val Ser Asp Ala Ser Phe Glu Glu Asp
    Rabbit        Val Lys Gln Ile Glu Ser Lys Ser Ala Phe Gln Glu Val Leu


Chlorobium    Ile Leu Asp Ser Asp Lys Ala Val Xle Val Asp Phe Trp Ala Ser  30
Chromatium    Val Xle Lys Ser Pro Asp Pro Val Leu Val Asp Tyr Trp Ala Asp
R. rubrum     Val Leu Lys Ala Asp Gly Pro Val Xle Val Asp Trp Ala Glu
    Rabbit    Asp Ser Ala Gly Asp Lys Leu Val Val Val Asp Phe Ser Ala Thr


Chlorobium    Trp Cys Gly Pro Cys Met Met Xle Gly Pro Val Ile Glu Gln Leu  45
Chromatium    Trp Cys Gly Pro Cys Lys Met Ile Ala Pro Val Leu Asp Glu Ile
R. rubrum     Trp Cys Gly Pro Cys Arg Gln Xle Ala Pro Ala Leu Glu Glu Leu
    Rabbit    Trp Cys Gly Pro Cys Lys Met Ile Lys Pro Phe Phe His Ala Leu


Chlorobium    Ala Asp Asp Tyr Glu Gly Lys Ala Ile Ile Ala Lys Xle Asn Val  60
Chromatium    Ala Asp Glu Tyr Ala Gly Arg Val Lys Xle Ala Lys Xle Asn Ile
R. rubrum     Ala Thr Ala Leu Gly Asp Lys Val Thr Val Ala Lys Ile Asn Ile
    Rabbit    Ser Glu Lys Phe Asn Asn Val Val Phe Ile Glu Val Asp Val Asp


Chlorobium    Asp Glu Asn Pro Asn Ile Ala Gly Gln Tyr Gly Xle Arg Ser Ile  75
Chromatium    Asp Glu Asn Pro Asn Thr Pro Pro Arg Tyr Gly Ile Arg Gly Ile
R. rubrum     Asp Glu Asn Pro Gln Thr Pro Ser Lys Tyr Gly Val Arg Gly Ile
    Rabbit    Asp Cys Lys Asp Ile Ala Ala Glu Cys Glu Val Lys Cys Met Pro


Chlorobium    Pro Thr Met Leu Ile Xle Lys Gly Gly Lys Val Val Asp Gln Met  90
Chromatium    Pro Thr Leu Met Leu Phe Arg Gly Gly Glu Val Glu Ala Thr Lys
R. rubrum     Pro Thr Leu Met Ile Phe Lys Asp Gly Gln Val Ala Ala Thr Lys
    Rabbit    Thr Phe Gln Phe Phe Lys Lys Gly Gln Lys Val Gly Glu Phe Ser


Chlorobium    Val Gly Ala Leu Pro Lys Asn Met Ile Ala Lys Lys Ile Asp Glu 105
Chromatium    Val Gly Ala Val Ser Lys Ser Gln Leu Thr Ala Phe Leu Asp Ser
R. rubrum     Ile Gly Ala Leu Pro Lys Thr Lys Leu Phe Glu Trp Val Glu Ala
    Rabbit    Gly Ala Asn Lys Glu Lys Leu Glu Ala Thr Ile Asn Glu Leu Leu


Chlorobium    His Ile Gly
Chromatium    Asn Xle
R. rubrum     Ser Val
```

Figure 11. Amino acid sequences (determined by tandem MS) of the thioredoxins isolated from *Chlorobium thiosulfatophilum, Chromatium vinosum, Rhodospirillum rubrum,* and rabbit bone marrow. Leucines and isoleucines are solidly underlined if differentiated (about 80% of occurrence); if undifferentiated (about 20%), they are denoted by Xle and underlined with a broken line.

3.3. Computer-Aided Interpretation of CID Spectra of Peptides

While the spectrum shown in Figure 7 is simple enough to be correctly interpreted in a relatively short time, a peptide of unknown sequence may require more time to work through and assign all peaks correctly. Having established the nature of the major fragmentation processes upon CID of the $(M + H)^+$ ions of a

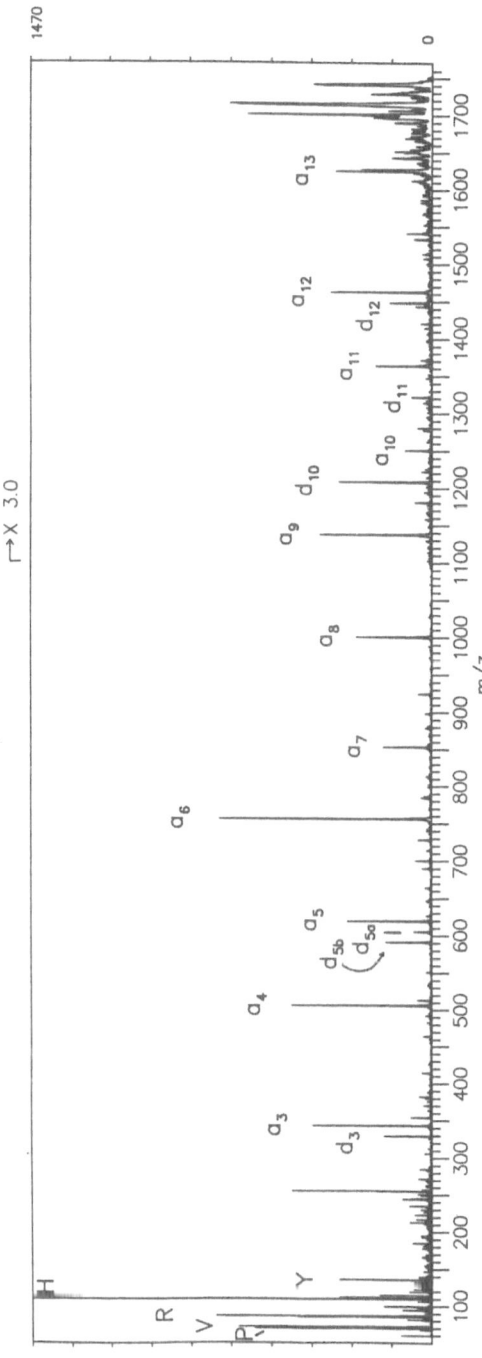

Figure 12. CID spectrum of 10 pmoles of the tetradecapeptide renin substrate (D-R-V-Y-I-H-P-F-H-L-L-V-Y-S) determined with an array detector.

large variety of peptides, it became possible to incorporate this knowledge into algorithms that systematically go through the spectrum and evaluate all possible structural elements and to present and rank the sequence or sequences that are compatible with the data.

We have developed two entirely different approaches. The first is an interactive process leading to a graphic display on a computer terminal, connecting all peaks in a CID spectrum that differ by the mass of an amino acid residue and labeling the pair by the one-letter code of the amino acid (Scoble *et al.*, 1987). The second algorithm is noninteractive and prints the most likely sequences and the assignment of the peaks corresponding to Scheme 1. These sequences are ranked by a score that represents the fraction of peaks (by intensity) assigned to the total summed intensity of all of the peaks in the spectrum; that is, a sequence that accounts for the entire spectrum would be assigned a score of 1.000 (Johnson and Biemann, 1988). Such algorithms not only eliminate the tedious search for pairs of peaks that differ by the mass of an amino acid residue, taking into account the various fragmentation processes depicted in Scheme 1, but also carry out an exhaustive search of all the possibilities for which the human interpreter would lack the patience. Most important is the speed of the search: it requires only a few minutes, which is similar to the time needed to acquire the data in the first place.

4. Conclusion

Having reduced the sample requirements of tandem MS by the use of the array detector, and having solved the data interpretation bottleneck by the computer algorithms described earlier, one can conclude that tandem MS is at least equally, if not significantly, more efficient than the conventional Edman method, even in its most sensitive version of the automated micro-gas phase sequencers. Thus, in situations in which a large number of sequences must be determined or checked in a short period of time, tandem MS will be more cost-efficient than the automated Edman sequencers, even though the capital investment is at least five times higher for a high-performance tandem MS than for a gas phase sequencer. The speed of the MS technique could be 10–100 times faster (depending on the organizational setup), resulting in from double to 20-fold efficiency in situations in which a large number of sequences must be determined as quickly as possible.

There are, of course, certain situations to which each approach is unique. The Edman method allows the sequencing of an N-terminal stretch of a protein of any size (unless its N-terminus is blocked), which is not possible with MS. On the other hand, the MS approach allows the sequencing of N-blocked peptides and can handle almost any modified amino acid, even if its structure is not known at the outset. Finally, MS is a general technique and is equally applicable to the investigation of other compounds types, such as carbohydrates.

ACKNOWLEDGMENTS. The author would like to thank J. E. Biller, C. E. Costello, J. A. Hill, S. A. Martin, I. A. Papayannopoulos, H. A. Scoble, and J. E. Vath for their contributions to the work described; S. Hopper and B. B. Buchanan for samples of the thioredoxins; and the National Institutes of Health for financial support (NIH Grants RR00317 and GM05472).

References

Barber, M., Bordoli, R. S., Sedgwick, R. D., and Tyler, A. N., 1981, Fast atom bombardment of solids (F.A.B.): A new ion source for mass spectrometry, *J. Chem. Soc., Chem. Comm.* **7**:325–327.

Biemann, K., and Scoble, H. A., 1987, Characterization by tandem mass spectrometry of structural modifications in proteins, *Science* **237**: 992–998.

Carr, S. A., and Roberts, G. D., 1986, Carbohydrate mapping by mass spectrometry: A novel method for identifying attachment sites of Asn-linked sugars in glycoproteins, *Anal. Biochem.* **157**: 396–406.

Chatton, B., Walter, P., Ebel, J. P., Lacroute, F., and Fasiolo, F., 1988, The yeast VAS1 gene encodes both mitochondrial and cytoplasmic valyl-tRNA synthetases, *J. Biol. Chem.* **263**:52–57.

Gibson, B. W., and Biemann, K., 1984, Strategy for the mass spectrometric verification and correction of the primary structures of proteins deduced from their DNA sequences, *Proc. Natl. Acad. Sci. USA* **81**: 1956–1960.

Hakansson, P., Kamensky, I., Sundqvist, B., Fohlman, J., Peterson, P., McNeal, C. J., and Macfarlane, R. D., 1982, [127]I-plasma desorption mass spectrometry of insulin, *J. Am. Chem. Soc.* **104**:2948–2949.

Hill, J. A., Martin, S. A., Biller, J. E., and Biemann, K., 1988, Use of a microchannel array detector in a four-sector tandem mass spectrometer, *Biomed. Environ. Mass Spectrom.* **17**:147–151.

Hill, J. A., Biller, J. E., Martin, S. A., Biemann, K., Yoshidome, K., and Sato, K., 1989a, Design considerations, calibration and applications of an array detector for a four sector tandem mass spectrometer, *Int. J. Mass Spectrom. Ion Processes* **92**:211–230.

Hill, J. A., Biller, J. E., Martin, S. A., Biemann, K., and Ishihara, M., 1989b, A flexible array detector of extended mass range for a high performance tandem mass spectrometer, *Proceedings 37th ASMS Conference on Mass Spectrometry and Allied Topics*, Miami, American Society of Mass Spectrometry, pp. 1077–1078.

Hopper, S., Johnson, R. S., Vath, J. E., and Biemann, K., 1989, Glutaredoxin from rabbit bone marrow. Purification, characterization and amino acid sequence determined by tandem mass spectrometry, *J. Biol. Chem.* **264**:20438–20447.

Johnson, R. S., and Biemann, K., 1987, The primary structure of thioredoxin from *Chromatium vinosum* determined by high performance tandem mass spectrometry, *Biochemistry* **26**:1209–1214.

Johnson, R. S., and Biemann, K., 1988, Computer program for the interpretation of CID mass spectra obtained from peptides of unknown sequence, Presented at the 36th ASMS Conference on Mass Spectrometry and Allied Topics, San Francisco, CA, pp. 1398–1399.

Johnson, R. S., Martin, S. A., Biemann, K., Stults, J. T., and Watson, J. T., 1987, Novel fragmentation process of peptides by collision induced decomposition in a tandem mass spectrometer: Differentiation of leucine and isoleucine, *Anal. Chem.* **59**:2621–2625.

Johnson, R. S., Mathews, W. R., Biemann, K., and Hopper, S., 1988a, Amino acid sequence of thioredoxin isolated from rabbit bone marrow determined by tandem mass spectrometry, *J. Biol. Chem.* **263**:9589–9597.

Johnson, R. S., Martin, S. A., and Biemann, K., 1988b, Collision induced fragmentation of (M + H)$^+$ ions of peptides. Side chain specific sequence ions, *Int. J. Mass Spectrom. Ion Processes* **86:**137–154.

Johnson, T. C., Yee, B. C., Carlson, D. E., Buchanan, B. B., Johnson, R. S., Mathews, W. R., and Biemann, K., 1988, Thioredoxin from *Rhodospirillum rubrum:* Primary structure and relation to thioredoxins from other photosynthetic bacteria, *J. Bacteriol.* **170:**2406–2408.

Mueller, D. R., Eckersley, M., and Richter, W. J., 1988, Hydrogen transfer reactions in the formation of "Y + 2" sequence ions from protonated peptides, *Org. Mass Spectrom.* **23:**217–222.

Offner, G. D., Sawlivich, W. B., Brecher, P., Costello, C. E., and Troxler, R. F., 1988, Characterization and amino acid sequence of a fatty acid binding protein from human heart, *Biochem. J.* **252:**191–198.

Reddy, V. A., Johnson, R. S., Biemann, K., Williams, R. S., Ziegler, F. D., Trimble, R. B., and Maley, F., 1988, Characterization of the glycosylation sites in yeast external invertase. I. *N*-linked oligosaccharide content of the individual sequons. *J. Biol. Chem.* **263:**6978–6985.

Sato, K., Asada, T., Ishihara, M., Kunihiro, F., Kammei, Y., Kubota, E., Costello, C. E., Martin, S. A., Scoble, H. A., Biemann, K., 1987, High performance tandem mass spectrometry: Calibration and performance of linked scans of a four-sector instrument, *Anal. Chem.* **59:**1652–1659.

Scoble, H. A., Biller, J. E., and Biemann, K., 1987, A graphics display-oriented strategy for the amino acid sequencing of peptides by tandem mass spectrometry, *Fresenius Z. Anal. Chem.* **327:**239–245.

Shimonishi, Y., Hong, Y-M., Kitagishi, T., Matsuo, T., Matsuda, H., Katakuse, I., 1980, Sequencing of peptide mixtures by Edman degradation and field-desorption mass spectrometry, *Eur. J. Biochem.* **112:**251–264.

Vath, J. E., Ph.D. Thesis, Microscale derivatization of peptides and glycolipids for mass spectrometry, Massachusetts Institute of Technology, 1990.

II

PROTEIN FOLDING AND STABILITY

Protein Folding and Stability

ROBERT L. BALDWIN

1. Goals

This section includes chapters on the process of protein folding and on factors that affect protein stability. The practical benefits of these studies are obvious, both for obtaining improved yields of proteins produced in expression systems and for improving the working properties of commercial enzymes. The basic motivation, however, for most scientists studying protein folding and stability is to help elucidate the mechanisms used to translate the linear amino acid sequence of a polypeptide into the three-dimensional structure of a protein. By studying the folding process, scientists set out to trap folding intermediates in order to determine their structures and find out what interactions stabilize these structures. By analyzing individual factors that affect stability, they can develop general methods for increasing protein stability, based on site-directed mutagenesis, and find out how individual interactions work together to determine the three-dimensional structure of a protein.

2. Folding of Fragments

Both the process of folding and factors that affect stability can be investigated by analyzing the folding of fragments. The C-peptide of ribonuclease A (RNase A) is used here to illustrate the procedure. This 13-residue peptide from the N-terminal end of RNase A corresponds to a helix-containing segment of the protein: residues 3–13 are H-bonded in a short α-helix. The initial finding that C-peptide forms a marginally stable α-helix in H_2O near 0°C (Brown and Klee,

ROBERT L. BALDWIN • Department of Biochemistry, Beckman Center, Stanford University Medical Center, Stanford, California 94305.

1971) was surprising, especially because similar studies of other protein fragments (Epand and Scheraga, 1968; Taniuchi and Anfinsen, 1969) had failed to show any evidence of helix formation and led to the generalization that short helices exist in aqueous solution only when stabilized by tertiary interactions with other segments. A later reinvestigation of C-peptide (Bierzynski *et al.*, 1982), which confirmed the conclusion of Brown and Klee (1971), produced the following paradox: although C-peptide shows clear evidence of partial helix formation, one can predict that it is impossible for any 13-residue peptide to form a substantial amount of α-helical structure in H_2O in a monomolecular reaction, if α-helix formation obeys the model of Zimm and Bragg (1959) and σ (the nucleation constant) is equal to 10^{-3} or less while the value of s (the helix growth parameter) is given separately for each amino acid, from the host-guest data provided by Sueki *et al.* (1984).

Further work suggested a resolution of this paradox, when it was found that two charged residues at either end of the helix, Glu 2^- and His 12^+, play special roles in stabilizing the helix (Shoemaker *et al.*, 1985; Rico *et al.*, 1986). It was shown that these residues participate in specific side-chain interactions, Glu 2^-. . .Arg 10^+ and Phe 8. . .His 12^+ (Rico *et al.*, 1986; Shoemaker *et al.*, 1987a) that can be explained by the X-ray structure of RNase A (see Fig. 1) if the structure of the 3-13 helix is preserved in the isolated peptide. Residues Glu 2^- and Arg 10^+ form a salt bridge (H-bonded ion pair) and the aromatic rings of Phe 8 and His 12 are stacked sideways in a conformation suggestive of an aromatic ring interaction (Blundell *et al.*, 1986; Bermejo *et al.*, 1986).

Making the Glu 2^-. . .Arg 10^+ salt bridge terminates the helix, because Glu 2^- must be in a nonhelical conformation to form this salt bridge. Thus, the Glu 2^-. . .Arg 10^+ interaction contributes to helix localization as well as stability. A third type of helix-stabilizing or destabilizing interaction was also found in C-peptide, when it was learned that the charge on the N-terminal residue has a major effect on helix stability (Shoemaker *et al.*, 1987b); Fairman *et al.*, 1989). This happens because the charge interacts with the nearby positive pole of the helix dipole. Four main-chain NH groups at the end of the helix are not H-bonded to CO groups, and each carries a partial positive charge. At the C-terminal end of the helix, each of four CO groups carries a partial negative charge, and the overall assembly of charges resembles an extended line dipole: hence the term "helix dipole." The charge on the N-terminal residue, which is close to the positive pole of the helix dipole, destabilizes the helix if it is positive (Lys^{2+} or Ala$^+$) and stabilizes the helix if it is negative (succinyl-Ala$^-$) (Shoemaker *et al.*, 1987b).

It was also possible to use this same type of system to test the effects of the helix–dipole interaction on the stability of reconstituted ribonuclease. The procedure was to vary the charge on the N-terminal residue in longer peptides containing residues 1–15 of RNase A (termed S-peptide 1-15). These 15-residue

Figure 1. Residues 1–13 of RNase A, showing the Glu 2⁻ . . . Arg 10⁺ salt bridge and the possible aromatic interaction between Phe 8 and His 12⁺. Other side chains have been removed. The coordinates of RNase A are from Wlodawer and Sjölin (1983) and are deposited in the Brookhaven Data Bank.

peptides bind strongly to the S-protein moiety (residues 21–124) to give reconstituted, enzymatically active ribonuclease. Varying the charge at residue 1 causes the thermal stability of reconstituted ribonuclease to change in the same manner as the stability of the isolated helix (Mitchinson and Baldwin, 1986).

Thus, on the one hand, C-peptide has provided a system for discovering and testing specific side-chain interactions, and on the other hand, it has shown how side-chain interactions contribute to helix localization and to protein stability. A conclusion of general interest concerning the mechanism of protein folding is that side-chain interactions between residues that are fairly close in sequence may be important in directing the folding process.

Covalently joined protein fragments have recently been used to model a more complex folding intermediate. Oas and Kim (1988) have used S–S bond formation to join covalently two fragments of pancreatic trypsin inhibitor (BPTI); one fragment (residues 43–58) contains a helix in BPTI, while the other fragment (residues 20–33) is from a β-sheet region. The individual fragments are nearly structureless, even at 0°C, but proximity enforced by covalent joining results in mutual stabilization of the secondary structures of both fragments, as shown by circular dichroism (CD) and nuclear magnetic resonance (NMR) studies. Mutter (1988) has introduced a synthetic strategy for bringing different peptide fragments into proximity by chemically cross-linking them to a common short "template" strand.

Fragment folding is proving useful even in the study of such basic interactions as the peptide H bond. A controversy exists in the older literature as to whether peptide H bonds make any significant contribution to protein stability, because competing H bonds to H_2O are nearly, if not equally, strong. These earlier studies are based on simple model compounds, such as urea (which forms weak dimers in H_2O; Schellman, 1955), N-methylacetamide (Klotz and Franzen, 1960), and δ-valerolactam (Susi et al., 1964). The helix formed by C-peptide unfolds with increasing temperature, showing that helix formation is enthalpy-driven (Bierzynski et al., 1982; Shoemaker et al., 1987b), and it is likely that the formation of peptide H bonds contributes most of the enthalpy change. Helices formed by short peptides of de novo design, which have simple repeating sequences and are stabilized by Glu^-, Lys^+ ion pairs, also show enthalpy-driven helix formation (Marqusee and Baldwin, 1987). Different side-chain interactions stabilize these two classes of peptide helix, which nevertheless show quite similar thermal unfolding curves. This finding suggests that there is a significant enthalpy of peptide H-bond formation in H_2O, in agreement with the hypothesis of Schellman (1955) and Hermans (1966). The thermodynamics of H-bond formation in proteins has also been studied by site-directed mutagenesis (Fersht et al., 1985). Using the binding of substrates to tyrosyl tRNA synthase as a model system, they found that the Gibbs free-energy change of making an H bond between neutral donor and acceptor groups is small but measurable (about -1 kcal/H bond).

Fragment folding studies may also prove useful in probing the role of the hydrophobic interaction in protein folding. Most workers agree with the conclusion of Kauzmann (1959) and Tanford (1962; cf. Nozaki and Tanford, 1971) that the hydrophobic interaction provides the major source of Gibbs free-energy driving folding, and that the free-energy change of burying a hydrophobic amino acid residue in the interior of a protein can be estimated semiquantitatively by transferring an amino acid side chain from H_2O to a suitable nonpolar solvent (e.g., n-octanol). There are two basic concerns about this procedure. One is the question of whether an organic liquid provides a reasonable model for the interior of a protein. The other is that two hydrophobic molecules, in H_2O, such as two benzene molecules, have a very weak tendency to associate into dimers (Stellner et al., 1983). In fact, the dimerization of benzene or other similar hydrocarbons is markedly stronger in the gas phase than in H_2O (Wood and Thompson, 1990). Yet the conventional view of the hydrophobic interaction is that H_2O forces two hydrophobic molecules together, because it is energetically difficult to accommodate both molecules separately in the H_2O structure; this requires making separate cavities and reorienting the H_2O molecules at the surface of each cavity. By studying leucine → alanine substitutions in model peptides that form dimeric α-helices (the coiled-coil structure), Hodges *et al.* (1988) have shown how the contribution of leucine–leucine interactions to stabilizing the coiled-coil structure can be analyzed, and this provides a promising approach, based on fragment folding, for analyzing the hydrophobic interaction in protein folding.

3. Trapping Folding Intermediates

The problem is discussed here from the perspective of studies on small, single-domain proteins. The basic difficulty in analyzing the kinetic pathway of folding is that, even after conditions have been found in which intermediates are populated, these intermediates exist only for very short times, that is, a few seconds or less. Methods capable of giving detailed information about protein structure, such as two-dimensional NMR and X-ray crystallography, require much longer times (at least a few hours). Various solutions to this problem have been proposed, including making ambient one-dimensional NMR observations of protein folding in cryosolvents (e.g., aqueous methanol) at subzero temperatures, where folding becomes extremely slow (Biringer and Fink, 1982). A few unusual small proteins show equilibrium intermediates in folding, notably α-lactalbumin (Kuwajima *et al.*, 1976), carbonic anhydrase (Wong and Tanford, 1973), β-lactamase (Robson and Pain, 1976), and apomyoglobin (Griko *et al.*, 1988). (Later references are given in the review by Ptitsyn, 1987.) These unusual proteins show a common unfolding behavior in which the native structure is not very stable and, when it is disrupted by a denaturant, an appreciable amount of secondary structure remains, as estimated by CD. In the folding of α-lactalbumin, a kinetic intermediate is formed rapidly that has properties corresponding to the equilibrium folding intermediate: in particular, both intermediates show the same unfolding curve when guanidinium chloride (GuHCl) is used to denature the protein (Ikeguchi *et al.*, 1986). This finding supports the hypothesis that the equilibrium intermediate is on the kinetic pathway of folding.

The equilibrium intermediates seen in the folding of these different proteins share common and striking properties, which have been reviewed recently by Ptitsyn (1987). They have high contents of secondary structure, as measured by CD, and are nearly as compact as the native protein, but they show little evidence of organized tertiary structure. Ptitsyn (1987) argues that the side chains do not adopt fixed conformations in this class of intermediates, which belong to "the molten globule state" in the terminology of Ohgushi and Wada (1983). Two mutants of β-lactamase have been found (Craig *et al.*, 1985) in which folding does not yield a nativelike structure, but instead a conformation resembling the molten globule state. Thus, one approach to determining the structures of folding intermediates is to analyze these uncommon equilibrium intermediates, and to find out if they belong on the kinetic pathway of folding.

It is also important to find methods of trapping or labeling kinetic intermediates in folding so that their structures can be determined at leisure, since at least some of these intermediates are likely to be on the kinetic pathway of folding. Creighton (1974) gave the first solution to this problem. Like most proteins that have S-S bonds, BPTI unfolds when it is reduced, and the regeneration of its S-S bonds is coupled to its refolding. Unlike many S-S-containing proteins, BPTI regenerates its S-S bonds in a reversible reaction, in which the same equilibrium curve for interconversion between the native and reduced forms is reached either by reoxidizing the reduced protein or by reducing the native protein (Creighton, 1977). A few 1-S-S and 2-S-S intermediates are well populated transiently during reoxidation. They can be trapped covalently, e.g., by reaction with iodoacetamide, and the different intermediates have been separated chromatographically. A preliminary one-dimensional NMR study suggests that the blocked intermediates contain some nativelike secondary structure (States *et al.*, 1987), and future two-dimensional NMR studies should define their structures. As noted earlier, fragment folding studies are giving valuable structural information about the structure of the major 30-51 S-S intermediate (Oas and Kim, 1988).

A recently reported noncovalent method pulse-labels kinetic folding intermediates in such a way that the structural locations of the labeled groups can be determined at leisure (Roder *et al.*, 1988; Udgaonkar and Baldwin, 1988). The method uses a multistage, stopped-flow, mix-and-quench apparatus to permit exchange between peptide NH groups and solvent (H_2O or D_2O), so that labeling during folding does not interfere with completion of the folding process. Using a procedure in which the unfolded protein is initially dissolved in D_2O, with all exchangeable NH groups deuterated, these authors initiate refolding by diluting out the denaturant (GuDCl) and adjusting the pH. Then, at various times after initiating folding, a pulse of exchange with H_2O is allowed to occur. The peptide NH groups, which have been converted to ND by unfolding the protein in D_2O, react with H_2O unless they are protected from exchange by structure in a folding intermediate. After folding is complete, ^{1}H-NMR is used to obtain the 2-D COSY spectrum of the folded protein in D_2O. Only the buried NH groups of the folded protein, which are stable to exchange with D_2O in the conditions of recording the COSY spectrum, can be studied. In intact proteins, many of the buried peptide NH groups are highly protected against exchange. Typically, protection is provided by H-bonding in α-helices and β-sheets, but accessibility to solvent is also required for exchange (Wagner, 1983; Englander and Kallenbach, 1984). Thus, this new experiment shows which peptide NH protons are protected against exchange by structure in a folding intermediate at various times during folding. The stability of the structure in the intermediates can also be probed by varying the pulse length during which exchange occurs or by increasing the pH of the exchange pulse, since exchange is typically base-catalyzed, even in folded proteins.

Results obtained by this new method are still preliminary, and the full potential remains to be evaluated. It is clear that the method gives detailed structural information about early events in folding, but it is not yet known whether late events can also be studied. In the folding of RNase A (S-S bonds intact), the entire β-sheet appears to be formed cooperatively at an early stage (Udgaonkar and Baldwin, 1988). In the folding of horse cytochrome c (FeIII), α-helices at the N- and C-termini of the protein are stabilized against exchange in an early folding event (Roder *et al.*, 1988), probably via a helix pairing reaction, since these helices are paired in folded cytochrome c.

The overall conclusion from these studies is that methods are now available for obtaining detailed structural information about intermediates in the folding process, as well as information about their stability. The expectation is that results obtained by these methods will contribute greatly to our understanding of the process by which the amino acid sequence determines the three-dimensional structure of a protein.

References

Bermejo, F. J., Rico, M., Santoro, J., Herranz, J. Gallego, E., and Nieto, J. L., 1986, Quantum-chemical calculations of a proposed Phe_n. . .His_{n+4} stabilizing interaction in peptide α-helices, *J. Mol. Structure* **142**:339–342.

Bierzynski, A., Kim, P. S., and Baldwin, R. L., 1982, A salt bridge stabilizes the helix formed by isolated C-peptide of RNase A, *Proc. Natl. Acad. Sci. USA* **79**:2470–2474.

Biringer, R. G., and Fink, A. L., 1982, Observation of intermediates in the folding of ribonuclease A at low temperature using proton nuclear magnetic resonance, *Biochemistry* **21**:4748–4755.

Blundell, T., Singh, J., Thornton, J., Burley, S. K., and Petsko, G. A., 1986, Aromatic interactions, *Science* **234**:1005.

Brown, J. E., and Klee, W. A., 1971, Helix-coil transition of the isolated amino terminus of ribonuclease, *Biochemistry* **10**:470–476.

Craig, S., Hollecker, M., Creighton, T. E., and Pain, R. H., 1985, Single amino acid mutations block a late step in the folding of β-lactamase from *Staphylococcus aureus*, *J. Mol. Biol.* **185**:681–687.

Creighton, T. E., 1974, Intermediates in the refolding of reduced pancreatic trypsin inhibitor, *J. Mol. Biol.* **87**:579–602.

Creighton, T. E., 1977, Energetics of folding and unfolding of pancreatic trypsin inhibitor, *J. Mol. Biol.* **113**:295–312.

Englander, S. W., and Kallenbach, N. R., 1984, Hydrogen exchange and structural dynamics of proteins and nucleic acids, *Q. Rev. Biophys.* **4**:521–655.

Epand, R. M., and Scheraga, H. A., 1968, The influence of long-range interactions on the structure of myoglobin, *Biochemistry* **7**:2864–2872.

Fairman, R., Shoemaker, K. R., York, E. J., Stewart, J. M., and Baldwin, R. L., 1988, Further studies of the helix dipole model: effects of a free α-NH$_3^+$ or α-COO-group on helix, *Proteins: Structure, Function, & Genetics* **5**:1–7.

Fersht, A. R., Shi, J.-P., Knill-Jones, J., Lowe, D. M., Wilkinson, A. J., Blow, D. M., Brick, P., Carter, P., Waye, M. M. Y., and Winter, G., 1985, Hydrogen bonding and biological specificity analyzed by protein engineering, *Nature* **314**:235–238.

Griko, Y. V., Privalov, P. L., Venjaminov, S. Y., and Kutyshenko, V. P., 1988. Thermodynamic study of the apomyoglobin structure, *J. Mol. Biol.* **202**:127–138.

Hermans, J., Jr., 1966, Experimental free energy and enthalpy of formation of the α-helix, *J. Phys. Chem.* **70**:510–515.

Hodges, R. S., Semchuk, P. D., Taneja, A. K., Kay, C. M., Parker, J. M. R., and Mant, C. T., 1988, Protein design using model synthetic peptides, *Peptide Res.* **1**:19–30.

Ikeguchi, M., Kuwajima, K., Mitani, M., and Sugai, S., 1986, Evidence for identity between the equilibrium unfolding intermediate and a transient folding intermediate: A comparative study of the folding reactions of α-lactalbumin and lysozyme, *Biochemistry* **25**:6965–6972.

Kauzmann, W., 1959, Factors in interpretation of protein denaturation, *Adv. Protein Chem.* **14**:1–63.

Klotz, I. M., and Franzen, J. S., 1960, The stability of interpeptide hydrogen bonds in aqueous solution, *J. Am. Chem. Soc.* **82**:5241.

Kuwajima, K., Nitta, K., Yoneyama, M., and Sugai, S., 1976, Three-state denaturation of α-lactalbumin by guanidine hydrochloride, *J. Mol. Biol.* **106**:359–373.

Marqusee, S., and Baldwin, R. L., 1987, Helix stabilization by Glu$^-$...Lys$^+$ salt bridges in short peptides of *de novo* design, *Proc. Natl. Acad. Sci. USA* **84**:8898–8902.

Mitchinson, C., and Baldwin, R. L., 1986, The design and production of semisynthetic ribonucleases with increased thermostability by incorporation of analogues with enhanced helical stability, *Proteins: Structure, Function, Genetics* **1**:23–33.

Mutter, M., 1988, Nature's rules and chemist's tools: A way for creating novel proteins, *Trends Biochem. Sci.* **13**:260–265.

Nozaki, Y., and Tanford, C., 1971, The solubility of amino acids and two glycine peptides in aqueous ethanol and dioxane solutions. Establishment of a hydrophobicity scale, *J. Biol. Chem.* **246**:2211–2217.

Oas, T. G., and Kim, P. S., 1988, A peptide folding intermediate, *Nature* **336**:42–48.

Ohgushi, M., and Wada, A., 1983, "Molten-globule state": A compact form of globular proteins with mobile side-chains, *FEBS Lett.* **164**:21–24.

Ptitsyn, O. B., 1987, Protein folding: hypotheses and experiments, *J. Protein Chem.* **6**:273–293.

Rico, M., Santoro, J., Bermejo, E. J., Herranz, J., Nieto, J. L., Gallego, E., and Jiménez, M. A., 1986, Thermodynamic parameters for the helix-coil thermal transition of ribonuclease S-peptide and derivatives from ^{1}H-NMR data, *Biopolymers* **25**:1031–1053.

Robson, B., and Pain, R. H., 1976, The mechanism of folding of globular proteins. Equilibria and kinetics of conformational transitions of penicillinase from *Staphylococcus aureus* involving a state of intermediate conformation, *Biochem. J.* **155**:331–344.

Roder, H., Elöve, G. A., and Englander, S. W., 1988, Structural characterization of folding intermediates in cytochrome c by H-exchange labelling and proton NMR, *Nature* **335**:700–704.

Schellman, J. A., 1955, The thermodynamics of urea solutions and the heat of formation of the peptide hydrogen bond, *Compt. Rend. Trav. Lab. Carlsberg Sér. Chim.* **29**:223–229.

Shoemaker, K. R., Kim, P. S., Brems, D. N., Marqusee, S., York, E. J., Chaiken, I. M., Stewart, J. M., and Baldwin, R. L., 1985, Nature of the charged-group effect on the stability of the C-peptide helix, *Proc. Natl. Acad. Sci. USA* **82**:2349–2353.

Shoemaker, K. R., Fairman, R., Kim, P. S., York, E. J., Stewart, J. M., and Baldwin, R. L., 1987a,

The C-peptide helix considered as an autonomous folding unit, *Cold Spring Harbor Symp. Quant. Biol.* **LII**:391–398.

Shoemaker, K. R., Kim, P. S., York, E. J., Stewart, J. M., and Baldwin, R. L., 1987b, Tests of the helix dipole model for stabilization of α-helices, *Nature* **326**:563–567.

States, D. J., Creighton, T. E., Dobson, C. M., and Karplus, M., 1987, Conformations of intermediates in the folding of the pancreatic trypsin inhibitor, *J. Mol. Biol.* **195**:731–739.

Stellner, K. L., Tucker, E. E., and Christian, S. D., 1983, Thermodynamic properties of the benzene-phenol dimer in dilute aqueous solution, *J. Sol. Chem.* **12**:307–313.

Sueki, M., Lee, S., Powers, S. P., Denton, J. B., Konishi, Y., and Scheraga, H. A., 1984, Helix-coil stability constants for the naturally occurring amino acids in H_2O.22. Histidine parameters from random poly (hydroxybutyl) glutamine co-L-histidine, *Marcomolecules* **17**:148–155.

Susi, H., Timasheff, S. N., and Ard, J. S., 1964, Near infrared investigation of interamide hydrogen bonding in aqueous solution, *J. Biol. Chem.* **239**:3051–3054.

Tanford, C., 1962, Contribution of hydrophobic interactions to the stability of the globular conformation of proteins, *J. Am. Chem. Soc.* **84**:4240–4247.

Taniuchi, H., and Anfinsen, C. B., 1969, An experimental approach to the study of the folding of staphylococcal nuclease, *J. Biol. Chem.* **244**:3864–3875.

Udgaonkar, J. B., and Baldwin, R. L., 1988, NMR evidence for an early framework intermediate on the folding pathway of ribonuclease A, *Nature* **335**:694–699.

Wagner, G., 1983, Characterization of the distribution of internal motions in the basic pancreatic trypsin inhibitor using a large number of internal NMR probes, *Q. Rev. Biophys.* **16**:1–57.

Wlodawer, A., and Sjölin, L., 1983, Structure of ribonuclease A: results of joint neutron and X-ray refinement at 2.0 A° resolution, *Biochemistry* **22**:2720–2728.

Wong, K-P. and Tanford, C., 1973, Denaturation of bovine carbonic anhydrase B by guanidine hydrochloride, *J. Biol. Chem.* **248**:8518–8523.

Wood, R. H., and Thompson, P. T., 1990, Differences between pair and bulk hydrophobic interactions, *Proc. Natl. Acad. Sci. USA* **87**:946–949.

Zimm, B. H., and Bragg, J. K., 1959, Theory of the phase transition between helix and random coil in polypeptide chains, *J. Chem. Phys.* **31**:526–535.

Genetic Analysis of Polypeptide Chain Folding and Misfolding in Vivo

JONATHAN KING, BEN FANE, CAMERON HAASE-PETTINGELL, ANNA MITRAKI, and ROBERT VILLAFANE

1. Introduction

Some 30 years have passed since the first protein structures were determined by X-ray crystallography, and since Anfinsen and co-workers established that the information for determining conformation could be contained solely within the polypeptide chain (see Anfinsen, 1973). The three-dimensional structure of more than 200 proteins is known to atomic resolution. For each of these proteins, and for thousands more, the amino acid sequence is also known exactly. Yet we are only beginning to be able to predict and understand how the sequence of the amino acids along polypeptide chains determines their three-dimensional conformation.

Two major reasons for the lag in solving this problem are now clear. The first is that polypeptide chains do not reach their native conformation directly, but must pass through a series of intermediate stages (Creighton, 1978; Kim and Baldwin, 1982). Thus, the amino acid sequence does not directly determine the structure of the native protein, but determines the conformation and sequence of formation of the kinetic intermediates. Solving the folding problem cannot be done simply by further inspecting more native structures; the conformation of the intermediates must be determined. This has been the focus of much of the work of the Baldwin group with ribonuclease (Chapter 3, this volume).

The second problem is the absence of information as to which amino acids

JONATHAN KING, BEN FANE, CAMERON HAASE-PETTINGELL, ANNA MITRAKI, and ROBERT VILLAFANE • Department of Biology, Massachusetts Institute of Technology, Cambridge, Massachusetts 02139.

in the chain direct the folding pathway, or determine the conformation. If intermediates are critical, then information in the sequence must specify the structure of the intermediates, as well as the stability and function of the native state (King, 1986). An examination of the sequences of homologous proteins, for example the hemoglobins, shows clearly that the residues at more than half of the positions can vary greatly, while the chains have the same final conformation (Bashford *et al.*, 1987).

Similarly, many of the successes of protein engineering depend on the fact that amino acids can be substituted at the active site, and the mutant polypeptide chain still folds up into the native structure. An often ignored conclusion of such work is that these active site residues are not important for determining conformation. The simplest way of understanding this is that only some of the positions are critical for determining the conformation.

2. The Protein Folding Grammar

One goal of the experimental work reviewed here is to help answer this question of which residues control the folding pathway. Three formal models for the distribution of the folding pathway information are shown in Figure 1. In the conventional model, every residue contributes to determining the conformation. This model underlines almost all algorithms for predicting secondary structure; each residue is assigned some propensity for forming a class of secondary structure and these values are added or multiplied according to a given rule. However, the underlying assumption that all residues contribute cannot be correct.

A comparison of the amino acid sequences of polypeptide chains such as the cytochromes and hemoglobins reveals the existence of many positions that can tolerate a great variety of amino acids while the chain folds into the same native structure. For example, the particularly thorough study by Bashford, Lesk, and Chothia (1987), comparing over 200 aligned hemoglobin sequences, reveals that some 60% of the sites vary in residue type.

Many of the recent dramatic successes in protein engineering depend on the existence of sequences that do not determine conformation. For example, the work of Fersht and colleagues in modifying the active site of tryptophan tRNA synthetase depends on the mutant polypeptide chains folding up into the native active protein—if this does not occur, the experiment cannot be done. Such results indicate that the conformation of the active site is determined not by the sequence of the active site, but elsewhere.

The second model is consistent with the comparative studies, in which only some residues or groups of residues control the folding. However, the work of

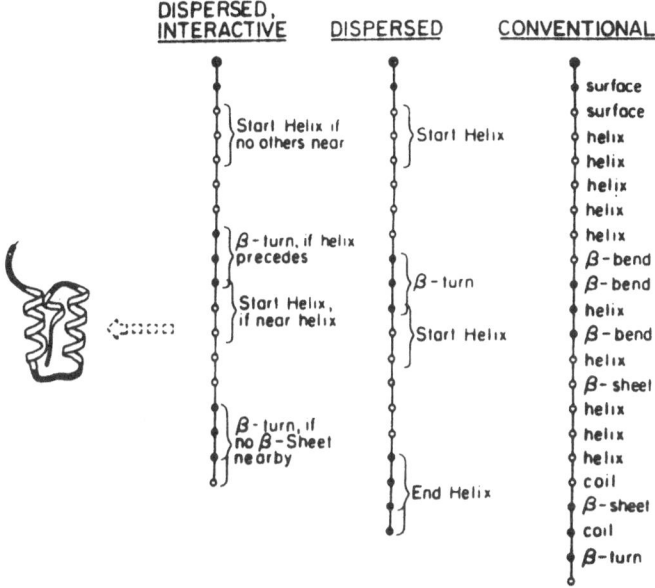

Figure 1. Folding grammars. The string of beads is an amino acid sequence that will form the helix–turn–helix conformation shown at the left. The "conventional" model at the right assumes that all residues in the sequence contribute to specifying chain conformation. The model in the center assumes that the specialized sequence directs the folding pathway, while the intervening residues are passive, presumably awaiting their later role in the activity of the native protein. The model at the left, which we prefer, assumes that dispersed sequences direct the folding pathway, but that these sequences are sensitive to their local environment, so that identical sequences would specify different conformations, depending on context.

Kabsch and Sander (1984), who have shown that identical pentapeptide sequences take different conformations in different proteins, indicates that this, too, cannot be correct.

The third interactive or context-dependent grammar is more realistic: dispersed sequences direct the folding pathway, but the information carried is sensitive to the local environment. In the case of helix formation, helix-initiating signals would be the polar clusters described by Presta and Rose (1988) or the charge residues interacting with the helix dipole described by Shoemaker *et al.* (1985). The residues are actually interacting in space and time, so that in another example residues could only take a beta turn configuration if they were in the presence of a preformed alpha helix. The same residues elsewhere in the sequence would assume a different local conformation.

Even this third model is simplistic, however, since it assumes contiguity of the essential residues. It is much more likely that spatially interacting residues are dispersed in the sequence, like the ion pairs formed by residues on the same side of the alpha helix (Marqusee and Baldwin, 1987).

We will focus on the following aspects of the folding problem: (1) how to identify those residues and groups of residues that direct the intracellular folding of newly synthesized chains released from the ribosome, and (2) the nature of the off-pathway interactions that must be avoided for productive maturation *in vivo*, and which have emerged as a practical problem in the biotechnology industry.

Our strategy has been to isolate mutations that specifically affect the intracellular folding pathway, rather than the native protein, and then characterize the mutant polypeptide chain to determine whether it has folded incompletely or incorrectly within the cell. Such mutants will exist only where there is indeed an intracellular folding pathway in which intermediates are well differentiated from the native structure. Genetic mapping and sequencing of such mutations identifies the critical residues in the folding pathway.

3. Intracellular Intermediates in Chain Folding and Assembly

The ability to identify residues affecting folding and not function requires an experimental system in which *in vivo*, or in crude extracts, one can trap and differentiate folding intermediates from mature protein. Such procedures have been developed for the thermostable tailspike of phage P22.

P22 is a dsDNA phage of *Salmonella typhimurium*. The tailspikes are the cell attachment organelles. The protein has an endorhamnosidase catalytic activity. This is, in fact, the mechanism of binding to and recognizing the O-antigen of the *Salmonella* cell surface. There are six tailspikes in a complete phage; each tailspike is itself a trimer of three polypeptide chains, the product of phage gene 9. Each polypeptide chain has 666 amino acids (Sauer *et al.*, 1982). No covalent modifications are associated with the maturation pathway.

The three-dimensional structure has not yet been solved; however, Tom Alber has obtained crystals diffracting to 2.4 Å, and we expect the structure within a year or two (personal communication). Characterization of the protein by Raman spectroscopy reveals that the secondary structure is dominated by beta-sheet (Sargent *et al.*, 1988). Given the fibrous nature of the protein, it is probably a cross-beta sandwich, in which short lengths of beta strand are adjacent and connected by turns, and two such segments form a sandwich. Though the sequence does not repeat, such elongated cross-beta structures are likely to have a conformational repeat in terms of a strand–turn–strand motif, or interstrand interactions.

Tailspikes are extremely stable, requiring temperatures over 80°C for thermal denaturation. They are resistant to most proteases. This presumably represents their function as the external structural protein of a virus, which has evolved to survive in the stomach, in the sewer, in the soil, and in the streets.

A particularly useful property is detergent resistance: under conditions in which all *Salmonella* proteins are denatured by SDS, tailspikes remain native.

The wild-type *in vivo* maturation pathway is depicted in Figure 2. Newly synthesized chains released from the ribosome form an early, partially folded intermediate, transforming to a species sufficiently structured for chain–chain recognition. These species interact to form the protrimer, in which the three chains are closely associated but not fully folded. The protrimer converts to the native tailspike, which is thermostable and protease and detergent resistant. Since additional folding is required after protrimer formation, there is no species corresponding to a "native" monomer.

Although the native protein is thermostable, early intermediates in the folding pathway are thermolabile (Goldenberg and King, 1981). As the temperature of maturation increases, the fraction of chains that fold successfully into native tailspikes decreases. The chains that do not successfully proceed through the folding and association pathway to the native tailspike are not degraded; they

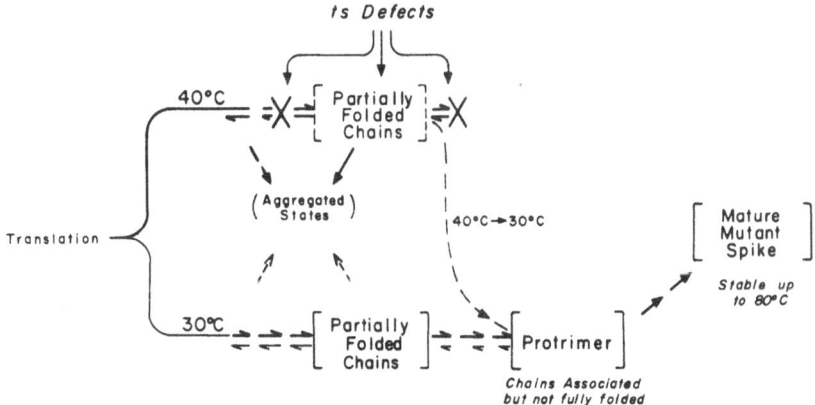

Figure 2. Pathway of intracellular folding and chain association for the P22 tailspike (Goldenberg and King, 1982; Goldenberg *et al.*, 1983). Newly synthesized chains released from ribosomes form a partially folded species. These convert to a species sufficiently structured for chain–chain recognition. These associate into the protrimer, which folds further to the native tailspike. The *tsf* mutations interfere with chain folding prior to the chain association step. Aggregates form as off-pathway species from early single-chain intermediates.

accumulate in an intact but nonnative state. If we examine the physical state of these chains, we find that they are very rapidly sedimenting aggregates; in short, they are typical inclusion bodies.

Most of the experiments described here are performed on proteins that are newly synthesized within cells, and the experiments are often performed with the crude cell extracts. Cells are infected with wild-type P22 phage or phage carrying a mutant of interest, and late after infection such cells fill up with phage components. Most of the experiments utilize radioactive amino acids, so that we observe polypeptide chains synthesized during a defined time period. Identification of chains in gel is by autoradiography or direct protein staining. In most experiments we block the assembly of the virus particle by introducing an additional mutation which prevents assembly of the shell. Under these conditions the DNA is not packaged, and the synthesis of the tailspike increases 10-fold over the wild-type level, so that it is a major protein within the infected cell.

Figure 3 shows some of the properties of the tailspike that are useful in gel analyses. In the presence of SDS the mature tailspike remains native and binds very little detergent. Its electrophoretic migration through an SDS polyacrylamide gel is therefore retarded. Incompletely or incorrectly folded tailspike chains form SDS/polypeptide chain complexes, which migrate according to the molecular mass of the polypeptide chain, at 72,000 Da. The second lane shows the migration of the protein in an SDS gel without boiling. The first lane has the same sample, but the sample was boiled before heating to denature the tailspike into its polypeptide chains. The faster mobility reflects the SDS/polypeptide chain complex. Lane c contains a lysate of bacteria infected with wild type phage, and chains that had not reached the native state at the time of cell lysis are collapsed into an SDS/polypeptide chain complex. Lane d contains a lysate of bacteria infected with ts H304 phage, which produces native tailspikes with altered electrophoretic mobility. A lysate of bacteria coinfected with wild type and ts H304 phage is shown at lane e. The resistance to detergent can thus be used to distinguish chains that are in the native conformation from those that have not yet reached that conformation.

4. Temperature-Sensitive Folding Mutations

If an amino acid substitution prevents the folding of a polypeptide chain, it should be an absolute lethal mutation. If it is an absolute lethal, the chain will not have or will not reach a stable native state. This will create very serious problems in arranging any collaboration with skilled protein biochemists. Most modern analytical technology assumes a stable native state, and most protein folding experiments measure some equilibrium constant between N and U. Instead of

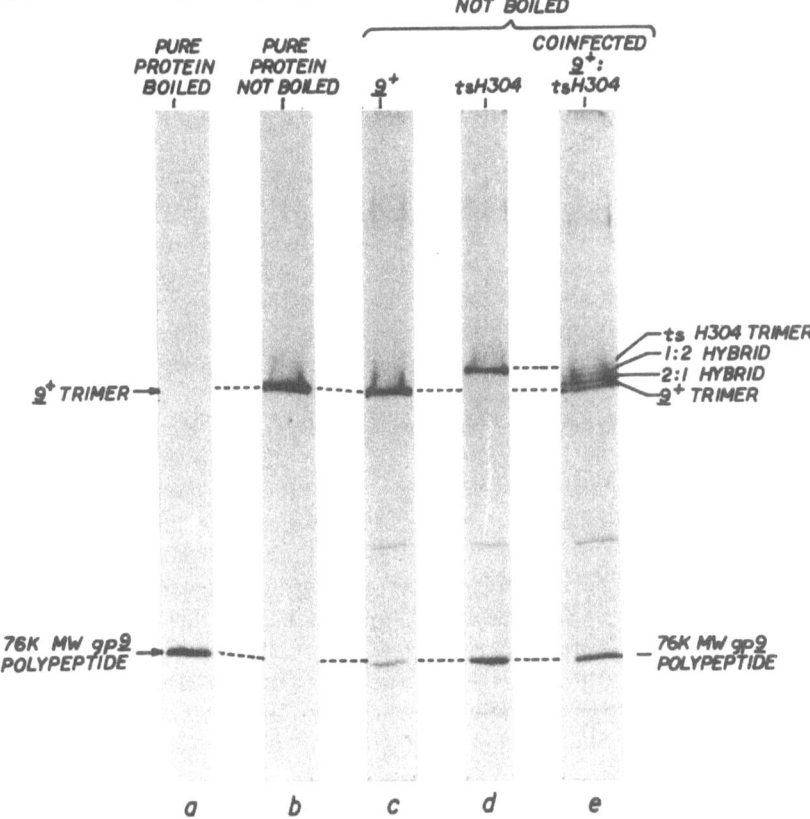

Figure 3. Formation of the hybrid tailspike. The tailspike accumulating in the lysate was electrophoresed through an acrylamide gel under nondenaturing conditions. The protein form in the wild-type 9+ lysate (lane c) is easily distinguished by electrophoretic mobility from the mutant protein accumulating in the ts H304 lysate shown in lane d. Cells that had been coinfected with 9+ and ts H304 phage so that both mutant and wild-type polypeptide chains can be synthesized in the same cytoplasm, display bands with mobility intermediate between mutant and wild type (lane e). These tracks are from a single slab gel, in which there was some bowing of the bands from one side of the gel to the other. The tracks have been rearranged for clarity of presentation, resulting in the discontinuous slant of some of the bands.

absolute lethal mutations, we chose to look for mutations that display conditional lethal defects in polypeptide chain folding.

In order to identify such residues we have isolated and characterized a large number of mutations whose character is opposite to that of traditional mutations. They have little or no effect on the native form of the protein; rather, they act on folding intermediates in the pathway. One set of these mutations includes the temperature-sensitive folding mutations.

The original characterization of temperature-sensitive mutants of phage T4 by Edgar and Lielausis (1964) reported that few of the mutations in the structural proteins actually rendered the particle thermolabile. Further characterization revealed that ts mutations fell into two classes: temperature-sensitive synthesis (TSS) and thermolabile protein (TL). Edgar and Lielausis suspected that many of the TSS mutations were causing folding defects, but they were not able to prove it with the techniques available at that time. One might think that a thermostable protein would not normally be an appropriate system for the isolation of ts mutants. These mutations are isolated by a simple criterion: the phage forms plaques at low temperature, but not at high temperature. Such mutations occur in most of the genes of the phage, and we use conventional genetics to localize the ones in gene 9 and then order them.

The tailspike gene is the locus of over 40 sites of temperature-sensitive mutations (Smith et al., 1980; King et al., 1986). These are of the TSS class for affecting the maturation process (Sadler and Novick, 1965). They are distinguished from the more familiar TL class, in which the mature mutant protein is rendered thermolabile (Hawkes et al., 1984). Since the TSS mutations in gene 9 affect the folding and subunit assembly pathway, they are termed temperature-sensitive folding (tsf). We recover tst mutants because the native protein is thermostable and an intermediate is thermolabile. Such mutants differentiate residues involved in stabilizing folding intermediates from residues involved in the stability and activity of the native protein (King, 1986;   et al., 1986).

At permissive temperatures (28°C), the tsf polypeptide chains fold into functional thermostable tailspikes. These mature mutant proteins are not thermolabile; their T_m for thermal inactivation is above 80°C, like the wild type. At restrictive temperature (38°C) the polypeptide chains are synthesized at the wild-type rate, but they fail to form the native tailspike and accumulate in an incompletely folded state, susceptible to detergent and proteases, and unreactive with antibody to the native state. These partially folded species that form at high temperature can enter the productive pathway if shifted to permissive temperature early in maturation (Smith and King, 1981). The tsf mutations destabilize an already thermolabile folding intermediate, shunting it off-pathway. This step occurs prior to chain association, in the monomeric state.

Figure 4 shows a genetic map of temperature-sensitive mutations in gene 9. Of 100 such genetically well-characterized mutations, about 40 have been sequenced, giving the precise location and amino acid substitution (Yu and King, 1984, 1988; Villafane and King, 1988). The mutations fall between positions 175 and 495. None have been recovered in the N-terminal region and none in the C-terminal region of the chains. Within the central region the mutants define about 30 positions.

The tsf mutations occur predominantly at sites of hydrophilic amino acids in the central 350 amino acids of the chain (Table I). Characterization of the purified

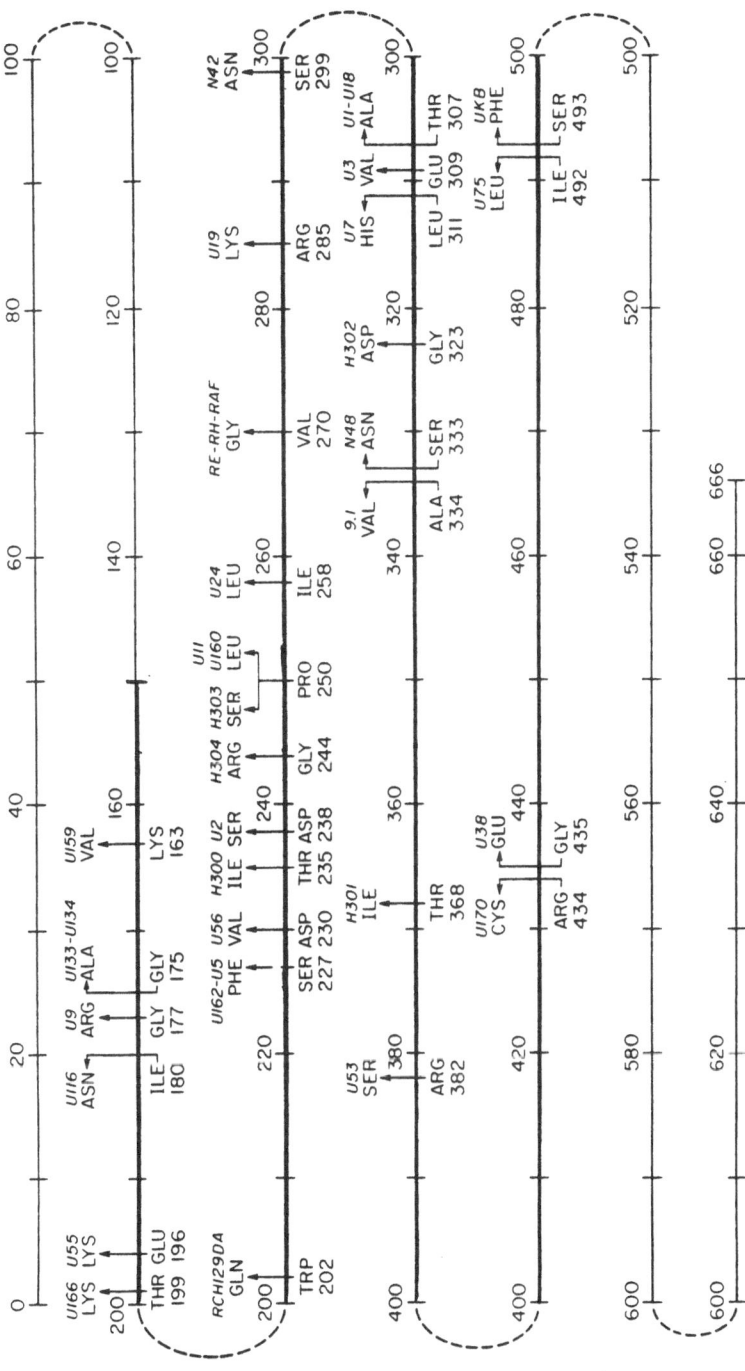

Figure 4. Map of the *tsf* mutations in gene 9. The region of the map containing the mutations is emphasized by a thicker line. Locations and amino acid substitutions for all of the *tsf* mutants whose sequence changes are currently known. The diagram is calibrated in amino acid position coordinates, with 666 amino acids in the complete tailspike chain (Sauer *et al.*, 1982); for example, in the tailspike mutant ts H304 at amino acid position 244, a glycine was substituted by an arginine residue.

mutant proteins, formed at permissive temperature, suggests that many are at the protein surface. This data is consistent with a model in which the central region of the chain forms a beta-sheet or cross-beta secondary structure as an early intermediate in the maturation pathway. The surface location and amino acid sequence at many of the mutants suggest that they may be marking beta turns (Yu and King, 1988).

5. Nonsense Mutations

In addition to temperature-sensitive mutations, more than 100 amber mutations have been isolated and characterized in gene 9 (Fane and King, 1987).

Table I. The Sites of the *tsf* Mutations in the Tailspike Gene

Allele codon number		Amino acid substitution	Local sequence								
tsU131	175	Gly>Ala	Ala .	Lys .	Phe .	Ile .	Gly .	Asp .	Gly .	Asn .	Leu
tsU134	175	Gly>Ala	"	"	"	"	"	"	"	"	"
tsU9	177	Gly>Arg	Phe .	Ile .	Gly .	Asp .	Gly .	Asn .	Leu .	Ile .	Phe
tsH304	244	Gly>Arg	Val .	Lys .	Phe .	Pro .	Gly .	Ile .	Glu .	Thr .	Leu
tsH302	323	Gly>Asp	Asn .	Tyr .	Val .	Ile .	Gly .	Gly .	Arg .	Thr .	Ser
tsU38	435	Gly>Glu	Leu .	Leu .	Val .	Arg .	Gly .	Ala .	Leu .	Gly .	Val
tsH303	250	Pro>Ser	Glu .	Thr .	Leu .	Leu .	Pro .	Pro .	Asn .	Ala .	Lys
tsU11	250	Pro>Leu	"	"	"	"	"	"	"	"	"
tsU160	250	Pro>Leu	"	"	"	"	"	"	"	"	"
tsU166	199	Thr>Lys	Met .	Glu .	Ser .	Thr .	Thr .	Thr .	Pro .	Trp .	Val
tsH300	235	Thr>Ile	Gly .	Tyr .	Gln .	Pro .	Thr .	Val .	Ser .	Asp .	Tyr
tsH301	368	Thr>Ile	Thr .	Trp .	Gln .	Gly .	Thr .	Val .	Gly .	Ser .	Thr
tsU18	307	Thr>Ala	Asp .	Gly .	Ile .	Ile .	Thr .	Phe .	Glu .	Asn .	Leu
tsU1	307	Thr>Ala	"	"	"	"	"	"	"	"	"
tsU5	227	Ser>Phe	Thr .	Leu .	Lys .	Gln .	Ser .	Lys .	Thr .	Asp .	Gly
tsU162	227	Ser>Phe	"	"	"	"	"	"	"	"	"
tsN42	299	Ser>Asn	Ala .	Asn .	Asn .	Pro .	Ser .	Gly .	Gly .	Lys .	Asp
tsN48	333	Ser>Asn	Gly .	Ser .	Val .	Ser .	Ser .	Ala .	Gln .	Phe .	Leu
tsU143	493	Ser>Phe	Gln .	Ile .	Tyr .	Ile .	Ser .	Gly .	Ala .	Cys .	Arg
tsU19	285	Arg>Lys	Gly .	Phe .	Leu .	Phe .	Arg .	Gly .	Cys .	His .	Phe
tsU53	382	Arg>Ser	Asn .	Leu .	Gln .	Phe .	Arg .	Asp .	Ser .	Val .	Val
tsU170	434	Arg>Cys	Asn .	Leu .	Leu .	Val .	Arg .	Gly .	Ala .	Leu .	Gly
tsU159	163	Lys>Val	Asp .	Phe .	Gly .	Gly .	Lys .	Val .	Leu .	Thr .	Ile
tsU34	163	Lys>Glu	"	"	"	"	"	"	"	"	"
tsU57	230	Asp>Val	Glu .	Ser .	Lys .	Thr .	Asp .	Gly .	Tyr .	Glu .	Pro
tsU56	230	Asp>Val	"	"	"	"	"	"	"	"	"
tsU2	238	Asp>Ser	Pro .	Thr .	Val .	Ser .	Asp .	Tyr .	Val .	Lys .	Phe
tsU55	196	Glu>Lys	Gly .	Val .	Phe .	Met .	Glu .	Ser .	Thr .	Thr .	Thr
tsU3	309	Glu>Val	Ile .	Ile .	Thr .	Phe .	Glu .	Asn .	Leu .	Ser .	Gly
tsmU8	344	Glu>Lys	Asn .	Gly .	Gly .	Phe .	Glu .	Arg .	Asp .	Gly .	Gly
tsU14	405	Glu>Lys	Asp .	Met .	Asn .	Pro .	Glu .	Leu .	Asp .	Arg .	Pro
ts9,1	334	Ala>Val	Ser .	Val .	Ser .	Ser .	Ala .	Gln .	Phe .	Leu .	Arg
tsU86	224	Leu>Ser	Val .	Val .	Ala .	Thr .	Leu .	Lys .	Gln .	Ser .	Lys
tsU7	311	Leu>His	Thr .	Phe .	Glu .	Asn .	Leu .	Ser .	Gly .	Asp .	Trp
tsU116	180	Ile>Asn	Asp .	Gly .	Asn .	Leu .	Ile .	Phe .	Thr .	Lys .	Leu

These define more than 60 additional sites of mutation. These are codons that have mutated to UAG (stop). A large number of these are in the N-terminal part of the gene. However, substituting amino acids in this region does not interfere with folding or chain association.

The insertion of a non-wild-type amino acid at many of the amber sites in the central region of the chain results in defective polypeptide chains. Some are temperature sensitive, and some have absolute defects. In general, these defects, like the *tsf* mutants, are in the folding and maturation of the chain, and not in the mature function (Fane and King, 1987).

The phenotypes of these mutants, with serine inserted at the amber codon,

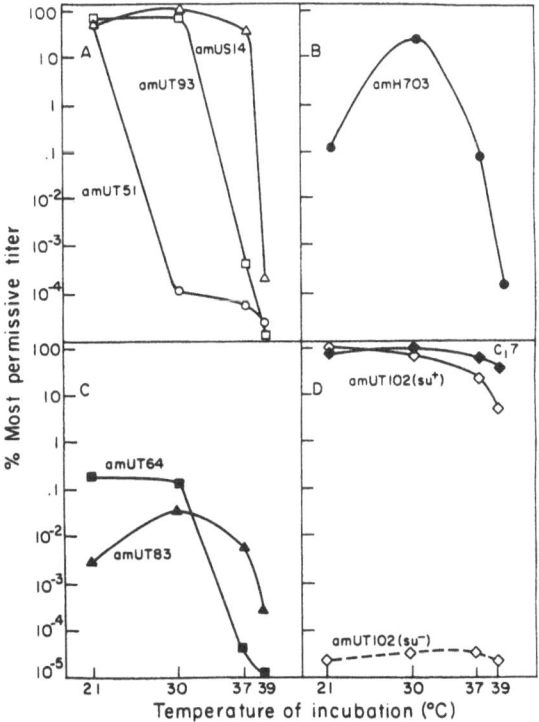

Figure 5. Efficiencies of plating of different gene 9 amber mutants on the serine inserting suppressing host (DB7154). The efficiencies of plating have been normalized to the titer of each mutant's most permissive host at 30°C. (A) The variety of ts phenotypes found among the amber mutants. (B) The complex cs/ts phenotype of amH703. Both defects are alleviated when this mutant is plated on DB7155 (glutamine). (C) The lethal phenotypes of two suppressed amber mutants. (D) The curve of wild-type *c1–7*, showing the temperature-dependent maturation of the wild-type tailspike. (D) also displays the curve of a tolerant amber *amUT102*; its efficiency of plating is host and temperature-independent. The dotted line in panel D depicts the titer of *amUT102* on the su- host (DB7136) as a function of temperature.

are shown in Fig. 5. For example, the top left panel shows mutants with different degrees of temperature sensitivity.

6. Intracellular Aggregation and Inclusion Body Formation

Early observations revealed that a significant fraction of the inactive tail-spike polypeptide chains accumulated in the form of a rapidly sedimenting aggregate (Goldenberg, 1981). These aggregates were not formed from the native protein, and correspond to the inclusion bodies described for proteins expressed from cloned genes (Marston, 1986).

To determine the physical fate of the mutant chains and to investigate the aggregation reaction, P22-infected *Salmonella* were labeled with a short pulse of C^{14}-amino acids during the period of late protein synthesis. After the pulse an aliquot of the culture was shifted to 28°C to serve as the permissive temperature control. This protocol allows the comparison of chains synthesized under the same conditions, but matured at different temperatures. Four minutes after the chase the infected cells were lysed and were then fractionated into low-speed supernatants and pellets. The pellets were resuspended and centrifuged again to minimize nonspecific trapping. After separation into supernatant and pellet fractions, samples were electrophoresed through an SDS gel without prior heating.

Figure 6 displays the distribution between supernatant and pellet of newly synthesized wild-type polypeptide chains released from the ribosomes at 28°C (bottom) and 38°C (top). The detergent-resistant native tailspikes migrate slowly near the tops of the gels. The detergent-sensitive, incompletely or incorrectly folded chains migrate as the SDS/polypeptide chain complex. Note that these experiments follow intact complete polypeptide chains; partially synthesized chains are not present in either band.

At the lower temperature (28°C), a substantial fraction of the newly synthesized wild-type polypeptide chains have matured to native tailspikes. They are found predominantly in the supernatant; however, SDS-sensitive, partially folded polypeptide chains are distributed between the soluble supernatant and aggregated pellet fractions. We define the chains that sediment into the low-speed pellets as aggregated.

At the higher temperature of maturation (38°C), a large fraction of the wild-type polypeptide chains appear in the aggregated state. Since the tailspike polypeptide chains found in the pellets are in the SDS-sensitive form, these chains were incompletely or incorrectly folded at the time of cell lysis and are not in the native SDS-resistant conformation (Goldenberg et al., 1982).

Similar results are found with *tsf* mutant–infected cells, except that no native tailspikes are formed at the high temperatures; all of the chains accumulate in the nonnative aggregated state (Smith and King, 1981; Haase-Pettingell and King, 1988).

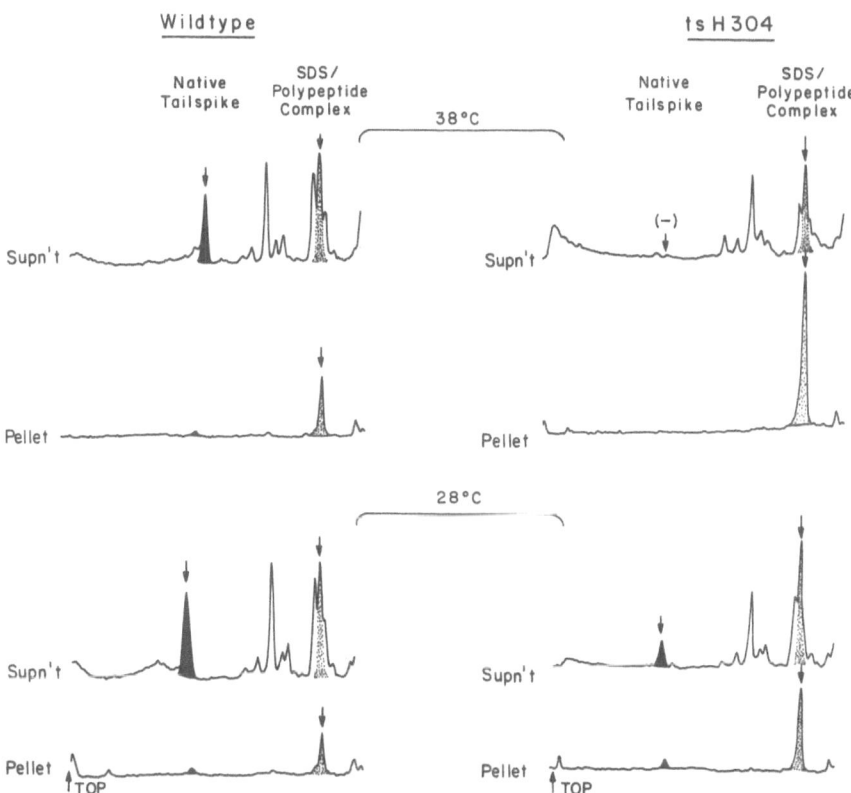

Figure 6. Densitometer tracing of an autoradiograph of a pellet supernatant separation. The tracings of autoradiograms show the distribution of newly synthesized polypeptide chains between supernatant and pellet fractions of phage infected cells.

Exponentially growing cultures of *Salmonella* bacteria were infected at 38°C with 9+ and 9− *tsfH304* phage strains, both carrying additional nonsense mutations in gene 5, to block capsid assembly, and in gene 13, to prolong lysis. At 40 min after infection, 2 μCi/ml of C^{14}-labeled amino acids was added to the infected cells. Incorporation was terminated by addition of excess cold amino acids at 41.5 min. At 43.5 a sample was shifted to 28°C. After 4.0 min of further incubation the cultures were chilled at 0°C. The infected cells were concentrated 20-fold by low-speed centrifugation and were then lysed by freezing and thawing. The lysed cells were fractionated into supernatant and pellet fractions by further centrifugation in a microfuge. The pellet was washed by resuspending and was centrifuged again. The supernatants were pooled and the pellet was resuspended to the volume of the pooled supernatants. Samples were electrophoresed through a 7½% SDS polyacrylamide gel, fixed, and dried, and an autoradiograph was prepared. The resulting autoradiograms were scanned with an LKB 2220 Ultroscan densitometer, and the intensity under the bands was integrated with the LKB 2220 Integrator. The native tailspikes in the wild-type infected cells served as the control.

7. Kinetics of Aggregate Formation

In order to determine the relationship between the soluble and aggregated forms of the tailspike polypeptide chains, we followed the kinetics of newly synthesized chains as a function of time after labeling at 38.5°C. One should note that the supernatant samples in Figure 6 contain nonnative but soluble forms of the tailspike polypeptide chains (SDS/polypeptide chain complex in the supernatant). Figure 7 shows the behavior of soluble and aggregated partially folded species in cells infected with wild-type and with two *tsf* mutants.

For both mutants, *tsfU9* and *tsfH304,* a substantial portion of the newly

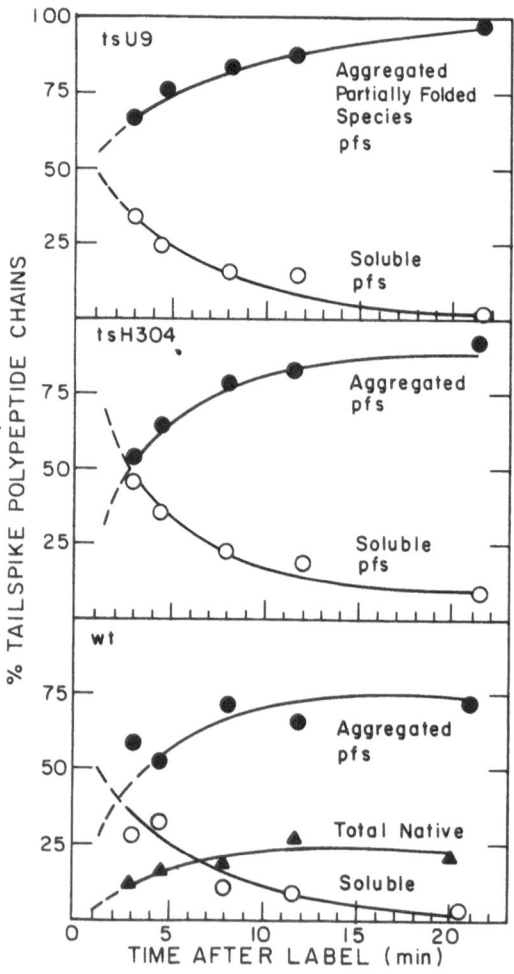

Figure 7. Time course of intracellular aggregation. Cells were infected and pulse-labeled as described in Figure 2, but the samples were labeled with 1 μCi/ml of C^{14}-amino acids. Samples were placed on ice at 3.5, 4.5, 7, 10, and 21.5 min after labeling. The infected cells were concentrated 5× by low-speed centrifugation and lysed by two cycles of freezing and thawing. The lysed cells were separated into supernatant and pellet fractions, the pellet was washed, and the samples were electrophoresed though a SDS polyacrylamide gel. Autoradiograms were prepared and analyzed as described in Figure 6.

Each value represents the percent of total C^{14}-labeled tailspike polypeptide chains present in the sample. O——O, soluble partially folded species (PFS); ●——● aggregate PFS; and △——△ native tailspike.

synthesized tailspike polypeptide chains was found in the supernatant (34% and 45%, respectively) at the earliest sampling point. As time progressed the soluble species disappeared, with a resulting increase in aggregated species. These results indicate that the aggregate is formed from the soluble, partially folded species. Since the conversion is essentially quantitative, this species appears to be a kinetic intermediate in aggregate formation. This result is similar to those reported in *in vitro* refolding experiments, in which aggregates form from intermediates in the refolding pathway (London *et al.*, 1974; Zettlmeissl *et al.*, 1979).

The wild-type, soluble partially folded species chase into two products, the native tailspike (25%) and the aggregated partially folded species (70%). Reducing the temperature of incubation during the maturation process increases the yield of the native protein and suppresses the off-pathway steps (Goldenberg *et al.*, 1982).

The aggregate present at the earliest time points is probably overestimated in these experiments. The fractionation into the supernatant and washed pellet requires about 2 hr. Although the aggregation reaction is slowed both by the cold and by the dilution associated with cell lysis, it does proceed after cell lysis. A majority of the soluble species are stable enough to endure the 2 hr of manipulations.

In the wild-type infection, the soluble, partially folded species include intermediates to both the aggregated and native states. In the mutant infected cells, the partially folded species is an intermediate solely in the aggregation pathway. These species will be referred to as partially folded intermediates (PFI).

These results demonstrate that aggregates of tailspike polypeptide chains in P22-infected cells form from species closely related to intermediates in chain folding and association. Increasing the temperature of maturation shifts the yield from native tailspike to nonnative aggregates. The simplest interpretation of these results is that the partially folded species that forms the aggregates is derived from the partial melting of the productive intermediate. The *tsf* mutations increase the yield of aggregate at higher temperatures by further destabilizing the thermolabile intermediate (Fig. 8).

Because of the high amount of beta structure in the native tailspike protein, the partially folded intermediates probably have partial beta-sheet conformations. These aggregates may represent an out-of-register form of the beta-sheet interactions that stabilize the native protein.

When the product of a cloned gene is expressed in a heterologous host, the cytoplasmic environment is generally not identical to the cytoplasm of the cell from which the cloned gene was derived. Although the native forms of the protein of interest may be stable in both cytoplasms, folding or subunit assembly intermediates are almost always less stable than their native states. For some proteins the altered conditions may destabilize folding or subunit association intermediates, resulting in off-pathway aggregation processes similar to those

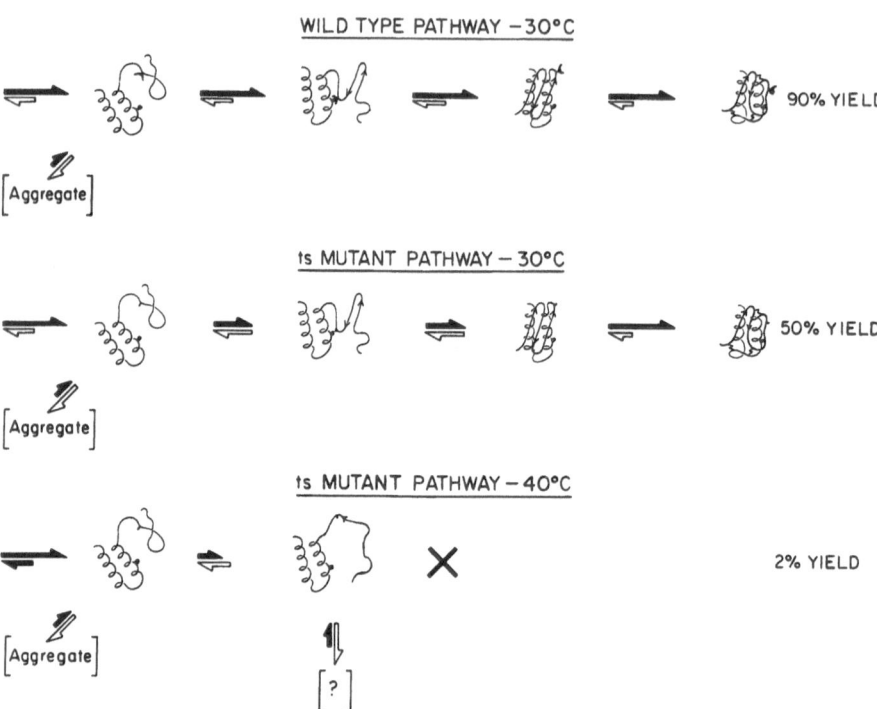

Figure 8. Model of the molecular basis of the defect conferred by a *tsf* mutation. The residue marked by the crescent plays a critical role in stabilizing a folding intermediate, but not the native form of the protein. The mutant chain can still fold at 30°C, but at higher temperatures folding intermediates are destabilized and aggregation occurs.

described for the tailspike. In these experiments, temperature was the critical variable. In other cases it may be ionic conditions, redox potential, or the presence or absence of a particular cofactor. Sorting out these features will generally require direct investigation of the maturation pathways *in vivo*.

8. *Isolation of Second-Site Suppressors of Folding Mutants*

A variety of different amino acids can be inserted at a single site in a polypeptide chain by growing phage carrying nonsense mutations in bacterial hosts that insert different amino acids at the nonsense codon. This results in the formation of mutant polypeptide chains. Starting with mutants that were temperature sensitive with the inserted amino acid, we isolated revertants that alleviated the folding defects, but still maintained the initial nonsense mutation. Such strains carry a second mutation that suppresses the defect caused at the initial site. The second-site suppressors can then be crossed out of the original strains and into other test strains, in order to examine the effect of the suppressors by themselves, and in combination with other mutations. Two well-defined suppressor mutations were isolated. These have been crossed into strains carrying a variety of well-characterized *tsf* folding mutations.

DNA fragments from phage carrying the suppressor mutations were cloned into M13 and sequenced by the Sanger dideoxy method. Suppressor 1 was a single base change corresponding to val 331 > ala. Suppressor 2 represented a single base change corresponding to ala 334 > val. The growth curves of the initial *tsf* mutant and the *tsf* mutant carrying the 334 suppressor mutation show that the second-site suppressor clearly alleviates the folding defect caused by the *tsf* mutation (Fane, 1988).

9. The Suppressing Amino Acid Substitutions Act at the Level of Polypeptide Chain Maturation

To determine if the suppression was occurring at the level of polypeptide chain folding and association, we examined the kinetics of tailspike maturation for appropriate mutants and suppressors, as previously described. At 38°C none of the chains carrying the *tsf* mutation was able to mature into the native tailspike form; when the suppressor mutation is present in the same chain, a substantially increased fraction of chains can reach the native structure (Fane, 1988).

10. The Suppressors Act in the Folding of the Single Chain, Not during Chain–Chain Association

The *tsf* mutations interfere with chain folding of the single chain prior to chain association. Nonetheless, it was possible that the suppressors could work through an effect on chain–chain association. We tested this by carrying out a mixed infection between the suppressor and the folding mutants, so that the folding defect and suppressing mutation were in different chains. By using a mutant with a charge change, we could identify the presence of each chain in native proteins. Both polypeptide chains formed native protein at permissive conditions. However, at restrictive conditions the presence of the suppressor could not rescue the *tsf* mutant chains, showing that the suppressor acts in *cis* and not *trans* to mutant amino acid substitutions.

11. The Suppressors Increase the Efficiency of Folding of the Wild-Type Chain

If the suppressor mutations increased the stability of the thermolabile intermediate or inhibited the off-pathway aggregation step, they would suppress many mutations acting on the early intermediate. The suppressor mutations were recombined away from the initial mutations and as isolated mutants. They were not distinguishable from wild type by plaque morphology. However, an investigations of the kinetics of folding and chain association revealed that the suppressors mature more efficiently than wild type at restrictive temperature; i.e., a higher proportion of suppressor chains than wild-type chains reach the native conformation (Fane, 1988).

The results reported here show that modest amino acid substitutions, i.e., val > ala and ala > val, in the center of the polypeptide chain correct the folding defects in the tailspike. These folding defects affect a single chain intermediate in the tailspike folding pathway, either destabilizing the intermediate or speeding up off-pathway aggregation (inclusion) reactions.

The suppressors could be acting in the following ways:

1. Through specific residue–residue interactions, for example, van der Waals packing, to correct packing or other defects in the critical intermediate that limit the yield of native tailspikes at high temperature.
2. By non-residue-specific interactions, which generally increase the stability of the thermolabile intermediate, thereby alleviating mutations that destabilize this intermediate.
3. By suppressing the off-pathway aggregation interaction that competes with the productive pathway. This suggests that residues 331/334 identify a site kinetically involved in intracellular aggregation.

References

Anfinsen, C. B., 1973, Principles that govern the folding of protein chains, *Science* **181**:223–230.

Bashford, D., Lesk, C., and Chothia, L., 1987, Determination of a protein fold. Unique feature of a globin amino acid sequence, *J. Mol. Biol.* **196**:199–216.

Creighton, T. E., 1978, Experimental studies of protein folding and unfolding, *Prog. Biophys. Mol. Biol.* **22**:231–298.

Edgar, R. S., and Lielausis, I., 1964, Temperature-sensitive mutants of bacteriophage T4D: Their isolation and genetic characterization, *Genetics* **49**:649–662.

Fane, B. A., 1988, Genetic analysis of the folding and maturation defects in the P22 tailspike endorhamnosidase, Ph.D. thesis, Massachusetts Institute of Technology.

Fane, B., and King, J., 1987, Identification of sites influencing the folding and subunit assembly of the P22 tailspike polypeptide chain using nonsense mutations, *Genetics* **117**:157–171.

Goldenberg, D., 1981, Genetic and biochemical analysis of the folding and subunit assembly of the bacteriophage P22 tail spike protein, Ph.D. thesis, Massachusetts Institute of Technology.

Goldenberg, D., and King, J., 1981, Temperature-sensitive mutants blocked in the folding of subunit assembly of the bacteriophage P22 tail spike protein. II. Active mutant proteins matured at 30°C, *J. Mol. Biol.* **145**:633–651.

Goldenberg, D., and King, J., 1982, Trimeric intermediate in the *in vivo* folding and subunit assembly of the tailspike endhorhamnosidase of bacteriophage P22, *Proc. Natl. Acad. Sci. USA* **79**:3403–3407.

Goldenberg, D. P., Berget, P. B., and King, J., 1982, Maturation of the tail spike endhorhamnosidase of Salmonella phage P22, *J. Biol. Chem.* **257**:7864–7871.

Goldenberg, D. P., Smith, D. H., and King, J., 1983, Genetic analysis of the folding pathway for the tail spike protein of phage P22, *Proc. Natl. Acad. Sci. USA* **80**:7060–7064.

Haase-Pettingell, C., and King, J., 1988, Formation of aggregates from a thermolabile *in vivo* folding intermediate in P22 tailspike maturation: A model for inclusion body formation. *J. Biol. Chem.* **263**:4977–4983.

Hawkes, R., Grutter, M. G., and Schellman, J., 1984, Thermodynamic stability and point mutations of bacteriophage T4 lysozyme, *J. Mol. Biol.* **175**:195–212.

Kabsch, W., and Sander, C., 1984, On the use of sequence homologies to predict protein structure: Identical pentapeptides can have completely different conformations, *Proc. Natl. Acad. Sci. USA* **81**:1075–1078.

Kim, P. S., and Baldwin, R. L., 1982, Specific intermediates in the folding reactions of small proteins and the mechanism of protein folding pathways, *Annu. Rev. Biochem.* **51**:459–489.

King, J., 1986, Genetic analysis of protein folding pathways, *Bio/technology* **4**:297–303.

King, J., Yu, M-h., Siddiqui, J., and Haase, C. A., 1986, Genetic identification of amino acid sequences influencing protein folding pathways, in: *Protein Engineering, Applications in Science, Medicine and Industry* (M. Inoue and R. Sarma, eds.) Academic Press, New York, pp. 275–291.

London, J., Skrzynia, C., and Goldberg, M., 1974, Renaturation of *E. coli* tryptophonase after exposure to 8 M urea. Evidence for the existence of nucleation centers, *Eur. J. Biochem.* **47**:409–415.

Marquesee, S., and Baldwin, R. L., 1988, Helix stabilization by Glu⁻. . .Lys⁺ salt bridges in short peptides of de novo design, *Proc. Natl. Acad. Sci. USA* **84**:8898–8902.

Marston, F. A. O., 1986, The purification of eukaryotic polypeptides synthesized in *Escherichia coli*, *Biochem. J.* **240**:1–12.

Presta, L., and Rose, G., 1988, Helix signal in protein, *Science* **240**:1632–1641.

Sadler, J. R., and Novick, A., 1965, The properties of repressor and the kinetics of its action, *J. Mol. Biol.* **12**:305–327.

Sargent, D., Benevides, J. M., Yu, M-h., King, J., and Thomas, Jr., G. J., 1988, Secondary structure and thermostability of the phage P22 tailspike: Analysis of Raman spectroscopy of the wild type protein and a temperature-sensitive folding mutant, *J. Mol. Biol.* **199**:491–502.

Sauer, R. T., Krovatin, W., Poteete, A. R., and Berget, P. B., 1982, Phage P22 tail protein: Gene and amino acid sequence, *Biochemistry* **21**:5811–5815.

Shoemaker, K. R., Kim, P. S., Brems, D. N., Marqusee, S., York, E. J., Chaiken, I. R., Stewart, J. M., and Baldwin, R. L., 1985, Nature of the charged-group effect on the stability of the C-peptide helix, *Proc. Natl. Acad. Sci. USA* **82**:2349–2353.

Smith, D. H., and King, J., 1981, Temperature-sensitive mutants blocked in the folding or subunit assembly of the bacteriophage P22 tail spike protein. III. Inactive polypeptide chain synthesized at 39°C, *J. Mol. Biol.* **145**:653–676.

Smith, D. H., Berget, P. B., and King, J., 1980, Temperature-sensitive mutants blocked in the folding or subunit assembly of the bacteriophage P22 tail-spike protein. I. Fine-structure mapping, *Genetics* **96**:331–352.

Villafane, R., and King, J., 1988, Nature and distribution of sites of temperature sensitive folding mutations in the gene for the P22 tailspike polypeptide chain, *J. Mol. Biol.* **204**:607–619.

Yu, M-H., and King, J., 1984, Single amino acid substitutions influencing the folding pathways of the phage P22 tail spike endhorhamnosidase, *Proc. Natl. Acad. Sci. USA* **81**:6584–6588.

Yu, M-H., and King, J., 1988, Surface amino acids as sites of temperature-sensitive folding mutations in the P22 tailspike protein, *J. Biol. Chem.* **263**:1424–1431.

Zettlmeissl, G., Rudolph, R., and Jaeniau, R., 1979, Reconstitution of lactic dehydrogenase. Noncovalent aggregation vs. reactivation. I. Physical properties and kinetics of aggregation, *Biochemistry* **18**:5567–5571.

5

Unfolding and Inactivation
Genetic and Chemical Approaches to the Stabilization of T4 Lysozyme and Human Interferon Gamma against Irreversible Thermal Denaturation

RONALD WETZEL, L. JEANNE PERRY,
MICHAEL G. MULKERRIN, and L. MICHAEL RANDALL

1. Introduction

One of the first criteria used for identifying and classifying proteins was the tendency of these "albuminoid" substances to respond to heating by coagulating in the manner of egg white albumin (see discussion in Fruton, 1972). By the early 1900s, thermally induced coagulation was being studied with purified proteins. By the 1930s, heat- and denaturant-induced coagulation was understood to involve two steps: a heat-dependent denaturation* step associated with an increase in solution viscosity and chemical accessibility of side-chain groups, and a heat-

---

*The word "denaturation" was originally (Chick and Martin, 1912) intended to apply to the first step in the coagulation process, by which point the still soluble protein molecule was recognized as being in a nonnative state. The word continues to be used to denote what we now understand to be an unfolding event, induced by temperature, solute, or pH changes, but it can also mean any transformation by which the protein is rendered nonnative, including chemical changes. In this chapter, the process in which the protein loses its natural folded structure is called "unfolding" or "reversible

---

RONALD WETZEL, L. JEANNE PERRY, MICHAEL G. MULKERRIN, and L. MICHAEL RAN-DALL • Biomolecular Chemistry Department, Genentech, Inc., South San Francisco, California 94080. *Present address for R. W.:* Macromolecular Sciences Department, SmithKline Beecham Pharmaceuticals, King of Prussia, Pennsylvania 19406. *Present address for L. J. P.:* Molecular Biology Institute, University of California at Los Angeles, Los Angeles, California 90024.

independent, but pH and salt-dependent, aggregation/precipitation step involving newly exposed groups (Wu, 1931; Mirsky and Pauling, 1936; Anson, 1938). A number of these coagulated proteins could be restored to the solubility and other properties of their native states by use of denaturing solvents followed by a return to native conditions (Neurath et al., 1944; Anson, 1945). The early observation of very high temperature coefficients for protein coagulation set this process apart from standard chemical reactions and was compared to the temperature dependence of state changes such as melting (Anson, 1938). Whereas the rates of chemical reactions might increase by a factor of two to three with a 10°C increase in temperature, the rate of coagulation was found to increase by factors of hundreds with a 10°C increase (Chick and Martin, 1912). The developing concepts of the nature of protein structure and stability in the first decades of this century depended greatly on such data.

Although the use of such techniques in the study of protein structural stability has long been supplanted by the study of *reversible** denaturation reactions, the process of irreversible* denaturation remains of immense practical interest. *In vitro*, proteins can lose structural integrity and activity upon storage as well as during use. Working biochemists are only too familiar with the sensitivity of their reagent enzymes to loss of function over time. Protein pharmaceuticals are especially sensitive to this process, since only a small degree of such damage is sufficient to raise concerns, for example about immunogenicity, even if potency is relatively unaffected. Stability under conditions of use can provide an especially important barrier to the use of proteins in industrial as well as laboratory applications, as most recently illustrated by the value of a thermally stable DNA polymerase in the polymerase chain reaction (Seiki and Gelfand, 1988). The irreversible denaturation of proteins plays many important roles *in vivo* as well. In humans, many genetic diseases are likely to involve the synthesis of defective proteins that spontaneously aggregate or exhibit low thermal stability (Daar et al., 1986; Kishi et al., 1987; Vulliamy et al., 1988). The amyloid deposits found in the brains of victims of Alzheimer's disease, Down syndrome, kuru, scrapie, and Creutzfeldt–Jakob disease are composed of specific precipitated proteins (Masters et al., 1985; Kang et al., 1987; Oesch et al., 1985; Robakis et al., 1986). Irreversible thermal denaturation of proteins also occurs in bacteria, in temperature-sensitive mutations (Goldenberg, 1988), and in the inclusion bodies formed during the biosynthesis of some recombinant proteins (Marston, 1986; Taylor et al., 1986).

---

denaturation." The words "reversible" and "irreversible" are used in the context of thermodynamic studies of reversible equilibria, as discussed by Tanford (1968). Thus, "irreversible" implies that the end state is no longer in equilibrium with the starting, native state under reaction conditions, but does not imply that native structure cannot be reestablished by subsequent "off-pathway" manipulations.

It is, of course, the onset of one or more of these irreversible processes, trapping protein molecules kinetically or thermodynamically, which interferes with a partially or completely unfolded protein's ability to return to a native folded state (Anson, 1938). Thus, irreversible processes must be overcome if the unfolding of a protein is to be made "thermodynamically"* reversible. With the exception of a few proteins, such reversible denaturation can be achieved only within a limited range of *in vitro* conditions. Such conditions are valuable, however, because the study of reversible unfolding allows the extraction of the thermodynamic parameters that describe the nature and relative magnitudes of the forces that give the protein its characteristic folding pattern and stability (Kauzmann, 1959; Tanford, 1968; Brandts, 1969; Ghelis and Yon, 1982; Privalov, 1982).

In protein denaturation, the word "irreversible" is normally used in the context of attempted thermodynamic measurements: if the protein loses activity, solubility, and/or structure after a simple heating and cooling cycle, it has "irreversibly" denatured (Tanford, 1968; Lumry and Biltonen, 1969). This does not preclude the possibility of reactivating the "irreversibly" denatured protein by some other treatment. It is often possible, for example, to reconstitute and reactivate such a protein by dissolving it in acid or a solute denaturant such as urea, followed by removal of the denaturant by dilution or dialysis under mild conditions (Neurath *et al.*, 1944; Anson, 1945; Lumry and Biltonen, 1969). If passage through denaturant alone can restore structure and activity, it is an indication that the inactivating structural change was mediated by noncovalent forces. Some "irreversible" inactivations that are chemical in nature can also be reversed with the right treatment; the best example is inactivation by formation of incorrect disulfide bonds, which can be reversed (sometimes with the help of denaturant, as described earlier) by reduction and reformation of any native disulfides (Haber and Anfinsen, 1962). For most other chemical irreversible inactivations, however, the chemistry that would allow specific restoration of structure and activity is not yet available.

The loss of the structural integrity of the globular protein in a preliminary unfolding step has long been perceived as playing a key role in most thermal inactivation mechanisms. Anson (1938) proposed two possible impediments to reversibility of denaturation: chemical damage caused by the denaturing conditions, and aggregation and precipitation of the denatured form.

Since the conditions required to break *covalent* bonds are already sufficient to unfold many proteins, it is not always clear whether unfolding is a necessary prerequisite for chemical inactivation (Zale and Klibanov, 1983). At the same time, it is well known that group-specific reagents often react with only a subset of the potentially reactive residues of a folded protein, and lack of reactivity is almost always attributed to the residue's being buried in the protein structure (Neurath *et al.*, 1944). Although chemically sensitive residues might be found on both the surface and in the interior of globular proteins, it is the internal ones whose modification is expected to be most devastating to packing integrity and therefore structural stability. Unfolding of the protein by heat or denaturant, however, can open the interior residues to attack by reagents (Anson, 1938, 1945; Neurath *et al.*, 1944). The relation of folding integrity to chemical lability is more established for the reaction of peptide bond hydrolysis catalyzed by enzymes. "Digestibility" was used as an early criterion for denaturation (unfolding) of a protein (Anson, 1938), and subsequent work has shown that proteins are much more susceptible to proteolysis when unfolded (Anfinsen and Scheraga, 1975).

Unfolded structures are also more susceptible than folded ones to *noncovalent* reactions like aggregation and adsorption. Although folded proteins in solution occupy at least a local minimum in the free energy surface, they can achieve lower free energies, and thus be favored in reactions, if they either refold into more stable solution structures, or associate into soluble aggregates, or leave solution by adsorption or precipitation. The hydrophobic and other interactions that function internally to give globular proteins their structural stability can also influence surface properties mediating misfolding, adsorption, and aggregation/precipitation reactions, when buried residues are exposed in an unfolded polypeptide. Early workers noted that unfolded proteins remained soluble until the pH was adjusted to the isoelectric point of the protein (Chick and Martin, 1912; Anson, 1938, 1945), which was consistent with a major involvement of apolar interactions in the aggregation process. There has also been some discussion of the possible involvement of more subtle and specific ionization effects in the aggregation of an unfolded protein (Lumry and Eyring, 1954).

Based on such considerations, it is possible to formulate a general mechanism to describe many different kinds of thermal irreversible inactivation reactions (Anson, 1938; Lumry and Eyring, 1954; Zale and Klibanov, 1983). In the following scheme, N is the native, folded state, U is an unfolded state that can interconvert with N, and I is a state from which neither N nor U can be regenerated under conditions of the experiment. I might be noncovalently denatured as a soluble aggregate, as precipitate, as protein adsorbed to the reaction vessel, or as an "incorrectly folded," soluble, monomeric, inactive form of the protein. I may also be the result of some covalent modification, such as oxidation of side chains of residues such as Met or Cys (the latter forming nonnative intra- or intermolecular disulfide bonds), β-elimination of native disulfide bonds, deamidation of Asn or Gln residues, or scission of the amide bonds of the peptide backbone either by chemical hydrolysis or by enzymatic cleavage by contaminating proteases or by autodigestion.

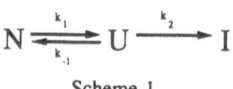

Scheme 1

Irreversible thermal denaturation and inactivation can thus be a complex reaction consisting of a number of processes occurring simultaneously. Indeed, in our work on T4 lysozyme (see Section 2) we have characterized the following thermally driven inactivating processes transpiring under similar conditions: adsorption to the glass or plastic reaction vessel, aggregation/precipitation, oxidation to disulfide-linked aggregates, thiol–disulfide interchange of disulfide-linked aggregates, and other covalent modifications. The attraction of this mechanism is that it suggests a simple solution to providing stability against many kinds of protein inactivation, even if the nature and mechanisms of the inactivating processes are not known. If the native, folded state N can be stabilized with respect to the unfolded state U, then at any given temperature in the transition region U will be less highly populated and therefore will less rapidly convert to I.

In this chapter we review our efforts to improve the stability of T4 lysozyme and interferon-gamma (IFN-γ) by exploiting an assumed relationship between thermal unfolding and thermal irreversible denaturation. The results share the intriguing aspect that, although both systems are shown to be capable of undergoing thermal inactivation consistent with the mechanism of Scheme 1, by far the most effective structural modifications in each case are those whose stabilizing power cannot be accounted for by their influence on folding stability. In both cases we discuss alternative hypotheses, some of which require invocation of an important unfolding intermediate in the inactivation mechanism.

2. T4 Lysozyme

The lability of T4 lysozyme when heated at neutral pH was reported as part of the initial characterization of the purified protein (Tsugita *et al.*, 1968). Streisinger and his colleagues found that after the enzyme was heated at 65°C at pH 6.5, nearly all the activity was gone after 5 min, as assessed in a turbidometric assay on cooled samples. As discussed in Section 1, the rapid inactivation observed under these conditions is due to several inactivation pathways acting independently and simultaneously on the sample. Irreversible thermal inactivation has been exploited to develop the major area of study of this protein, the effect of mutation on folding stability. Cells harboring T4 phage producing inactive lysozyme no longer lyse efficiently; selection schemes based on this "halo assay" (Streisinger *et al.*, 1961) have led to the isolation of both thermolabile (Remington *et al.*, 1978) and thermostable (Alber and Wozniak, 1985) lysozyme sequence variants, which have served as the foundation for the ongoing studies on the relationship between structure and thermodynamic stability in this system (Alber *et al.*, 1986, 1987; Matthews *et al.*, 1987; Matsumura *et al.*, 1988).

Until recently, irreversible denaturation at neutral pH and above imparted a barrier that restricted reversible unfolding experiments to working at low pH conditions, where thermal and denaturant induced unfolding is reversible (Elwell and Schellman, 1975). T4 lysozyme undergoes a cooperative, reversible unfolding transition when heated at pH 2, with a Tm of 37°C (Elwell and Schellman, 1977). The Tm for the reversible unfolding at pH 3 is 55°C (Elwell and Schellman, 1977). Analysis of such data, obtained by spectropolarimetry, reveals that the protein achieves a stable three-dimensional folding pattern in the same way as other globular proteins, through the relatively small, favorable free energy of stabilization derived from the sum of large enthalpy and entropy terms (Hawkes et al., 1984; Schellman et al., 1981). Recently, Becktel and Baase (1987) found conditions that allow reversible unfolding of the molecule up to pH 7. Consistent with other globular proteins, the free energy of stabilization of T4 lysozyme is at a minimum (most stabilizing) in the neutral pH region, with an associated Tm of 67°C at pH 5.5 and 64–65°C at pH 6.5 (Becktel and Baase, 1987; Wetzel et al., 1988). The fact that Tsugita et al. (1968) found the molecule to be labile at 65°C is not surprising in light of these Tm values and the expected relationship between reversible unfolding and irreversible denaturation/ inactivation outlined in the introduction.

In 1982, based in part on the demonstration by Tsugita et al. (1968) of T4 lysozyme's thermal lability in vitro, we chose this protein as a model system to investigate the ability of an engineered disulfide bond to stabilize the protein. The results of these studies have been reported (Perry and Wetzel, 1984, 1986, 1987; Mulkerrin et al., 1986; Wetzel et al., 1988) and reviewed (Wetzel, 1986, 1988) elsewhere from the point of view of the role of disulfide bonds in protein structure and stability. This chapter will present some of these results, along with previously unpublished work, in terms of mechanisms of protein inactivation and approaches to protein stabilization.

2.1. Sulfhydryl Oxidation

T4 lysozyme contains two cysteine residues, one at position 54 in the N-terminal lobe and the other at position 97 in the C-terminal lobe. These cysteines should be incapable of forming a disulfide bond in the native state, as suggested by the three-dimensional structure determined crystallographically (Remington et al., 1978; Fig. 1) and are found in the free sulfhydryl form in the protein isolated from phage extracts of Escherichia coli. Since sulfhydryl oxidation is expected when proteins are heated in the presence of air, from our earliest experiments we routinely added EDTA to buffers to chelate contaminating divalent metal ions known to catalyze the reaction between molecular oxygen and sulfhydryl groups (Torchinsky, 1981). Preliminary results (Perry and Wetzel, 1984) suggested that oxidation was, in fact, being effectively suppressed, since under these conditions the wild-type protein undergoes heat inactivation no more slowly in the presence of mercaptoethanol than in its absence (Fig. 2). This

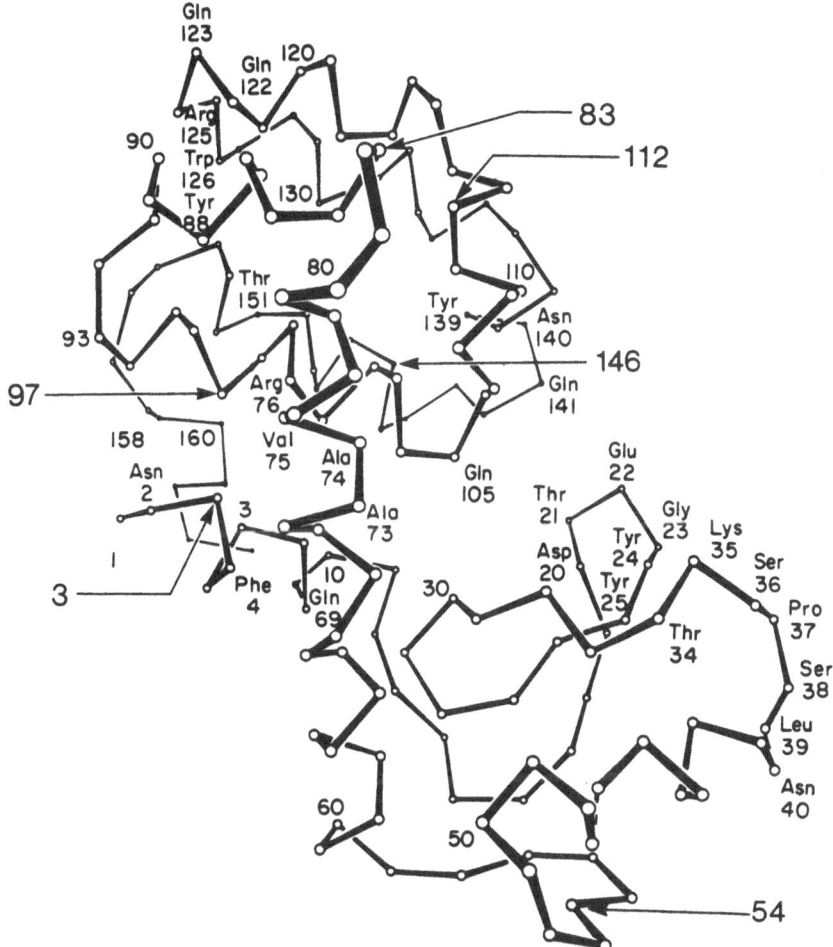

Figure 1. X-ray crystal structure of T4 lysozyme with locations of key amino acid residues indicated. Adapted from Remington *et al.*, 1978.

experiment was deceptive, however, in terms of the efficiency of EDTA in inhibiting air oxidation. Under the conditions used in these early experiments, the rates and extents of other irreversible inactivation reactions (aggregation and adsorption) were accelerated by the presence of high phosphate concentration and low T4 lysozyme concentration (see Section 2.3). Later these parameters were changed, resulting in a decrease in the rates of thermal adsorption and noncovalent aggregation. What we did not appreciate until later was that by thus reducing these rates we allowed the formerly minor component of thiol oxidation

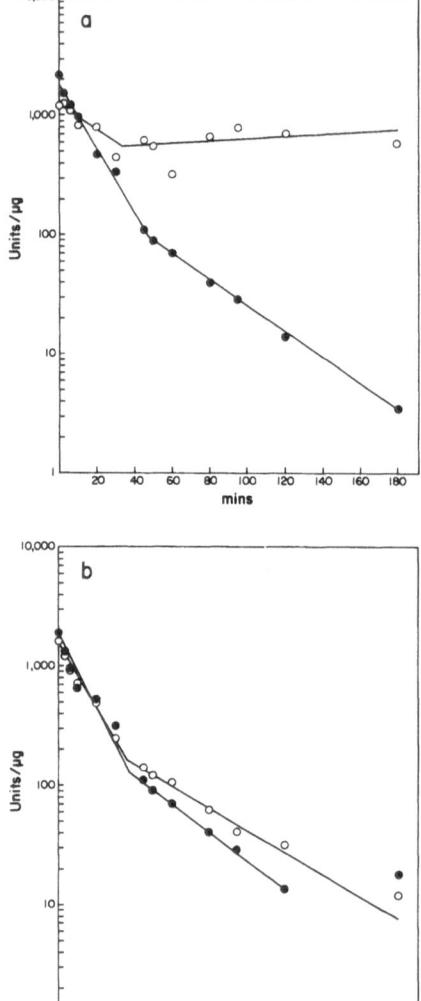

Figure 2. Kinetics of inactivation of T4 lysozyme derivatives heated at 67°C at 3 μg/ml in 0.1 M potassium phosphate, pH 6.5, 0.1 M NaCl, 1 mM EDTA. Samples were removed from heat at timed intervals, cooled, diluted, and assayed by the turbidometric assay at 20°C. (a) Wild-type (●) and oxidized [I3C-97C] (○) T4 lysozymes, no further additions; (b) wild-type (●) and reduced [I3C] (○) T4 lysozymes with added 10 mM β-mercaptoethanol. From Perry and Wetzel, 1984, copyright 1984 by AAAS.

to become a major component of the reaction, and that this oxidation was not completely suppressed by EDTA.

These revised conditions, standard for most of the work discussed here, are as follows: 30 μg/ml T4 lysozyme, 5–10 mM potassium phosphate, 1 mM EDTA, 0.2 M KCl, pH 6.5. Under these conditions sulfhydryl oxidation plays a major role in the irreversible inactivation of the wild-type enzyme at 65–70°C.

This is shown by the ability of added mercaptoethanol to significantly slow the inactivation rate, and by the ability of mercaptoethanol plus denaturant to restore more of the lost activity than does denaturant alone (Perry and Wetzel, 1987). The products of inactivation are disulfide-linked oligomers, as demonstrated by mobility in nonreducing SDS-PAGE (Perry and Wetzel, 1987). This thermal inactivation by thiol oxidation appears to require prior thermal unfolding of the molecule, since a plot of inactivation rate vs. temperature shows a rapid increase in rate around 65°C, which is just where the tertiary and secondary structures of the protein are known to be lost under these conditions (Becktel and Baase, 1987; Wetzel *et al.*, 1988). Although this correlation might be coincidental, the case is strengthened by the behavior of some T4 lysozyme sequence variants heated under the same conditions. Figure 3 shows the inactivation rate/temperature curves for the wild-type protein and two mutants. All three enzymes become highly susceptible to inactivation over a small temperature range. The tem-

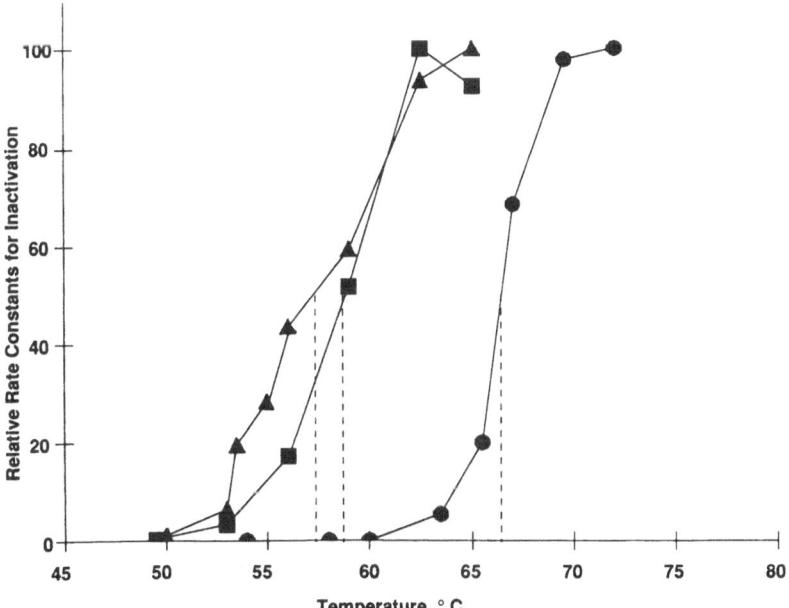

Figure 3. Dependence of rate constants for irreversible inactivation for T4 lysozyme derivatives on incubation temperature. Kinetics were determined as described in the legend to Figure 1, except that proteins at 30 μg/ml were heated in a buffer of 5 mM phosphate, pH 6.5, 1 mM EDTA, 0.2 m KCl. Since maximum inactivation rates for each protein differed, rates for each protein are displayed as values relative to that molecule's rate maximum. Lines to baseline indicate temperatures at which rate is half-maximal for wild-type (●, 67°C), reduced I3C/C54T/A146T (■, 58°), and reduced I3C/C54T/R96H (▲, 57°).

perature of half-maximal inactivation rate is different for each protein, however, and the values obtained agree well with the Tm value measured for wild type and the Tm values calculated, assuming additivity (Becktel *et al.*, 1986), from the T_m effects of the individual mutations (Wetzel *et al.*, 1988). (It should be noted that thiol oxidation is not the only thermal inactivation reaction occurring under these conditions, but it appears to account for 25–50% of the total inactivation. The curves in Fig. 3 thus reflect overall rates, but at the same time would be expected to be sensitive to significant oxidative inactivation occurring at temperatures substantially below the Tm, which is not observed.)

In these experiments, no effort was made to exclude oxygen. Although rigorous elimination of dissolved oxygen would stabilize the protein toward thermal, oxidative inactivation, this is not always technically feasible. It is a bit surprising that EDTA was not more effective at suppressing the oxidation. Although the combination of EDTA and added reducing agent is effective at stabilizing T4 lysozyme against thermal, oxidative processes, in some cases such an approach may backfire, since the interaction of reducing agent with metal ions and oxygen can generate peroxide which might itself be an effective inactivating agent (Torchinsky, 1981). (However, we have not investigated the chemical integrity of T4 lysozyme heated in the presence of EDTA and mercaptoethanol.) Whether or not chemical damage is a factor, it is clear that a more general, buffer-independent cure for sensitivity to oxidative inactivation would be to modify the protein itself. This can be done in at least two ways, as discussed in the following.

Since thermal inactivation by oxidation is dependent on prior unfolding, one way to stabilize T4 lysozyme to this process in the 65–70° range would be to build mutations into the protein that would provide additional stability to reversible unfolding. Such mutations are known (Alber and Wozniak, 1985; Matthews *et al.*, 1987; Wetzel *et al.*, 1988), and together, given additivity of effect (Becktel *et al.*, 1986; Matsumura *et al.*, 1986), might be expected to provide the molecule with an additional 5–10° to the T_m; such a mutant would be expected to exhibit a sigmoidal curve like those in Fig. 3, but with a temperature of half-maximal rate correspondingly higher than that of the wild type.

In the case of sulfhydryl oxidation, however, another obvious curative step would be to remove the labile part of the unfolded molecule, in this case the reactive sulfhydryl groups. In fact, when the two Cys residues are replaced by other amino acids, the new protein exhibits a significantly lower inactivation rate than wild type at 70°C (Fig. 5), and none of the lost activity appears to be due to oxidation (Perry and Wetzel, 1987). This might be viewed as a simple example of a general mechanism of protein stabilization, elaborated in the rest of this chapter, which focuses on the properties of unfolded states rather than on the stability of the folded state.

It is likely that one reason for previously encountered difficulties in obtaining reversible unfolding of T4 lysozyme at conditions above pH 4–5 has in fact been cysteine oxidation, a reaction favored at basic pHs (Torchinsky, 1981). However, even a T4 lysozyme without cysteines is inactivated at an appreciable rate at pH 6.5 (Fig. 5), in reactions driven by noncovalent forces. These will be discussed in later sections.

2.2. Thiol/Disulfide Interchange

When disulfide-bond-containing proteins are unfolded with urea, the presence of a free cysteine residue in the protein or small amounts of a thiol compound can induce gel formation (Huggins *et al.*, 1951; Frensdorff *et al.*, 1953). Similar effects are observed in the irreversible thermal denaturation of such proteins (Warner and Levy, 1958) and can play major roles in the properties of food proteins (Kinsella, 1982). More subtle effects have been described in which a protein can isomerize to a new disulfide arrangement under conditions where the molecule retains globular structure and nativelike behavior (McKenzie *et al.*, 1972; Stroupe and Foster, 1973). These scrambling reactions are primarily a result of thiol/disulfide interchange, a process in which the deprotonated form of a thiol attacks a disulfide to displace one of the components, generating a new disulfide and a new thiolate. Because of the dependence of the reaction on the thiolate form, thiol/disulfide interchange is highly favored in the basic pH range (Torchinsky, 1981).

When the Ile residue at position 3 of T4 lysozyme is replaced with a Cys by directed mutagenesis, the new residue can be oxidatively cross-linked to the Cys at position 97 to form a disulfide bond. This protein retains 50% of its starting enzymatic activity after being heated for 3 hr at pH 6.5 and 65°C. In contrast, the wild-type protein exhibits only about 0.1% of its starting activity after the same treatment (Fig. 2). Although this result was dramatic and was in the expected direction, it was clear that the mechanism by which the effect was achieved was complicated: the mutant lysozyme decays by a biphasic process, with an initial rapid decay to 50% starting activity followed by a very slow phase producing little or no further inactivation (Fig. 2).

This disulfide-cross-linked mutant also contained an unpaired Cys at position 54. It seemed possible that at least part of the biphasic behavior in the inactivation kinetics was due to this residue (Perry and Wetzel, 1984). This was supported by an experiment which showed that treatment of the cross-linked mutant with the thiol reagent iodoacetic acid stabilized the protein to heating as compared with its untreated precursor (Perry and Wetzel, 1986). When the product of heat inactivation of the disulfide mutant with Cys unblocked was analyzed by SDS-PAGE, a "ladder" of covalent oligomers of T4 lysozyme was observed (Perry and Wetzel, 1986). The most prevalent form in the reaction mixture is a dimer, which could arise either by air oxidation of the Cys54 residues of two monomers to each other (see Section 2.1), or from the attack of Cys54 of one monomer on the 3-97 disulfide of another in a disulfide interchange reaction. Although the linkages of the dimers produced by each reaction would be expected to be different (54–54' for the oxidation, 54–3' or 54–97' for interchange), the dimer in the reaction mixture was not isolated or characterized. Some of the inactivation must be due to thiol/disulfide interchange, however, since higher oligomers are also present in the reaction mixture.

Just as for air oxidation of wild-type T4 lysozyme, the disulfide-cross-linked mutant can also be stabilized to inactivation *via* disulfide interchange by replacing the unpaired Cys54 with another amino acid (Perry and Wetzel, 1986). Although these cross-linked double mutants can still be inactivated by heat, they decay very slowly at 65°C (Fig. 5), requiring much higher temperatures to produce inactivation, as described in more detail in later sections.

How does the process of thiol–disulfide interchange depend on the folding integrity of the protein? Although the disulfide-cross-linked single mutant of T4 lysozyme, containing an unpaired Cys54, was not observed to differ from the wild type in stability in the 4–25°C range, the dependence of inactivation rate on temperature was not studied for this mutant. It is possible that thiol–disulfide interchange inactivation may not always strictly adhere to the standard unfolding-linked mechanism of inactivation: As stated earlier, isomeric disulfide forms of some proteins can be generated under conditions well below the temperatures at which they undergo cooperative melting (McKenzie *et al.*, 1972; Stroupe and Foster, 1973). Thus, thiol–disulfide interchange might participate in the unfolding reaction itself, in the reverse reaction of the well-studied thiol–disulfide-mediated folding of proteins (Creighton, 1978; Freedman and Hillson, 1980). Denaturation presumably becomes irreversible in reactions facilitated by disulfide interchange when incorrectly disulfide-bonded monomers, and/or disulfide-linked aggregates, ultimately undergo noncovalent aggregation and/or precipitation.

2.3. Adsorption

Although thiol–disulfide interchange is an important inactivation mechanism under some conditions, a more significant contribution to the fast phase of the inactivation kinetics shown in Fig. 2 turned out to be loss of protein by adsorption. Thus, a cross-linked double mutant (I3C-97C/C54V), even though it is stable to thiol–disulfide interchange and thiol oxidation because of its lack of free sulfhydryls, also exhibits an initial rapid inactivation phase on heating under the same conditions. The amount of activity lost in the rapid phase, however, drops from about 40% to less than 5% when the starting lysozyme concentration is increased from 3 to 30 μg/ml (R. Wetzel, unpublished observations), and the amount of activity lost at a concentration of 3 μg/ml decreases as the diameter of the glass test tube increases (consistent with a decreasing surface/volume ratio), as shown in Figure 4A. This suggests that there are a finite number of adsorption-competent sites on the walls of the glass tubes in which samples are heated. Pretreatment of the glass with trimethylchlorosilane promoted rather than inhibited the adsorption process (Fig. 4A). Adsorption was also more serious in polypropylene tubes. These facts suggest that this adsorption process is mediated predominantly by hydrophobic interactions.

The fact that this disulfide-cross-linked T4 lysozyme is stable to many of the other inactivation processes normally acting on the wild type made it possible to more clearly observe these adsorption effects, but this mode of inactivation also clearly affects non-cross-linked molecules (Fig. 4B). We have not investigated the temperature dependence of the adsorption reaction for any T4 lysozyme. Surface adsorption has important consequences for the practical use of proteins (Brash and Horbett, 1987) and can reasonably be said to be important *in vivo* as well, in the interaction of proteins with membranes. In at least one case surface properties of a protein have been shown to correlate well with folding stability (Kato and Yutani, 1988; Horbett 1988).

2.4. Aggregation

With adsorption and disulfide-associated effects suppressed, T4 lysozyme continues to be sensitive to inactivation when heated at or above its Tm in the neutral pH region. Figure 5 shows that the inactivation rate for a non-cross-linked cysteine-free mutant [T4 lysozyme(C54V/C97S)], when heated at a concentration of 30 μg/ml, is significantly slower than that of the wild type, but even slower to inactivate is the cross-linked mutant. Figure 6 shows that the temperature at which the inactivation rate is half maximal for the oxidation-stable non-cross-linked mutant agrees well with the Tm for the same protein measured under conditions where unfolding is reversible (Wetzel *et al.*, 1988). This suggests that the major remaining inactivation process is also dependent upon the folded state of the molecule.

This process apparently involves only conformational effects, since inactivated protein can be reactivated by the action of denaturants. Figure 7 shows that almost all of the activity lost when C54V/C97S is heated at 70°C can be restored if the inactivated sample is treated with guanidine hydrochloride and then diluted to low denaturant concentration. The thermal irreversible denaturation clearly involves aggregation, since light-scattering particles are generated (Fig. 8) with about the same kinetics as the directly observed inactivation (Fig. 5).

This aggregation-related inactivation reaction is essentially totally blocked by the insertion of a disulfide bond. Figure 5 shows that the inactivation rate of a 3–97 cross-linked molecule is very slow under conditions at which C54V/C97S inactivates rapidly. The nature of the residual inactivation of the cross-linked

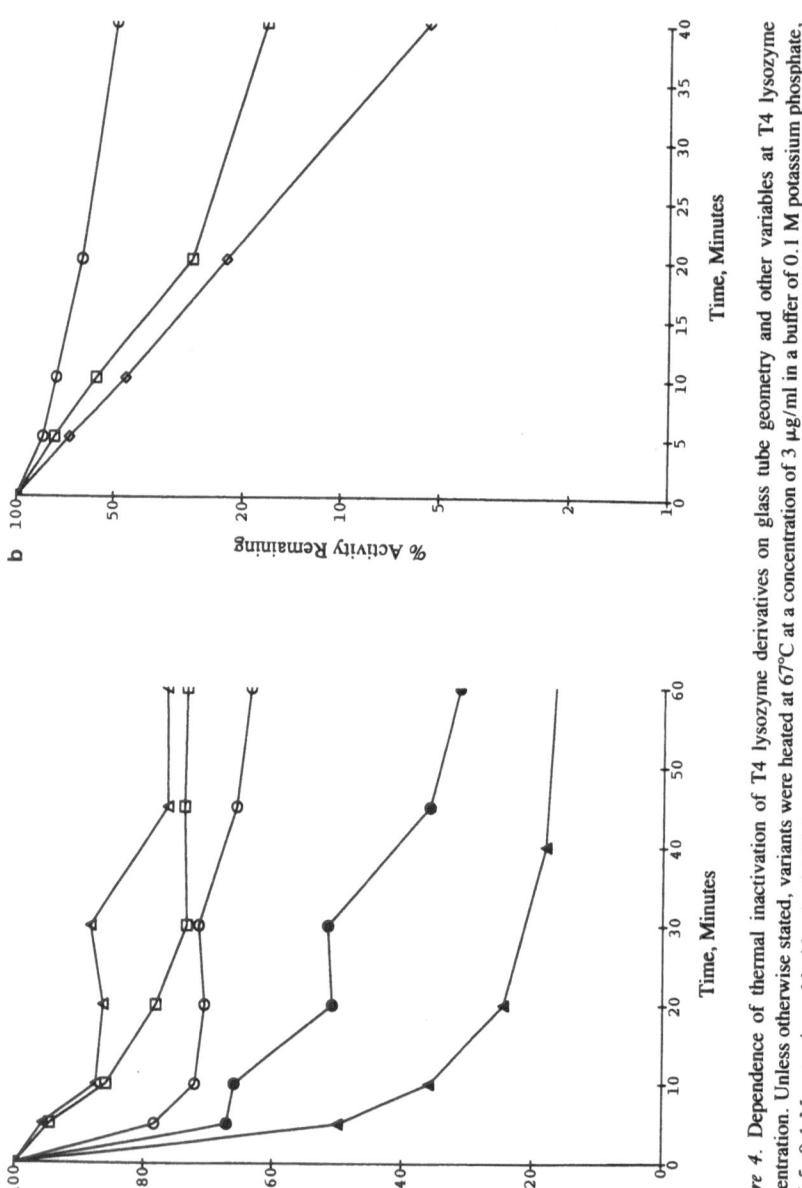

Figure 4. Dependence of thermal inactivation of T4 lysozyme derivatives on glass tube geometry and other variables at T4 lysozyme concentration. Unless otherwise stated, variants were heated at 67°C at a concentration of 3 μg/ml in a buffer of 0.1 M potassium phosphate, pH 6.5, 0.1 M potassium chloride, 1 mM EDTA, in glass tubes of varying dimensions, and were removed from heat at defined intervals, cooled, and assayed at 20°C. (A) Disulfide-cross-linked T4 lysozyme [I3C-97C/C54T] in 6 × 50 mm tubes as described (○) or with sodium rather than potassium as cation (●), in 10 × 75 mm tubes as described (□) or after tubes had been treated with trimethylchlorosilane (▲), and in 12 × 75 mm tubes (△) as described. (B) Non-cross-linked T4 lysozyme [C54T] in potassium salts as listed and in glass tubes measuring 6 × 50 mm (◇), 10 × 75 mm (□), and 12 × 75 mm (○).

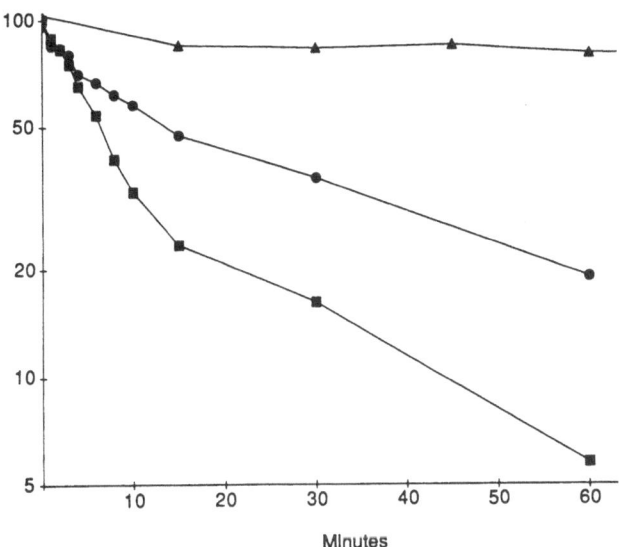

Figure 5. Inactivation kinetics of T4 lysozyme derivatives heated at pH 6.5 and 70°C. Buffers and other conditions as described in legend to Figure 3. I3C-97C/C54V (▲), C54V/C97S (●), wild type (■).

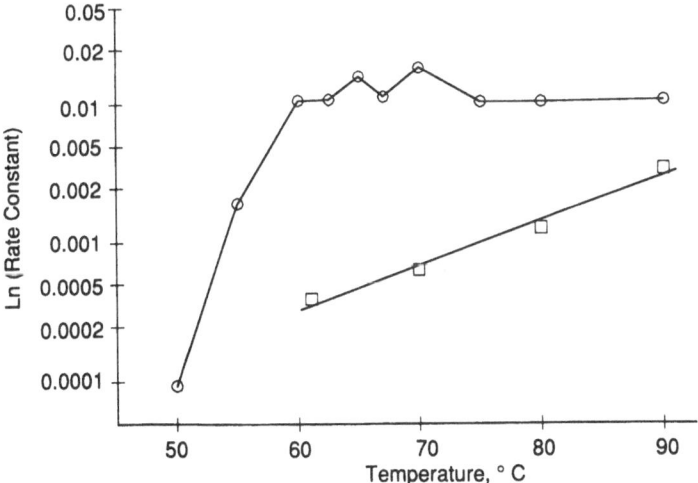

Figure 6. Dependence of rate constants for irreversible inactivation for T4 lysozyme derivatives on incubation temperature. Reactions conditions were as described in the legend to Figure 3. Variants studied were non-cross-linked C54V/C97S (○) and cross-linked I3C-97C/C54V (□). Data for the latter replotted according to the Arrhenius equation yield an E_A of 18 kcal/mole.

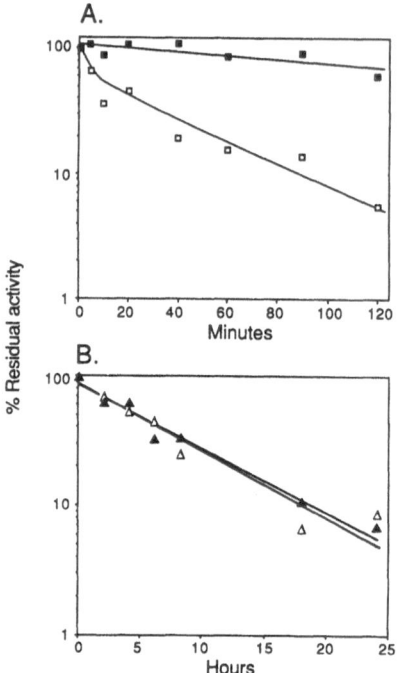

Figure 7. Ability of a denaturation/renaturation cycle to restore activity to heat-inactivated T4 lysozyme derivatives. Proteins were heated at 70°C under the conditions described in the legend to Figure 3, then cooled. One set of duplicates for each protein was diluted into native buffer and assayed as normally (□, △). A separate set of duplicates was worked up by addition of solid guanidine hydrochloride to give an 8M solution which, after standing at room temperature 30 min, was diluted into native buffer, incubated to allow refolding, then further diluted and assayed (■, ▲). (A) Behavior of the non-cross-linked C54V/C97S variant. (B) Behavior of the cross-linked I3C-54C/C54V molecule. Note the differing time scales for A and B, reflecting the differing stabilities to inactivation. (From Wetzel *et al.,* 1988.)

variant will be discussed in the next section. The implications of this result for the roles of disulfide bonds in proteins have been discussed elsewhere (Wetzel *et al.,* 1988; Wetzel, 1988). In the context of this chapter, however, the central point is that, like cysteine oxidation, another major mode of inactivation of T4 lysozyme can be eliminated, not by changing the Tm of the protein, but by altering the kinetic or solution properties of an unfolded form. Evidence for this statement will now be presented.

The folding stability of T4 lysozyme is, in fact, somewhat improved by the 3–97 cross-link. In spite of some destabilization by the Ile3→Cys substitution (which decreases the Tm of T4 lysozyme by 1° at pH 2; W. J. Becktel and W. A. Baase, unpublished results), the 3–97 cross-link is worth 1.1–1.2 kcal/mole in free energy of stabilization compared to wild type, corresponding to an improvement in Tm of 3°C at pH 6.5 (Wetzel *et al.,* 1988).

This small stabilizing effect on the stability of the folded molecule, however, is apparently not responsible for the dramatic stability to inactivation provided by the disulfide shown in Figs. 5 and 6. Thus, the residual inactivation observed for the cross-linked molecule is qualitatively different from the inactivation of non-cross-linked lysozyme. For example, Fig. 6 shows that the thermal dependence of the log of the inactivation rate for the cross-linked protein is

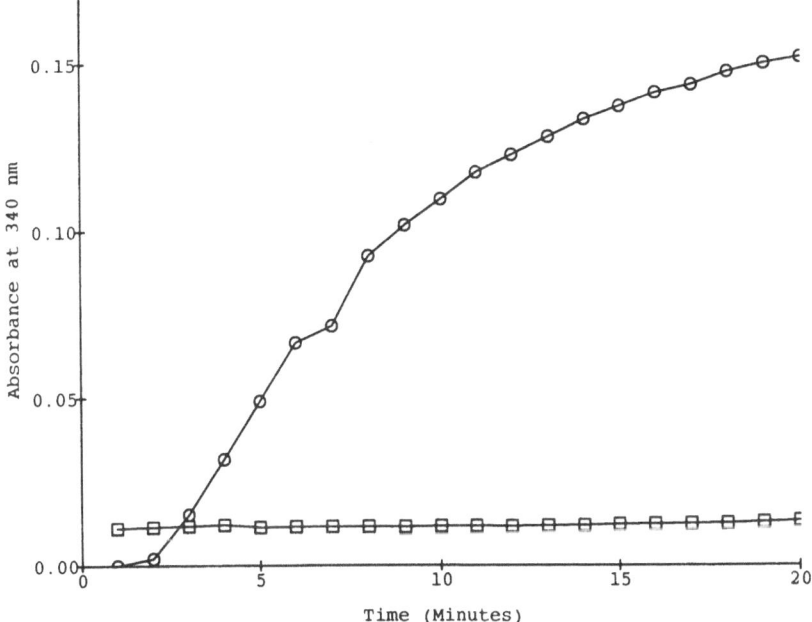

Figure 8. Absorbance change at 340 nm associated with light scattering on heating of lysozyme derivatives. Wild type (○) and cross-linked I3C-97C/C54V (□) T4 lysozymes were heated in glass cuvettes at 70°C at pH 6.5 in the buffer described in the legend to Fig. 3 and the absorbance at 340 nm monitored with time. The C54V/C97S mutant gives results similar to the wild type.

linear with a relatively small slope, while the temperature dependence of the rate for the non-cross-linked lysozyme is biphasic, undergoing a large increase over the small temperature range 50–60 and little further change above 60°C (a behavior qualitatively similar to that of other non-cross-linked T4 lysozymes and to other globular proteins as well). [In fact, it is interesting that the presence or absence of the 3–97 cross-link determines whether T4 lysozyme in its thermal inactivation resembles a classical thermal coagulation, with a rate increase in the hundreds for a temperature increase of 10°C (Chick and Martin, 1912; Anson, 1938), or a typical chemical reaction, with a rate increase of 2–3 for a 10°C temperature increase.] In addition, Fig. 7 shows that, while activity from inactivated non-cross-linked T4 lysozyme can be recovered by denaturation/ renaturation, this treatment has no effect on the inactivated cross-linked molecule. A more direct indication of the nature of the disulfide's effect is found in Fig. 9. This experiment shows that 3–97 cross-linked lysozymes that are relatively unstable to reversible unfolding, because of the introduction of additional temperature-sensitive mutations, are as stable to irreversible inactivation as the rela-

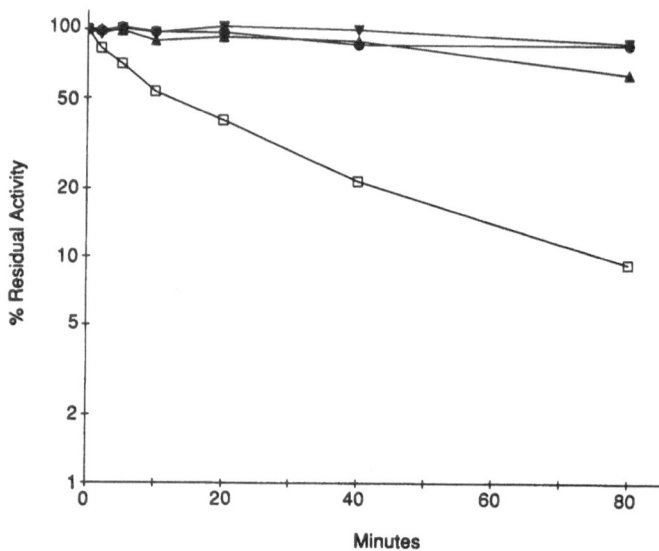

Figure 9. Thermal inactivation of a series of T4 lysozyme variants. Proteins were heated at 73°C. and assayed as described in the legend to Figure 3. The three disulfide-cross-linked mutants with their independently measured Tm values are I3C-97C/C54T (69° C, ●), I3C-97C/C54V/R96H (60° C, ▲), and I3C-97C/C54T/A146T (62° C,▼). The non-cross-linked control, which was heated with the addition of 1 mM DTT, is C54T (66°C, □). Adapted from Wetzel *et al.*, 1988.

tively high-melting I3C-97C/C54T. In fact, all three cross-linked molecules are much more stable to inactivation than a non-crosslinked lysozyme control, despite the fact that the latter is more stable to reversible unfolding than two of the three cross-linked mutants.

    If the disulfide is not providing stability to irreversible denaturation by decreasing the extent to which the protein is unfolded at reaction conditions, how might it be acting? Although the mechanism of Scheme 1 does not allow for many other explanations, the modified mechanism of Scheme 2 does. That T4 lysozyme may undergo thermal unfolding through an unfolding intermediate is supported by two studies on the denaturant unfolding of the molecule (Desmadril and Yon, 1984). Furthermore, studies on other proteins suggest the important role played by unfolding intermediates, as opposed to fully unfolded forms as demanded by Scheme 1, in their irreversible denaturation and aggregation both *in vitro* (London *et al.*, 1974; Yano and Irie, 1975; Ikai *et al.*, 1978; Rudolph *et al.*, 1979; Zettlmeissl *et al.*, 1979; Zetina and Goldberg, 1980; Ghelis and Yon, 1982; Fish *et al.*, 1985; Horowitz and Criscimagma, 1986; Brems *et al.*, 1987, 1988; Mitraki *et al.*, 1987; Brems, 1988; Sanchez-Ruiz *et al.*, 1988) and *in vivo* (Haase-Pettingell and King, 1988). Transient folding/unfolding intermediates are being detected and studied at increasing levels of detail (Udgaonkar and Baldwin, 1988; Roder *et al.*, 1988).

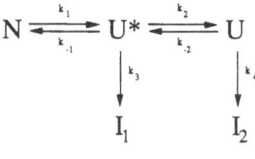

Scheme 2

In the context of this expanded general mechanism for irreversible denaturation, one attractive hypothesis for the role of the disulfide in T4 lysozyme is that it alters the structure of U^* or U to the extent that these unfolded forms are less susceptible to the irreversible, inactivating steps, effectively decreasing the rate constants k_3 and k_4. For example, a cross-link might reduce the total exposure of normally buried hydrophobic residues in an unfolded state, thus reducing its tendency to aggregate. Reduction of disulfide bonds consistently yields molecules that achieve much higher viscosities after solvent-induced unfolding than do their cross-linked forms (Tanford, 1968). The behavior of the cross-linked lysozyme in SDS (Perry and Wetzel, 1986) and urea-gradient (Fig. 10) gels, and in reverse-phase high-performance liquid chromatography (HPLC) (Perry and Wetzel, 1984), supports this idea of a more compact, less hydrophobic state.

Another possibility is that the cross-link substantially changes some of the rate constants for interconversion of the native and various unfolded states, effectively decreasing the steady-state concentration of U^*. This in turn would decrease the rate of inactivation. For example, the disulfide might greatly decrease the rate constant for unfolding, k_1. However, by monitoring unfolding in the far UV circular dichroism spectrum we have established that the rate of unfolding of the cross-linked protein at 65°C is too fast to account for the reduction in the inactivation kinetics (M. G. M and R. W., unpublished data). In fact, the urea gradient gel shown in Fig. 10 suggests that the added disulfide *increases* the rate of urea-induced unfolding, compared to wild type (see Goldenberg and Creighton, 1984, for interpretation). Another way to decrease the steady-state concentration of U^* would be to increase the rate of its decomposition to N and U. Data from denaturant-induced unfolding experiments suggests that the disulfide does indeed have this property. It has been shown that in guanidine hydrochloride-induced unfolding at low temperatures, while non-cross-linked T4 lysozyme unfolds with two-phase kinetics (consistent with a highly populated intermediate), T4 lysozyme with a 3–97 cross-link unfolds in a single kinetic phase with a rate comparable to that of the fast phase of non-crosslinked T4 lysozymes (Anderson *et al.*, 1990).

Thus there is some evidence to support either of the two hypotheses for the mode of action of the 3–97 cross-link, and at present there is no more direct data to address whether either, or both, are responsible. One mechanism that can be

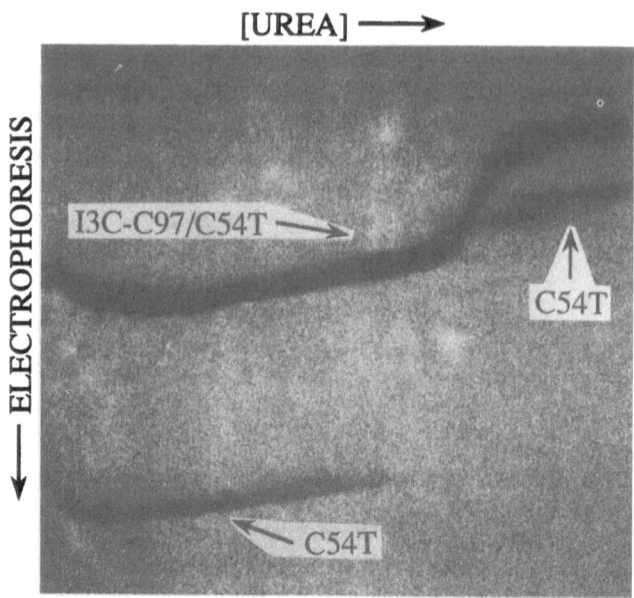

Figure 10. Urea gradient gel of T4 lysozyme derivatives. The gel was poured and run as described for basic proteins in Goldenberg and Creighton (1984). Native T4 lysozymes (C54T) (150 μg) was loaded first and the gel run for 4 hr, after which 150 μg of the native cross-linked I3C-97T/C54T was loaded and the run continued as described. Identification of bands was confirmed by running each protein on separate gels; both were run here on the same gel to avoid effects from differences in the urea gradients. Besides folding kinetics (see text), the gel also indicates the smaller hydrodynamic radius of the urea-denatured cross-linked protein compared to the urea-denatured non-cross-linked protein.

eliminated is that the disulfide facilitates correct refolding and blocks aggregation as the sample is cooled prior to being assayed. This cannot explain the observations that (1) the inactivating process is clearly occurring at the elevated temperature, as shown by the dependence of the thermal inactivation on heating time (Fig. 2), (2) generation of light-scattering particles is coincident with this inactivation (Fig. 8), and (3) altered cooling regimens have no effect on the extent of inactivation (R. Wetzel, unpublished results).

One way to address the mechanism of action of the disulfide, as well as to explore the generality of this method of protein stabilization, is to observe how the stabilizing effect varies as the location of the cross-link is changed. The 3–97 cross-link can be thought of as spanning the N- and C-terminal lobes of the protein as well as a considerable portion of the primary sequence. In contrast, a disulfide between positions 83 and 112 spans a much shorter piece of polypeptide and is located entirely within the highly helical C-terminal domain (Fig. 1).

Figure 11 shows that the reduced, non-cross-linked four-replacement mutant inactivates similarly to other non-cross-linked T4 lysozymes, while the oxidized mutant, containing a cross-link between positions 83 and 112, is about as stable as a 3–97 cross-linked molecule.

2.5. Chemical Inactivation

After oxidation, thiol–disulfide interchange, adsorption, and conformation-related aggregation reactions have been suppressed, T4 lysozyme continues to exhibit some irreversible loss of activity in response to heating in the neutral pH range, as shown, for example, in Figs. 5 and 6 for T4 lysozyme (I3C-97C/C54V). This lost activity cannot be restored by denaturation/ renaturation (Fig. 7), even in the presence of reducing agent (Wetzel *et al.*, 1988). Further, altered mobilities of peptides in HPLC maps of pepsin digests indicate changes in the chemical structure of the protein occurring in parallel with the loss of activity (R. Wetzel, unpublished data). The rates of this inactivation exhibit a simple logarithmic dependence on temperature (Fig. 6), yielding an Arrhenius E_A of about 18 kcal/mole, similar to value of 24 kcal/mole measured for the inactivation of hen egg white lysozyme (Shugar and Syruczek, 1954), and in the range typical for chemical reactions like ester and amide hydrolysis. These observations suggest

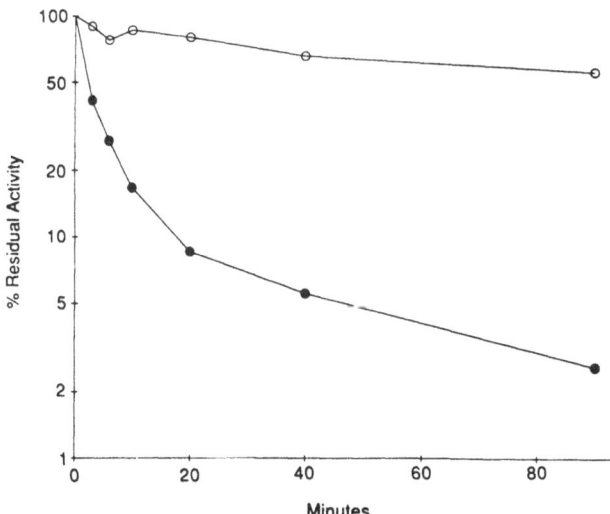

Figure 11. Thermal inactivation of oxidized, cross-linked (○) and reduced, non-cross-linked (●) forms of T4 lysozyme [C54V/C97S/K83C-A112C] at 70°C. and pH 6.5 under the standard conditions described in the legend to Figure 3. The reduced mutant was heated in the presence of 1 mM DTT to preserve it in the non-cross-linked state during heating.

that the remaining inactivation processes involve chemical reactions of the polypeptide. These processes have been well documented for ribonuclease (Zale and
Klibanov, 1986) and hen egg white lysozyme (Ahern and Klibanov, 1985), and
include side-chain deamidation, peptide bond hydrolysis, and β-elimination of
disulfide bonds (Volkin and Klibanov, 1987). The overall rate of inactivation of the
cross-linked T4 lysozyme is consistent with the rates of chemical inactivation
observed in these other systems. In fact, the rate of inactivation of T4 lysozyme
(I3C-97C/C54T) is essentially the same as that of hen egg white lysozyme
measured under the same conditions (Mulkerrin *et al.*, 1986).

The relationship of the rate of chemical inactivation of T4 lysozyme
(I3C-97C/C54T) to its folded state cannot be addressed with the data shown in
Fig. 6. Thus, the lack of appreciable inactivation of the cross-linked variant at
60°C can be sufficiently explained by treating this inactivation as a simple chemical reaction described by the Arrhenius equation. That is, linear extrapolation of
the high temperature data to 60°C predicts the observed low rate, without considering the mutant's unfolding transition at 66°C. Since T4 lysozyme is already
unfolded under conditions in which many chemical inactivation reactions become measurable, it is not a good model system to address this question. However, the dependence of chemical inactivation rates on protein unfolding has been
observed for ribonuclease (Zale and Klibanov, 1983), which has a higher T_m.
Although this dependence of chemical inactivation rates on unfolding may prove
to be a general feature of proteins that exhibit relatively high T_m values, there are
also cases in which tertiary structure can actually promote chemical modification
of proteins. Kossiakoff (1988) has shown that certain deamidation reactions are
promoted by a particular local folding motif, so that an asparagine in the same
sequence context is either favored or disfavored in deamidation, depending on
local structure. Since kinetics have not been determined, the practical importance
of such deamidation processes is not clear. The observation does, however,
introduce the possibility that some chemical reactions may occur preferentially in
the folded state in solution, in contrast to the mechanisms shown in Schemes 1
and 2.

2.6. Conclusions from T4 Lysozyme Experiments

Perhaps the most important observation coming out of these studies is that a
single disulfide bond can impart a dramatic increase in stability to a protein
susceptible to thermal inactivation by conformation-related aggregation. It is not
clear whether this may prove to be a general method for stabilizing proteins to
such irreversible denaturation processes. In this regard, it is encouraging that two
disulfides, the 3–97 and 83–112, independently stabilize T4 lysozyme despite
their different locations. In addition, there are at least two examples of normally
disulfide-bonded proteins that become susceptible to irreversible thermal denaturation when their disulfides are broken (Plunkett and Ryan, 1980; Morehead

et al., 1984). Furthermore, hen lysozyme, which becomes significantly less ordered at room temperature when its disulfides are broken by reduction (White, 1982), also exhibits a substantial decline in solubility under native conditions (Bewley and Li, 1969), whereas it undergoes reversible thermal unfolding under native conditions with its disulfides intact (Imoto *et al.*, 1972). The poor solubility in native conditions of reduced proteins in general has been noted as well (Orsini *et al.*, 1975). Notwithstanding the above discussion, it is clear from experiments with the protease subtilisin that engineered disulfide bonds are not a panacea for thermal instability (Wells and Powers, 1986; Pantoliano *et al.*, 1987; Mitchinson and Wells, 1989), and that we are far from understanding the factors that determine when and how added cross-links will be effective (Wetzel, 1987, 1988).

By screening for phage mutants that no longer lyse infected cells at a restrictive temperature just above 40°C, a number of T4 lysozyme mutants that exhibit thermal sensitivity have been isolated (Remington *et al.*, 1978). Despite the relatively low screening temperature, however, the purified lysozymes from these mutants exhibit Tm values no more than 10°C lower than the Tm of the wild type, 65°C. If all modes of irreversible inactivation require prior unfolding of T4 lysozyme, how can growth at 42°C distinguish between protein Tm values of 55° and 65°C? One attractive possibility is that proteases may exploit some local chain motion (Fontana *et al.*, 1986) in the lysozyme to initiate proteolytic decomposition, a possibility supported by recent *in vitro* experiments (Schellman, 1986). Whatever the basis for this *in vivo* effect, it is clear that the model for the relationship between reversible unfolding and irreversible denaturation shown in Scheme 1 is incapable of explaining the behavior of T4 lysozyme in the cell, jut as it provides an inadequate model for explaining the results of experiments to introduce, by design, improved stability *in vitro.*

By definition, processes that render a protein's denaturation irreversible provide constraints to the ability to undertake thermodynamic analysis of its reversible unfolding. Mutations such as those described here, which bestow reversibility on the system, thus enable more rigorous analysis. This is illustrated by recently published characterization of the low-temperature unfolding of disulfide-cross-linked mutants of T4 lysozyme (Chen and Schellman, 1989; Chen *et al.*, 1989). One can imagine moving other mutations into the disulfide background in order to obtain better reversibiity under wider conditions to more rigorously compare a series of mutants.

3. Interferon-γ

Human interferon-γ (HuIFN-γ) is a noncovalent dimer (Yphantis and Arakawa, 1987) of identical subunits of 143 amino acids. The protein has no cysteine residues, but when expressed by mammalian cells it does contain carbohydrate at two positions (Rinderknecht *et al.*, 1984). Synthesized in *E. coli*, the molecule lacks the carbohydrate but acquires an added N-terminal methionine, making a 144-amino-acid subunit. Both sequence-based predictions (Zoon and Wetzel, 1984) and circular dichroism (CD) measurements (Hsu and Arakawa, 1985) indicate a high alpha-helix content. Beyond this secondary structure feature, the molecule bears little (DeGrado *et al.*, 1982) or no apparent sequence or structural homology with the type I interferons. The experiments described here were performed with a molecule 140 amino acids long that lacks the C-terminal four amino acids but begins with the initiator Met; this molecule has essentially the same activity and stability as the full-length sequence.

Classically, IFN-γ was distinguished from the (type I) α and β interferons by the sensitivity of the former to exposure to pH 2, and for this reason the stability of the protein to acid treatment has been extensively characterized (Arakawa *et al.*, 1985; Hsu and Arakawa, 1985). There is little information on the stability of the protein in the neutral pH range. Attempts to study the reversible thermal unfolding of the molecule were initially unsuccessful due to aggregation (M. G. Mulkerrin, unpublished observations). Aggregation was also a major constraint in devising a well-behaved formulation for storage of the vialed protein as a pharmaceutical (C. Ward, personal communication). We have recently discovered a dramatic pH dependence to the thermal aggregation, which has made possible reversible unfolding studies of the molecule (Mulkerrin and Wetzel, 1989). The differential effects of pH on reversible and irreversible denaturation, described below, demonstrate the inadequacy of the mechanism of Scheme 1 as a basis for rational improvement of HuIFN-γ stability. Preliminary results of genetic experiments on HuIFN-γ, also described, may provide candidates for molecules of improved properties and may also unveil relationships between *in vitro* and *in vivo* stabilities of HuIFN-γ.

3.1. In Vitro Experiments

Although thermal unfolding of HuIFN-γ is irreversible at pH 6, it is reversible at pH 5. Unfolding is also reversible at pH 6 if 0.25 M guanidine hydrochloride is included in the buffer. Figure 12 shows a thermal denaturation profile of the molecule, determined by monitoring circular dichroism ellipticity at 222 nm at pH 6 and 0.25 M guanidine hydrochloride, then correcting the data to 0 M denaturant. The Tm at pH 6 calculated in this way is 53.5°C. The variation of Tm with respect to pH (Table I, values with 0.25 M guanidine hydrochloride, uncorrected) is typical of globular proteins, being highest in the neutral pH range.

The irreversible denaturation of HuIFN-γ can be conveniently followed by the development of light-scattering particles in the spectrophotometer. At the point when light scattering is no longer increasing with heating time, no activity and little UV absorbing material remain in the centrifugation supernate. On the

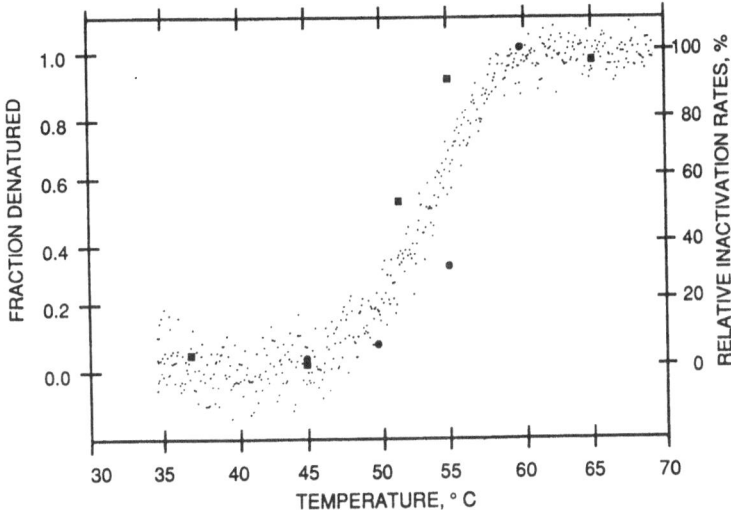

Figure 12. Effects of temperature on various measurements of thermal denaturation of HuIFN-γ. The data points (·) represent mole fractions of unfolded protein, each derived from an individual measurement of the circular dichroism of HuIFN-γ, corrected to 0 M guanidine hydrochloride from the 0.25 M used in the unfolding experiment, from the same sample of protein taken through the melt at 0.1°C/min at pH 6. Also plotted are relative rates of generation of light scattering by a sample of HuIFN-γ heated at pH 6 (●) and the fractional amount of activity lost in samples of HuIFN-γ in PBS containing 0.5% serum albumin after heating 90 min (■). (From Mulkerrin and Wetzel, 1989.)

other hand, activity can be recovered by a denaturation/renaturation cycle (M. G. Mulkerrin, unpublished), as expected for material inactivated *via* a noncovalent aggregation.

Also plotted in Fig. 12 are the kinetics of light scattering with respect to temperature at pH 6, and the temperature-dependence of recovery of immunochemical activity after heating at pH 6.8 for 1 hr. The agreement between the thermodynamic data, on the one hand, and the kinetics of aggregation and inactivation, on the other, suggests that partial or complete unfolding precedes

Table I. Thermal Denaturation of HuIFN-γ[a]

pH	T_m(°C)	pH	T_m(°C)
4.75	30.5	5.75	42.5
5.00	35.5	6.00	41.0
5.22	38.4	6.25	41.8
5.50	40.5		

[a]Samples include 0.25 M guanidine hydrochloride.

inactivation. This is also suggested by the correspondence between the temperature-dependent time lag to onset of generation of light scattering and the half-life of the slow phase observed in unfolding under reversible conditions (Mulkerrin and Wetzel, 1989). Finally, this is suggested by the fact that a truncated, 125-amino-acid-long variant of HuIFN-γ, HuIFN-γ[A124K/Δ(125-143)], has a higher Tm (52°C, compared to 45°C for wild type at pH 5) and requires a higher temperature for equivalent inactivation compared to full-length HuIFN-γ (M. G. Mulkerrin and R. Wetzel, unpublished results).

As the pH is varied from 6 to 5, the inactivation rate (light-scattering kinetics) drops to essentially zero. This behavior suggests that the protonation of some groups with pKas in this range has a stabilizing effect on HuIFN-γ. However, since the inactivation mechanism at pH 6 involves prior unfolding, as discussed earlier, and since unfolding itself is faciliated at pH 5 compared to pH 6, there is a dilemma. The stabilizing effect of shifting pH from 6 to 5 on irreversible denaturation must involve something other than pH stabilization of the folded structure itself. One possibility is an effect of protonation on the properties (solubility, kinetic lifetime, etc.) of some unfolded state.

Since nuclear magnetic resonance experiments show that both His residues titrate in the pH range 5–6 (M. G. Mulkerrin and P. Johnston, unpublished results), we evaluated the stability properties of HuIFN-γ treated with His-directed reagents. Ruthenium pentamine, which is specific for His residues (Matthews *et al.*, 1978; Yocum *et al.*, 1982), produces a derivative that does not generate light scattering on heating at pH 6, 65°C, a temperature well above the Tm of both wild-type and adduct (Mulkerrin and Wetzel, 1989). This suggests that the ruthenium derivative also produces its stabilizing effect by altering a property of some unfolded state of HuIFN-γ.

It is not clear how protonation or ruthenium modification stabilizes HuIFN-γ to irreversible thermal denaturation. The conventional view is that folded and unfolded states can become labile to aggregation when the molecule possesses a net charge near zero, which presumably allows other interactions, such as hydrophobic forces, to predominate (see Introduction). The HuIFN-γ version studied here, however, has a net charge of about +9 at pH 6, and thus would not be expected to be especially sensitive to the addition of further positive charge in its aggregation properties at this pH. Nonetheless, both stabilizing effects—the transition from pH 6 to 5, and the ruthenium modification—would be expected to increase the net positive charge, which is qualitatively consistent with such an interpretation. It may be significant that the charge distribution of the HuIFN-γ molecule is highly skewed with respect to the polypeptide chain, with the C-terminal 16 residues accounting for +8 of the net charge at pH 6. Thus, the N-terminal 125 residues of the polypeptide undergo a charge change from +1 to +3 when the two His residues within it are protonated (assuming negligible protonation of Asp and Glu residues) in the transition from neutral pH 6 to pH 5. This may have a significant effect on the aggregation properties of the unfolded molecule, despite the highly charged C-terminal tail.

As discussed for T4 lysozyme, these effects could also conceivably be due to a retardation of the unfolding rate. However, the unfolding kinetics for HuIFN-γ at 50°C, pH 5, although slow, are not slow enough to account for the nearly complete lack of aggregation under these conditions (Mulkerrin and Wetzel, 1989). Another possibility, parallel to the earlier discussion on T4 lysozyme, is that the pH change affects the tendency of an unfolding intermediate to aggregate. What little data is available supports this hypothesis, since thermal unfolding kinetics under reversible conditions are biphasic (M. G. Mulkerrin and R. Wetzel, unpublished results), which suggests the involvement of an unfolding intermediate.

Whatever the molecular details of these stabilizing treatments, however, both clearly produce their effects by altering the properties of some unfolded form of the protein. These results make an interesting contrast to those of Zale and Klibanov (1983), who showed that the pH dependence of the thermal-irreversible inactivation of ribonuclease, which is entirely due to chemical reactions, can be quantitatively accounted for by the pH dependence of its thermal-reversible unfolding equilibrium.

3.2. Genetic Experiments

The approach to protein stabilization described above, proposing a class of structural changes whose stabilizing effect rests on some improved property of an unfolded state, may have some general validity. Even so, it is not immediately clear what sorts of general methods might be available to accomplish this goal. Cysteine oxidation of an unfolded protein can be effectively blocked by mutagenic replacement of cysteine residues. As discussed earlier, disulfide bonds might have some general ability to retard aggregation of thermally unfolded proteins. What about those proteins that might not be amenable to or respond to cross-linking? What about oligomeric proteins like HuIFN-γ?

It may be possible to take a genetic approach to protein stability to attack such problems in some proteins. For example, we have constructed a collection of randomly generated mutants of HuIFN-γ expressed in the cytoplasm of *E. coli* (Wetzel *et al.*, 1990). The wild-type protein and many of the mutants are expressed in large amounts (about 10–20% of the cellular protein) in the cells. At the same time, a significant number of the mutants produce no detectable protein in a native lysate. Using a series of simple microtiter-format immunochemical screens, we have identified two classes of low-expression mutants: those in which HuIFN-γ does not stably accumulate, and those in which the protein is essentially deposited entirely in inclusion bodies and is thus not available in a native lysis procedure (R. Wetzel and C. Veilleux, unpublished results). In addition, we have identified a class of mutants that yields much higher than wild-type levels of HuIFN-γ in a native lysate. This class seems to be due to the preferential accumulation of these proteins in the cytoplasm and very little in inclusion bodies. This is in contrast to the wild-type and most normally expressed mutants in our expression system, in which the protein is partitioned over both fractions but is found mostly in inclusion bodies.

Inclusion bodies in bacteria are densely packed deposits of protein which are highly insoluble in native buffer when isolated from cells, but which can be solubilized by denaturant. They have been observed to be formed in response to the synthesis in the cell of abnormal forms or abnormally high levels (Prouty *et al.*, 1975; Georgiou *et al.*, 1986) of normal bacterial proteins, and more recently in cells producing heterologous gene products (Wetzel and Goeddel, 1983; Marston, 1986). Their density characteristics suggest that the makeup of these particles can vary with the protein being expressed, ranging from relatively disordered to more ordered structures (Taylor *et al.*, 1986). This suggests that they might be formed from structures ranging from random-coil, unfolded conformations, on the one hand, to native structures of limited solubility, on the other. A second source of the ordered class of inclusion bodies might be partially folded intermediates; Haase-Pettingell and King (1988) have presented convincing evidence that the defect in a series of "temperature sensitive folding" mutants in the P22 phage tailspike protein is an increase in the ability of a folding intermediate to be deposited into inclusion bodies.

Although inclusion bodies of HuIFN-γ tend to be relatively highly ordered (Taylor *et al.*, 1986), the mutants have not been characterized well enough to know whether the inclusion bodies formed in the biosynthesis of HuIFN-γ are composed of an aggregated folding intermediate. If this is the case, however, it may be that mutant proteins that are not deposited in inclusion bodies may have improved thermal stability to irreversible thermal denaturation *in vitro* as well.

The other class of expression mutants containing no detectable levels of HuIFN-γ are also of interest. Some of these mutants probably arise from frame shift and nonsense mutations, but others probably arise from proteolysis of polypeptide chains that are incompletely folded due to inadequate thermodynamic stability or to barriers in folding kinetics, as discussed earlier. Some mutants containing HuIFN-γ exclusively in inclusion bodies may also be explained by increased sensitivity of the soluble form to proteolysis. Analysis of the sequence changes in such mutants should give further information on the folding stability of HuIFN-γ.

Stability in the cell or test tube might ultimately be due to improved stability of the folded state to unfolding or aggregation, to the reduced kinetic lifetime of an unfolding/folding intermediate, or to the reduced aggregation tendency or enhanced proteolytic susceptibility of a folding/unfolding intermediate or of the thermally unfolded state. It would thus be interesting to compare *in vitro* and *in vivo* properties of a series of such mutants.

4. General Conclusions

Although almost all globular proteins are composed of the same 20 amino acids and are constructed by the same kinds of covalent and noncovalent interactions, they vary enormously in their stabilities to irreversible denaturation, just as they do in the thermodynamic stabilities of their folded states and in their activities and functions. As discussed here and elsewhere, a great part of a protein's stability toward irreversible thermal denaturation derives from its folding stability. It is also clear, however, that other aspects of a protein's structure, and the medium in which it finds itself, can make major contributions to stability.

4.1. Modes of Thermal-Irreversible Denaturation

At very cold temperatures some proteins can exhibit the phenomenon of "cold denaturation" (Brandts, 1964; Pace and Tanford, 1968; Nojima *et al.*, 1977; Privalov *et al.*, 1986; Griko *et al.*, 1988; Chen and Schellman, 1989; Chen *et al.*, 1989). Although this little-understood process seems to be derived from the general nature of the forces stabilizing globular proteins—the existence of a minimum in the temperature dependence of the free energy of stabilization—it is usually not observed because of the practical limit of solvent freezing. It is possible, however, that cold denaturation plays a role in the lability of some proteins to lyophilization—for example, due to the possibility of supercooling during the initial freezing step.

At relatively mild conditions, where a protein is stably folded, proteins may still be labile. Harsh chemicals may have access to the interior due to their small size, and/or due to loosening of protein conformation, by normal breathing modes or by "unzipping" of the structure as surface accessible residues are modified. Similarly, proteases may cleave folded proteins at loops that become transiently accessible due to local fluctuations (Fontana *et al.*, 1986), thereby making the remaining structure more susceptible to further proteolytic breakdown. Intramolecular thiol–disulfide interchange may compromise the folded structure and folding stability of a protein containing both disulfides and free sulfhydryls, leading to structures that might further denature in other ways, such as through the formation of disulfide-linked oligomers or noncovalent aggregates. Kossiakoff (1988) has described a structural motif that promotes deamidation of Asn residues in the folded, crystalline state under mild conditions. Folded proteins can, of course, also be precipitated by pH or salt effects, but normally these precipitates readily redissolve in native buffer of a more favorable pH.

Some chemical reactions, such as thiol oxidation promoted by ambient levels of molecular oxygen, can require an unfolded state of the protein, as shown for T4 lysozyme's sensitivity to disulfide-mediated oligomerization by oxidation. Many proteins undergo aggregation and other conformation-related modes of irreversible denaturation when exposed to temperatures in the range of their cooperative unfolding transitions. Although this aggregation has generally been attributed to the poor solution properties of the thermally unfolded state, it is possible that in many cases the aggregating species is a poorly behaved unfolding intermediate. Almost all of the examples cited here, which give some evidence for the involvement of a folding intermediate in aggregation under denaturing conditions, are either multidomain proteins (Zetina and Goldberg, 1980; Horowitz and Criscimagma, 1986; Mitraki *et al.*, 1987), like T4 lysozyme, or multisubunit proteins (London *et al.*, 1974; Yano and Irie, 1975; Rudolph *et al.*, 1979; Zettlmeissl *et al.*, 1979; Haase-Pettingell and King, 1988), like HuIFN-γ. The fact that the well-characterized example of bovine growth hormone (Brems *et al.*, 1987, 1988; Brems, 1988) appears to be an exception (Abdel-Mequid *et al.*, 1987) makes it especially interesting.

As discussed earlier, an upper temperature limit to protein stability is generally provided by the stabilities of amide bonds to hydrolysis and of disulfide bonds to β-elimination and scrambling (Zale and Klibanov, 1986). Although at least some of these processes can be promoted in an unfolded protein (Zale and Klibanov, 1983), for most proteins the reaction conditions required are so harsh that they would be expected, in any case, to promote the unfolding of the most stable proteins. Thus, in order for amide hydrolysis and disulfide decomposition to set a practical limit to protein stability, the protein must be stable to other inactivating mechanisms which occur at lower temperatures. For example, triose phosphate isomerase, which has a Tm of 60°C (Casal *et al.*, 1987), undergoes a critical deamidation in an asparagine important for subunit stability when heated at neutral pH and 100°C. Mutagenic replacement of this residue with threonine allowed retention of a functional dimer interface that at the same time survived exposure to 100°C better than wild type. This stabilization, however, could only be demonstrated by heating the enzyme in denaturing levels of guanidine hydrochloride (followed by refolding in native buffer at normal temperature), since the enzyme heated in native buffer is inactivated at 60°C by an aggregation process (Ahern *et al.*, 1987). Because the Tm values of most proteins are well below the temperatures required to do chemical damage, unfolding-related modes of irreversible denaturation in general will have to be solved first in order to achieve practical improvement in stability of such proteins (Tomazic and Klibanov, 1988).

4.2. Strategies for Stabilizing Proteins

As reviewed in the preceding section, there are many ways in which proteins can be irreversibly denatured and inactivated upon exposure to heat or other harsh treatments. It is reasonable to expect that the prospects for stabilizing a particular protein to a particular stress will become more favorable the more one knows about the processes by which irreversible denaturation occurs in that system. Many, but not all, of these processes are ruled by mechanisms that include at least partial unfolding of the protein's globular structure. Thus, attacking the folding stability of the protein is a reasonable approach and in many cases can be expected to give modest gains in stability in the temperature range around the Tm of the protein. Other options should also be explored, however, since some inactivating processes do not directly depend on the folded state of the protein. Even for those that do, dramatic stabilization can sometimes be achieved through structural changes that appear to operate on some parameter other than the free energy of stabilization, as illustrated in the experiments described in this chapter. This approach to stabilizing proteins to unfolding has been utilized in facilitating the refolding of some proteins involving disulfide formation, in which the solubilities of reduced, denatured states and refolding intermediates were improved by polyalanylation (Epstein and Anfinsen, 1962; Freedman and Sela, 1966; Orsini *et al.*, 1975).

It is possible that attempts to stabilize proteins to some processes by improving folding stability may actually backfire. For example, in attempting to improve a protein that gives unacceptable aggregation after a lyophilization/ resuspension cycle, given the relationship between unfolding and aggregation discussed earlier, one might be tempted to look for a protein variant with a higher Tm than the wild type. However, since cold denaturation may be the process responsible for the aggregation, stability to this process and not to heat denaturation may be the important parameter. Nojima *et al.* (1977) have pointed out that it is theoretically possible for proteins exhibiting increased stability to thermal unfolding to be either more or less stable to cold unfolding, depending on the thermodynamic mechanism of stabilization. In addition, it is possible that increasing the folding stability of some proteins may lead to undesirable decreases in functional specific activity.

Although it may be useful to consider approaches directed at the solution or kinetic stability of an unfolded form, a significant drawback is the limited empirical and theoretical basis for such an approach. For this reason, a genetic approach is particularly attractive. Genetic selections or screens for thermally stable mutant proteins have in the past generated molecules whose *in vivo* stability correlated with an improved folding stability measured *in vitro* (Goldenberg, 1988). However, this relationship may not always, strictly speaking, be the basis for the selection (see discussion on T4 lysozyme and P22 tailspike protein). As discussed in this chapter, properly chosen and interpreted selections or screens may produce protein variants that can be expected to exhibit stability to irreversible denaturation, regardless of changes in folding stability, when tested *in vitro*.

Some genetic diseases can be attributed to the loss of activity in a particular protein that is ultimately due to a decrease in protein stability. Other disorders are accompanied by, and possibly caused by, a buildup of denatured protein. There are intriguing similarities in the appearance, nonrandom composition, and solubility characteristics of bacterial inclusion bodies (Prouty *et al.*, 1975) and amyloid deposits (Benson and Wallace, 1989). While in some cases a reduction in the folding stability of the protein may be sufficient to explain the phenotype, future work may reveal a class of abnormal proteins whose irreversible inactivation and/or precipitation is governed by factors other than Tm.

References

Abedel-Mequid, S. S., Shieh, H-S., Smith, w. W., Dayringer, H. E., Violand, B. N., and Bentle, L. A., 1987, Three dimensional structure of a genetically engineered variant of porcine growth hormone, *Proc. Natl. Acad. Sci. USA* **84**:6434–6437.

Ahern, T. J., and Klibanov, A. M., 1985, The mechanism of irreversible enzyme inactivation at 100°C, *Science* **228**:1280–1284.

Ahern, T. J., Casal, J. I., Petsko, G. A., and Klibanov, A. M., 1987, Control of oligomeric enzyme thermostability by protein engineering, *Proc. Natl. Acad. Sci. USA* **84**:675–679.

Alber, T., and Wozniak, J. A., 1985, A genetic screen for mutations that increase the thermal stability of phage T4 lysozyme, *Proc. Natl. Acad. Sci. USA* **82**:747–750.

Alber, T., Gruetter, M. G., Gray, T. M., Wozniak, J. A., Weaver, L. H., Chen, B-L., Baker, E. N., and Matthews, B. W., 1986, Structure and stability of mutant lysozymes from bacteriophage T4, in: *Protein Structure, Folding and Design* (D. L. Oxender, ed.), pp. 307–318, Alan R. Liss, New York.

Alber, T., Dao-pin, S., Nye, J. A., Muchmore, D. C., and Matthews, B. W., 1987, Temperature-sensitive mutations of bacteriophage T4 lysozyme occur at sites with low mobility and low solvent accessibility in the folded protein, *Biochemistry* **26**:3754–3758.

Anderson, W. D., Fink, A. L., Perry, L. J., and Wetzel, R., 1990, Effect of an engineered disulfide bond on the folding of T4 lysozyme at low temperatures, *Biochem.* **29**:3331–3337.

Anfinsen, C. B., and Scheraga, H. A., 1975, Experimental and theoretical aspects of protein folding, *Adv. Prot. Chem.* **29**:205–300.

Anson, M. L., 1938, The coagulation of proteins, in: *The Chemistry of the Amino Acids and Proteins* (C. L. A. Schmidt, ed.), pp. 407–428, Charles C Thomas, Springfield, IL.

Anson, M. L., 1945, Protein denaturation and the properties of protein groups, *Adv. Prot. Chem.* **2**:361–386.

Arakawa, T., Allton, N. K., and Hsu, Y-R., 1985, Preparation and characterization of recombinant DNA-derived human interferon-γ, *J. Biol.Chem.* **260**:14435–14439.

Becktel, W. J., and Baase, W. A., 1987, Thermal denaturation of bacteriophage T4 lysozyme at neutral pH, *Biopolymers* **26**:619–623.

Becktel, W. J., Baase, W. A., Chen, B. L., Muchmore, D. C., Schellman, C. G., and Schellman, J. A., 1986, The thermodynamics of protein denaturation—Single and multiple amino acid variants of bacteriophage-T4 lysozyme, *Biophys. J.* **49**:572a.

Benson, M. D., and Wallace, M. R., 1989, Amyloidosis, in: *The Metabolic Basis of Inherited Disease* (C. R. Scriver, A. L. Beaudet, W. S. Sly, and D. Valle, eds.), pp. 2439–2460.

Bewley, T. A., and Li, C. H., 1969, The reduction of protein disulfide bonds in the absence of denaturants, *Int. J. Prot. Res.* **1**:117–124.

Brandts, J. F., 1964, The thermodynamics of protein denaturation. I. The denaturation of chymotrypsinogen, *J. Amer. Chem. Soc.* **86**:4291–4301.

Brandts, J. F., 1969, Conformational transitions of proteins in water and in aqueous mixtures, in: *Structure and Stability of Biological Macromolecules* (S. Timasheff and G. Fasman, eds.), pp. 213–290, Marcel Dekker, New York.

Brash, J. L., and Horbett, T. A., eds., 1987, *Proteins at Interfaces,* American Chemical Society, Washington, DC.

Brems, D. N., 1988, Solubility of different folding conformers of bovine growth hormone, *Biochemistry* 27:4541–4546.

Brems, D. N., Plaisted, S. M., Dougherty, Jr., J. J., and Holzman, T. F., 1987, The kinetics of bovine growth hormone folding are consistent with a framework model, *J. Biol. Chem.* 262:2590–2596.

Brems, D. N., Plaisted, S. M., Havel, H. A., and Tomich, C.-S. C., 1988, Stabilization of an associated folding intermediate of bovine growth hormone by site-directed mutagenesis, *Proc. Natl. Acad. Sci. USA* 85:3367–3371.

Casal, J. I., Ahern, T. J., Davenport, R. C., Petsko, G. A., and Klibanov, A. M., 1987, Subunit interface of triosephosphate isomerase: Site-directed mutagenesis and characterization of the altered enzyme, *Biochemistry* 26:1258–1264.

Chen, B., and Schellman, J. A., 1989, Low-temperature unfolding of a mutant of phage T4 lysozyme. I. Equilibrium studies, *Biochemistry* 28:685–691.

Chen, B., Baase, W. A., and Schellman, J. A., 1989, Low-temperature unfolding of a mutant of phage T4 lysozyme. II. Kinetic investigations, *Biochemistry* 28:691–699.

Chick, H., and Martin, C. J., 1912, On the "heat coagulation" of proteins. Part IV. The conditions controlling the agglutination of proteins already acted on by hot water, *J. Physiol.* 45:261–295.

Creighton, T. E., 1978, Experimental studies of protein folding and unfolding, *Prog. Biophys. Mol. Biol.* 33:231–297.

Daar, I. O., Artymiuk, P. J., Phillips, D. C., and Maquat, L. E., 1986, Human triose-phosphate isomerase deficiency: A single amino acid substitution results in a thermolabile enzyme, *Proc. Natl. Acad. Sci. USA* 83:7903–7907.

DeGrado, W. F., Wasserman, Z. R., and Chowdhry, V., 1982, Sequence and structural homologies among type I and type II interferons, *Nature* 300:379–381.

Desmadril, M., and Yon, J. M., 1984, Evidence for intermediates during unfolding and refolding of a two-domain protein, phage T4 lysozyme: Equilibrium and kinetic studies, *Biochemistry* 23:11–19.

Elwell, M., and Schellman, J., 1975, Phage T4 lysozyme; physical properties and reversible unfolding, *Biochim. Biophys. Acta* 386:309–323.

Elwell, M., and Schellman, J., 1977, Stability of phage T4 lysozymes: 1. Native properties and thermal stability of wild type and two mutant lysozymes, *Biochim. Biophys. Acta* 494:367–383.

Epstein, C. J., and Anfinsen, C. B., 1962, Reversible reduction of disulfide bonds in trypsin and ribonuclease coupled to carboxymethyl cellulose, *J. Biol. Chem.* 237:2175–2179.

Fish, W. W., Danielsson, A., Nordling, K., Miller, S. H., Lam, C. F., and Bjork, I., 1985, Denaturation behavior of antithrombin in guanidinium chloride. Irreversibility of unfolding caused by aggregation, *Biochemistry* 24:1510–1517.

Fontana, A., Fassina, G., Vita, C., Dalzoppo, D., Zamai, M., and Zambonin, M., 1986, Correlation between sites of limited proteolsis and segmental mobility in thermolysin. *Biochemistry* 25:1847–1851.

Freedman, M. H., and Sela, M., 1966, Recovery of antigenic activity upon reoxidation of completely reduced polyalanyl rabbit immunoglobulin G, *J. Biol.Chem.* 241:2383–2396.

Freedman, R. B., and Hillson, D. A., 1980, Formation of disulfide bonds, in: *The Enzymology of Post-translational Modification of Proteins, Vol. 1* (R. B. Freedman and H. C. Hawkins, eds.), pp. 157–212, Academic Press, Orlando, FL.

Frensdorff, H. K., Watson, M. T., and Kauzmann, W., 1953, The kinetics of protein denaturation. IV. The viscosity and gelation of urea solutions of ovalbumin, *J. Am. Chem. Soc.* 75:5157–5166.

Fruton, J. S., 1972, *Molecules and Life; Historical Essays on the Interplay of Chemistry and Biology,* pp. 87–95, Wiley, New York.

Georgiou, G., Telford, J. N., Shuler, M. L., and Wilson, D. B., 1986, Localization of inclusion bodies in *Escherichia coli* overproducing β-lactamase or alkaline phosphatase, *Appl. Environ. Microbiol.* **52**:1157–1161.

Ghelis, C., and Yon, J., 1982, *Protein Folding*, Academic Press, New York.

Goldenberg, D. P., 1988, Genetic studies of protein stability and mechanisms of folding, *Annu. Rev. Biophys. Biophys. Chem.* **17**:481–507.

Goldenberg, D. P., and Creighton, T. E., 1984, Gel electrophoresis in studies of protein conformation and folding, *Anal. Biochem.* **138**:1–18.

Griko, Yu.V., Privalov, P. L., Sturtevant, J. M., and Venyaminov, S.Yu., 1988, Cold denaturation of staphylococcal nuclease, *Proc. Natl. Acad. Sci. USA* **85**:3343–3347.

Haase-Pettingell, C. A., and King, J., 1988, Formation of aggregates from a thermolabile *in vivo* folding intermediate in P22 tailspike maturation; a model for inclusion body formation, *J. Biol. Chem.* **263**:4977–4983.

Haber, E., and Anfinsen, C. B., 1962, Side-chain interactions governing the pairing of half-cystine residues in ribonuclease, *J. Biol. Chem.* **237**:1839–1844.

Hawkes, R., Gruetter, M. G., and Schellman, J., 1984, Thermodynamic stability and point mutations of bacteriophage T4 lysozyme, *J. Mol. Biol.* **175**:195–212.

Horbett, T. A., 1988, Molecular origins of the surface activity of proteins, *Prot. Eng.* **2**:172–174.

Horowitz, P., and Criscimagma, N. L., 1986, Low concentrations of guanidinium chloride expose apolar surfaces and cause differential perturbation in catalytic intermediates of rhodanese, *J. Biol. Chem.* **261**:15652–15658.

Hsu, Y-R., and Arakawa, T., 1985, Structural studies on acid unfolding and refolding of recombinant human interferon-γ, *Biochemistry*, **24**:7959–7963.

Huggins, C., Tapley, D. F., and Jensen, E. V., 1951, Sulfhydryl-disulfide relationships in the induction of gels in proteins by urea, *Nature* **167**:592–593.

Ikai, A., Tanaka, S., and Noda, H., 1978, Reactivation kinetics of guanidine denatured bovine carbonic anhydrase B, *Arch. Biochem. Biophys.* **190**:39–45.

Imoto, T., Johnson, L. N., North, A. C. T., Phillips, D. C., and Rupley, J. A., 1972, Vertebrate lysozymes, in: *The Proteins* (P. D. Boyer, ed.), pp. 665–868, Academic Press, New York.

Kang, J., Lemaire, H-G., Unterbeck, A., Salbaum, J. M., Masters, C. L., Grzeschik, K-H., Multhaup, G., Beyreuther, K., and Mueller-Hill, B., 1987, The precursor of Alzheimer's disease amyloid A4 protein resembles a cell-surface receptor, *Nature* **325**:733–736.

Kato, A., and Yutani, K., 1988, Correlation of surface properties with conformational stabilities of wild-type and six mutant tryptophan synthase α-subunits substituted at the same position, *Prot. Engl.* **2**:153–156.

Kauzmann, W., 1959, Some factors in the interpretation of protein denaturation, *Adv. Prot. Chem.* **14**:1–63.

Kim, P. S., and Baldwin, R. L., 1982, Specific intermediates in the folding reactions of small proteins and the mechanism of protein folding, *Annu. Rev. Biochem.* **51**:459–489.

Kinsella, J. E., 1982, Relationships between structure and functional properties of food proteins, in: *Food Proteins* (P. F. Fox and J. J. Condon, eds.), pp. 51–103, Applied Science Pubs., London.

Kishi, H., Mukai, T., Hirono, A., Fujii, H., Miwa, S., and Hori, K., 1987, Human aldolase A deficiency associated with a hemolytic anemia: Thermolabile aldolase due to a single base mutation, *Proc. Natl. Acad. Sci. USA* **84**:8623–8627.

Kossiakoff, A. A., 1988, Tertiary structure is a principal determinant to protein deamidiation, *Science* **240**:191–194.

London, J., Skrzynia, C., and Goldberg, M. E., 1974, Renaturation of *Escherichia coli* tryptophanase after exposure to 8 M urea, *Eur. J. Biochem.* **47**:409–415.

Lumry, R., and Biltonen, R., 1969, Thermodynamic and kinetic aspects of protein conformations in relation to physiological function, in: *Structure and Stability of Biological Macromolecules* (S. Timasheff and G. Fasman, eds.), pp. 65–212, Marcel Dekker, New York.

Lumry, R., and Eyring, H., 1954, Conformation changes of proteins, *J. Phys. Chem.* **58**:110–120.

Marston, F. A. O., 1986, The purification of eukaryotic polypeptides synthesized in *Escherichia coli, Biochem. J.* **240**:1–12.

Masters, C. L., Simms, G., Weinman, N. A., Multhaup, G., McDonald, B. L., and Beyreuther, K., 1985, Amyloid plaque core protein in Alzheimer disease and Down's syndrome, *Proc. Natl. Acad. Sci. USA* **82**:4245–4249.

Matsumura, M., Yasumura, S., and Aiba, S., 1986, Cumulative effect of intragenic amino-acid replacements on the thermostability of a protein, *Nature* **323**:356–358.

Matsumura, M., Becktel, W. J., and Matthews, B. W.,1988, Hydrophobic stabilization in T4 lysozyme determined directly by multiple substitutions of Ile 3, *Nature* **334**:406–410.

Matthews, B. W., Nicholson, H., and Becktel, W. J., 1987, Enhanced protein thermostability from site-directed mutations that decrease the entropy of unfolding, *Proc. Natl. Acad. Sci. USA* **84**:6663–6667.

Matthews, C. R., Erickson, P. M., Van Vliet, D. L., and Petersheim, M., 1978, Synthesis of pentaammineruthenium-histidine complexes in ribonuclease-A, *J. Am. Chem. Soc.* **100**:2260–2262.

McKenzie, H. A., Ralston, G. B., and Shaw, D. C., 1972, Location of sulfhydryl and disulfide groups in bovine β-lactoglobulins and effects of urea, *Biochemistry* **11**:4539–4547.

Mirsky, A. E., and Pauling, L., 1936, On the structure of native, denatured, and coagulated proteins, *Proc. Natl. Acad. Sci. USA* **22**:439–447.

Mitchinson, C., and Wells, J. A., 1989, Protein engineering of disulfide bonds in subtilisin BPN′ *Biochemistry* **28**:4807–4815.

Mitraki, A., Betton, J-M., Desmadril, M., and Yon, J. M., 1987, Quasi-irreversibility in the unfolding-refolding transition of phosphoglycerate kinase induced by guanidine hydrochloride, *Eur. J. Biochem.* **163**:29–34.

Morehead, H., Johnson, P. D., and Wetzel, R., 1984, Roles of the 29-138 disulfide bond of subtype A of human α-interferon in its antiviral activity and conformational stability, *Biochemistry* **23**:2500–2507.

Mulkerrin, M. G., and Wetzel, R., 1989, The pH dependence of the reversible and irreversible thermal denaturation of gamma interferon *Biochemistry* **28**:6556–6561.

Mulkerrin, M. G., Perry, L. J., and Wetzel, R., 1986, Stability and solution structure of a disulfide cross-linked T4 lysozyme, in: *Protein Structure, Folding and Design* (D. L. Oxender, ed.), pp. 297–305, Alan R. Liss, New York.

Neurath, H., Greenstein, J. P., Putnam, F. W., and Erickson, J. O., 1944, The chemistry of protein denaturation, *Chem. Rev.* **34**:157–265.

Nojima, H., Ikai, A., Oshima, T., and Noda, H., 1977, Reversible thermal unfolding of thermostable phosphoglycerate kinase. Thermostability associated with mean zero enthalpy change, *J. Mol. Biol.* **116**:429–442.

Oesch, B., Westaway, D., Walchli, M., McKinldey, M. P., Kent, S. B. H., Aebersold, R., Barry, R. A., Tempst, P., Teplow, D. B., Hood, L. E., Prusiner, S. B., and Weissman, C., 1985, A cellular gene encodes scrapie PrP 27-30 protein, *Cell* **40**:735–746.

Orsini, G., Skrzynia, C., and Goldberg, M. E., 1975, The renaturation of reduced polyalanylchymotrypsinogen and chymotrypsinogen, *Eur. J. Biochem.* **59**:433–440.

Pace, N. C., and Tanford, C., 1968, Thermodynamics of the unfolding of β-lactoglobulin A in aqueous urea solutions between 5 and 55°, *Biochemistry* **7**:198–208.

Pantoliano, M. W., Ladner, R. C., Bryan, P. N., Rollence, M. L., Wood, J. F., and Poulos, T. L., 1987, Protein engineering of subtilisin BPN′: Enhanced stabilization through the introduction of two cysteines to form a disulfide bond, *Biochemistry* **26**:2077–2082.

Perry, L. J., and Wetzel, R., 1984. Disulfide bond engineered into T4 lysozyme: Stabilization of the protein toward thermal inactivation, *Science* **226**:555–557.

Perry, L. J., and Wetzel, R., 1986, Unpaired cysteine-54 interferes with the ability of an engineered disulfide to stabilize T4 lysozyme, *Biochemistry* **25**:733–739.

Perry, L. J., and Wetzel, R., 1987, The role of cysteine oxidation in the thermal inactivation of T4 lysozyme, *Protein Engineering* **1**:101–105.

Plunkett, G., and Ryan, C. A., 1980, Reduction and carboxamidomethylation of the single disulfide bond of proteinase inhibitor I from potato tubers, *J. Biol Chem.* **255**:2752–2755.

Privalov, P. L., 1982, Stability of proteins: Proteins which do not present a single cooperative system, *Adv. Prot. Chem.* **35**:1–104.

Privalov, P. L., Griko, Yu.V., Venyaminov, S.Yu., and Kutyshenko, V. P., 1986, Cold denaturation of myoglobin, *J. Mol. Biol.* **190**:487–498.

Prouty, W. F., Karnovsky, M. J., and Goldberg, A. L., 1975, Degradation of abnormal proteins in *Escherichia coli*, *J. Biol. Chem.* **250**:1112–1122.

Remington, S. J., Anderson, W. F., Owen, J., Ten Eyck, L. F., Grainger, C. T., and Matthews, B. W., 1978, Structure of the lysozyme from bacteriophage T4: An electron density map at 2.4 Å resolution, *J. Mol. Biol.* **118**:81–98.

Rinderknecht, E., O'Connor, B. H., and Rodriquez, H., 1984, Natural human interferon-γ: Complete amino acid sequence and determination of sites of glycosylation, *J. Biol. Chem.* **259**:6790–6797.

Robakis, N. K., Sawh, P. R., Wolfe, G. C., Rubenstein, R., Carp, R. I., and Innis, M. A., 1986, Isolation of a cDNA clone encoding the leader peptide of prion protein and expression of the homologous gene in various tissues, *Proc. Natl. Acad. Sci. USA* **83**:6377–6381.

Roder, H., Elove, G. A., and Englander, S. W., 1988, Structural characterization of folding intermediates in cytochrome c by H-exchange labelling and proton NMR, *Nature* **335**:700–704.

Rudolph, R., Zettlmeissl, G., and Jaenicke, R., 1979, Reconstitution of lactic dehydrogenase. Noncovalent aggregation vs. reactivation. 2. Reactivation of irreversibly denatured aggregates, *Biochemistry* **18**:5572–5575.

Sanchez-Ruiz, J. M., Lopez-Lacomba, J. L., Cortijo, M., and Mateo, P. L., 1988a, Differential scanning calorimetry of the irreversible thermal denaturation of thermolysin, *Biochemistry* **27**:1648–1652.

Schellman, C. G., 1986, Proteolysis as a probe for motility in mutant bacteriophage T4 lysozymes, *Biophys. J.* **49**:493a.

Schellman, J. A., Lindorfer, M., Hawkes, R., and Gruetter, M., 1981, Mutations and protein stability, *Biopolymers* **20**:1989–1999.

Seiki, R. K., and Gelfand, D. H., 1988, Primer-directed enzymatic amplification of DNA with a thermostable DNA polymerase, *Science* **239**:487–494.

Shugar, D., and Syruczek, E., 1954, Kinetics of heat inactivation of lysozyme and the influence of various buffers and manganese ions, *Bull. Acad. Polon. Sci. Cl. II* **2**:73–78.

Streisinger, G., Muhai, F., Dreyer, W., Miller, B., and Horiuchi, S., 1961, Mutations affecting lysozyme of phage T4, *Cold Spring Harbor Symp. Quant. Biol.* **26**:25–30.

Stroupe, S. D., and Foster, J. F., 1973, Further studies of the sulfhydryl-catalyzed isomerization of bovine mercaptoalbumin, *Biochemistry* **12**:3824–3830.

Tanford, C., 1968, Protein denaturation, *Adv. Prot. Chem.* **23**:121–282.

Taylor, G., Hoare, M., Gray, D. R., and Marston, F. A. O., 1986, Size and density of protein inclusion bodies, *Biotechnology* **4**:553–557.

Tomazic, S. J., and Klibanov, A. M., 1988, Why is one *Bacillus* α-amylase more resistant against irreversible thermoinactivation than another? *J. Biol. Chem.* **263**:3092–3096.

Torchinsky, Yu.M., 1981, *Sulfur in Proteins*, p. 55, Pergamon Press, Oxford.

Tsugita, A., Inouye, M., Terzaghi, E., and Streisinger, G., 1968, Purification of T4 lysozyme, *J. Biol. Chem.* **243**:391–397.

Udgaonkar, J. B., and Baldwin, R. L., 1988, NMR evidence for an early framework intermediate on the folding pathway of ribonuclease A, *Nature* 335:694–699.

Volkin, D. B., and Klibanov, A. M., 1987, Thermal destruction processes in proteins involving cystine residues, *J. Biol. Chem.* 262:2945–2950.

Vulliamy, T. J., D'Urso, M., Battistuzzi, G., Estrada, M., Foulkes, N. S., Martini, G., Calabro, V., Poggi, V., Giordano, R., Town, M., Luzzatta, L., and Persico, M. G., 1988, Diverse point mutations in the human glucose-6-phosphate dehydrogenase gene cause enzyme deficiency and mild or severe hemolytic anemia, *Proc. Natl. Acad. Sci. USA* 85:5171–5175.

Warner, R. C., and Levy, M., 1958, Denaturation of bovine plasma albumin. II. Isolation of intermediates and mechanism of the reaction at pH 7, *J. Am. Chem. Soc.* 80:5735–5744.

Wells, J. A., and Powers, D. B., 1986, *In vivo* formation and stability of engineered disulfide bonds in subtilisin, *J. Biol. Chem.* 261:6564–6570.

Wetzel, R., 1986, Investigation of the structural role of disulfides by protein engineering: A study with T4 lysozyme, in; *Protein Engineering: Applications in Science, Medicine and Industry* (Inouye, M., and Sarma, R., eds.), pp. 257–274, Academic Press, New York.

Wetzel, R., 1987, Gordian knot loosened somewhat, *Prot. Engin.* 1:79–80.

Wetzel, R., 1988, Harnessing disulfide bonds using protein engineering, *Trends Biochem. Sci.* 12:478–482.

Wetzel, R., and Goeddel, D. V., 1983, Synthesis of polypeptides by recombinant DNA means, in: *The Peptides: Analysis, Synthesis, Biology, Vol. 5* (E. Gross and J. Meienhofer, eds.), pp. 1–64, Academic Press, New York.

Wetzel, R., Perry, L. J., Baase, W. A., and Becktel, W. J., 1988, Disulfide bonds and thermal stability in T4 lysozyme, *Proc. Natl. Acad. Sci. USA* 85:401–405.

Wetzel, R., Perry, L. J., Veilleux, C., and Chang, G., 1990, Mutational analysis of the carboxyl terminus of human interferon-γ, *Protein Engineering* 3:(in press).

White, F. H., Jr., 1982, Studies on the relationship of disulfide bonds to the formation and maintenance of secondary structure in chicken egg white lysozyme, *Biochemistry* 21:967–977.

Wu, H., 1931, Studies on denaturation of proteins XIII. A theory of denaturation, *Chin. J. Physiol.* 5:321–344.

Yano, Y., and Irie, M., 1975, Renaturation of yeast inorganic pyrophosphatase denatured in urea and guanidine hydrochloride, *J. Biochem.* 78:1001–1011.

Yocum, K. M., Shelton, J. B., Shelton, J. R., Schroeder, W. A., Worosila, G., Isied, S. S., Bordignon, E., and Gray, H. B., 1982, Preparation and characterization of a pentaamineruthenium III derivative of horse heart ferricytochrome c, *Proc. Natl. Acad. Sci. USA* 79:7052–7055.

Yphantis, D. A., and Arakawa, T., 1987, Sedimentation equilibrium measurements of recombinant DNA derived human interferon γ, *Biochemistry* 26:5422–5427.

Zale, S. E., and Klibanov, A. M., 1983, On the role of reversible denaturation (unfolding) in the irreversible thermal inactivation of enzymes, *Biotech. Bioeng.* 25:2221–2230.

Zale, S. E., and Klibanov, A. M., 1986, Why does ribonuclease irreversibly inactivate at high temperatures? *Biochemistry* 25:5432–5444.

Zetina, C. R., and Goldberg, M. E., 1980, A comparative study of the thermal inactivation of the isolated and associated domains within the β2 subunit of *Escherichia coli* tryptophan synthetase, *J. Biol. Chem.* 255:4381–4385.

Zettlmeissl, G., Rudolph, R., and Jaenicke, R., 1979, Reconstitution of lactic dehydrogenase. Noncovalent aggregation vs. reactivation. 1. Physical properties and kinetics of aggregation, *Biochemistry* 18:5567–5571.

Zoon, K. C., and Wetzel, R., 1984, Comparative structures of mammalian interferons, in: *Handbook of Experimental Pharmacology, Vol. 71* (P. E. Came and W. A. Carter, eds.), pp. 79–100, Springer, Berlin.

III

PRINCIPLES OF RECEPTOR
DESIGN AND REGULATION

Protein–Tyrosine Kinases and Their Substrates
Old Friends and New Faces

TONY HUNTER, KATHLEEN L. GOULD,
RICHARD A. LINDBERG, JILL MEISENHELDER,
DAVID S. MIDDLEMAS, and DAVID P. THOMPSON

1. Introduction

Protein phosphorylation, through its ability to increase or decrease the activity of proteins, plays a central role in cellular regulation. Most proteins known to be regulated by phosphorylation are substrates for both a protein kinase and a protein phosphatase. Indeed, the ready reversibility of protein phosphorylation makes it ideally suited for rapid responses. Protein phosphorylation is used extensively in the internal cellular response pathways to external stimuli. Receptor-mediated recognition of hormones, growth factors, and neurotransmitters is commonly transduced across the plasma membrane through the activation of intracellular protein kinases. Many of the surface receptors for peptide mitogens are themselves membrane-spanning protein kinases consisting of an external ligand-binding domain and a cytoplasmic protein kinase domain (for review, see Yarden and Ullrich, 1988). The EGF, PDGF, and CSF-1 receptors are examples of this type. In these cases signal transduction is a direct result of activation of the receptor protein kinase upon binding of the growth factor.

TONY HUNTER, KATHLEEN L. GOULD, JILL MEISENHELDER, and DAVID S. MIDDLEMAS • Molecular Biology and Virology Laboratory, The Salk Institute, San Diego, California 92138. *RICHARD A. LINDBERG and DAVID P. THOMPSON* • Molecular Biology and Virology Laboratory, The Salk Institute, San Diego, California 92138; and Department of Biology, University of California at San Diego, La Jolla, California 92093.

Other surface receptors are coupled to response systems that release second messengers. For instance, many external stimuli are coupled through G proteins to phospholipase C (PLC), which upon activation hydrolyzes membrane phosphoinositides to yield diacylglycerol (DAG) and inositol trisphosphate (IP_3) (for review, see Neer and Clapham, 1988). DAG is a direct activator of protein kinase C; IP_3 causes release of intracellular Ca^{2+} from the endoplasmic reticulum, which in turn can stimulate a series of calmodulin-dependent protein kinases. Some hormones are coupled to adenyl cyclase through the agency of a G protein, and the resultant cAMP activates cAMP-dependent protein kinase (Neer and Clapham, 1988). Other factors increase cGMP levels by activating guanyl cyclase, and thus stimulate cGMP-dependent protein kinase. Finally, a recent finding suggests that protein phosphatases may also play a direct role in signal transduction. The leukocyte common antigen (also known as T200 or CD45), a T- and B-cell surface protein, has a large cytoplasmic domain that shows striking relatedness to one of the phosphotyrosine phosphatases (Charbonneau *et al.*, 1988). Although there is no direct evidence that CD45 is a receptor, by analogy with the growth factor receptor protein kinases it seems possible that binding of a ligand to the external domain of CD45 could activate its cytoplasmic phosphatase domain.

Each of the primary signal-transducing proteins kinases can act pleiotropically to phosphorylate a series of substrates. Commonly included among these substrates are other protein kinases, which can in turn phosphorylate additional protein kinases, thus creating a protein kinase cascade. A cascade of this sort has the important property of amplifying the initial signal. In addition to such linear pathways, protein kinases are able to phosphorylate elements in other protein-kinase-mediated pathways, thus creating a protein kinase network.

Many of the signal-transducing protein kinases are protein–tyrosine kinases; this is true of all of the growth factor receptor protein kinases (Hunter and Cooper, 1985; Yarden and Ullrich, 1988). It is not clearly understood why tyrosine phosphorylation has been singled out for this purpose, because *a priori* there seems to be little reason to think that tyrosine phosphorylation is any more effective than serine and threonine phosphorylation in changing protein activity. The answer may well lie in the timing of the evolution of protein–tyrosine kinases from protein–serine kinases. Although protein–serine kinases are well documented in prokaryotes, the existence of protein–tyrosine kinases has not been convincingly demonstrated in these cells. They are scarce in unicellular eukaryotes, if they exist at all. Protein–tyrosine kinases appear to have emerged with the first metazoans and have grown in number in step with the increasing complexity of multicellular organisms. For this reason tyrosine phosphorylation may have evolved to serve as a cell–cell signaling mechanism, either through direct cell–cell contact or through the recognition of soluble factors. The first idea that protein–tyrosine kinase receptors might signal cell–cell contact is given credence by the fact that some growth factors are synthesized in membrane-anchored forms that give rise to humoral factors through processing.

In addition to serving as growth factor receptors, many of the other protein-tyrosine kinases are found localized to the inner face of the plasma membrane, where they are concentrated at sites of cell–cell junctions and cell–substratum attachment (Hunter and Cooper, 1985). these include protein kinases in the *src*, *fps*, and *abl* families. The true function of these protein–tyrosine kinases is not known, although several membrane regulatory functions have been proposed. One attractive possibility is that they act as subunits of surface receptors that lack their own protein kinase domain. Recent evidence to support this idea comes from the discovery that the T-cell surface proteins, CD4 and CD8, are tightly associated with p56^{lck}, which is a protein–tyrosine kinase in the *src* family (Rudd *et al.*, 1988; Veillette *et al.*, 1988). CD4 and CD8 are believed to be critical for T-cell recognition by other cells in the immune system, and therefore in one sense act as receptors.

The number of distinct vertebrate protein–tyrosine kinases now stands at more than 30 (Table I). These enzymes can be subdivided into families based on sequence, structural, and functional similarities; Figure 1 shows a typical example of the structure of a protein–tyrosine kinase from each family. In some cases further diversity is generated through alternative splicing from a single primary transcript. Many of the known protein–tyrosine kinases were first identified as the products of oncogenes, which were derived from the cellular gene (protoon-cogenes) encoding the corresponding normal cellular protein–tyrosine kinase; for instance, three members of the *src* family were discovered as retroviral oncogenes. Similarly, the archetypal members of the *fps* and *abl* families are retroviral oncogenes. Two members of the growth factor receptor family came to light as retroviral oncogenes, while several potential additional members were identified as oncogenes through the transfection of tumor cell DNA. The latter proteins are classified in this group because their predicted structures are analogous to those of the bona fide receptors. While this certainly implies that they act as surface receptors, no soluble ligands have been identified for any of these proteins. Indeed, extensive but unsuccessful efforts have been made to find a soluble ligand for the c-*erb*B2 (*neu*) protein. This failure would be consistent with the possibility that the ligands for these proteins are not circulating factors, but rather anchored cell surface molecules, and that normally these receptors are involved in monitoring cell–cell interaction. Nevertheless, from the examples of the EGF, PDGF, and CSF-1 receptors, it is clear that at least one of the mitogenic signal pathways uses tyrosine phosphorylation.

About half of all known oncogenes encode protein–tyrosine kinases. From intensive study of several of the protein–tyrosine kinase oncogenes, we have come to understand a good deal about the mechanisms of oncogenic activation of normal protein–tyrosine kinases. The emerging principle is that activation re-

Table I. Protein–Tyrosine Kinase Families

---

src family
    pp60^{c-src}—two forms generated by alternate splicing
    pp62^{c-yes}
    p61^{c-fgr}
    p59^{fyn}
    p56^{lck}
    p56^{lyn}
    p59^{hck}
    tkl protein (possibly chicken *lck*)
fps family
    p92^{c-fes}/p98^{c-fps}
    NCP94 (FER or *flk* protein)
abl family
    p150^{c-abl}—two different N-termini generated by alternate splicing
    arg protein
Growth factor receptors

EGF receptor	PDGF receptor
c-*erb*B2 (*neu* protein)	CSF-1 receptor
	c-*kit* protein
Insulin receptor	*ret* protein
IGF-1 receptor	
met protein	*eph* protein
trk protein	*eck* protein
*trk*B protein	*elk* protein
c-*ros* protein	
c-*sea* protein	

Others
    p75
    ltk protein
    flg protein
    flt protein

---

quires mutational events within the protein coding region that result in the loss of negative regulation. (The specific example of oncogenic activation of pp60^{c-src} will be discussed later.) From the plethora of protein–tyrosine kinase oncogenes, it is clear that deregulating protein–tyrosine kinase activity is a powerful way to transform cells. One characteristic of transformed cells is their uncontrolled growth. Since tyrosine phosphorylation is the basis of a major mitogenic pathway, one can readily imagine how a deregulated protein–tyrosine kinase could continuously phosphorylate substrates in this pathway and thus promote unbridled growth. If the deregulated protein–tyrosine kinase is normally part of such a pathway, it could simply phosphorylate its conventional substrates in an unscheduled manner. Because of the broad substrate specificity of most protein–

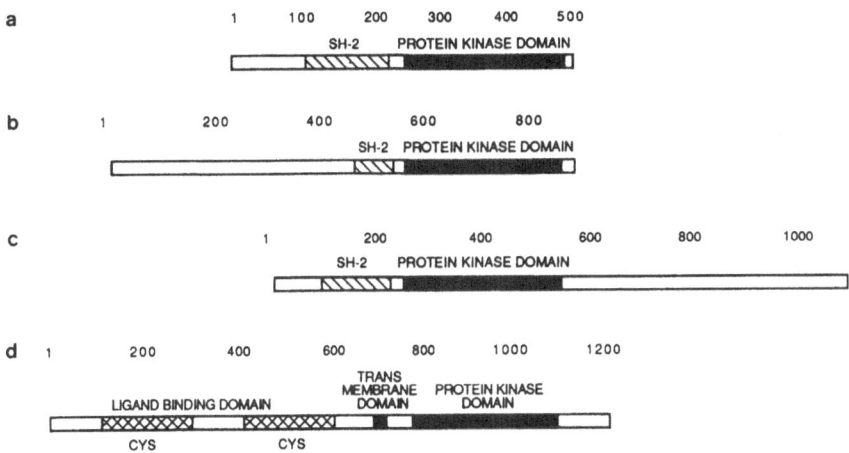

Figure 1. Structures of prototypic members of protein–tyrosine kinase families. The structures of the c-*src* (a), c-*fps* (b), and c-*abl* (c) proteins and the EGF receptor (d) are shown as examples of these four families of protein–tyrosine kinases. Each protein is depicted as a box, with the amino acid numbers given above. The catalytic domain is shown as a filled box, and the SH-2 region, when present, is shown as a hatched box. For the EGF receptor the transmembrane domain is a stippled box, and the cysteine-rich regions in the external domain are cross-hatched.

tyrosine kinases, even protein–tyrosine kinases not involved in signal transduction may have the potential to be converted into oncoproteins by suitable mutagenic manipulations, which could result in both deregulation and relocalization. To understand the role of tyrosine phosphorylation in growth control and oncogenesis, one will obviously need to identify the substrates for the different protein–tyrosine kinases. This issue will be addressed later.

Since there are already over 30 known vertebrate protein–tyrosine kinases, it might seem superfluous to try to identify additional enzymes. However, we have undertaken this task for several reasons. First, in order to fully understand the protein kinase regulatory network one must identify all of its component protein kinases. Second, their propensity for oncogenic activation means that new protein–tyrosine kinase genes are potential protooncogenes. Third, it is important to find out whether all protein–tyrosine kinases are related, or whether there might be enzymes of this sort based on a different sequence principle.

2. Isolation of Novel Protein–Tyrosine Kinases

Although protein kinases range widely in size, all known examples show sequence relationships over a region of about 300 amino acids, which proves to be the catalytic domain (Fig. 1) (Hunter and Cooper, 1985). Taking advantage of this relationship, several strategies have been developed for the isolation of new protein kinase sequences. New members of families have frequently been obtained by low-stringency screening of cDNA genomic libraries; the *lyn, fyn,* and *hck* genes were identified in this manner as members of the *src* family. Although the low-stringency approach sometimes yields protein kinases in a different family, a more global strategy for identifying protein–tyrosine kinases has used an approach that was first employed successfully to identify a series of novel protein–serine kinases (Hanks, 1987). In this homology probing strategy, cDNA libraries are screened sequentially with degenerate oligonucleotide probes designed to hybridize to two or more regions that are highly conserved in the catalytic domain of a specific subset of protein kinases.

We have applied this homology probing method to a HeLa-cell-derived Okayama/Berg library and two rat λ phage cDNA libraries. The oligonucleotide probes were designed to hybridize to regions that code for the conserved sequences HRDLAAR and P(I/V)KW(T/M)APE, which are ~40 residues apart in the catalytic domain, and to be selective for protein–tyrosine kinase cDNAs rather than protein–serine/threonine kinase cDNAs. The first oligonucleotide was designed to detect receptorlike protein–tyrosine kinases but not *src*-family protein–tyrosine kinases, whereas the APE-region probe represents a composite of all protein–tyrosine kinases. Among the positive cDNA clones sequenced from the HeLa cell library, two appeared to be bona fide novel protein–tyrosine kinases. The first of the two, *eck* (for epithelial *c*ell *k*inase), was found with the HRDLAAR probe, and has the receptor configuration in that region. It also has the DFG sequence common to all protein kinases, an autophosphorylation site tyrosine, and an APE sequence, but it differs from any known protein kinase in less conserved regions. The *eck* cDNA hybridized to a 4.8-kb mRNA, which is expressed at the highest levels in cell lines of epithelial origin. Analysis of rat tissue RNAs also showed that expression is greatest in tissues that contain proliferating epithelia; skin, intestine, lung, and ovary had the highest levels of *eck* mRNA, with nine other organs showing markedly less expression.

The deduced amino acid sequence for the kinase domain shows that *eck* is most closely related to the *eph* (Hirai *et al.,* 1987) and the very recently described *elk* (Letwin *et al.,* 1988) protein–tyrosine kinases. In the catalytic domain there is ~60% identity between *eck* and either *eph* or *elk,* whereas *eck* has ~30–45% identity to other protein–tyrosine kinases. Like *eph* and other receptor protein–tyrosine kinases, *eck* has a 24-residue hydrophobic transmembrane domain followed by several basic residues. This stop-transfer sequence is situated 61 amino acids N-terminal to the kinase domain. The 97-amino-acid sequence C-terminal to the *eck* kinase domain does not contain obvious similarities with any other protein–tyrosine kinases except *eph* and *elk.* Although the entire structure of the *eck* external domain has not been determined, the *eck* protein has at least

300 amino acids upstream of the transmembrane domain, including a cysteine-rich region that is also present in *eph. eph, elk,* and *eck* constitute a new family of receptor protein–tyrosine kinases. Based on the cell type in which it is expressed, *eck* appears to be a receptor protein–tyrosine kinase that may be involved in proliferation or differentiation of epithelial cells. Antisera to the *eck* coding region are being raised to identify the *eck* gene product and to prove that it is a protein–tyrosine kinase.

The second novel protein–tyrosine kinase cDNA clone isolated from the HeLa cell library appears to encode a soluble protein rather than a receptorlike protein. The rat homologue of this cDNA was also isolated from a PC12 cell brain cDNA library. This protein kinase is of interest because it lacks a tyrosine in the position within the catalytic domain where all other protein–tyrosine kinases are found to "autophosphorylate." This protein kinase shares homology with the *src, fes,* and *abl* kinases in a region to the N-terminal side of the catalytic domain, which has been termed SH-2 and which has a regulatory role (Sadowski *et al.,* 1986; Pawson, 1988). We have also isolated several *flg* protein–tyrosine kinase cDNA clones, and a novel *trk*-related cDNA clone from a rat cerebellar brain cDNA library. We are currently extending this approach by making use of the polymerase chain reaction technique to amplify the sequences lying between the two oligonucleotide target regions, which will then be cloned.

The second approach we have used for identifying novel protein–tyrosine kinase cDNA clones is to screen λgt11 cDNA expression libraries with anti-phosphotyrosine antibodies. In this strategy, the detection of positive clones relies on enzymatic function rather than any degree of sequence homology. The rationale for using this approach is as follows: Since *Escherichia coli* lacks detectable phosphotyrosine in proteins, a eukaryotic cDNA expressed in *E. coli* that encodes an active protein–tyrosine kinase could be detected if it were able to either autophosphorylate or phosphorylate bacterial proteins. The fact that it is feasible to detect an active protein–tyrosine kinase in *E. coli* with antiphosphotyrosine antibodies was demonstrated by Kipreos *et al.* (1987), who were able to select temperature-sensitive mutants of v-*abl* by screening v-*abl*-expressing *E. coli* with antiphosphotyrosine antibodies. A low frequency of positives might be expected, because the approach would rely not only on the complete catalytic domain being expressed, but also on its being able to form a functional enzyme. In addition, enzymatic activity might require the absence of sequences that would normally have a negative regulatory function. In principle, this approach is less prejudiced than strategies relying on sequence homology, because it selects only for protein–tyrosine kinase function.

We screened a human fibroblast λgt11 library with affinity-purified anti-phosphotyrosine antibodies and obtained several positive clones. The first clone we characterized encodes residues 130 to the C-terminus of a previously described protein–tyrosine kinase, *lyn* (Yamanashi *et al.,* 1987), a region including the entire catalytic domain (Lindberg *et al.*, 1988). Upon inducing expression of the *lyn* cDNA insert in an inducible plasmid expression vector, both the *lyn* protein itself and various bacterial proteins were found to be phosphorylated on tyrosine by immunoblotting with antiphosphotyrosine antibodies. Our success validates the approach of using antiphosphotyrosine antibodies for the cloning of active protein–tyrosine kinases. Other groups have also used expression screening with antiphosphotyrosine antibodies to isolate novel protein–tyrosine kinase cDNAs. For example, *elk* and *flk* (which appears to be identical to FER) were obtained from a rat brain cDNA library (Letwin *et al.*, 1988), and *flg, yes,* and a series of novel protein–tyrosine kinases were obtained from a chick fibroblast cDNA library in this manner (E. Pasquale, personal communication).

After the identification of a novel protein–tyrosine kinase, one is faced with the problem of determining its function. The pattern of expression in different tissues and cell types may give some clues. In the case of *eck,* for instance, its expression in epithelial cells suggests that it could function as a receptor for an epithelial cell growth factor. Another approach we have taken is to create a protein kinase phylogenetic tree by paired computer-assisted comparison of the sequences of a series of protein kinase catalytic domains (Hanks *et al.*, 1988). Protein kinases known to have similar functions prove to cluster within the tree. For instance, the protein kinases that are regulated by second messengers, such as cAMP, DAG, and Ca^{2+}/calmodulin, form a separate branch. The protein–tyrosine kinases families depicted in Table I are all readily discernible as separate limbs. The *eck, eph,* and *elk* genes form a branch arborizing from the *src, fps,* and *abl* limb. A new protein–tyrosine kinase sequence can readily be interpolated into this tree. More information would undoubtedly be gained if this type of analysis were extended to noncatalytic domains.

It is hard to guess how many protein–tyrosine kinases are encoded by the mammalian genome. The rate of discovery of protein–tyrosine kinases has been exponential for the past few years (Hunter, 1987b) and shows no sign of abating. Their number may well be over a hundred, and could be considerably higher. Not all of them will be expressed in a single cell type, but it is clear that many can be present simultaneously. Interestingly, the large number of protein–tyrosine kinases is not reflected in extensive protein–tyrosine phosphorylation. This is due both to the strict negative regulation of most protein–tyrosine kinases and to a high level of phosphotyrosine phosphatase activity.

3. Regulation of Protein–Tyrosine Kinase Activity: Phosphorylation of pp60$^{c\text{-}src}$

The activity of protein–tyrosine kinases can be regulated in a variety of fashions. They are generally inactive in the cell; for instance, the basal activity of the growth factor receptor protein–tyrosine kinases is low, being stimulated severalfold upon binding ligand. Another important mode of regulation involves phosphorylation. This can occur via either autophosphorylation or transphosphorylation. A prime example of a protein–tyrosine kinase that is regulated by phosphorylation is pp60$^{c\text{-}src}$.

pp60$^{c\text{-}src}$ is a member of a family of closely related, membrane-associated protein–tyrosine kinases, which includes *src, yes, fgr, fyn, lyn, lck,* and *hck* (Cooper, 1990). These protein–tyrosine kinases are all ~60 kDa, and share about 75% sequence identity over the C-terminal 450 residues. Although they lack a transmembrane domain, all of them have an N-terminal myristoylated glycine residue, which is essential for membrane association. Apart from this glycine, the N-terminal 80 residues of these proteins are unrelated. This unique region may confer individual specificities on these enzymes. However, it also possible that these proteins serve analogous functions in different cells. Each protein has a distinct pattern of expression, with many being found only in specific hematopoietic lineages. pp60$^{c\text{-}src}$ is found at the highest levels in platelets and certain neurons, but it is almost universally expressed. The actual functions of the *src* family protein–tyrosine kinases are not known, but their localization to the inner face of cytoplasmic membranes, including the plasma membrane, has suggested that these proteins may have a signal-transducing function. As indicated by the tight association of p56^{lck} with CD4 and CD8 (Rudd *et al.,* 1988; Veillette *et al.,* 1988), these protein–tyrosine kinases might act as subunits for ligand-binding proteins lacking their own catalytic domain.

One essential feature of this family of protein–tyrosine kinases is that their enzymatic activity is negatively regulated by tyrosine phosphorylation. pp60$^{c\text{-}src}$ is phosphorylated by a unknown protein–tyrosine kinase at the most C-terminal tyrosine, Tyr 527, which lies in the tail beyond the catalytic domain (Cooper *et al.,* 1986). This phosphorylation suppresses the catalytic activity of pp60$^{c\text{-}src}$ by an as yet undefined mechanism (Courtneidge, 1985; Cooper and King, 1986). When isolated from the cell, pp60$^{c\text{-}src}$ is nearly fully phosphorylated at Tyr 527 and is therefore largely in an inactive state. Presumably the stimulus that turns on pp60$^{c\text{-}src}$ enzymatic activity either activates the Tyr 527 phosphatase or inhibits the Tyr 527 protein kinase. A conserved tyrosine is present at this position in each of the *src* family protein kinases, and it seems probable that they are all negatively regulated by phosphorylation at this residue (Cooper, 1988).

The regulation of these normal protein–tyrosine kinases has an important bearing on their oncogenic activation. Consider the conversion of pp60$^{c\text{-}src}$ into pp60$^{v\text{-}src}$ (for review, see Hunter, 1987a). Although chicken pp60$^{c\text{-}src}$ and pp60$^{v\text{-}src}$ are similar in structure over their N-terminal 514 amino acids, apart from a few scattered point mutations, the two proteins diverge completely at their C-termini. From residue 515 the C-terminal 19 amino acids of pp60$^{c\text{-}src}$ have been replaced in pp60$^{v\text{-}src}$ by 12 unrelated amino acids. The obvious consequence of removal of the C-terminus is the elimination of the regulatory tyrosine.

Indeed, as predicted by the negative regulation model, pp60$^{v\text{-}src}$ has much greater protein kinase activity than pp60$^{c\text{-}src}$. Conversion of pp60$^{c\text{-}src}$ into a transforming protein can be achieved by simply mutating Tyr 527 to the nonphosphorylatable residue phenylalanine (Hunter, 1987a). This confirms that the major effect of the deletion of the C-terminus is removal of this phosphorylation site. Although some of the other point mutations in the body of the v-*src* gene have been shown to be sufficient to induce transforming activity in their own right, it seems likely that the primary event in the conversion of the c-*src* gene into a transforming gene is the inactivation through deletion or mutation of the critical regulatory residues at the C-terminus.

Another tyrosine phosphorylation event that regulates pp60$^{c\text{-}src}$ activity is phosphorylation of Tyr 416, which lies in the catalytic domain. Tyr 416 is the major autophosphorylation site for pp60$^{c\text{-}src}$ *in vitro*, but *in vivo* Tyr 416 is not detectably phosphorylated in pp60$^{c\text{-}src}$ (Gould *et al.*, 1985). In contrast, Tyr 416 is a major phosphorylation site in pp60$^{v\text{-}src}$. Thus, phosphorylation of Tyr 416 reflects increased pp60^{src} phosphotransferase activity. Phosphorylation of Tyr 416 probably occurs normally by an intramolecular mechanism, but under some circumstances it can be phosphorylated by other protein–tyrosine kinases. Site-directed mutation of Tyr 416 to phenylalanine in pp60$^{v\text{-}src}$ reduces its potency as a transforming protein about twofold (Kmiecik *et al.*, 1988). This implies that autophosphorylation can be an autogenous activation mechanism. Similar autogenous activation is seen upon autophosphorylation with some of the growth factor receptor protein–tyrosine kinases.

Polyoma virus middle T antigen, the viral transforming protein, binds to and activates pp60$^{c\text{-}src}$ (Courtneidge and Smith, 1983; Bolen *et al.*, 1984). In this complex pp60$^{c\text{-}src}$ can be phosphorylated on Tyr 90 or 92 *in vitro* (J. S. Brugge, personal communication). This may simply reflect the increased pp60$^{c\text{-}src}$ activity, since we have identified Tyr 90 as a minor site of autophosphorylation that is used at high concentrations of ATP. Moreover, since mutations in this region of pp60$^{c\text{-}src}$ can activate it for transformation (Hunter, 1987a), there is reason to think that phosphorylation of Tyr 90 could influence pp60$^{c\text{-}src}$ activity.

There are other phosphorylation events that might modulate the activity of pp60$^{c\text{-}src}$. For example, protein kinase C phosphorylates pp60$^{c\text{-}src}$ on Ser 12. Normally, Ser 12 is unphosphorylated, but it is rapidly and stoichiometrically phosphorylated when cells are treated with 12-tetradecanoyl-phorbol-13-acetate (TPA) to activate protein kinase C (Gould *et al.*, 1985). Rapid Ser 12 phosphorylation is also observed when quiescent mouse NIH3T3 fibroblasts are treated with PDGF, PGF$_{2\alpha}$, serum, Na$_3$VO$_4$, bombesin, or basic FGF (Gould and Hunter, 1988). All of these agents are mitogenic and are known to induce the turnover of phosphatidylinositol (PI) in NIH3T3 cells, which leads to the production of DAG and the activation of protein kinase C. We have been unable to discern any effects of this modification on the activity or localization of pp60^{c-src}. A nearby serine residue, Ser 17, is phosphorylated by the cAMP-dependent protein kinase (Collett *et al.*, 1979; Patschinsky *et al.*, 1986). However, unlike Ser 12, Ser 17 is constitutively phosphorylated to a high level in both pp60^{c-src} and pp60^{v-src}. As for Ser 12, no obvious consequence of Ser 17 phosphorylation has been reported. Moreover, there is no dramatic phenotypic effect of site-directed mutation of either Ser 17 (Cross and Hanafusa, 1983; Hirota *et al.*, 1988) or Ser 12 (P. Yaciuk and D. Shalloway, personal communication) to alanine.

Treatment of resting fibroblasts with platelet-derived growth factor (PDGF) leads to the appearance of a more slowly migrating form of pp60^{c-src} (p60$^+$), accounting for about 5% of the total population (Ralston and Bishop, 1985; Gould and Hunter, 1988). p60$^+$ is modified in the N-terminal 18 kDa and, unlike pp60^{c-src} in resting cells, contains phosphotyrosine in this N-terminal region. The appearance of p60$^+$ is correlated with a two- to threefold increase in pp60^{c-src} protein kinase activity toward an exogenous substrate. The appearance of p60$^+$ is a very rapid event, occurring within 2 min of PDGF addition to quiescent NIH3T3 cells (Gould and Hunter, 1988). Both the PDGF AA and BB dimers induce the formation of p60$^+$ as efficiently as the natural PDGF AB dimer, but neither the formation of p60$^+$ nor the enhancement of pp60^{c-src} protein kinase activity has been observed with any other mitogen.

p60$^+$ is highly phosphorylated and contains four to five phosphates per mole of pp60^{c-src} (Gould and Hunter, 1988). Analysis of N-terminal fragments of ^{32}P-labeled p60$^+$ by peptide mapping shows the presence of a novel phosphoserine-containing and a novel phosphotyrosine-containing peptide in the N-terminal 18 kDa, as well as the expected Ser 17 and Ser 12 peptides (Gould and Hunter, 1988). These novel sites have not yet been mapped, nor have the protein kinases that phosphorylate them been identified. The tyrosine might be phosphorylated by the PDGF receptor, another protein–tyrosine kinase, or pp60^{c-src} itself by autophosphorylation. Despite the increase in *in vitro* protein kinase activity, there is no detectable difference in the level of phosphorylation of Tyr 527 in p60$^+$ compared with the main form of pp60^{c-src} in PDGF-treated cells, nor is there any increase in the level of Tyr 416 phosphorylation, which is often an indication of increased pp60^{src} activity. Thus, the elevated pp60^{c-src} protein–tyrosine kinase activity in PDGF-treated cells is apparently not due to a relief of negative regulation at Tyr 527. Overall, it seems likely that the combined phosphorylation events that give rise to p60$^+$ are also responsible for the increase in PTK activity. It is tempting to speculate that the activation of pp60^{c-src} in PDGF-treated cells is important for the mitogenic response, but since a similar modification of pp60^{c-src} is not seen with other mitogens, it may be involved in some response unique to PDGF.

There is an interesting parallel between the results obtained with $p60^{+}$ in PDGF-treated cells and $pp60^{c\text{-}src}$ in mitotic cells, where a novel threonine phosphorylation in the N-terminal 18 kDa correlates with the appearance of a more slowly migrating form of $pp60^{c\text{-}src}$ and an increase in protein–tyrosine kinase activity (Chackalaparampil and Shalloway, 1988). In addition, elevated $pp60^{c\text{-}src}$ protein–tyrosine kinase activity is detected in neuronal cells (Brugge *et al.*, 1985; Lynch *et al.*, 1986; Cartwright *et al.*, 1987, 1988). It is not known how the activity is increased, but there are a number of possibilities. Neurons contain a neuronal-specific form of $pp60^{c\text{-}src}$ that has an additional 6-amino-acid exon inserted at position 118. $pp60^{c\text{-}src}$ in neurons is phosphorylated at yet another distinct serine site (Brugge *et al.*, 1987; Cartwright *et al.*, 1987), which appears to be Ser 75. Finally, $pp60^{c\text{-}src}$ from neuronal cells appears to be underphosphorylated (Gould and Hunter, 1989), which could result in a low stoichiometry of phosphorylation at Tyr 527 and contribute to its increased activity. All of these observations suggest that the N-terminal 18 kDa of $pp60^{c\text{-}src}$ is likely to be a critical site for regulation by phosphorylation. However, there is no direct evidence in any of these three situations that $pp60^{c\text{-}src}$ has actually increased protein kinase activity in the cell.

From the aforementioned it is apparent that $pp60^{c\text{-}src}$ is subject to "multisite" phosphorylation (summarized in Fig. 2). At least nine sites of phosphorylation have been identified, and these are recognized by a minimum of six distinct protein kinases. So far only three of these phosphorylations have been shown to have functional consequences at the level of phosphotransferase activity. Nevertheless, the other sites are conserved in $pp60^{c\text{-}src}$ in birds and mammals and could well be important. These sites could regulate substrate accessibility, as well as interaction with regulatory proteins. Some of the phosphorylation sites occur in the SH-2 region, which is known to have a regulatory role and is believed to interact with other cellular proteins (DeClue *et al.*, 1987). It is interesting to note that Ser 12 and Ser 17 are located in the unique N-terminal region of $pp60^{c\text{-}src}$ and could therefore modulate some activity unique to $pp60^{c\text{-}src}$. In other proteins subject to multisite phosphorylation, such as glycogen synthase, many of the phosphorylation sites are interactive (Cohen, 1988). The same could be true for $pp60^{c\text{-}src}$.

This detailed study of a single protein–tyrosine kinase has revealed a very complex pattern of phosphorylation. It seems likely that most other protein–tyrosine kinases will prove to be equally complex. In principle, this cross-talk between different protein kinases allows the establishment of a multidimensional protein kinase network. In many ways one can draw an analogy between this network and an electronic circuit (Hunter, 1987b). The challenge will be to work out the logic of the connections between the different proteins kinases and deduce how the signal output changes as the input is varied.

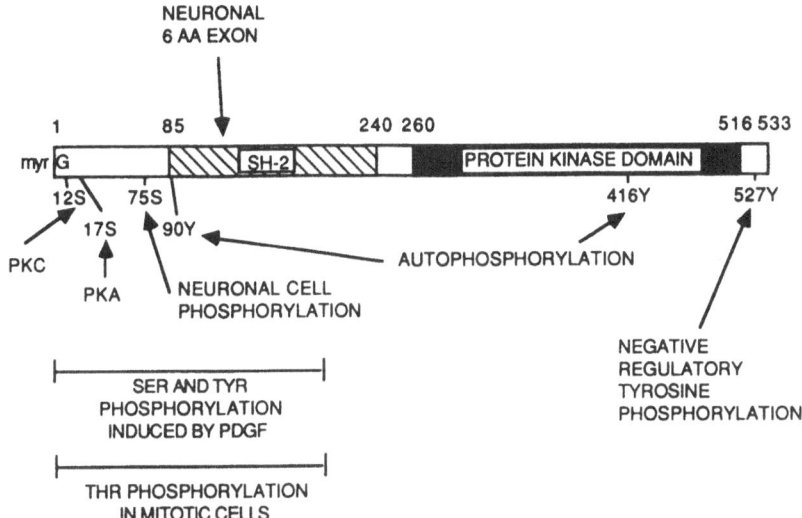

Figure 2. Phosphorylation sites in pp60$^{c\text{-}src}$. Chicken pp60$^{c\text{-}src}$ is depicted as a box, using the same conventions as in Fig. 1. Amino acid numbers are above the box for the beginning and end of the SH-2 region and catalytic domains. Sites of phosphorylation are below the box, with an indication of the protein kinases responsible where they are known (PKC = protein kinase C; PKA = cAMP-dependent protein kinase). The sites of serine and tyrosine phosphorylation in PDGF-treated cells and the site of threonine phosphorylation in mitotic cells are not mapped precisely, but they are known to lie within the first 200 residues.

4. Protein–Tyrosine Kinase Substrates

To understand how tyrosine phosphorylation elicits mitogenesis and cellular transformation, one must identify the substrates for the protein–tyrosine kinases in question and then determine how phosphorylation alters their activity. Given that most protein kinases are pleiotropic, one would expect there to be multiple distinct primary targets. In this way the multifarious nature of the responses to mitogens and oncogenic protein–tyrosine kinases could be explained by the existence of target proteins in many different cellular pathways.

The main problem in identifying protein–tyrosine kinase substrates is that phosphoserine (90%) and phosphothreonine (10%) constitute the bulk of phosphate linked to protein in mammalian cells, with phosphotyrosine accounting for about 0.05% in most normal cells. Even in a cell transformed by a deregulated protein–tyrosine kinase oncoprotein, the level of phosphotyrosine rarely rises above 0.5%. Three main methods are used to detect phosphotyrosine-containing proteins in intact cells. First, use can be made of the relatively high resistance of the phosphate ester linkage of phosphotyrosine to alkali compared to that of the more abundant phosphoserine. Two-dimensional gel analysis of ^{32}P-labeled cells followed by alkali treatment of the fixed gel reveals a pattern of alkali-resistant phosphoproteins containing phosphothreonine and phosphotyrosine (Cooper and Hunter, 1981). Phosphoamino acid analysis of the resistant proteins indicates which proteins contain phosphotyrosine. A fair number of substrates have been identified in this fashion (Table II). Second, immunoprecipitation of candidate substrates from ^{32}P-labeled cells followed by SDS gel electrophoresis and phosphoamino acid analysis of the gel band has been successful in identifying substrates, particularly in the case of cytoskeletal proteins (Table II). Third, the recent widespread use of antiphosphotyrosine antibodies for immunoblotting or immunoprecipitation has revealed potential substrates in considerably greater numbers than had previously been detected by other methods.

A plethora of phosphotyrosine-containing proteins have been identified in cells transformed by oncogenic protein–tyrosine kinases. For instance, well over 50 newly phosphorylated proteins can be detected in cells transformed by pp60^{v-src} upon immunoblotting one-dimensional SDS gels with antiphosphotyrosine antibodies (Kamps and Sefton, 1988). At the moment very little is known about most of these proteins. Table II lists the substrates on which at least some

Table II. Substrates for Protein–Tyrosine Kinases[a]

Substrate	Oncogenic protein– tyrosine kinases	Receptor protein– tyrosine kinases	Detection method
Talin	+	?	Immunoprecipitation
Fibronectin receptor	+	?	Immunoprecipitation
Vinculin	+	−	Immunoprecipitation
p81 (ezrin)	+	+	Two-dimensional gel
p50	+	?	Coprecipitation with pp60^{v-src}
p42	+	+	Two-dimensional gel
Enolase	+	−	Two-dimensional gel
p36 (calpactin I)	+	(+)	Two-dimensional gel
p35 (calpactin II)	−	+	*In vitro* substrate
Lactate dehydrogenase	+	−	Immunoprecipitation
Phosphoglycerate mutase	+	−	Two-dimensional gel
T-cell receptor ζ chain	?	?	Immunoprecipitation
Calmodulin	+	−	Immunoprecipitation

[a]The substrates are all proteins that can be detected phosphorylated on tyrosine in intact cells. They are listed in order of decreasing molecular weight. The columns indicate whether these proteins are phosphorylated by one or more of the retroviral protein–tyrosine kinases and/or by one of the growth factor receptor protein–tyrosine kinases. In addition to these proteins, most of the protein–tyrosine kinases themselves can be shown to contain phosphotyrosine in intact cells as a result of either autophosphorylation or transphosphorylation.

preliminary characterization has been carried out. Among these are several proteins whose phosphorylation is clearly gratuitous. These include the three glycolytic enzymes, enolase, lactate dehydrogenase, and phosphoglycerate mutase (Cooper *et al.*, 1983). The stoichiometry of tyrosine phosphorylation is low (1–5% mole P.Tyr/mole protein) for all three, and there is no measurable difference in the activities of these enzymes in $pp60^{v-src}$-transformed cells. Furthermore, none of these enzymes is rate-limiting in glycolysis, so that even if phosphorylation changed their activity this would not result in an altered rate of glycolysis.

Another set of substrates form part of the cytoplasmic submembraneous or cortical skeleton. These include p36 (also termed calpactin I or lipocortin II) (Radke and Martin, 1979) and p35 (calpactin II or lipocortin I) (Fava and Cohen, 1984), which are members of a burgeoning family of Ca^{2+}-dependent phospholipid binding proteins (Crompton *et al.*, 1988). These proteins are built up of four or eight copies of a 75-amino-acid repeat, which is the minimal Ca^{2+}/phospholipid binding domain. p81, a component of the microvillar core (Gould *et al.*, 1986), and three elements of the focal contact structure, vinculin (Sefton *et al.*, 1981), talin (Pasquale *et al.*, 1986; DeClue and Martin, 1987), and the β subunit of the fibronectin receptor (Hirst *et al.*, 1986), are also localized to the cortical skeleton. Most of the relevant protein–tyrosine kinases are associated with either the membrane or the submembraneous skeleton and are in an ideal position to phosphorylate these proteins. This raises the question of whether these proteins are phosphorylated adventitiously due to a physical juxtapositioning, or whether their phosphorylation has a functional consequence. Certainly the phosphorylation of proteins in the focal contact structure could result in the diminished cellular adhesiveness and cell-rounding characteristic of transformed cells. However, it should be remembered that these are also characteristics of cells transformed by oncogenes that do not encode protein–tyrosine kinases.

Faced with trying to discern which of the multitude of phosphotyrosine-containing proteins revealed by global screening methods are likely to be critical substrates, we have turned again to a directed search. In part this has been prompted by the increasing availability of antibodies against regulatory proteins, which have been generated as a result of molecular cloning. For the signal-transducing protein–tyrosine kinases it seems reasonable that many of the primary substrates will be localized to the submembrane locale. A number of obvious candidates are listed in Table III. We have focused on PLC, a key enzyme in the PI cycle, which is stimulated by a wide range of agonists and which through the generation of DAG and IP_3 leads to the activation of protein kinase C and a variety of Ca^2-dependent enzymes.

Multiple PLC isozymes are known, and at least five PI-specific enzymes have been purified. Rhee and his colleagues have purified three such enzymes from bovine brain (Ryu *et al.*, 1988); termed PLC-I, PLC-II, and PLC-III, they

Table III. Candidate Substrates for Signal-Transducing Protein Kinases[a]

Candidate substrate	Specific protein	Phosphorylation
Phosphatidylinositol kinases	Type I PI kinase (p85)	↑ P.tyr and P.ser
Phospholipases C and A$_2$	PLC-I	P.ser (no change)
	PLC-II (PLC-148)	↑ P.tyr and P.ser
	PLC-III	P.ser (no change)
	PLA$_2$	?
G proteins	?	
Ion channels	?	
Transport proteins	?	
Extracellular matrix receptors	FN receptor β subunit	↑ P.tyr
Protein–serine kinases	CDC2Hs	P.tyr and P.thr
	c-*raf* protein	↑ P.ser and P.tyr
	MAP-2 protein kinase	↑ P.tyr and P.thr
Protein phosphatases	Protein phosphatase 1	↑ P.tyr

[a]The candidate substrates are families of proteins. The individual members that have been shown to be phosphorylated are indicated, as are the phosphoamino acids that are present.

have molecular masses of 150,000, 145,000, and 85,000, respectively. cDNA clones for all three enzymes have been isolated from a rat brain library (Suh *et al.*, 1988b,c). PLC-II has also been cloned by Knopf and his co-workers, who called it PLC-148 (Stahl *et al.*, 1988). A central region of PLC-II has sequence similarity with the SH-2 region of pp60[c-src], as well as with regions in the v-*crk* (Mayer *et al.*, 1988) and p21[c-ras] GTPase activator (GAP) (Vogel *et al.*, 1988) proteins. This suggests that there may be a regulatory network mediated via the SH-2 regions of these proteins that connects the different signal pathways in which they are involved.

In collaboration with Sue Goo Rhee (National Institutes of Health) we have used monoclonal antibodies specific for PLC-I, PLC-II, and PLC-III (Suh *et al.*, 1988a) to determine whether any of these enzymes is a protein–tyrosine kinase substrate both in PDGF- or epidermal growth factor (EGF)-treated cells and in pp60[v-src]-transformed cells. PLC-I and PLC-III are expressed at very low levels in most fibroblastic cells. There was no marked change in their level of phosphorylation under any circumstance; PLC-I contained predominantly phosphoserine, whereas PLC-III contained both phosphoserine and phosphothreonine. In contrast to PLC-I and PLC-III, the phosphorylation of PLC-II was markedly stimulated when quiescent NIH3T3 cells were treated with PDGF or when human epidermoid A431 carcinoma cells were treated with EGF. With PDGF, there was a 10- to 50-fold increase in the level of phosphorylation within 30 sec, and this high level of phosphorylation persisted for up to 1 hr. Only a low level of phosphoserine was detected in PLC-II from resting cells, but phosphoserine and phosphotyrosine were present in equal amounts following PDGF treatment. Peptide mapping studies showed that at least three new phosphotyrosine-containing peptides were present in PLC-II from PDGF-treated cells. PLC-II phosphorylation was also rapidly stimulated on serine and tyrosine in EGF-treated A431 cells, and similar novel sites of tyrosine phosphorylation were found. Several of the same sites could be phosphorylated *in vitro* when purified bovine PLC-II was treated with purified EGF receptor. These sites of phosphorylation are currently being mapped within PLC-II; it would be intriguing if they lie within the central SH-2-like region. Our results extend those of Wahl *et al.* (1988), who recently found that EGF treatment leads to an increase in the amount of PLC activity that can be recovered from cells by immunoabsorption with antiphosphotyrosine antibodies.

PDGF is known to activate PI turnover in NIH3T3 cells. The mechanism of activation is unknown, but since PDGF-induced PI turnover is insensitive to treatment with pertussis toxin, which leads to the inactivation of some G proteins, it appears that activation of PLC is not mediated through a G protein (Letterio *et al.*, 1986). This is in contrast to mitogens, such as bombesin, where stimulation of PI turnover is blocked by pertussis toxin (Letterio *et al.*, 1986). Thus, it is tempting to speculate that PDGF could stimulate the PDGF receptor to phosphorylate PLC-II directly, and that this in turn would increase PLC-II activity. The effects of tyrosine phosphorylation on PLC-II activity are currently being assessed. In contrast to PDGF, $pp60^{v-src}$ causes only a weak elevation in PI turnover (Martins *et al.*, 1989). Interestingly, we found that PLC-II from $pp60^{v-src}$-transformed cells had very low levels of phosphotyrosine.

Evidence is accumulating that other members of the candidate list are also substrates for protein–tyrosine kinases. For instance, it has been shown that type I PI kinase is phosphorylated on tyrosine in both PDGF-treated (Kaplan *et al.*, 1987) and polyoma virus-transformed cells (Courtneidge and Heber, 1987), and that this is correlated with increased PI kinase activity. PI is normally phosphorylated in two steps to PI-4,5-diphosphate (PIP_2), which is the preferred substrate for the PI-specific PLCs. Instead, type I PI kinase generates PI-3-P, which does not lie directly on this pathway, but could itself be a novel second messenger or give rise to one (Whitman *et al.*, 1988). As mentioned earlier, the β subunit of the fibronectin receptor is phosphorylated on tyrosine in $pp60^{v-src}$-transformed cells. This phosphorylation is said to decrease the affinity of the receptor both for talin on the inside of the cell and for fibronectin on the outside. This could certainly have the consequence of reducing cellular adhesiveness.

Clearly events other than those occurring at the cell surface can be induced by mitogens and transforming proteins. Among such events are the induction of gene expression and increased protein synthesis. Signal pathways that originate at the cell surface must be able to traverse the cytoplasm and reach the nucleus. As discussed at the outset, one type of signal pathway with this property would involve a protein kinase cascade. Recently, two cytoplasmic protein–serine kinases have been found to be phosphorylated on tyrosine and activated in response to mitogen treatment. These are the c-*raf* protein (D. K. Morrison and T. M. Roberts, personal communication) and the MAP-2 protein kinase, which is activated following insulin treatment (Ray and Sturgill, 1988). The CDC2Hs protein kinase (D. Beach and J. Y. J. Wang, personal communication), which is involved in cell cycle regulation is also phosphorylated on tyrosine. These findings offer encouragement that it will ultimately prove possible to unravel the signal transduction pathways that utilize protein phosphorylation.

ACKNOWLEDGMENTS. We thank Steve Hanks for his help with the cloning of protein–tyrosine kinases with degenerate oligonucleotides and Mark Kamps and Bart Sefton for their gift of affinity-purified antiphosphotyrosine antibodies. We are deeply indebted to Sue Goo Rhee (NIH) for his generous supply of anti-PLC monoclonal antibodies and purified PLC-II. This work was supported by USPHS Grant CA-39780 from the NCI to T. H., and by a grant from the American Business Foundation for Cancer Research. R. A. L. and D. P. T. were supported by an NIH predoctoral training grant to the University of California, San Diego. D. S. M. was supported by an NRSA Fellowship from NINCDS.

References

Bolen, J. B., Thiele, C. J., Israel, M., Yonemoto, W., Lipsich, L. A., and Brugge, J. S., 1984, Enhancement of cellular *src* gene product tyrosyl kinase activity following polyoma virus infection, *Cell* **38**:767–777.

Brugge, J. S., Cotton, P. C., Queral, A. E., Barrett, J. N., Nonner, D., and Keane, R. W., 1985, Neurones express high levels of a structurally modified activated form of pp60^{c-src}, *Nature* **316**:554–557.

Brugge, J., Cotton, P., Lustig, A., Yonemoto, W., Lipsich, L., Coussens, P., Barrett, J. N., Nonner, D., and Keane, R., 1987, Characterization of the altered form of the c-*src* gene product in neuronal cells, *Genes Dev.* **1**:287–296.

Cartwright, C. A., Simantov, R., Kaplan, P. L., Hunter, T., and Eckhart, W., 1987, Alterations in pp60^{c-src} accompany differentiation of neurons from rat embryo striatum, *Mol. Cell. Biol.* **7**:1830–1840.

Cartwright, C. A., Simantov, R., Cowan, W. M., Hunter, T., and Eckhart, W., 1988, pp60^{c-src} expression in the developing rat brain, *Proc. Natl. Acad. Sci. USA* **85**:3348–3352.

Chackalaparampil, I., and Shalloway, D., 1988, Altered phosphorylation and activation of pp60^{c-src} during fibroblast mitosis, *Cell* **52**:801–810.

Charbonneau, H., Tonks, N. K., Walsh, K. A., and Fischer, E. A., 1988, The leukocyte common antigen (CD45): A putative receptor-linked protein–tyrosine phosphatase, *Proc. Natl. Acad. Sci. USA* **85**:7182–7186.

Cohen, P., 1988, Protein phosphorylation and hormone action, *Proc. Roy. Soc. Lond. B* **234**:115–144.

Collett, M. S., Erikson, E., and Erikson, R. L., 1979, Structural analysis of the avian sarcoma virus transforming protein: Sites of phosphorylation, *J. Virol.* **29**:770–781.

Cooper, J. A., 1990, The *src*-family of protein-tyrosine kinases, in: *Peptides and Protein Phosphorylation* (B. Kemp and P. Alewood, eds.), CRC Press, Orlando, FL, pp. 85–113.

Cooper, J. A., and Hunter, T., 1981, Changes in protein phosphorylation in Rous sarcoma virus transformed cells, *Mol. Cell. Biol.* **1**:165–178.

Cooper, J. A., and King, C. S., 1986, Dephosphorylation or antibody binding to the carboxy terminus stimulates pp60^{c-src}, *Mol. Cell. Biol.* **6**:4467–4477.

Cooper, J. A., Reiss, N. A., Schwartz, R. J., and Hunter, T., 1983, Three glycolytic enzymes are phosphorylated at tyrosine in cells transformed by Rous sarcoma virus, *Nature* **302**:218–223.

Cooper, J. A., Gould, K. L., Cartwright, C. A., and Hunter, T., 1986, Tyr^{527} is phosphorylated in pp60^{c-src}: Implications for regulation, *Science* **231**:1431–1433.

Courtneidge, S. A., 1985, Activation of the pp60^{c-src} kinase by middle T antigen binding or by dephosphorylation, *EMBO J.* **4**:1471–1477.

Courtneidge, S. A., and Heber, A., 1987, An 81 kd protein complexed with middle T antigen and pp60^{c-src}: A possible phosphatidylinositol kinase, *Cell* **50**:1031–1037.

Courtneidge, S. A., and Smith, A. E., 1983, Polyoma virus transforming protein associates with the product of the cellular *src* gene, *Nature* **303**:435–439.

Crompton, M. R., Moss, S. E., and Crumpton, M. J., 1988, Diversity in the lipocortin/calpactin family, *Cell* **55**:1–3.

Cross, F. R., and Hanafusa, H., 1983, Local mutagenesis of Rous sarcoma virus: The major sites of tyrosine and serine phosphorylation of pp60^{src} are dispensable for transformation, *Cell* **34**:597–607.

DeClue, J. E., and Martin, G. S., 1987, Phosphorylation of talin at tyrosine in Rous sarcoma virus transformed cells, *Mol. Cell Biol.* **7**:371–378.

DeClue, J. E., Sadowski, I., Martin, G. S., and Pawson, T., 1987, A conserved domain regulates interactions of the v-*fps* protein-tyrosine kinase with the host cell, *Proc. Natl. Acad. Sci. USA* **84**:9064–9068.

Fava, R. A., and Cohen, S., 1984, Isolation of a calcium-dependent 35-kilodalton substrate for the epidermal growth factor receptor/kinase from A-431 cells, *J. Biol. Chem.* **259**:2636–2645.

Gould, K. L., and Hunter, T., 1988, Platelet-derived growth factor induces multisite phosphorylation of pp60^{c-src} and increases its protein–tyrosine kinase activity, *Mol. Cell. Biol.* **8**:3345–3356.

Gould, K. L., and Hunter, T., 1989, AtT20 cells express modified forms of pp60^{c-src}, *Mol. Endocrinol.* **3**:79–88.

Gould, K. L., Woodgett, J. R., Cooper, J. A., Buss, J. E., Shalloway, D., and Hunter, T. 1985, Protein kinase C phosphorylates pp60^{src} at a novel site, *Cell* **42**:849–857.

Gould, K. L., Cooper, J. A., Bretscher, A., and Hunter, T., 1986, The protein-tyrosine kinase substrate, p81, is homologous to a chicken microvillar core protein, *J. Cell. Biol.* **102**:660–669.

Hanks, S. K., 1987, Homology probing: Identification of cDNA clones encoding members of the protein–serine kinase family, *Proc. Natl. Acad. Sci. USA* **84**:388–392.

Hanks, S. K., Quinn, A. M., and Hunter, T., 1988, The protein kinase family; Conserved features and deduced phylogeny of the catalytic domains, *Science* **241**:42–52.

Hirai, H., Maru, Y., Hagiwara, K., Nishida, J., and Takaku, F., 1987, A novel putative tyrosine kinase receptor encoded by the *eph* gene, *Science* **238**:1717–11719.

Hirota, Y., Kato, J.-Y., and Takeya, T., 1988, Substitution of Ser-17 of pp60^{c-src}: Biological and biochemical characterization in chicken embryo fibroblasts, *Mol. Cell. Biol.* **8**:1826–1830.

Hirst, R., Horwitz, A., Buck, C., and Rohrschneider, L., 1986, Phosphorylation of the fibronectin receptor complex in cells transformed by oncogenes that encode tyrosine kinases, *Proc. Natl. Acad. Sci. USA* **83**:6470–6474.

Hunter, T., 1987a, A tail of two *src*'s: Mutatis mutandis, *Cell* **49**:1–4.

Hunter, T., 1987b, A thousand and one protein kinases, *Cell* **50**:823–829.

Hunter, T., and Cooper, J. A., 1985, Protein-tyrosine kinases, *Annu. Rev. Biochem.* **54**:897–930.

Kamps, M. P., and Sefton, B. M., 1988, Identification of multiple novel polypeptide substrates of the v-*src*, v-*yes*, v-*fps*, v-*ros*, and v-*erb*-B oncogenic tyrosine protein kinases utilizing antisera against phosphotyrosine, *Oncogene* **2**:305–315.

Kaplan, D. R., Whitman, M., Schaffhausen, B., Pallas, D. C., White, M., Cantley, L., and Roberts, T. M., 1987, Common elements in growth factor stimulation and oncogenic transformation: 85 kd phosphoprotein and phosphatidylinositol kinase activity, *Cell* **50**:1021–1029.

Kipreos, E. T., Lee, G. J., and Wang, J. Y. J., 1987, Isolation of temperature-sensitive mutants of v-*abl* oncogene by screening with antibodies for phosphotyrosine, *Proc. Natl. Acad. Sci. USA* **84**:1345–1349.

Kmiecik, T. E., Johnson, P. J., and Shalloway, D., 1988, Regulation by the autophosphorylation site in overexpressed pp60$^{c\text{-}src}$, *Mol. Cell. Biol.* **8**:4541–4546.

Letterio, J. J., Coughlin, S. R., and Williams, L. T., 1986, Pertussis toxin-sensitive pathway in the stimulation of c-*myc* expression and DNA synthesis by bombesin, *Science* **234**:1117–1119.

Letwin, K., Yee, S-P., and Pawson, T., 1988, Novel protein-tyrosine kinase cDNAs related to *fps*, *fes* and *eph* cloned using anti-phosphotyrosine antibody, *Oncogene* **3**:621–627.

Lindberg, R. A., Thompson, D. P., and Hunter, T., 1988, Identification of cDNA clones that code for protein-tyrosine kinases by screening expression libraries with antibodies against phosphotyrosine, *Oncogene* **3**:629–633.

Lynch, S. A., Brugge, J. S., and Levine, J. M., 1986, Induction of altered c-*src* gene product during differentiation of embryonal carcinoma cells, *Science* **234**:873–876.

Martins, T. J., Sugimoto, Y., and Erikson, R. L., 1989, Dissociation of inositol trisphosphate from diacylglycerol production in Rous sarcoma virus-transformed fibroblasts, *J. Cell Biol.* **108**:683–691.

Mayer, B. J., Hamaguchi, M., and Hanafusa, H., 1988, A novel viral oncogene with structural similarity to phospholipase C, *Nature* **332**:272–275.

Neer, E. J., and Clapham, D. E., 1988, Roles of G protein subunits in transmembrane signalling, *Nature* **333**:129–134.

Pasquale, E. B., Maher, P. A., and Singer, S. J., 1986, Talin is phosphorylated on tyrosine in chicken embryo fibroblasts transformed by Rous sarcoma virus, *Proc. Natl. Acad. Sci. USA* **83**:5507–5511.

Patschinsky, T., Hunter, T., and Sefton, B. M., 1986, Phosphorylation of the transforming protein of Rous sarcoma virus: direct demonstration of phosphorylation of serine 17 and identification of an additional site of tyrosine phosphorylation in pp60$^{v\text{-}src}$ of Prague Rous sarcoma virus, *J. Virol.* **59**:73–81.

Pawson, T., 1988, Non-catalytic domains of cytoplasmic protein–tyrosine kinases: Regulatory elements in signal transduction, *Oncogene* **3**:491–495.

Radke, K., and Martin, G. S., 1979, Transformation by Rous sarcoma virus: effects of *src* gene on the synthesis and phosphorylation of cellular polypeptides, *Proc. Natl. Acad. Sci. USA* **76**:5212–5216.

Ralston, R., and Bishop, J. M., 1985, The product of the protooncogene c-*src* is modified during the cellular response to platelet-derived growth factor, *Proc. Natl. Acad. Sci. USA* **82**:7845–7849.

Ray, L. B., and Sturgill, T. W., 1988, Insulin-stimulated microtubule-associated protein kinase is phosphorylated on tyrosine and threonine *in vivo*, *Proc. Natl. Acad. Sci. USA* **85**:3753–3757.

Rudd, C. E., Trevillyan, J. M., Dasgupta, J. D., Wong, L. L., and Schlossman, S. F., 1988, The CD4 receptor is complexed in detergent lysates to a protein-tyrosine kinase (pp58) from human T lymphocytes, *Proc. Natl. Acad. Sci. USA* **85**:5190–5194.

Ryu, S. H., Suh, P-G., Cho, K. S., Lee, K. Y., and Rhee, S. G., 1987, Bovine brain cytosol contains three immunologically distinct forms of inositol-phospholipid-specific phospholipase C, *Proc. Natl. Acad. Sci. USA* **84**:6649–6653.

Sadowski, I., Stone, J. C., and Pawson, T., 1986, A non-catalytic domain conserved among cytoplasmic protein–tyrosine kinases modifies the kinase function and transforming action of Fujinami sarcoma virus P130*gag-fps*, *Mol. Cell. Biol.* **6**:4396–4408.

Sefton, B. M., Hunter, T., Ball, E. H., and Singer, S. J., 1981, Vinculin: A cytoskeletal target of the transforming protein of Rous sarcoma virus, *Cell* **24**:165–174.

Stahl, M. L., Ferenz, C. R., Kelleher, K. L., Kriz, R. W., and Knopf, J. L., 1988, Sequence similarity of phospholipase C with the non-catalytic region of *src, Nature* **332**:269–272.

Suh, P-G., Ryu, S. H., Choi, W. C., Lee, K-Y., and Rhee, S. G., 1988a, Monoclonal antibodies to three phospholipase C isozymes from bovine brain, *J. Biol. Chem.* **263**:14497–14504.

Suh, P-G., Ryu, S. H., Moon, K. H., Suh, H. W., and Rhee, S. G., 1988b, Cloning and sequence of multiple forms of phospholipase C, *Cell* **54**:161–169.

Suh, P-G., Ryu, S. H., Moon, K. H., Suh, H. W., and Rhee, S. G., 1988c, Inositol phospholipid-specific phospholipase C: Complete cDNA and protein sequences and sequence homology to tyrosine kinase-related oncogene products, *Proc. Natl. Acad. Sci. USA* **85**:5419–5423.

Veillette, A., Bookman, M. A., Horak, E. M., and Bolen, J. R., 1988, The CD4 and CD8 T cell surface antigens are associated with the internal membrane tyrosine-protein kinase p56*lck, Cell* **55**:301–308.

Vogel, U. S., Dixon, R. A. F., Schaber, M. D., Diehl, R. E., Marshall, M. S., Scolnick, E. M., Sigal, I. S., and Gibbs, J. S., 1988, Cloning of bovine GAP and its interaction with oncogenic *ras* p21, *Nature* **335**:90–93.

Wahl, M. I., Daniel, T. O., and Carpenter, G., 1988, Antiphosphotyrosine recovery of phospholipase C activity after EGF treatment of A-431 cells, *Science* **241**:968–970.

Whitman, M., Downes, C. P., Keeler, M., Keller, T., and Cantley, L., 1988, Type I phosphatidyl-inositol kinase makes a novel inositol phospholipid, phosphatidylinositol-3-phosphate, *Nature* **332**:644–646.

Yamanashi, Y., Fukushige, S., Sukegawa, J., Miyajima, N., Matsubara, K., Yamamoto, T., and Toyoshima, K., 1987, The *yes*-related cellular gene *lyn* encodes a possible tyrosine kinase similar to p56*lck, Mol. Cell. Biol.* **7**:237–243.

Yarden, Y., and Ullrich, A., 1988, Growth factor receptor tyrosine kinases, *Annu. Rev. Biochem.* **57**:443–478.

IV

THE GUANINE NUCLEOTIDE BINDING PROTEIN FAMILY

G Proteins
A Family of Signal-Transducing Molecules

EVA J. NEER

1. The Family of G Proteins

The signals from many receptors for hormones and neurotransmitters are sent through the plasma membrane by a set of heterotrimeric GTP-binding proteins composed of a GTP-binding α subunit and a $\beta\gamma$ subunit that is a nondissociable dimer made up of two different polypeptides (for recent reviews, see Gilman, 1987; Stryer and Bourne, 1986; Lochrie and Simon, 1988; Neer and Clapham, 1988). The receptors on the cell surface are extremely specific for their ligands, which have very low affinities for heterologous receptors. Thus, there is very little cross-talk due to agonist binding at an inappropriate receptor. However, the next steps in the cascade appear to be less precise. Within a single cell, several separate receptors can often regulate each final pathway. For example, in the heart, acetylcholine and adenosine receptors appear to regulate the same K^+ channel (Kurachi *et al.*, 1986).

In the last 3 years, there has been an explosive increase in information regarding the structures of receptors, G proteins, and some of the enzymes and ion channels that are regulated by this transmembrane signaling system. These studies have shown a surprising sequence conservation within each class of protein, but a much larger number of isoforms than one might guess. This apparent redundancy raises important questions about how the cell ensures that hormonal signals are precisely transmitted from the receptor to the correct effector without inappropriate cross-talk. Part of the specificity lies in the structure of the G proteins. For example, only one category of G protein, G_s, stimulates

EVA J. NEER • Department of Medicine, Brigham and Women's Hospital and Harvard Medical School, Boston, Massachusetts 02115.

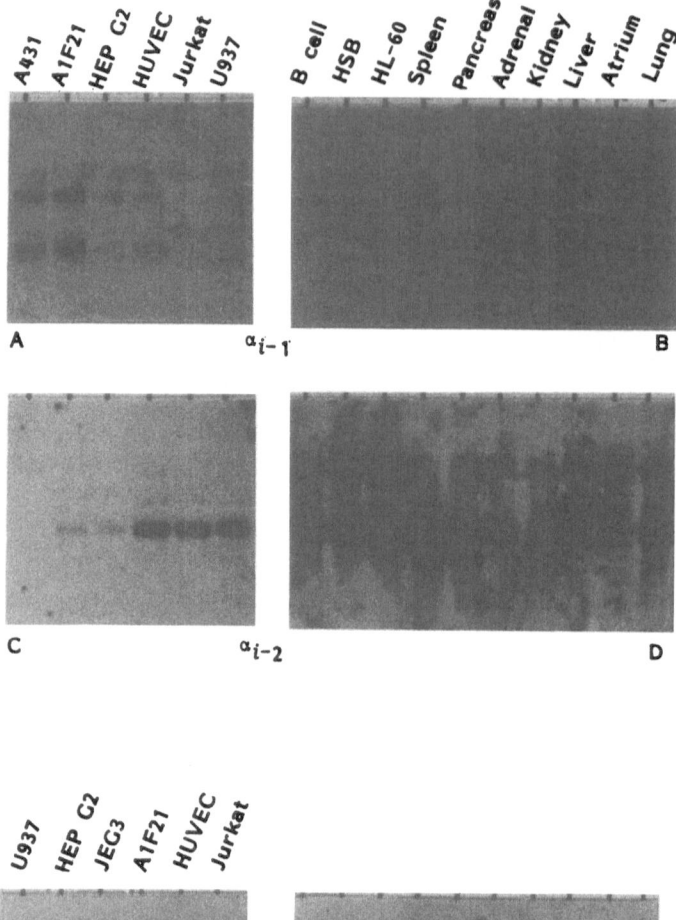

Figure 1. mRNA analysis from human fetal tissue and from human cell lines (Kim et al., 1988). Total cellular RNA was extracted from the human fetal tissues and cell lines. Blots were hybridized with probes for α_{i-1}, α_{i-2}, and α_{i-3} as follows: (A and B) EcoR1-BAM fragment containing nucleotides 867–1344 of the published α_{i-1} sequence and additional 3' untranslated region corresponding to the larger mRNA. The predominant mRNA bands are at 3.9 and 2.2 kb. (C and D) Full-length EcoR1 insert of human α_{i-2}. The predominant mRNA band is at 2.7 kb. (E) A 600-bp fragment derived from

adenylyl cyclase, although all the isoforms of G_s can apparently do so (Graziano et al., 1987). Another major category of G proteins is the G_i/G_o family, whose members regulate channels and phosphoinositol turnover and inhibit adenylyl cyclase (see review cited earlier). The members of this group are very closely related, with about 70% sequence identity between α_o and the three isoforms of α_i and 90% or greater identity among α_{i-1}, α_{i-2}, and α_{i-3} (Jones and Reed, 1987; Kim et al., 1988; Lochrie and Simon, 1988). It has been very difficult to assign specific functions to these isoforms, which often behave indistinguishably when reconstituted with receptors or effectors (for example, see Florio and Sternweis, 1985; Yatani et al., 1988).

2. Distribution of G Protein α Subunits

The high degree of cross-talk found in reconstitution experiments would be of no consequence if the G proteins were, in fact, distributed in different cells. In the family of the transducins, G proteins that couple visual pigments to a cyclic GMP phosphodiesterase in the retina, there is, in fact, a precise cellular distribution. The α subunit of transducin 1 is located entirely in the rods, while that of transducin 2 is found exclusively in the cones (Lerea et al., 1986). Outside the visual system, however, there does not seem to be a strict cellular specificity for expression of G-protein subtypes. Figure 1 shows that many human tissues and cell lines express mRNA for more than one type of α subunit of the α_i family. The mRNA from several permanent and primary human cell lines was probed with the cDNA specific for each of the α_i subtypes. It is apparent from the figure that the cells analyzed contain multiple subtypes of α_i protein. Both α_{i-2} and α_{i-3} are expressed in all the cell types we have analyzed. However, the amount of α_{i-2} mRNA appears to be more variable among different cells than α_{i-3}, which seems to be the most evenly distributed. The α_{i-1} protein has the most limited distribution and is, in general, absent from cells of myelocytic lineage such as T cells, B cells, and HSB cells (Kim et al., 1988).

The α subunit of G_i and G_o can be identified by the pertussis toxin reaction. Pertussis toxin transfers ADP-ribose from NAD to the α subunits of this class of G proteins. This reaction allows specific radioactive labeling of the G proteins if [^{32}P]NAD is used as a substrate. The heterogeneity of G proteins in a single cell

---

the 5' flanking region of the α_{i-3} gene as described in the text. (F) The Hind III-Hind III fragment of α_{i-3} cDNA. The predominant mRNA band is at 2.8 kb. The blots were washed at 65°C in 0.2 SSC and 0.1% SDS. The cell lines are A431 (epithelial carcinoma); A1F21 (human dermal fibroblast); HepG-2 (hepatoma); HUVEC (human endothelial cells); Jurkat (T cell); U937 (monocyte); B cell (Epstein–Barr virus transformed peripheral blood B cells); HSB (T cell); HL60 (monocyte).

revealed by analysis of mRNA is corroborated by two-dimensional gel electrophoresis of pertussis toxin substrates in several clonal cell lines. Several examples are shown in Figure 2. In each of the cell types, several ADP-ribosylated protein spots are detected by this method. It is possible that some of these spots may be attributed to posttranslational modification of the protein. However, correlation of the position of the ADP-ribosylated spots found in crude membranes with the position of ADP-ribosylated pure α subunits suggests that at least the majority of the spots represent different α subunit isoforms.

Although cells may contain several different types of α subunits, they certainly do not contain them in equal proportions. A dramatic example of this is found in the brain, where the 39-kDa pertussis toxin substrate (called α_{39} or α_o) is extremely abundant (Neer *et al.*, 1984; Sternweis and Robishaw, 1984), comprising about 0.5% of the total brain particulate protein. Immunocytochemical studies performed in collaboration with Dr. Solomon Snyder and his colleagues showed that α_o immunoreactivity was highly localized to particular regions of the brain (Worley *et al.*, 1986). Therefore, the average concentration in the whole brain underestimates the amount of α_o that is present in certain cells. In these

Endothelial cell

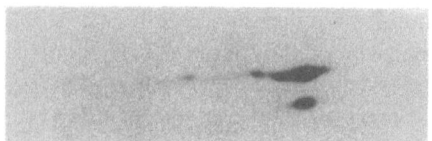

A431 carcinoma

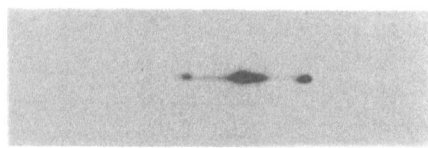

Jurkat T cell

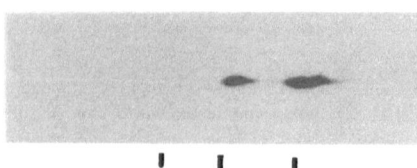

pl   5.37 5.31 5.26

Figure 2. Membranes from human umbilical vein endothelial cells (A), A431 human carcinoma cells (B), or Jurkat, a transformed human T-cell line (C), were ADP-ribosylated by pertussis toxin as described by Huff *et al.* (1985). The membranes were solubilized and analyzed by isoelectric focusing followed by SDS-polyacrylamide gel electrophoresis as described by O'Farrell *et al.* (1977). Isoelectric focusing was done using ampholytes of pH 5–7. Markers of known isoelectric points were used to calibrate the gels.

cells, the concentration of α_o may approach 5% of the brain particulate protein. It is not yet clear what function α_o serves in the central nervous system, although it has been shown to regulate ion channels (Hescheler *et al.*, 1987; VanDongen *et al.*, 1988).

3. Regulation of α_o Levels

One of the puzzling features about the G proteins is that the levels of mRNA for the various α subunits seem to bear no relationship to the quantity of α subunit protein present in the cell. Thus, α_o is by far the most abundant G-protein α subunit in the brain, yet the message is extremely difficult to detect; in contrast, the protein α_s is one of the least abundant α subunits in brain, and its mRNA is relatively abundant. We explicitly compared relative mRNA and protein levels for α_o in rat brain, neonatal cardiocytes, and pituitary cells. The results are given in Table I. The relative amount of α_o mRNA detected on Northern blots of polyA-selected mRNA was approximately equal in all three cell types. Probing the same polyA-selected mRNA Northern blot with a probe

Table I. Stability of α_0 Protein in GH_4 Cells, Cardiocytes, and Brain

	Relative α_0 mRNA levels[a]	Relative α_0 protein[b]	Half-life of α_0 protein (hr)[c]
Brain	1	1	—
GH_4 cells	1	0.2	$28 \pm 7 \ (n = 4)$
Cardiocytes	1	0.2	$>72 \ (n = 4)$

[a]10 ug of polyA-containing RNA from GH_4 cells or rat cardiocytes and rat brain was loaded onto an agarose gel, transferred to nitrocellulose following denaturing agarose gel electrophoresis, and hybridized using as a probe [^{32}P]-labeled rat α_0 cDNA (EcoR1-EcoR1 fragment, α_0 cDNA was a kind gift from Dr. Randall Reed, Johns Hopkins University, Baltimore, MD). The autoradiograms were analyzed by densitometry; areas under the 4.4-kb α_0 mRNA peak are given relative to the amount of α_0 mRNA in brain.
[b]GH_4 cells, neonatal rat ventricle, or adult rat brain was analyzed by Western blotting as described by Huff *et al.* (1985) using an anti-α_0 polyclonal antibody raised in rabbits. Equal amounts of total cellular protein were loaded for each sample analyzed. The amount of immunoreactive α_0 protein was quantitated by densitometry. Areas under the peaks are given relative to the amount in brain.
[c]The time course of degradation was measured by quantitative immunoprecipitation of α_0 from cells pulsed for 6–20 hr with [^{35}S]methionine to label the α_0 protein, followed by a chase period of up to 80 hr. Decay of α_0 from GH_4 cells followed first-order kinetics. The half-time for decay could not be determined in cardiocytes, since there was only a modest decrease in immunoprecipitated [^{35}S]α_0 over the time of the chase (Silbert *et al.*, 1990).

for α_{i-2} showed that all three contained large amounts of easily detectable α_i to message, although heart and GH_4 cells contained two to three times as much α_i to message as did the brain. These results indicate that the low level of α_o mRNA is not due to degradation of the mRNA analyzed, nor to incomplete transfer of the mRNA.

Despite the uniformity of mRNA levels, the steady-state level of α_o protein is very different between brain and peripheral tissues, as we have previously demonstrated (Huff *et al.*, 1985). However, the amount of α_o protein in neonatal cardiocytes and GH_4 cells is approximately equal (Table I).

In order to analyze the factors that regulate the levels of α_o protein, we determined the degradation rate of α_o in cultured neonatal cardiocytes and GH_4 cells. The cells were metabolically labeled with [^{35}S]methionine for 6–20 hr, and the methionine was then chased with the unlabeled amino acid. A loss of label from α_o was determined by quantitative immunoprecipitation using a polyclonal α_o antibody (previously described in Huff *et al.*, 1985). A decay of label in the immunoprecipitated protein followed first-order kinetics and allowed calculation of a half-time for degradation of α_o, as indicated in Table I. The α_o protein in GH_4 cells decays with a half-time of 28 ± 7 ($n = 4$) hr, while the degradation rate of α_o in cultured cardiocytes is extremely slow, with a half-time in excess of 72 hr ($n = 4$). Since we have previously shown that the steady-state level of the protein in the two cell types is approximately equal, we conclude from the results of studies of the degradation rate of α_o that the rate of synthesis of the protein in the two cell types must be significantly different. Since the mRNA levels are approximately equal, these observations suggest that translational control may be important in setting the α_o levels.

4. G Proteins as Bifurcating Signals

A central feature of the mechanism of action of the membrane-bound heterotrimeric G proteins is that activation involves dissociation of the guanine nucleotide liganded α subunit from $\beta\gamma$ according to the following equilibrium:

$$\alpha\beta\gamma + G \rightleftarrows \alpha_G + \beta\gamma$$

where G is a guanine nucleoside triphosphate. The dissociation of the heterotrimer into α and $\beta\gamma$ subunits raises the possibility that both α and $\beta\gamma$ subunits may be signaling molecules generated by a hormonal stimulus.

There is ample evidence that the α subunit, in its guanine nucleoside trisphosphate liganded form, can be an activator of effectors (such as adenylyl cyclase (Northup *et al.*, 1983), the muscarinic K^+ channel (Codina *et al.*, 1987; Logothetis *et al.*, 1988), and cyclic GMP phosphodiesterase in the retina (Fung *et al.*, 1981). The clear evidence that the guanine nucleoside trisphosphate liganded α subunit can activate enzymes and ion channels does not rule out the possibility that βγ may also be a regulatory protein. Several examples have now been demonstrated in which βγ seems to function as a positive regulator. Jelsema and Axelrod (1987) showed that the βγ subunit of transducin can activate phospholipase A_2 in rod outer segment membranes. Logothetis *et al.* (1987, 1988) showed that the βγ subunit can activate the muscarinic K^+ channel in patch-clamp experiments using chick and rat cardiac atrial membranes. Activation of the K^+ channel by βγ subunits is probably not due to a direct interaction of βγ with the K^+ channel, but appears to be mediated by activation of phospholipase A_2 (Kim *et al.*, 1989). Antibodies to phospholipase A_2 can block activation of the channel by the βγ subunit. However, they do not block activation of the channel by muscarinic agonists. This is expected, since Logothetis *et al.* (1988) have previously shown that the effects of α and βγ on the channel are not additive, and that the channel can be fully activated by either subunit. Muscarinic agonists would also release activated α subunits which would activate the channel independently of any effect on phospholipase A_2. In regulation of the muscarinic K^+ channel, therefore, both α and βγ seem to be positive effectors. Whiteway *et al.* (1989) have recently shown that in the yeast *Saccharomyces cerevisiae* proteins very similar to βγ act as a positive regulator on signal transduction through the mating phenomenon receptor. Thus, it may generally be true that in eukaryotic organisms α and βγ can each set in motion a train of responses, although the mechanisms by which they act may be different.

5. Organizing Principles

The studies we have presented so far suggest that the cell is faced with a complicated problem in devising ways to ensure that signals from hormone receptors are transmitted to the correct effector, since receptors and effectors are coupled through a family of G proteins that are extremely similar structurally. However, even though many of the subunits may act interchangeably *in vitro* in the cell, a variety of mechanisms may operate to enhance specificity (see Table II).

One extremely important factor is the stoichiometry of the various α subunits. For example, in cardiac atria the β-adrenergic receptor interacts with a heterotrimer of $\alpha_s\beta\gamma$. Activation of the receptor by an agonist should cause dissociation of α from the βγ subunit, and the βγ subunit might then be expected to activate the K^+ channel. In fact, activation of the K^+ channel by β-adrenergic receptors is extremely weak. One explanation is that the amount of $\alpha_s\beta\gamma$ in cardiac cells might be considerably lower than the amount of $\alpha_i\beta\gamma$, so that the amount of βγ released from that source would be small in comparison to the

Table II. Factors that Modify Receptor–G
Protein–Effector Interaction

1. Stoichiometry of receptors, G proteins, and effectors
2. Differential subunit affinity
3. Compartmentation within membrane domains
4. Association with other proteins

amount of βγ that could be released from other α components. Thus, although βγ released from many receptors could, in principle, activate the K⁺ channel, only receptors that could release large amounts of βγ might, in fact, do so.

Another potential regulatory factor is the relative affinity of different α subunits for βγ. All α subunits do not have the same affinity for βγ, and some may dissociate more readily than others (Huff *et al.*, 1985). Therefore, one pool of αβγ hetertrimers may be less effective as a source of βγ than another. Finally, receptor number can affect the amount of G protein activated. If only a small number of receptors is present, only a small amount of G protein might be activated. These G proteins would stimulate effectors with a high affinity for the subunits, but would not stimulate effectors with a low affinity.

Considerations of stoichiometry and relative subunit affinity are particularly important if all of the members of the hormone responsive system can move freely in the plane of the membrane. Free diffusion of some components in the plasma membrane have been demonstrated in some cells (for example, Tolkovsky *et al.*, 1982; see also review by Neer and Clapham, 1988). However, there is also suggestive evidence that hormone receptors, G proteins, and perhaps effectors may be restricted to membrane domains. Kinetic studies suggest that discrete pools of cyclic AMP in cells may originate from particular populations of adenylyl cyclase that can be differentiated from the total adenylyl cyclase (Dufau *et al.*, 1978; Buxton and Brunton, 1983). The components of the hormone responsive system have been shown to be asymmetrically distributed in apical and basolateral membranes of polarized epithelia (Schlatz *et al.*, 1975; Dominguez *et al.*, 1987). The cytoskeleton may influence the ability of hormones to stimulate adenylyl cyclase and of detergents to solubilize the G proteins (Rasenick *et al.*, 1981; Insel and Kennedy, 1978). Elements of the cytoskeleton may also associate directly with G-protein subunits (Carlson *et al.*, 1986). Subcellular compartmentation does not rule out the possibility of subunit dissociation, since the components might well dissociate, but simply not be free to travel very far in the plane in the bilayer.

Finally, there may be additional proteins which interact with the G-protein receptors and/or effectors, and which enhance the specificity or rates of the reaction. Recently, a cytosolic protein (Trahey and McCormick, 1987) has been identified that acts as a nucleotide exchange factor for the *ras* protein. Further work has shown that there are multiple proteins of different molecular weights that seem to perform this function for other *ras*-related proteins (Garrett *et al.*, 1989).

The various mechanisms we have outlined are not exclusive of each other, and each may apply to different parts of the same cells or have greater or lesser importance in different signaling systems. The challenge for the future will be to define how the G proteins actually function *in situ*, and how the general principles developed through analysis of purified proteins can be applied to living cells.

References

Buxton, I. L., and Brunton, L. L., 1983, Compartments of cyclic AMP and protein kinase in mammalian cardiomyocytes, *J. Biol. Chem.* **258**:10233–10239.

Carlson, K. E., Woolkalis, M. J., Newhouse, M. G., and Manning, D. R., 1986, Fractionation of the β subunit common to guanine nucleotide-binding regulatory proteins with the cytoskeleton, *Mol. Pharmacol.* **30**:463–468.

Codina, Y., Yatani, A., Grenet, D., Brown, A. M., and Birnbaumer, L., 1987, The α subunit of the GTP-binding protein G_k opens atrial potassium channels, *Science* **238**:1288–1291.

Dominguez, P., Velasco, G., Barros, F., and Lazo, P. S., 1987, Intestinal brush border membranes contain regulatory subunits of adenylyl cyclase, *Proc. Natl. Acad. Sci. USA* **84**:6965–6969.

Dufau, M. L., Horner, J. A., Hayashi, K., Tsuruhara, T., Conn, P. M., and Catt, K. J., 1978, Actions of choleragen and gonadotropin in isolated Leydig cells, *J. Biol. Chem.* **253**:3721–3729.

Florio, V. A., and Sternweis, P. C., 1985, Reconstitution of resolved muscarinic cholinergic receptors with purified GTP-binding proteins, *J. Biol. Chem.* **260**:3477–3483.

Fung, B. K. K., Hurley, J. B., and Stryer, L., 1981, Flow of information in the light-triggered cyclic nucleotide cascade of vision, *Proc. Natl. Acad. Sci. USA* **78**:152–56.

Garrett, M. D., Self, A. J., van Oers, C., and Hall, A., 1989, Identification of distinct cytoplasmic targets for ras/R-ras and rho regulatory proteins, *J. Biol. Chem.* **264**:10–13.

Gilman, A. G., 1987, G-proteins: Transducers of receptor-generated signals, *Annu. Rev. Biochem.* **56**:615–650.

Graziano, M. P., Casey, P. J., and Gilman, A. G., 1987, Expression of cDNAs for G-proteins in *E. coli*, *J. Biol. Chem.* **262**:11375–11381.

Hescheler, J., Rosenthal, W., Trautwein, W., and Schultz, G., 1987, The GTP-binding protein G_o regulates neuronal calcium channels, *Nature (London)* **325**:445–447.

Huff, R. M., Axton, J. M., and Neer, E. J., 1985, Physical and immunological characterization of a guanine nucleotide-binding protein purified from bovine cerebral cortex, *J. Biol. Chem.* **260**:10864–10871.

Insel, P. A., and Kennedy, M. S., 1978, Colchichine potentiates β-adrenoreceptor-stimulated cyclic AMP in lymphoma cells by an action distal to the receptor, *Nature* **273**:471–473.

Jelsema, C. L., and Axelrod, J., 1987, Stimulation of phospholipase A_2 activity in bovine rod outer segments by the βγ subunits of transducin and its inhibition by the α subunit, *Proc. Natl. Acad. Sci. USA* **84**:3623–3627.

Jones, D. T., and Reed, R. R., 1987, Molecular cloning of five GTP-binding protein cDNA species from rat olfactory neuroepithelium, *J. Biol. Chem.* **262:**14241–14249.

Kim, D., Lewis, D. L., Graziadei, Neer, E. J., Bar-Sagi, D., and Clapham, D. E., 1989, G-protein βγ-subunits activate the cardiac muscarinic K$^+$-channel via phospholipase A$_2$, *Nature* **337:**557–560.

Kim, S. Y., Ang, S-L., Bloch, D., Bloch, K., Kawahara, Y., Lee, R. L., Tolman, C., Seidman, J. G., and Neer, E. J., 1988, Identification of cDNA encoding an additional α subunit of a human GTP-binding protein, *Proc. Natl. Acad. Sci. USA* **85:**4153–4157.

Kurachi, Y., Nakajima, T., and Sugimoto, T., 1986, On the mechanism of activation of muscarinic K$^+$ channels by adenosine in isolated atrial cells: involvement of GTP-binding proteins, *Pflügers Arch.* **407:**264–274.

Lerea, C. L., Somers, D. E., Hurley, J. B., Klock, I. B., and Bunt-Milam, A. M., 1986, Identification of specific transducin α subunits in retinal rod and cone photoreceptors, *Science* **234:**77–80.

Lochrie, M. A., and Simon, M. I., 1988, G-protein multiplicity in signal transduction systems, *Biochemistry* **27:**4957–4965.

Logothetis, D. E., Kurachi, Y., Galper, J., Neer, E. J., and Clapham, D. E., 1987, The βγ subunits of GTP-binding proteins activate the muscarinic K$^+$ channel in heart, *Nature* **325:**321–326.

Logothetis, D. E., Northup, J., Neer, E. J., and Clapham, D. E., 1988, Specificity of action of guanine nucleotide-binding regulatory protein subunits on the cardiac muscarinic K$^+$ channels, *Proc. Natl. Acad. Sci.* **85:**5814–5818.

Neer, E. J., and Clapham, D. E., 1988, Roles of G-proteins in transmembrane signalling, *Nature (London)* **333:**129–134.

Neer, E. J., Lok, J. M., and Wolf, L. G., 1984, Purification and properties of the inhibitory guanine nucleotide regulatory unit of brain adenylate cyclase, *J. Biol.Chem.* **259:**14222–14229.

Northup, J. K., Smigel, M. D., Sternweis, P. C., and Gilman, A. G., 1983, The subunits of the stimulatory regulatory component of adenylate cyclase, *J. Biol. Chem.* **258:**11369–11376.

O'Farrell, P. Z., Goodman, H. M., and O'Farrell, P. H., 1977, High-resolution two-dimensional electrophoresis of basic as well as acidic proteins, *Cell* **12:**1133–1142.

Rasenick, M. M., Stein, P. J., and Bitensky, M. W., 1981, The regulatory subunit of adenylate cyclase interacts with cytoskeletal components, *Nature* **294:**560–562.

Schlatz, L. J., Schwartz, I. L., Kinne-Saffran, E., and Kinne, R., 1975, Distribution of parathyroid hormone-stimulated adenylate cyclase in plasma membranes in cells of the kidney cortex, *J. Mem. Biol.* **24:**131–144.

Silbert, S., Michel, T., Lee, R., and Neer, E. J., 1990, Differential degradation rates of the G protein α$_o$ in cultured cardiac and pituitary cells, *J. Biol. Chem.* **265:**3102–3105.

Sternweis, P. C., and Robishaw, J., 1984, Isolation of two proteins with high affinity for guanine nucleotides from membranes of bovine brain, *J. Biol. Chem.* **259:**13806–13813.

Stryer, L., and Bourne, H. R., 1986, G-proteins: A family of signal transducers, *Annu. Rev. Cell Biol.* **2:**391–419.

Tolkovsky, A. M., Braun, S., and Levitski, A., 1982, Kinetics of interaction between the β-receptor, the GTP regulatory protein, and the catalytic unit of turkey erythrocyte adenylate cyclase, *Proc. Natl. Acad. Sci. USA* **79:**213–217.

Trahey, M., and McCormick, F., 1987, A cytoplasmic protein stimulates normal N-ras p21 GTPase, but does not affect oncogene mutants, *Science* **238:**542–545.

VanDongen, A. M., Codina, J., Olate, J., Mattera, R., Joho, R., Birnbaumer, L., and Brown, A. M., 1988, Newly identified brain potassium channels gated by the guanine nucleotide binding protein G$_o$, *Science* **242:**1433–1437.

Whiteway, M., Hougan, L., Dignard, D., Thomas, D. Y., Bell, L., Saari, G. C., Grant, F. J.,

O'Hara, P., and MacKay, V. L., 1989, The STE4 and STE18 genes of yeast encode potential β and γ subunits of the mating factor receptor-coupled G protein, *Cell* **56**:467–477.

Worley, P. F., Baraban, J. M., Van Dop, C., Neer, E. J., and Snyder, S., 1986, G_o, a guanine nucleotide-binding protein: Immunohistochemical localization in rat brain resembles distribution of second messenger systems, *Proc. Natl. Acad. Sci. USA* **83**:4561–4565.

Yatani, A., Mattera, R., Codina, J., Graf, R., Okabe, K., Padrell, E., Iyengar, R., Brown, A. M., and Birnbaumer, L., 1988, The G protein-gated atrial K^+ channel is stimulated by three distinct G_i alpha-subunits, *Nature* **336**:680–682.

8

Model of Signal Transduction by G Proteins
Roles of α Subunits and $\beta\gamma$ Dimers in Regulation of Ionic Channels and Adenylyl Cyclase

LUTZ BIRNBAUMER, ATSUKO YATANI, RAVI IYENGAR, JOHN D. HILDEBRANDT, JUAN CODINA, and ARTHUR M. BROWN

1. Introduction

About 80% of all hormone and neurotransmitter receptors interact with signal transducing GTP-binding proteins (G proteins) to modulate cell activity. G proteins are $\alpha\beta\gamma$ heterotrimers, and although not unequivocally proven, there is general agreement that receptors act by catalyzing their activation by GTP to give α^*-GTP plus a $\beta\gamma$, at which point they regulate the activity of effector systems, and that they lose their regulatory activity after GTP is hydrolyzed to GDP by the α subunit and subunit reassociation to give an inactive αGDP$\beta\gamma$ trimer. There is also agreement that effector systems, such as adenylyl cyclase, phosphodiesterases, and phospholipases, are "readout systems" that monitor and report through activity changes on the activation state of the G protein. This chapter discusses

LUTZ BIRNBAUMER • Departments of Cell Biology and Molecular Physiology and Biophysics, Baylor College of Medicine, Houston, Texas 77030.   *ATSUKO YATANI and ARTHUR M. BROWN* • Department of Molecular Physiology and Biophysics, Baylor College of Medicine, Houston, Texas 77030.   *RAVI IYENGAR* • Department of Pharmacology, Mount Sinai School of Medicine, New York, New York 10029.   *JOHN D. HILDEBRANDT* • Worcester Foundation for Experimental Biology, Shrewsbury, Massachusetts 01545.   *JUAN CODINA* • Department of Cell Biology, Baylor College of Medicine, Houston, Texas 77030.

the molecular basis of signal transduction by G proteins: the pros and cons of whether G proteins dissociate into α plus βγs in the normal phospholipid environment and, if so, what can be said about the role that this dissociation reaction plays in the mode of action of receptors; and the α vs. βγ controversy that has arisen in two areas of G protein action: stimulation of K^+ channels and inhibition of adenylyl cyclase. The chapter ends with a discussion of the roles of βγ dimers in signal transduction as perceived by the authors.

2. Mechanism of Receptor-Mediated Activation of G Proteins

Two models may explain how receptors activate G proteins (Fig. 1). One is a passive nucleotide exchange model (Model I). It assumes that the sole effect of receptors is to promote the opening of the nucleotide binding site so that bound GDP can be spontaneously replaced by GTP. The other is an active subunit dissociation model (Model II). It assumes that, in addition to nucleotide exchange, receptors are required for activation of the G protein by GTP. In the first model the G-GTP to G*GTP reaction is spontaneous; in the second the "starring" occurs only if the G protein is complexed with receptor.

Seven basic sets of receptor-mediated signal transduction data must be accounted for:

1. Receptors act as catalysts of G-protein activation and have the ability to "move" from one molecule to another. It can be calculated that one receptor has the ability to activate more than one G protein. This has been shown in intact membranes for the action of β-adrenergic receptors acting on G_s in turkey erythrocytes (Fig. 2; Tolkovsky and Levitzki, 1978) and for the action of light-activated rhodopsin on transducin in retinal rod outer segments (Fung et al., 1981), and with purified components incorporated into phospholipid vesicles (Asano et al., 1984). In the latter study, 1 mole of receptor promoted the binding of GTPγS to ca. 10 moles of G_s.

2. Receptors that are coupled by G proteins exist in at least two agonist affinity states, one with high affinity and the other with low affinity for agonist (Rodbell et al., 1971; Maguire et al., 1976; Kent et al., 1980). These receptor states correlate with being, or not being, associated with a G protein. The high-affinity state associated with G protein is found in membranes in the absence of guanine nucleotides (GTP, GTP analogs, GDP, or GDP analogs; Fig. 3A,C) and is dependent for its formation on the presence of a functional G protein (Fig. 3A,B). The free, low-affinity state receptor forms upon addition of guanine nucleotides (di- or triphosphates; Fig. 3C, for exception see below) as well as upon removal of G protein by inactivation (Kim and Neubig, 1985) or receptor purification (Shorr et al., 1981). Coincorporation of pure low-affinity receptor and pure active G protein into phospholipid vesicles results in G-protein-depen-

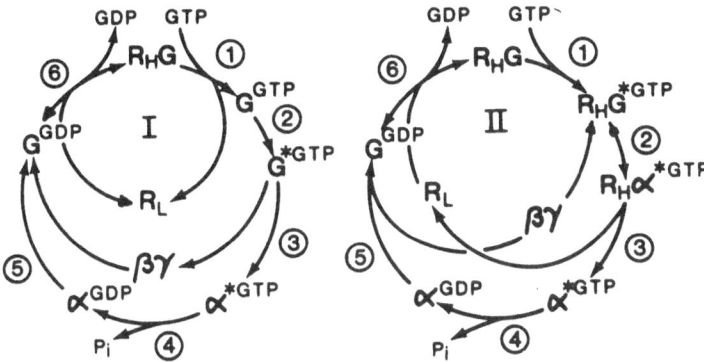

Figure 1. Two models for the catalysis of G-protein activation by hormone-receptor complex. In Model I, activation by nucleotide exchange (left side of figure), the role of the receptors is solely to promote nucleotide exchange. In Model II, activation by subunit dissociation (right side of figure), the role of the receptors is both to promote nucleotide exchange and to stabilize a GTP-dependent "activated" form of the G protein. In both models the G protein undergoes a cyclical dissociation-reassociation reaction and oscillates between GDP, nucleotide-free, and GTP states. The cycle is driven energetically forward by the capacity of the G protein to hydrolyze GTP (Cassel and Selinger, 1976; Godchaux and Zimmerman, 1979; Brandt and Ross, 1986; Cerione *et al.*, 1984). In both models the receptor has high affinity for the agonist (R_H) when associated with the nucleotide-free trimeric $\alpha\beta\gamma$ form of the G protein and has low affinity for the agonist (R_L) when it is free. In both models the receptor has higher affinity for the trimeric $\alpha\beta\gamma$ form of G than the G-GDP, thus accounting for the finding that GDP and G-GDP analogs promote the R_H to R_L transition (Fig. 3), and in both $\beta\gamma$ dimers are required for the interaction of α with R. Thus, both models are identical in reactions 4, 5, and 6, but differ fundamentally in reaction 1. Model I assumes that reaction 1 is symmetrical to reaction 6, with receptor merely allowing the GDP–GTP exchange to occur passively, and that reactions 2 and 3 are thermodynamic consequences of the formation of G-GTP. Further, although the dissociation of G* to α* plus $\beta\gamma$ is indicated (reaction 3), it is not an obligatory feature if only the catalytic nature of receptor action must be accounted for. Model II, on the other hand, assumes that no G* forms unless it is "aided" by receptor, and hence that receptor should have even higher affinity for the G*-GTP state than the nucleotide-free state of G. As a consequence, this model is absolutely dependent on reaction 2 (subunit dissociation) to account for the catalytic nature of receptor action. Thermodynamic properties do not allow R both to stabilize the G* state and to dissociate from it. Reaction 3 of Model II states further that the α*GTP loses its ability to stay associated with receptor and decomposes further into free activated α*GTP plus free receptor, thus accounting for the fact that under "working" conditions (saturation by both GTP and hormone, and hence sustained regulation of effector) only a small proportion of receptors are found in their high-affinity, G-protein-associated state. Thus, in Model II, the G-protein cycle is driven forward not only by the GTPase, but also, and obligatorily so, by the subunit dissociation reaction.

dent reformation of high affinity agonist binding (Cerione *et al.*, 1984; Tota *et al.*, 1987).

3. As purified from tissues and membranes, unactivated G proteins are obtained mostly as G–GDP complexes (Godchaux and Zimmerman, 1979; Ferguson *et al.*, 1986), and agonist–receptor complexes promote GDP–GTP ex-

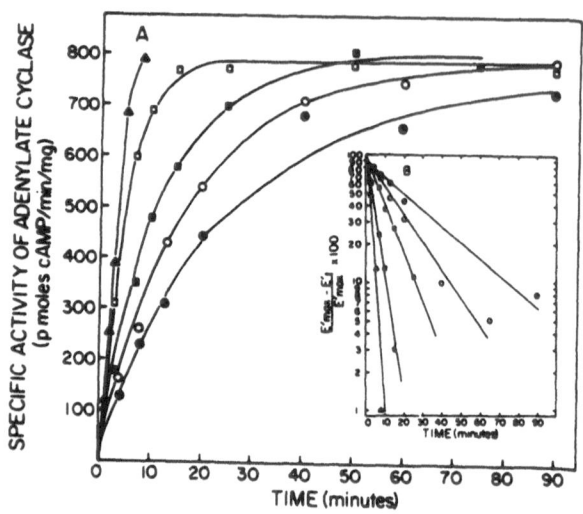

Figure 2. Receptor number determines the rate of activation of G_s by guanine nucleotide. Turkey erythrocyte membranes with 1.12 pmole/mg β-adrenergic receptors (▲) were alkylated, leading to loss of 42 (□), 70 (■), 87 (□), and 91 (●)% of receptors. The corresponding fractional rates of activation of G_s by GMP-P(NH)P, 0.374 min^{-1} in nonalkylated membranes, were reduced by 42%, 52.4%, 87.3%, and 92%, respectively. This indicates that receptors act catalytically rather stoichiometrically. A stoichiometric mode of action predicts that as receptors are lost, fewer G_s molecules are activated.

change (Cassel and Selinger, 1978; Michel and Lefkowitz, 1982; Murayama and Ui, 1984).

4. Purified G proteins subjected to "activation" treatments, such as GTPγS plus Mg^{2+} or AlF_4^- (GDP) plus Mg^{2+}, dissociate into α* plus βγ (Howlett and Gilman, 1980; Northup *et al.*, 1983a; Codina *et al.*, 1984a; Fung, 1983), and isolated α* molecules are potent regulators of effectors such as adenylyl cyclase (Northup *et al.*, 1983a; Smigel, 1986, rod outer-segment phosphodiesterase (Fung *et al.*, 1981), and various ionic channels (Codina *et al.*, 1987a,b; Yatani *et al.*, 1988a,b; Mattera *et al.*, 1989a,b).

5. High concentrations of βγ dimers inhibit activation of G_s as seen in membranes (Katada *et al.*, 1984a,b; see Fig. 7C) and reconstituted phospholipid vesicles (Cerione *et al.*, 1986), and interfere with activation of G_p, the G protein responsible for activation of membrane-bound phospholipase C, as inferred from *Xenopus* oocyte injection studies (Moriarty *et al.*, 1988). βγ dimers also delay the rate at which purified G_s is activated by GTPγS (Birnbaumer *et al.*, 1985a,b; Northup *et al.*, 1983b).

6. In liver membranes as well as most other adenylyl cyclase systems, except turkey erythrocyte membranes, the rate-limiting step in the activation of

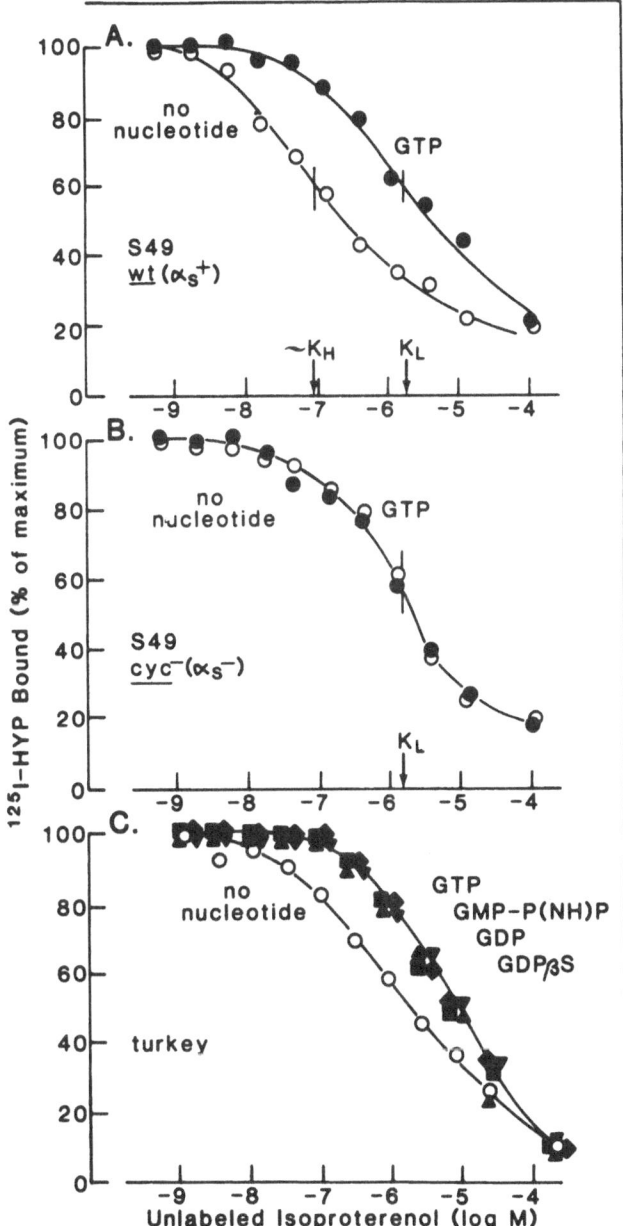

Figure 3. High-affinity, GTP-sensitive agonist binding to β-adrenoceptors is dependent on the presence of an active G protein (A vs. B), and low-affinity binding can be induced with either GTP and analogs or GDP and analogs. (A and B) Binding to S49 cell membranes; (C) binding to turkey erythrocyte membranes. All reactions contained 10 mM Mg^{2+}. For details see Birnbaumer *et al.* (1985a,b).

G_s by GTP and its analogs [GTPγS and GMP-P(NH)P] is not the binding of the nucleotide, but the G_s to G_s^* transition, as seen by the fact that saturating concentrations of each nucleotide activate G_s at distinct rates (Fig. 4; Birnbaumer *et al.*, 1980; Iyengar *et al.*, 1980). Recent studies by Ross and collaborators on the reaction rates and interactions of nucleotides, receptors, and G protein in reconstituted vesicles also showed that receptors act not only to "open" the nucleotide binding site, thereby releasing GDP and facilitating GTP binding, but also to accelerate the subsequent step that changes G_s to the G_s^* conformation, as judged by rate of GTPγS binding (Asano *et al.*, 1984; Brandt and Ross, 1986).

7. In both platelet and renal proximal tubule membranes, the addition of GTP promotes an increase in the proportion of receptors in the high-affinity state (Grandt *et al.*, 1982; Watanabe *et al.*, 1986).

Although the most recent reviews (e.g., Gilman, 1987; Weiss *et al.*, 1988)

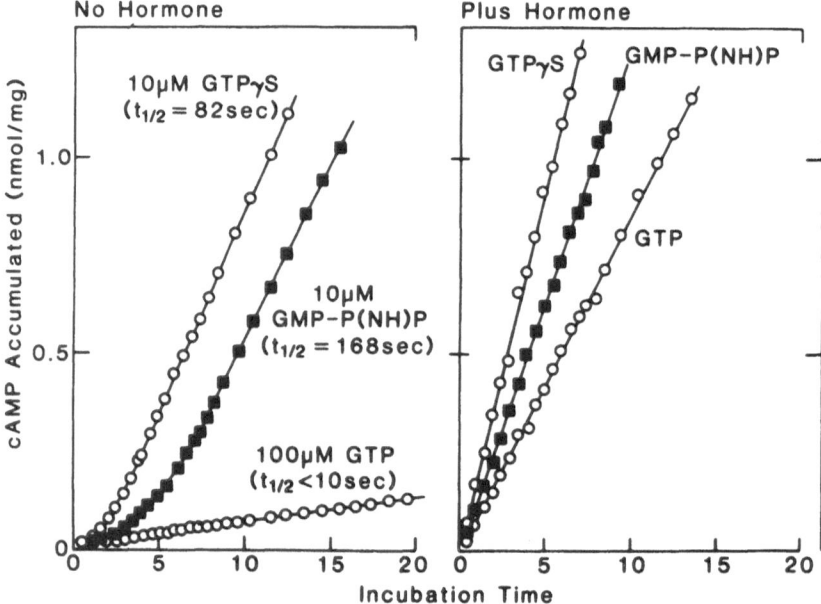

Figure 4. Differential rates of activation of liver membrane G_s by saturating concentrations of GTP, GTPγS, and GMP-P(NH)P (A) and stimulation by glucagon (B). Adapted from studies with liver membranes (Birnbaumer *et al.*, 1980; Iyengar *et al.*, 1980). Note that the rate-limiting step in the activation of GTP analogs cannot be GDP dissociation, since the same $t_{1/2}$ would be predicted, and the action of receptor cannot be restricted to promoting the nucleotide exchange reaction. Similar data have been obtained in several other membrane systems, including rabbit corpus luteum and S49 *wt*, MDCK, and GH$_4$C$_1$ cells, but not with turkey erythrocyte membranes. In the latter case GDP dissociation is rate-limiting, so the effects of receptor on the rate of activation, if any, cannot be assessed by this approach.

propose that receptors act according to Model I, which accounts for the first five data sets. Model II would seem to fit the data better, since only it can account for all seven data sets. Thus, contrary to the prediction of Model I, which does not allow for an effect of hormone receptor on the G to G* transition, the measured acceleration by hormone of the activation of G_s by GTP and its analogs in all membranes studied indicates that the steps affected by the hormone–receptor complex include the "activation" of G_s to G_s^*, and that therefore an RG–RG* reaction must exist. Likewise, there is no provision in Model I for a GTP-induced increase in high-affinity forms of receptors, which Model II readily allows for by accepting variable degrees of reversibility in reaction in the Rα* to R plus α* reaction, and assuming that effectors may be regulated by both α* and Rα*.

Several aspects of the regulatory cycle of G protein and its modulation by hormone receptor need to be studied. As shown in Figure 1, the RG to RG*-GTP transition is the sum of several individual reactions, including the intermediary formation of an RG–GTP complex. Abundant data are available to indicate that Mg^{2+} ion plays a crucial role, for which no place has yet been made in the model. Further, even though it is clear that $\beta\gamma$ dimers need to dissociate from α* before R, it is not clear which of the receptor-regulated reactions, G to G* or G* to α* plus $\beta\gamma$, is the actual rate-limiting step.

Since, upon activation, G proteins presumably generate two products, α* and $\beta\gamma$ dimers, the question is whether the signal transduction message of the dissociated G protein is carried only by the α subunit (as assumed in Fig. 1), or whether the $\beta\gamma$ dimer also participates as a bearer of a signal. This last question has been addressed experimentally by more than one laboratory. The answers differ and are the topic of the next section of this chapter.

3. Which G-Protein Subunit Mediates the Effects of Receptors?

The overall implicit assumption in the description of different G proteins is that their α subunits are the key to specificity in deciding which receptors and which effectors they regulate. This applies especially to effectors, which are used to name G proteins, e.g., G_s, G_i, G_p. There are, however, two completely independent sets of studies that might suggest otherwise. One relates to the mechanism by which receptors regulate a class of ionic channels, the so-called cardiac muscarinic K^+ channels, and the other relates to the mechanism by which hormones inhibit adenylyl cyclase.

3.1. Regulation of Ionic Channels: α or $\beta\gamma$, and What Does Each Do to Regulation of Channel Activity?

The existence of a K^+ channel in atrial heart cells that is under control of a G protein akin to the control of adenylyl cyclase or a phospholipase by G proteins was predicted by electrophysiological studies (Soejima and Noma, 1984; Pfaffinger *et al.*, 1985). The name G_k was coined to define the putative PTX-sensitive G protein responsible for the regulation of a K^+ channel (Breitweiser and Szabo, 1985). In late 1986, research carried out independently by us at Baylor College of Medicine (Yatani *et al.*, 1987a) and by Clapham, Neer, and colleagues at Harvard (Logothetis *et al.*, 1987a) discovered regulation of the atrial inwardly rectifying "muscarinic" K^+ channel by a guanine nucleotide binding G protein, measured by voltage clamp techniques under cell-free conditions in inside-out membrane patches under conditions that excluded mediation by phosphorylation or formation of a soluble second messenger (cAMP, Ca, IP3). Both groups agreed that a G protein regulates this type of channel, but at that time concluded the opposite with respect to which subunit was responsible for the regulation of the channel. In Houston, we observed that GTPγS-activated human erythrocyte (hRBC) G_i [G_i^*, a mixture of GTPγS-α_i (α_i^*) plus βγ], but not GTPγS-activated G_s (G_s^*, the equivalent mixture of α_s^* plus βγ), was active in stimulating the K^+ channel, from which we concluded that K^+ channel regulation was mediated by the α_i^*. The active G protein was named G_k, and its α subunit α_k. G_k was active at very low concentrations, 1–10 pM. Maximal effects were obtained with 50–100 pM. In Boston, Neer, Clapham, and colleagues found that bovine brain α_i subunits, as well as α_o subunits, treated with GTPγS and referred to as GTPγS-α_{41} and GTPγS-α_{39}, did not stimulate the channel, but that βγ dimers did. Furthermore, GDP-α_{41} and GDP-α_{39} inhibited the effects of βγ dimers. This led them to conclude that the mechanism by which muscarinic receptors activate atrial K^+ channels involves the formation of βγ dimers which either directly or indirectly stimulate channel activity.

The reasons that led to such disparate conclusions can be found, we believe, in the quality and purity of the purified/activated G proteins and G-protein subunits, and in the various detergent-related artifacts that may interfere with the recording of single channel currents from isolated inside-out membrane patches by the patch clamp technique.

In addressing the question of whether α subunits activate the channel, experiments carried out subsequently by us (Codina *et al.*, 1987a,b; Kirsch *et al.*, 1988) and others (Cerbai *et al.*, 1988), as well as more recently by Logothetis *et al.* (1988), proved α subunits to be active in the absence of βγ dimers. In our hands α subunits act with approximate EC50s of 10–40 pM (Kirsch *et al.*, 1988). In addition, hRBC α_k^* also stimulates a similar acetylcholine- and somatostatin-responsive K^+ channel present in inside-out membrane patches of the prolactin-secreting rat pituitary GH_3 cell line (Fig. 5A; Yatani *et al.*, 1987b). Amino acid sequence analysis of purified hRBC α_k established that it is encoded by the so-called α_i-3 gene (Codina *et al.*, 1988). Since ongoing studies in our laboratory allowed us to express the human α_i-3 subunit in *E. coli* using recombinant DNA techniques, we sought to further confirm the identity of α_k by testing its possible effect on K^+ channel activity (Mattera *et al.*, 1989a). As shown in Figure 5B, recombinant GTPγS-activated α_i-3, but not recombinant GTPγS-activated α_s, does indeed activate atrial K^+ channels. To our surprise, when we tested recombinant forms of α_i-1 and α_i-2, they were as active as recombinant α_i-3 (Yatani *et al.*, 1988b). This suggested strongly that there are not one but three G_k proteins. Tests with GTPγS-activated α subunits from subsequently purified native bovine brain G_i-1 (Fig. 5C) and human erythrocyte G_i-2 (Fig. 5D) confirmed the existence of G_k isoforms (Yatani *et al.*, 1988b).

Based on these studies, there is now agreement that G-protein-gated K^+ channels are regulated by α subunits of the stimulating G protein. However, there are still discrepancies about which of the G proteins are active and which are not. In the same article in which Logothetis *et al.* corrected their original statement relating to the activity to α subunit by showing positive results with their bovine brain GTPγS-α_{39} (α_o^*) and our hRBC α_i-3 (called by them GTPγS-α_{40}), they also reported "consistent" lack of effect with a preparation of bovine brain GTPγS-α_{41} and inhibition of GTPγS-stimulated K^+ channel activities by GDP-α_{41} (Logothetis *et al.*, 1988). Both these results are unexpected and point to the difficulties the biochemist has in these types of studies. Based on our finding, any GTPγS-α_i subunit (including GTPγS-α_{41}) should have worked in the hands of Logothetis *et al.* (1988). The fact that GTPγS-α_{41} did not suggests the presence of an inhibitory contamination, as we had suggested upon reading their original results (Birnbaumer and Brown, 1987). It is, of course, not clear where the inhibition is occurring. It could be at the level of the G protein or the channel. The "reversal" (inhibition for us) of GTPγS-activated single K^+ channel currents observed by Logothetis *et al.* (1988) with the same α_i preparation prior to GTPγS treatment (GDP-α_{41}), although in agreement with the idea that the protein comes with an inhibitor, does not clarify its site of action either. Unfortunately, none of the standard control experiments testing directly for presence of an inhibitor in the bovine brain α_{41} were carried out or reported on. Boiling, a manipulation requested from us by several referees, is not necessarily a stringent test either, for the interference could well be due to the presence of thermolabile phospholipase(s) or proteases. What one would like to see, for example, is a positive effect of muscarinic stimulation or GTPγS stimulation, or GTPγS-α_{40} in a patch that is unresponsive to putatively inactive GTPγS-α_{41}. In the meantime, the suggestion that inhibition of K^+ channel activity by addition of high concentrations of GDP-α_{41} represents removal of activating βγ dimers from their regulatory site(s), or of a physiologically relevant inhibitor, requires supportive evidence. Confirming our contention that G proteins gate K^+ channels *via* their α subunits, a monoclonal antibody (MAb A4) raised against the α subunit of transducin, and which also interacts with α_i-3 but not βγ dimers, was found to block muscarinic activation of the channels, and its blocking activity could be titrated by exogenously added GTPγS-α_k^* activity (Yatani *et al.*, 1988c; Fig. 5E).

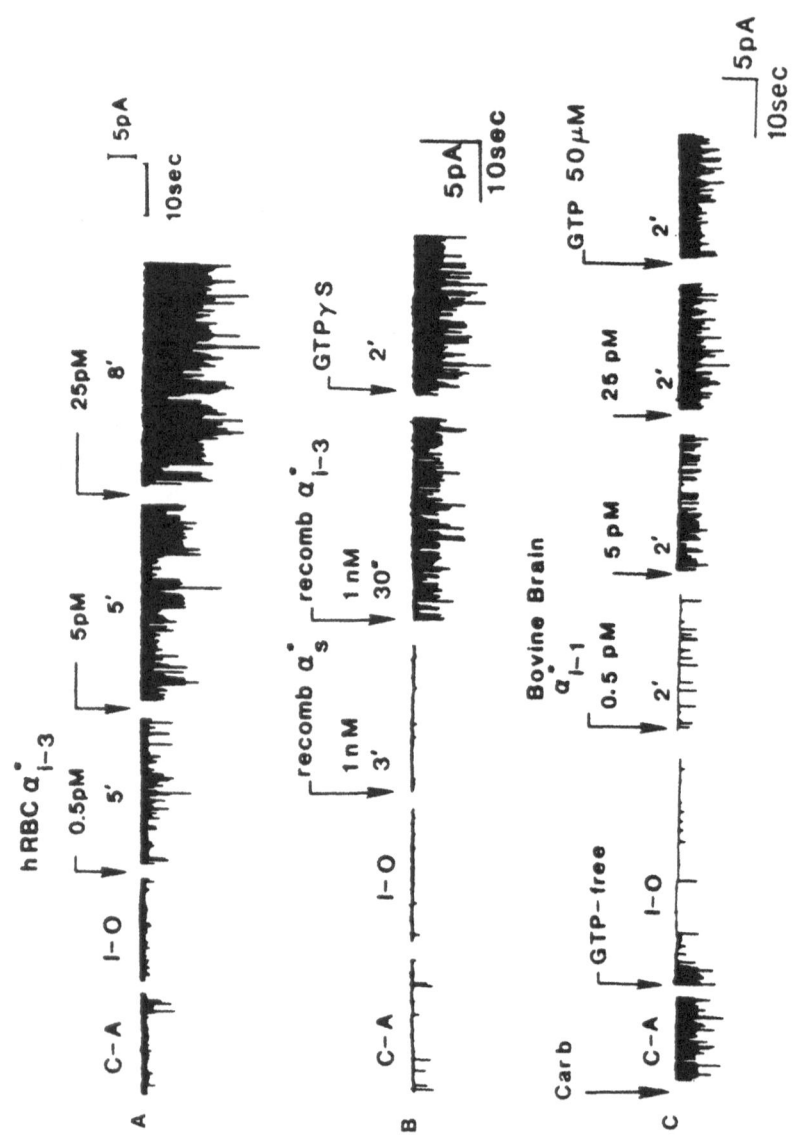

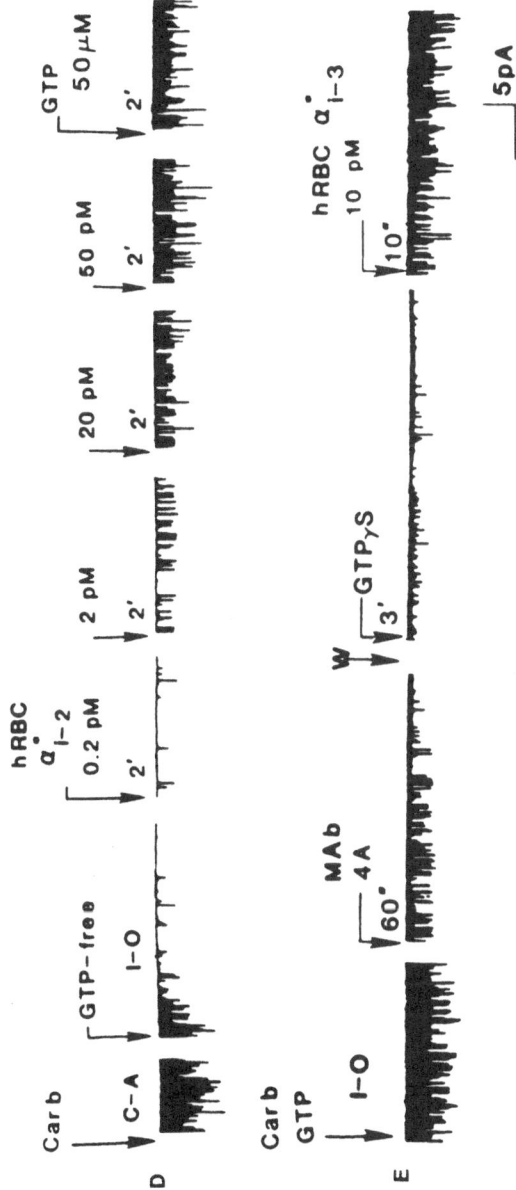

Figure 5. Representative examples of the effects of native and recombinant PTX-sensitive α subunits on GH₃ cell or atrial G-protein-gated K⁺ channels. In all cases α subunits were preactivated with GTPγS, as denoted by the asterisk. Recombinant molecules have a nine-amino-acid amino-terminal extension (Mattera *et al.*, 1989a; Yatani *et al.*, 1988b). (A) Stimulation of GH₃ cell single-channel K⁺ currents by human erythrocyte α_{i}-3 (hRBC α_{i}^{*}-3); (B) stimulation of guinea pig atrial muscarinic K⁺ channel by recombinant α_{i}-3 but not by recombinant α_{s}; (C) stimulation of atrial K⁺ channels by bovine brain α_{i}-1 (α_{i}^{*}-1); (D) stimulation of guinea pig atrial K⁺ channels by erythrocyte α_{i}-2 (hRBC α_{i}^{*}-2); (E) inhibition of muscarinic K⁺ currents by monoclonal anti-α antibody 4A (MAb 4A). Single K⁺ channel currents were recorded from the cell-attached (*C-A*) and inside-out (*I-O*) configurations as noted.

While the dispute between us and Logothetis *et al.* about whether α subunits are active in regulating "muscarinic" (G-protein-gated) K^+ channels has arrived, at least in part, at the position that αs are active, the discrepancy about what βγ dimers may do has not. The effects of βγ dimers have been tested on a variety of functions, including not only K^+ channels, but also G_s regulation of adenylyl cyclase, calmodulin-stimulated adenylyl cyclase, adenylyl cyclase proper, deactivation of AlF_4^-- and GTPγS-activated G_s, and phospholipase A_2. As shown in Fig. 6, in all instances the concentrations required to obtain effects with βγ dimers are higher by a minimum of 10-fold, an average of 100-fold, and, in the case of reversal of GTPγS-stimulated $α_s$ and stimulation of phospholipase A_2, by as much as 10,000-fold the concentrations required of α subunits to regulate effector functions. Some of the effects of βγ dimers may be significant, while others may not; some may be true effects of the added βγ molecules, while others may need confirmation for they could also be the result of contaminants. Some examples are the stimulation of phospholipase A_2 activity by close to 1 μM βγ prepared from rod outer segments, and stimulation of K^+ channels by 1–10 nM βγ dimers dissolved with the aid of the detergent CHAPS, which could be due to substances coadded with the βγ dimers, including tightly bound detergent, and the inhibition of calmodulin-stimulated adenylyl cyclase, which may well be a nonspecific effect of the hydrophobic character of βγ interacting nonspecifically with the hydrophobic pocket of calmodulin.

If α subunits are able to act at an average of 100-fold lower concentrations than those at which βγ dimers can mimic their real or perceived actions (Fig. 6), it is difficult to see how the effects of half-maximally effective concentrations of hormones and neurotransmitters could be exerting their action *via* the βγ aspect of the receptor-activated G protein.

The presumed stimulation of K^+ channels by βγ dimers is one case in which it is unlikely to be of physiological significance because of the concentrations required; it is possible that it is not even due to the βγ dimers, but rather to a combination of the hydrophobic βγ added with the detergent used to keep it dissolved. Thus, Logothetis *et al.* (1988) observed stimulatory effects with 1–10 nM of bovine brain or human placenta βγ dimers (reported threshold 200 pM) using buffers with CHAPS. The same questions of purity that applied to the inhibitory effects of bovine brain $α_{41}$ preparations also apply to the stimulatory effects of the βγ preparation; that is, might the effect be due to a contaminating phospholipase or protease? For example, we have found that low concentrations of trypsin stimulate atrial K^+ channels, while high concentrations inactivate it. Also, in our hands (Kirsch *et al.*, 1988), but not in those of Logothetis *et al.* (1988), CHAPS has marked nonspecific effects on K^+ channels, including the stimulation of a channel with the same mean open times and slope conductance as the one stimulated by muscarinic ligands, GTPγS, or GTPγS-$α_i^*$ (Kirsch *et al.*, 1988). Isenberg and collaborators have also noted stimulatory effects of

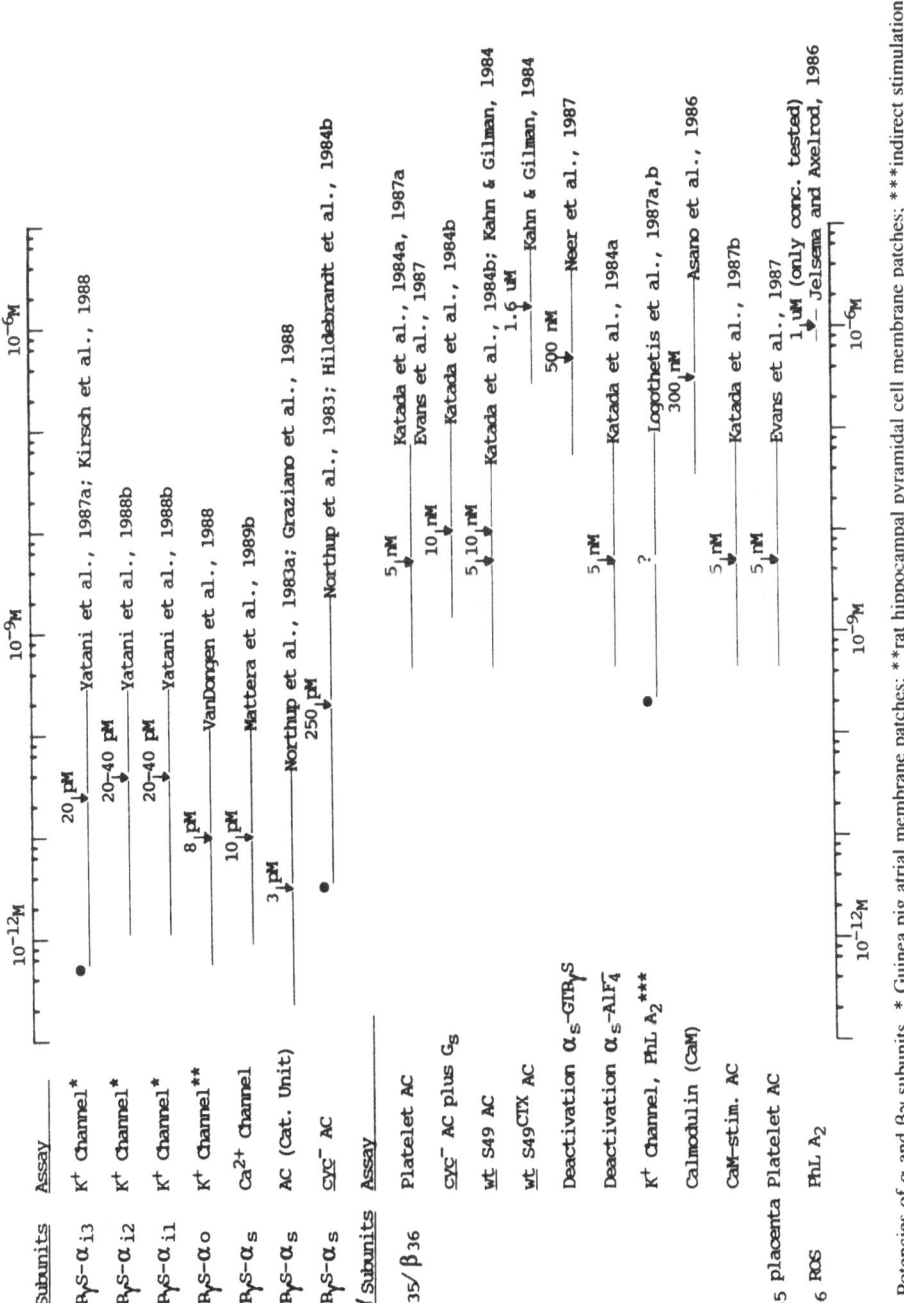

Figure 6. Potencies of α and βγ subunits. * Guinea pig atrial membrane patches; **rat hippocampal pyramidal cell membrane patches; ***indirect stimulation of K⁺ channel via phospholipase A (PhLA) metabolite; ●, lowest concentration reported with activity; ↓, approximate EC_{50}.

CHAPS on K^+ currents (personal communication), although not as much as we saw. The possibility exists, therefore, that the effects noted by Logothetis *et al.* are just those of hydrophobic $\beta\gamma$ dimers "delivering" CHAPS to the membrane patch. In recent unpublished work, presented at the 1988 meeting of the Society of General Physiologists held in Woods Hole (Massachusetts), Dr. Clapham inhibited the effect of CHAPS-$\beta\gamma$ with an antibody to phospholipase A_2 (PhL A_2) and mimicked muscarinic activation of the K^+ channel with 10 μM free arachidonic acid. The suggestion was made that $\beta\gamma$ dimers, rather than acting directly on the K^+ channel, may be acting by activating membrane-bound PhL A_2. The observed stimulation would then be mediated either by arachidonic acid directly or by one of its metabolites, in analogy to what may be the mode of action of phenylalanylmethionylarginylphenylalanylamide (FMRF-amide) in aplysia (Piomelli *et al.*, 1987). The concentrations of $\beta\gamma$ required for this effect remain the same (high), and all the purity/detergent arguments raised with respect to K^+ channel regulation remain in place for PhL A_2 regulation.

Our experience with $\beta\gamma$ dimers has been not only different, but indeed opposite, to that of Logothetis *et al.* We use Lubrol PX as detergent to maintain $\beta\gamma$ dimers monodispersed (Codina *et al.*, 1984a,b; Hildebrandt *et al.*, 1984a), at concentrations at which it has no effect on its own and does not interfere with muscarinic (acetylcholine-receptor-mediated), GTPγS, or hRBC α_k^* stimulation of single-channel atrial K^+ currents (Fig. 7A–D). We never observed intrinsic stimulatory effects of $\beta\gamma$ dimers (Fig. 7A; Codina *et al.*, 1987a,b). This was not due to the presence of Lubrol, as indicated by the fact that $\beta\gamma$-unresponsive patches remained α-responsive (not shown). Contrary to any of the predictions based on results reported by Logothetis *et al.*, we find that $\beta\gamma$-dimers inhibit basal and muscarinic receptor-stimulated K^+ channel activities. Thus, adding $\beta\gamma$ dimers (free of PTX substrate), purified from human placenta by Tony Evans (Genentech; Evans *et al.*, 1987), inhibits the K^+ channel activity induced by GTP or by carbachol plus GTP (Fig. 7C,D). It is noteworthy that inhibition of the spontaneous GTP-induced, agonist-independent activity appears to occur faster (compare Fig. 7C to 7D).

The mechanism of ionic channel regulation by α subunits needs further study. For example, stimulation could be either the result of a direct G protein–ionic channel interaction or a membrane-delimited phenomenon, in which the G protein generates a lipid-soluble local second messenger, which then directly or indirectly modulates the activity of the ionic channel. The second mode of action would be true, for example, if channel gating were due to arachidonic acid or one of its metabolites, as Piomelli *et al.* (1988) suggest for FMRF-amide regulation of aplysia K^+ currents. Although we tend to favor the direct mode of action, which resembles the way that adenylyl cyclase and rod cell phosphodiesterase are regulated, there are no data that rule out an indirect mechanism of action.

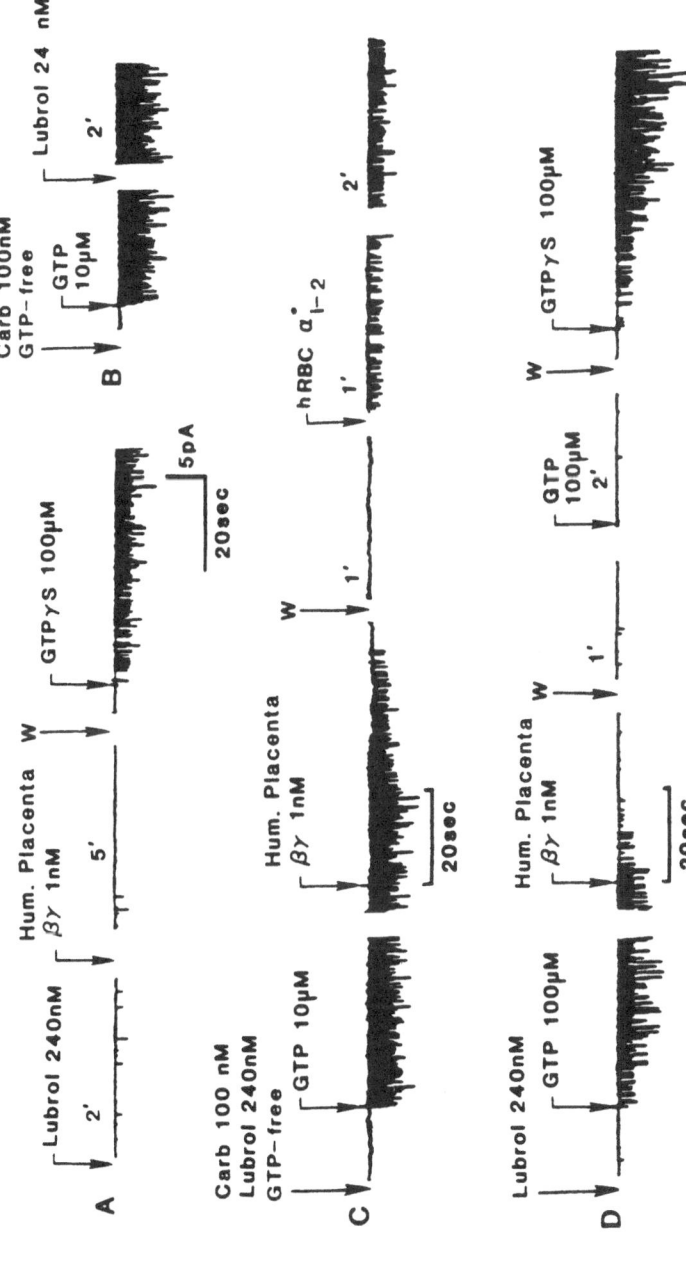

Figure 7. Effects of βγ dimers muscarinic K⁺ currents when presented to guinea pig atrial membrane patches in 240 nM Lubrol PX. (A) Lack of effect of βγ dimers in eliciting stimulation of atrial K⁺ channel activity and lack of a stimulatory effect of Lubrol; (B) Lack of effect of Lubrol in interfering with receptor-mediated GTP dependent stimulation of the atrial G-protein-gated K⁺ channel; (C) inhibition by βγ dimers of GTP- and carbachol-stimulated atrial K⁺ channels; (D) inhibition by βγ dimers of basal (GTP) dependent atrial K⁺ channel activity. In each case, similar results were obtained with bovine brain and human placenta βγ dimers.

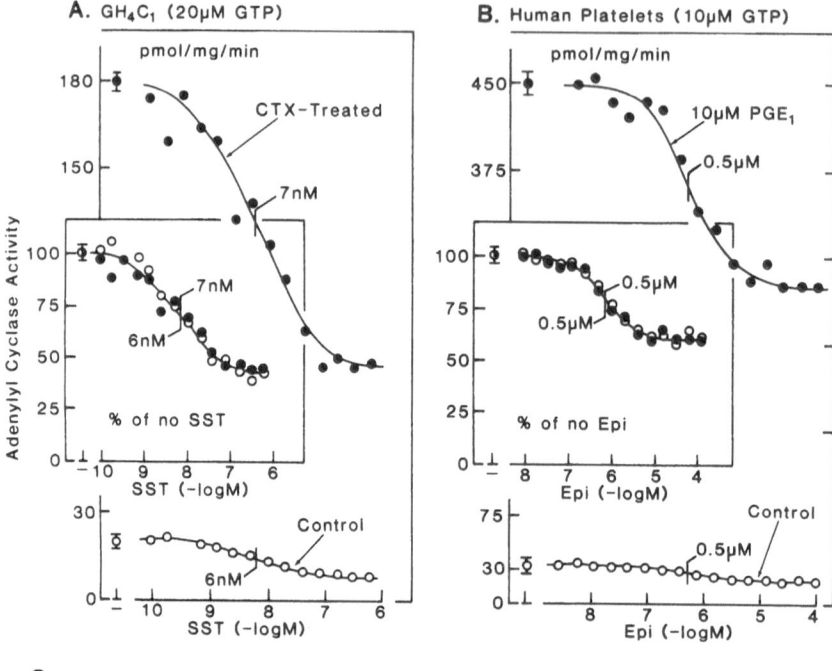

A. GH₄C₁ (20µM GTP)

B. Human Platelets (10µM GTP)

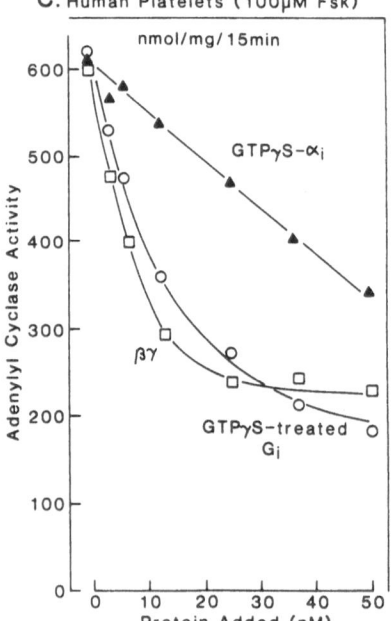

C. Human Platelets (100µM Fsk)

Figure 8. (A and B) Inhibitory regulation of adenylyl cyclases: Independence from state and type of stimulation of G_s and independence from presence of G_s. (A) Dose–response curves for inhibition of basal and PGE_1-stimulated adenylyl cyclase by epinephrine (α_2-adrenerghic effect) in human platelet membranes (from Hildebrandt and Birnbaumer, unpublished); (B) dose–response curves for inhibition of basal and cholera toxin stimulated adenylyl cyclase by somatostatin (SST) in homogenates of GH_4C_1 cells. (Adapted from Toro *et. al.*, 1987.) (Insets): Normalized activities. (C) Effects of added G_i or G_i components on human platelet adenylyl cyclase. (Adapted from Katada *et al.*, 1984a.) Note that all experiments are without forskolin.

3.2. Inhibitory Regulation of Adenylyl Cyclase: α_i or $\beta\gamma$?

Different studies carried out between 1982 and 1985 agreed that hormonal inhibition of adenylyl cyclase involves a pertussis toxin (PTX)-sensitive G_i, but reached opposing conclusions as to which aspect of the G_i mediates the action of the inhibitory hormones. One set of studies showed that there are inhibitory somatostatin receptors (Jakobs *et al.*, 1983) in membranes from *cyc*- S49 cell membranes (then suspected and now known to be devoid of α_s subunits), and that the adenylyl cyclase of these membranes shows normal inhibitory regulation, not only in response to somatostatin plus GTP (Jakobs *et al.*, 1983; Katada *et al.*, 1984b), but also in response to GTP or GTP analogs alone, measurable in both the presence and absence of the activity-enhancing diterpene forskolin (Hildebrandt *et al.*, 1982). Based on this, we proposed that hormonal inhibition, like hormonal stimulation of adenylyl cyclase, is mediated by a G protein (G_i) distinct from G_s (Hildebrandt *et al.*, 1983), that the effect of G_i occurs independently of the presence of G_s, and that the effect was mediated by the α_i subunit of the $\alpha\beta\gamma$ complex (Hildebrandt *et al.*, 1983; Hildebrandt and Birnbaumer, 1983).

The other set of studies provided two important observations: (1) It was found that even though adding GTPγS-activated liver G_i (mixture of GTPγS-α_i^* plus $\beta\gamma$) inhibits adenylyl cyclase in, for example, platelet membranes, the effect could be quantitatively mimicked and accounted for by the added $\beta\gamma$ dimers. About 5 nM $\beta\gamma$ dimers was needed to obtain half-maximal effects. (2) In these studies (Katada *et al.*, 1984a), as well as others (Roof *et al.*, 1985; Katada *et al.*, 1987a), resolved GTPγS-α_i^* complexes had only weak intrinsic inhibitory activity and then only at significantly higher concentrations than those required to inhibit G_s-driven adenylyl cyclase by $\beta\gamma$. For example, in one case 40 nM GTPγS-α_i^* was needed to obtain 50% of maximal effect (Katada *et al.*, 1984a; Fig. 8C), and it is possible that this might have been caused by locally released GTPγS, which inhibits *cyc*- adenylyl cyclase with an EC50 of 11 nM (Hildebrandt and Birnbaumer, 1983). In contrast, about 1000 times lower concentrations of GTPγS-activated α subunits are required to stimulate adenylyl cyclase and ionic channels (Fig. 6). Thus, the physiological meaning of the effects of high concentrations of GTPγS-activated α_i is questionable, at best.

Based on the fact that $\beta\gamma$ dimers inhibit adenylyl cyclase in membranes and that α_i subunits do not, or do so only poorly, Gilman and collaborators have proposed that hormonal inhibition of adenylyl cyclase is mediated primarily by $\beta\gamma$ dimers acting by law of mass action to prevent dissociation of the $\beta\gamma\alpha_s^*$-GTP complex, and hence α_s action (Gilman, 1987). There appear to be two shortcomings to this interpretation, however: (1) It is based on the failure to obtain effects with GTPγS-activated α_i subunits, that is, on a negative result. Such failure is certainly no proof that the effects should not exist, especially in view of the finding that *cyc*- S49 cells show the same inhibitory regulation by somatostatin receptors as *wt* S49 cells. (2) The interpretation does not account for the kinetics of action of inhibitory hormones vis-à-vis the concurrent absence or presence of stimulatory regulation of G_s that is not competitive. Thus, one would expect that the effectiveness of "$\beta\gamma$", i.e., an inhibitory hormone, would be reduced when G_s is being activated by a stimulatory hormone receptor, which, according to the basic premise proposed for this mechanism, has to cause subunit dissociation and hence decrease the affinity of α_s for $\beta\gamma$. Likewise, one would expect reduced effectiveness of an inhibitory hormone receptor, acting putatively through $\beta\gamma$ dimers, after ADP ribosylation of G_s with cholera toxin, since this manipulation is known to reduce the affinity of α_s for $\beta\gamma$. Yet, neither of these predictions for a $\beta\gamma$-regulated system holds true (Fig. 8A,B; Toro *et al.*, 1987; Hildebrandt and Birnbaumer, unpublished). In addition, the concentrations of $\beta\gamma$ dimers that mimic G_i action, while lower (5–10 nM) than those of activated α_i subunits, are still 200-500 times higher than those at which bona fide effects of α subunits on effectors are observed.

$\beta\gamma$ dimers inhibit GTP-regulated K^+ channels more rapidly than GTP plus agonist stimulated single K^+ channel currents (Fig. 7C,D), which agrees with similar results obtained previously by incorporating a $\beta\gamma$ dimer generating system, composed of hRBC G_i plus photoactivated rhodopsin, into phospholipid vesicles containing fixed amounts of G_s, β-adrenoceptor, and resolved catalytic unit of adenylyl cyclase (Cerione *et al.*, 1985). In these experiments $\beta\gamma$ dimers were more effective in inhibiting the basal GTP-driven adenylyl cyclase activity than in inhibiting the activity in the presence of GTP plus isoproterenol and were, in fact, a requirement for phenotypic expression of a properly regulated neurotransmitter stimulated adenylyl cyclase system (Cerione *et al.*, 1985; Birnbaumer, 1987).

3.3. What Is the Role of $\beta\gamma$ Dimers?

Based on data on ionic channel regulation and the mode by which inhibitory regulation of adenylyl cyclase occurs in *cyc-* membranes, the role of $\beta\gamma$ dimers cannot be that of receptor lambans, i.e., the bearer of the primary message of receptor occupation and hence the mediator of receptor action. This does not mean, of course, that $\beta\gamma$ dimers play a secondary role in signal transduction. Using the subunit dissociation model as a basis for analyzing the available data (Fig. 1), $\beta\gamma$ dimers can be predicted to have at least three key roles in signal transduction: (1) to provide the free energy required for receptors to act as catalytic regulators of G-protein activation by dissociating the receptor–G protein–GTP complex; (2) to allow for reactivation of the α-GDP complex by recombining with the α subunit to reform the holoprotein; (3) to act as inhibitor of ligand-independent activation of G proteins by GTP.

Experimental results show that the facts support these predictions. (1) The α-GDP complex of transducin does not interact with rhodopsin unless it is supplemented with βγ dimers (Kanaho *et al.*, 1984; Watkins *et al.*, 1981; Navon and Fung, 1987), and the α subunits of G_o cannot promote high affinity agonist binding to muscarinic receptors in reconstituted vesicles unless βγ dimers are present as well (Florio and Sternweis, 1985). Thus, βγs are essential for recycling α subunits. (2) As shown earlier in phospholipid vesicles (Cerione *et al.*, 1985), and now again in inside-out membrane patches (Yatani, Okabe, Evans, Codina, Birnbaumer, and Brown, submitted; Fig. 7), the absence of or insufficient levels of βγ dimers lead to "noise" in the various G-protein-coupled receptor–effector systems, as seen in the absence of added receptor ligands. Thus, βγs improve the signal-to-noise ratio in signal transduction by selectively blocking the ligand independent activations. (3) Addition of excess βγ dimers inhibits formation of activated effector-competent G proteins (assumed to be α^* -GTP and $R\alpha^*GTP$), as shown for the inhibition of adenylyl cyclase in isolated membranes (Katada *et al.*, 1984a,b) and, more indirectly, for the inhibition of phospholipase C activation in *Xenopus* oocytes (Moriarty *et al.*, 1988). These experiments, together with the finding that the addition of GDP-α_i complexes to isolated *wt* S49 membranes (but not to *cyc-* membranes) results in an increase in activated α_s (Katada *et al.*, 1984b), are perhaps the best experimental data available to support the hypothesis that G-protein subunit dissociation occurs naturally in the plasma membrane environment, and is not an artifact of the test tube.

In addition to the roles of βγ dimers, there are other questions that need investigation. For example, the assumption is made that regulation of effector is possible only after βγ has dissociated, but actually it is not known at what point the α subunit becomes competent to associate with effector. It is also not known at what point the α subunit acquires the capacity to hydrolyze GTP. There is no information about the rates at which subunit dissociation may have to proceed to account for rates of activation and deactivation of G proteins. The assumption is generally made that the intrinsic turnoff reaction of the system is GTP hydrolysis. However, direct measurements of GTP hydrolysis by purified G proteins (G_o, G_i, transducin) have given activities in the order of 3–6/min, which would seem too low to account for the actual rates at which, for example, the photoreceptor cell stops signaling to its neighboring cell after light is turned off, or the G_K-gated K^+ channel ceases activity when GTP is merely washed out (Yatani, Brown, and Birnbaumer, unpublished). Indeed, K^+ channel turnoff predicts a GTPase activity of at least 100/min, to account for the rate of signal extinction.

Further tests on the validity of assumptions and the possible dynamic aspects of subunit interactions in signal transduction will require perturbation in the relative abundance of the interacting components. While not easily possible in the test tube, this is now possible with the use of recombinant DNA techniques. These afford the creation of new cell lines that overexpress one or the other of the components of the cycle. Studies with such "new" cell lines are bound to be the basis of future insights into the dynamics of signal transduction.

ACKNOWLEDGMENTS. This work was supported by grants from the National Institutes of Health, the American Heart Association, and the Welch Foundation.

References

Asano, T., Pedersen, S. E., Scott, C. W., and Ross, E. M., 1984, Reconstitution of catecholamine-stimulated binding of guanosine 5'-O-(3-thiotriphosphate) to the stimulatory GTP-binding protein of adenylate cyclase, *Biochemistry* 23:5460–5467.

Asano, T., Ogasawara, N., Kitajima, S., and Sano, M., 1986, Interaction of GTP-binding proteins with calmodulin, *FEBS Lett.* 203:135–138.

Birnbaumer, L., 1987, Which G protein subunits are the active mediators in signal transduction, *Trends Pharmacol. Sci.* 8:209–211.

Birnbaumer, L., and Brown, A. M., 1987, G protein opening of K^+ channels, *Nature* 327:21–22.

Birnbaumer, L., Swartz, T. L., Abramowitz, J., Mintz, P. W., and Iyengar, R., 1980, Transient and steady state kinetics of the interaction of nucleotides with the adenylyl cyclase system from rat liver plasma membranes: Interpretation in terms of a simple two-state model, *J. Biol. Chem.* 255:3542–3551.

Birnbaumer, L., Codina, J., Sunyer, T., Cerione, R. A. Rosenthal, W., Hildebrandt, J. D., Caron, M. G., Lefkowitz, R. J., and Sekura, R. D., 1985a, Structural and functional properties of N_s and N_i, the regulatory components of adenylyl cyclase, in: *Pertussis Toxin and its Effects* (R. D. Sekura, J. Moss, and M. Vaughan, eds.), pp. 77–104, Academic Press, San Diego.

Birnbaumer, L., Hildebrandt, J. D., Codina, J., Mattera, R., Cerione, R. A., Sunyer, T., Rojas, F. J., Caron, M. G., Lefkowitz, R. J., and Iyengar, R., 1985b, "Structural basis of adenylate cyclase stimulation and inhibition by distinct guanine nucleotide regulatory proteins," in: *Molecular Mechanisms of Signal Transduction* (P. Cohen and M. D. Houslay, eds.), pp. 131–182, Elsevier/North Holland Biomedical Press, Amsterdam.

Brandt, D. R., and Ross, E. M., 1986, Catecholamine-stimulated GTPase cycle. Multiple sites of regulation by beta-adrenergic receptor and Mg^{2+} studied in reconstituted receptor-G_s vesicles, *J. Biol. Chem.* 261:1656–1664.

Breitweiser, G. E., and Szabo, G., 1985, Uncoupling of cardiac muscarinic and beta-adrenergic receptors from ion channels by a guanine nucleotide analogue, *Nature* 317:538–540.

Cassel, D., and Selinger, Z., 1976, Catecholamine-stimulated GTPase activity in turkey erythrocyte membranes, *Biochim. Biophys. Acta* 252:538–551.

Cassel, D., and Selinger, Z., 1978, Mechanism of adenylate cyclase activation through the beta-adrenergic receptor: Catecholamine-induced displacement of bound GDP by GTP, *Proc. Natl. Acad. Sci. USA* 75:4155–4159.

Cerbai, E., Klöckner, and Isenberg, G., 1988, The α subunit of the GTP binding protein activates muscarinic potassium channels of the atrium, *Science* 240:1782–1783.

Cerione, R. A., Codina, J., Benovic, J. L., Lefkowitz, R. J., Birnbaumer, L., and Caron, M. G., 1984, The Mammalian $beta_2$-adrenergic receptor: Reconstitution of the pure receptor with the pure stimulatory nucleotide binding protein (N_s) of the adenylate cyclase system, *Biochemistry* 23:4519–4525.

Cerione, R. A., Staniszewski, C., Caron, M. G., Lefkowitz, R. J., Codina, J., and Birnbaumer, L., 1985, A role for N_i in the hormonal stimulation of adenylate cyclase, *Nature* 318:293–295.

Cerione, R. A., Staniszewski, C., Gierschik, P., Codina, J., Somers, R., Birnbaumer, L., Spiegel, A. M., Caron, M., and Lefkowitz, R. J., 1986, Mechanism of guanine nucleotide regulatory protein-mediated inhibition of adenylate cyclase. Studies with isolated subunits of transducin in a reconstituted system, *J. Biol. Chem.* **261**:9514–9520.

Codina, J., Hildebrandt, J. D., Birnbaumer, L., and Sekura, R. D., 1984a, Effects of guanine nucleotides and Mg on human erythrocyte Ni and N_s, the regulatory components of adenylyl cyclase, *J. Biol. Chem.* **259**:11408–11418.

Codina, J., Hildebrandt, J. D., Sekura, R. D., Birnbaumer, M., Bryan, J., Manclark, C. R., Iyengar, R., and Birnbaumer, L., 1984b, N_s and N_i, the stimulatory and inhibitory regulatory components of adenylyl cyclases. Purification of the human erythrocyte proteins without the use of activating regulatory ligands, *J. Biol. Chem.* **259**:5871–5886.

Codina, J., Yatani, A., Grenet, D., Brown, A. M., and Birnbaumer, L., 1987a, The *alpha* subunit of G_k opens atrial potassium channels, *Science* **236**:442–445.

Codina, J., Grenet, D., Yatani, A., Birnbaumer, L., and Brown, A. M., 1987b, Hormonal regulation of pituitary GH_3 cell K^+ is mediated by its *alpha* subunit, *FEBS Lett.* **216**:104–106.

Codina, J., Olate, J., Abramowitz, J., Mattera, R., Cook, R. G., and Birnbaumer, L., 1988, *Alpha*$_i$-3 cDNA encodes the *alpha* subunit of G_k, the stimulatory G protein of receptor-regulated K^+ channels, *J. Biol. Chem.* **263**:6746–6750.

Evans, T., Fawzi, A., Fraser, E. D., Brown, M. L., and Northup, J. K., 1987, Purification of a beta$_{35}$ form of the *beta–gamma* complex common to G-proteins from human placental membranes, *J. Biol.Chem.* **262**:176–181.

Ferguson, K. M., Higashijima, T., Smigel, M., and Gilman, A. G., 1986, The influence of bound GDP on the kinetics of guanine nucleotide binding to G proteins, *J. Biol. Chem.* **261**:7393–7399.

Florio, V. A., and Sternweis, P. C., 1985, Reconstitution of resolved muscarinic cholinergic receptors with purified GTP-binding proteins, *J. Biol. Chem.* **260**:3477–3483.

Fung, B. K-K., 1983, Characterization of transducin from bovine retinal rod outer segments. I. Separation and reconstitution of the subunits, *J. Biol. Chem.* **256**:10495–10502.

Fung, B. K-K., Hurley, J. B., and Stryer, L., 1981, Flow of information in the light-triggered cyclic nucleotide cascade of vision, *Proc. Natl. Acad. Sci. USA* **78**:152–156.

Gilman, A. G., 1987, G proteins: Transducers of receptor-generated signals, *Annu. Rev. Biochem.* **56**:615–649.

Godchaux, W., III, and Zimmerman, W. F., 1979, Membrane-dependent guanine nucleotide binding and GTPase activities of soluble protein from bovine rod cell outer segments, *J. Biol. Chem.* **254**:7874–7884.

Grandt, R., Aktories, K., and Jakobs, K. H., 1982, Guanine nucleotides and monovalent cations increase agonist affinity of prostaglandin E_2 receptors in hamster adipocytes, *Mol. Pharmacol.* **22**:320–326.

Graziano, M. P., Freissmuth, M., and Gilman, A. G., 1988, Expression of $G_{s\alpha}$ in *Escherichia coli:* Purification and properties of two forms of the protein, *J. Biol. Chem.* **264**:409–418.

Hildebrandt, J. D., and Birnbaumer, L., 1983, Inhibitory regulation of adenylyl cyclase in the absence of stimulatory regulation. Requirements and kinetics of guanine nucleotide induced inhibition of the *cyc*- S49 adenylyl cyclase, *J. Biol. Chem.* **258**:13141–13147.

Hildebrandt, J. D., Hanoune, J., and Birnbaumer, L., 1982, Guanine nucleotide inhibition of *cyc*-S49 mouse lymphoma cell membrane adenylyl cyclase, *J. Biol. Chem.* **257**:14723–14725.

Hildebrandt, J. D., Sekura, R. D., Codina, J., Iyengar, R., Manclark, C. R., and Birnbaumer, L., 1983, Stimulation and inhibition of adenylyl cyclases is mediated by distinct proteins, *Nature* **302**:706–709.

Hildebrandt, J. D., Codina, J., Risinger, R., and Birnbaumer, L., 1984a, Identification of a *gamma* subunit associated with the adenylyl cyclase regulatory proteins N_s and N_i, *J. Biol.Chem.* **259**:2039–2042.

Hildebrandt, J. D., Codina, J., and Birnbaumer, L., 1984b, Interaction of the stimulatory and inhibitory regulatory proteins on the adenylyl cyclase system with the catalytic component of *cyc*- S49 cell membranes, *J. Biol. Chem.* **259**:13178–13185.

Howlett, A. C., and Gilman, A. G., 1980, Hydrodynamic properties of the regulatory component of adenylate cyclase, *J. Biol. Chem.* **260**:2861–2866.

Iyengar, R., Abramowitz, J., Riser, M, and Birnbaumer, L., 1980, Hormone receptor-mediated stimulation of the rat liver plasma membrane adenylyl cyclase system: Nucleotide effects and analysis in terms of a two-state model for the basic receptor-affected enzyme, *J. Biol. Chem.* **255**:3558–3564.

Jakobs, K. H., Aktories, K., and Schultz, G., 1983, A nucleotide regulatory site for somatostatin inhibition of adenylate cyclase in S49 lymphoma cells, *Nature* **303**:177–178.

Jelsema, C. L., and Axelrod, J., 1987, Stimulation of phospholipase A_2 activity in bovine rod outer segments by the beta–gamma subunits of transducin and its inhibition by the alpha subunit, *Proc. Natl. Acad. Sci. USA* **84**:3623–3627.

Kahn, R. A., and Gilman, A. G., 1984, ADP-ribosylation of G_s promotes the dissociation of its alpha and beta subunits, *J. Biol. Chem.* **259**:6235–6240.

Kanaho, Y., Tsai, S-C., Adamik, R., Hewlett, E. L., Moss, J., and Vaughan, M., 1984, Rhodopsin-enhanced GTPase activity of the inhibitory GTP-binding protein of adenylate cyclase, *J. Biol. Chem.* **259**:7378–7381.

Katada, T., Bokoch, G. M., Northup, J. K., Ui, M., and Gilman, A. G., 1984a, The inhibitory guanine nucleotide-binding regulatory component of adenylate cyclase. Properties and function of the purified protein, *J. Biol. Chem.* **259**:3568–3577.

Katada, T., Bokoch, G. M., Smigel, M. D., Ui, M., and Gilman, A. G., 1984b, The inhibitory guanine nucleotide-binding regulatory component of adenylate cyclase. Subunit dissociation and the inhibition of adenylate cyclase in S49 lymphoma *cyc*- and wild type membranes, *J. Biol.Chem.* **259**:3586–3595.

Katada, T., Oinuma, M., Kusakabe, K., and Ui, M., 1987a, A new GTP-binding protein in brain tissues serving as the specific substrate of islet-activating protein, pertussis toxin, *FEBS Lett.* **213**:353–358.

Katada, T., Kusakabe, K., Oinuma, M., and Ui, M., 1987b, A novel mechanism for the inhibition of adenylate cyclase via inhibitory GTP-binding proteins. Calmodulin-dependent inhibition of the cyclase catalyst by the beta–gamma subunits of GTP-binding protein, *J. Biol. Chem.* **262**:11897–11900.

Kent, R. S., DeLean, A., and Lefkowitz, R. J., 1980, A quantitative analysis of beta-adrenergic receptor interactions: Resolution of high and low affinity states of the receptor by computer modeling of ligand binding data, *Mol. Pharmacol.* **17**:14–23.

Kim, M. H., and Neubig, R. R., 1985, Parallel inactivation of α_2-adrenergic agonist binding and N_i by alkaline treatment, *FEBS Lett.* **192**:321–325.

Kirsch, G., Yatani, A., Codina, J., Birnbaumer, L., and Brown, A. M., 1988, The *alpha* subunit of G_k activates atrial K^+ channels of embryonic chick, neonatal rat and adult guinea pig, *Am. J. Physiol.* **254**:H1200–H1205.

Logothetis, D. E., Kurachi, Y., Galper, J., Neer, E. J., and Clapham, D. E., 1987a, The $\beta\gamma$ subunits of GTP-binding proteins activate the muscarinic K^+ channel in heart, *Nature* **325**:321–326.

Logothetis, D. E., Kurachi, Y., Galper, J., Neer, E. J., and Clapham, D. E., 1987b, G Protein opening of K^+ channels: Reply to Birnbaumer and Brown, *Nature* **327**:21–22.

Logothetis, D. E., Kim, D., Northup, J. K., Neer, E. J., and Clapham, D. E., 1988, Specificity of action of guanine nucleotide-binding regulatory protein subunits on the cardiac muscarinic K^+ channel, *Proc. Natl. Acad. Sci. USA* **85**:5814–5818.

Maguire, M. E., Van Arsdale, P. M., and Gilman, A. G., 1976, An agonist-specific effect of guanine nucleotides on binding to the beta adrenergic receptor, *Mol. Pharmacol.* **12**:335–339.

Mattera, R., Yatani, A., Kirsch, G. E., Graf, R., Olate, J., Codina, J., Brown, A. M., and

Birnbaumer, L., 1989a), Recombinant α_i-3 subunit of G protein activates G_k-gated K^+ channels, *J. Biol. Chem.* **264**:465–471.

Mattera, R., Graziano, M. P., Yatani, A., Zhou, Z., Graf, R., Codina, J., Birnbaumer, L., Gilman, A. G., and Brown, A. M., 1989b, Individual splice variants of the α subunit of the G protein G_s activate both adenylyl cyclase and Ca^{2+} channels, *Science* **243**:804–807.

Michel, T., and Lefkowitz, R. J., 1982, Hormonal inhibition of adenylate cyclase. Alpha$_2$-adrenergic receptors promote release of [^{3}H]guanylylimidodiphosphate from platelet membrane, *J. Biol. Chem.* **257**:13557–13563.

Moriarty, T. M., Gillo, B., Carty, D. J., Premont, R. G., Landau, E. M., and Iyengar, R., 1988, Beta–gamma subunits of GTP binding proteins inhibit muscarinic stimulation of phospholipase C, *Proc. Natl. Acad. Sci. USA* **85**:8865–8869.

Murayama, T., and Ui, M., 1984, [^{3}H]GDP release from rat and hamster adipocyte membranes independently linked to receptors involved in activation or inhibition of adenylate cyclase. Differential susceptibility to two bacterial toxins, *J. Biol. Chem.* **259**:761–769.

Navon, S. E., and Fung, B. K-K., 1987, Characterization of transducin from bovine retinal rod outer segments. Participation of the amino-terminal region of T_α in subunit interaction, *J. Biol. Chem.* **262**:15746–15751.

Neer, E. J., Wolf, L. G., and Gill, D. M., 1987, The stimulatory guanine-nucleotide regulatory unit of adenylate cyclase from bovine cerebral cortex, *Biochem. J.* **241**:325–336.

Northup, J. K., Smigel, M. D., Sternweis, P. C., and Gilman, A. G., 1983a, The subunits of the stimulatory regulatory component of adenylate cyclase. Resolution of the activated 45,000-dalton (alpha) subunit, *J. Biol. Chem.* **258**:11369–11376.

Northup, J. k., Sternweis, P. C., and Gilman, A. G., 1983b, The subunits of the stimulatory regulatory component of adenylate cyclase. Resolution, activity and properties of the 35,000 dalton (b) subunit, *J. Biol. Chem.* **258**:11361–11368.

Pfaffinger, P. J., Martin, J. M., Hunter, D. D., Nathanson, N. M., and Hille, B., 1985, GTP-binding proteins couple cardiac muscarinic receptors to a K channel, *Nature* **317**:536–538.

Piomelli, D., Volterra, A., Dale, N., Siegelbaum, S. A., Kandel, E. R., Schwartz, J. H., and Belardetti, F., 1987, Lipoxygenase metabolites of arachidonic acid as second messengers for presynaptic inhibition of *Aplysia* sensory cells, *Nature* **328**:38–43.

Rodbell, M., Krans, H. M. M., Pohl, S. L., and Birnbaumer, L., 1971, The glucagon-sensitive adenyl cyclase system in plasma membranes. IV. Binding of glucagon: Effect of guanyl nucleotides, *J. Biol. Chem.* **246**:1872–1876.

Roof, D. J., Applebury, M. L., and Sternweis, P. C., 1985, Relationships within the family of GTP-binding proteins isolated from bovine central nervous system, *J. Biol. Chem.* **260**:16242–16249.

Shorr, R. G. L., Lefkowitz, R. J., and Caron, M. G., 1981, Purification of the *beta*-adrenergic receptor. Identification of the hormone binding subunit, *J. Biol. Chem.* **256**:5820–5826.

Smigel, M. D., 1986, Purification of the catalyst of adenylate cyclase, *J. Biol. Chem.* **261**:1976–1982.

Soejima, M., and Noma, A., 1984, Mode of regulation of the ACh-sensitive K-channel by the muscarinic receptor in rabbit atrial cells, *Pflügers Arch.* **400**:424–431.

Tolkovsky, A. M., and Levitzki, A., 1978, Mode of coupling between the β-adrenergic receptor and adenylate cyclase in turkey erythrocytes, *Biochemistry* **17**:3795–3810.

Toro, M-J., Montoya, E., and Birnbaumer, L., 1987, Inhibitory regulation of adenylyl cyclases. Evidence against a role for beta/gamma complexes of G proteins as mediators of G_i-dependent hormonal effects, *Mol. Endocrinol.* **1**:669–676.

Tota, M. R., Kahler, K. R., and Schimerlik, M. I., 1987, Reconstitution of the purified porcine atrial muscarinic acetylcholine receptor with purified porcine atrial inhibitory guanine nucleotide binding protein, *Biochemistry* **26**:8175–8182.

VanDongen, A., Codina, J., Olate, J., Mattera, R., Joho, R., Birnbaumer, L., and Brown, A. M.,

1988, Newly identified brain potassium channels gated by the gaunine nucleotide binding (G) protein G_o, *Science* **242:**1433–1437.

Watanabe, T., Umegaki, K., and Smith, W. L., 1986, Association of solubilized prostaglandin E_2 receptor from renal medulla with a pertussis toxin-reactive guanine nucleotide regulatory protein, *J. Biol. Chem.* **261:**13430–13439.

Watkins, P. A., Moss, J., and Vaughan, M., 1981, ADP ribosylation of membrane proteins from human fibroblasts. Effect of prior exposure of cells to choleragen, *J. Biol. Chem.* **256:**4895–4899.

Weiss, E. R., Kelleher, D. J., Woon, C. W., Soparkar, S., Osawa, S., Heasley, L. E., and Johnson, G. L., 1988, Receptor activation of G proteins, *FASEB J.* **2:**2841–2848.

Yatani, A., Codina, J., Brown, A. M., and Birnbaumer, L., 1987a, Direct activation of mammalian atrial muscarinic K channels by a human erythrocyte pertussis toxin-sensitive G protein, G_k, *Science* **235:**207–211.

Yatani, A., Codina, J., Sekura, R. D., Birnbaumer, L., and Brown, A. M., 1987b, Reconstitution of somatostatin and muscarinic receptor mediated stimulation of K^+ channels by isolated G_k protein in clonal rat anterior pituitary cell membranes, *Mol. Endocrinol.* **1:**283–289.

Yatani, A., Imoto, Y., Codina, J., Hamilton, S. L., Brown, A. M., and Birnbaumer, L., 1988a, The stimulatory G protein of adenylyl cyclase, G_s, also stimulates dihydropyridine-sensitive Ca^{2+} channels. Evidence for direct regulation independent of phosphorylation by cAMP-dependent protein kinase or stimulation by a dihydropyridine agonist, *J. Biol. Chem.* **263:**9887–9895.

Yatani, A., Mattera, R., Codina, J., Graf, R., Okabe, K., Padrell, E., Iyengar, R., Brown, A. M., and Birnbaumer, L., 1988b, The G protein-gated atrial K^+ channel is stimulated by three distinct $G_i\alpha$-subunits, *Nature* **336:**680–682.

Yatani, A., Hamm, H., Codina, J., Mazzoni, M. R., Birnbaumer, L., and Brown, A. M., 1988c, A monoclonal antibody to the α subunit of G_k blocks muscarinic activation of atrial K^+ channels, *Science* **241:**828–831.

Structural Homologies in G-Binding Proteins

BRIAN F. C. CLARK, MICHAEL JENSEN,
MORTEN KJELDGAARD, and SØREN THIRUP

1. Introduction

The common property of G-binding proteins is, by definition, the ability to bind guanine nucleotides. In this review we shall make our task easier by concentrating on the domain or subunits responsible for carrying out this function. This is not, however, intended to preempt the issue: a common function of such a general nature, exercised in so many different biological organisms and biochemical contexts, does not automatically imply a high, or even a detectable, degree of homology among the proteins that proffer it. Indeed, the G-proteins provide a spectrum of degrees of kinship that range from the intimate to the unrecognizable. Satisfyingly, this spectrum correlates largely with similarity of function, at least within the general classes of G-protein. Nevertheless, between these classes it remains a matter for conjecture whether the difference between just-detectable homology and no detectable homology, significant as it may be in the statistical sense, is significant in the subjective sense of "telling us anything about evolution."

As of March 1988 there were some 8000 proteins of known sequences, but the crystal structure been determined for only some 300 of these. Compared with this, the G-proteins are lagging behind: of some 100 or more sequenced G-protein genes (over 150 have been identified), only two structures have been published. However, the importance of this group of proteins and the prospect of

BRIAN F. C. CLARK, MICHAEL JENSEN, MORTEN KJELDGAARD, and SØREN THIRUP
• Division of Biostructural Chemistry, Department of Chemistry, Aarhus University, 8000 Aarhus C, Denmark.

comparative data have given impetus to several groups, including our own, that are now working on further crystal structures.

G-proteins bind GTP, and almost all of them convert it to GDP. In most cases, the γ-phosphate group is released as inorganic phosphate ion; in some, such as the enzyme GTP:AMP phosphotransferase and many kinases, it is used to phosphorylate another molecule. Therefore, quantitative criteria for the comparative analysis of the function/structure relationships in G-proteins will include the equilibria and rates of binding of GTP and GDP. For those G-proteins whose function is hydrolytic, the same aspects of inorganic phosphate binding will also be relevant, but interest will center above all on the rate of conversion of GTP to GDP in the binding pocket. This is partly because the hydrolysis of the triphosphate is an important step in the reaction: in protein synthesis, alteration in this rate results in a modulation of the accuracy of translation as governed by the "proofreading" step, while in the case of signal-transmitting G-proteins, the hydrolysis of GTP is the switching-off step, the rate of which helps to determine both the amplification factor of the response and its decay time. But another, unifying reason for interest in this step is that comparison of a large number of enzymes whose function is a single, relatively simple reaction may be expected, within the next few years, to lead to a better understanding of the effects of changes in structure (mutations) upon enzymic activity, and thus to a semiempirical grasp of the central teleonomic issue of biochemistry: *why* enzymes are *how* they are.

In the following paragraphs, we shall therefore summarize the current knowledge of G-binding proteins in decreasing order of dimensionality. First we examine and compare the three-dimensional (tertiary) structures of the two G-proteins for which they are known. We then take a comparative look at the sequences ("one-dimensional structures") of a much wider range of G-proteins in a short section that leans heavily on the work of W. Merrick's group. By linear interpolation, the secondary structures of proteins are frequently misdescribed as "two-dimensional"; while not associating ourselves with this facile geometry, we shall include a brief comparison of secondary structures, both known and predicted.

2. Structural Homology in Three Dimensions

2.1. Elongation Factor Tu (EF-Tu)

The function of this protein is basically to form a three-way adaptor linking aminoacyl tRNA, a GTP molecule as a source of energy, and the ribosome; however, many secondary functions, such as protection of the aminoacyl group against hydrolysis, have been attributed to it (for a general review, see Clark,

1980). The purpose of this three-way adaptor is not simply to bind the other species together or (for aa-tRNA and the ribosome) to strengthen their binding; the EF-Tu molecule provides a mechanical linkage between the irreversible hydrolysis of GTP and conformational changes of the programmed ribosome whose molecular details are still unknown. When the complex [GTP·EF-Tu·aa-tRNA·ribosome] is formed (see Fig. 1), the probability of GTP hydrolysis rises as compared with that in the free complex [GTP·EF-Tu·aa-tRNA]. If GTP is hydrolyzed, the EF-Tu, complexed with the resulting GDP, separates off and the aa-tRNA is left bound to the ribosome, committed to the next step of the protein elongation cycle. The waiting time for GTP hydrolysis must therefore be short enough to allow protein synthesis to proceed at an acceptable rate, but not so short as to be virtually instantaneous: if GTP hydrolysis is too prompt, then *any* aminoacyl-tRNA can be incorporated, including those that do not match the messenger codon, and this situation would make the translation of the mRNA highly inaccurate.

Thus, EF-Tu is a three-way adaptor with both mechanical coupling and a timer function, and the latter is connected intimately with the hydrolysis of GTP—and therefore with the fine structure of the GTP-binding site. A complete description of the function in terms of structure will have to account for at least the following four states of the molecule: (1) ternary complex, with GTP and aa-tRNA; low GTP hydrolysis rate; (2) quaternary complex, with the above plus the ribosome; high GTP hydrolysis rate; (3) posthydrolytic, bound to GDP and with a greatly lowered affinity for the ribosome and for aa-tRNA; (4) in complex with EF-Ts, which opens the G-binding pocket like an oyster and exchanges GDP for fresh GTP. At the moment it is not known whether the catalytic groups that determine the rate of hydrolysis of the bound GTP are all from EF-Tu, or whether

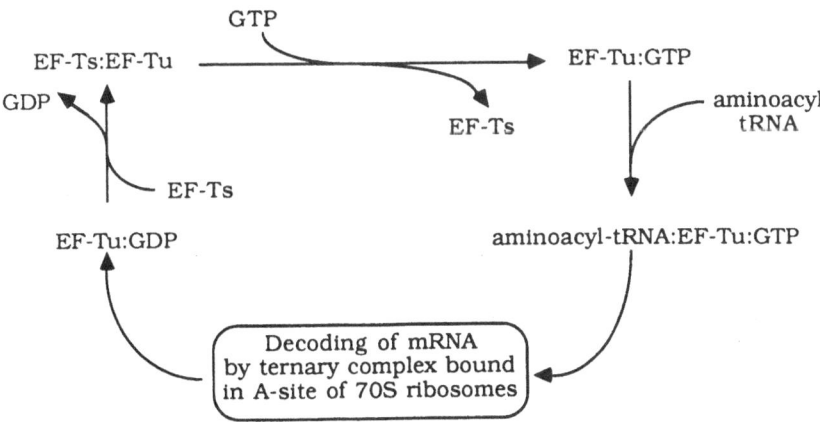

Figure 1. Scheme to show the EF-Tu cycle in the elongation step.

some are ribosomal. But functional studies show that at least some are on EF-Tu, since it can, under certain conditions, hydrolyze GTP in the absence of ribosomes.

EF-Tu has a structure with three domains (la Cour *et al.*, 1985; Jurnak, 1985), as shown in Fig. 2. The N-terminal domain I contains the G-binding site. It is also the best resolved; domains III (C-terminal) and especially II (central) show more thermal disorder and poorer definition in electron density maps, and these will not be considered further here, as they do not represent elements of structure common to all classes of G-protein. However, their lack of clarity has been a major obstacle to the accurate determination of the structure of this protein.

A recent report (Nyborg and la Cour, 1989) has described in detail the progress in Århus in elucidating the structure of EF-Tu. The findings of another group (in Riverside, California) are the same in all important details concerning the G-domain (Jurnak, 1985, and personal communication). Here we summarize the principal structural features of the G-domain I, as shown in Fig. 3.

This domain consists of six strands of β-structure. Five of these are parallel and the sixth, the end strand in the sheet, runs antiparallel to the rest. This strand is linked to the next by a simple, well-defined turn (residues $T_{71}PT$), while the

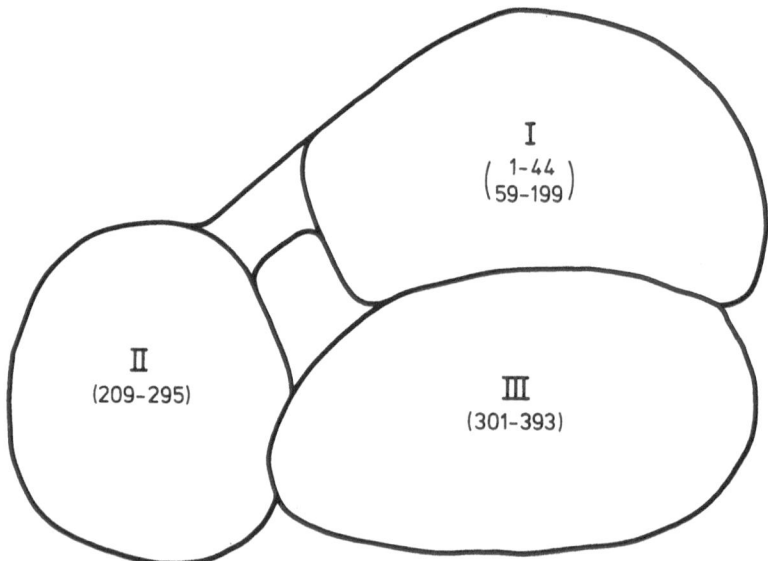

Figure 2. Overall domain structure of EF-Tu. The position of binding of GDP in domain I is indicated. The different degrees of structural compactness are indicated by the terms "tight," "loose," and "floppy," applied to domains I, III, and II, respectively.

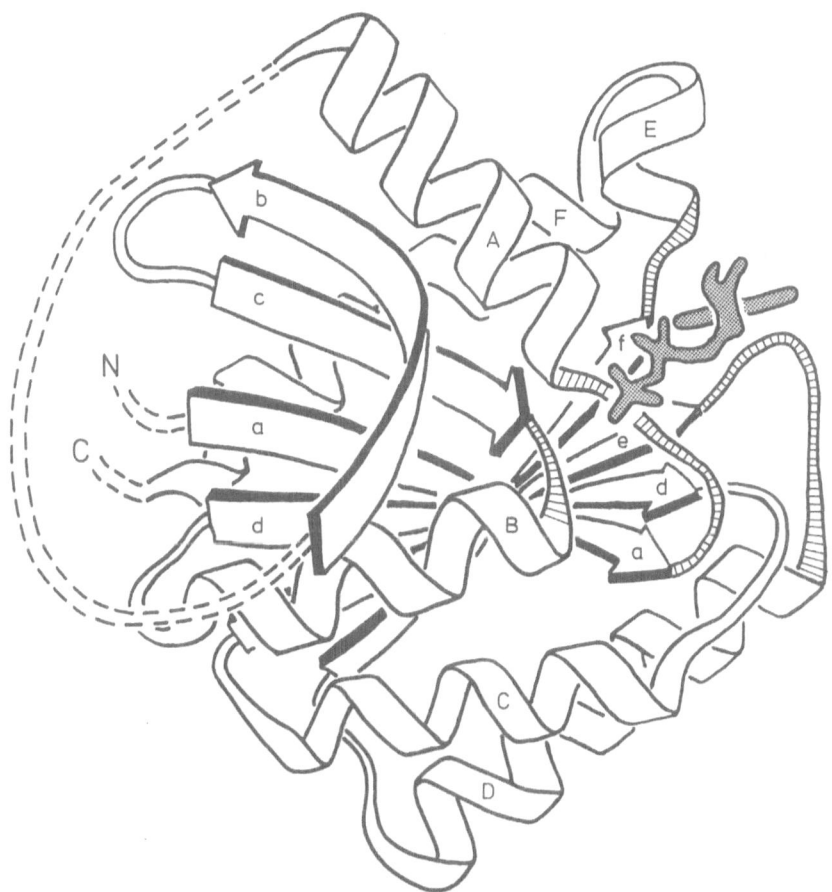

Figure 3. The G-domain of EF-Tu. The GDP molecule is shown and the four loops that bind it are shaded. A stretch is missing (residues 45–58); this is removed enzymically to facilitate crystallization.

others are connected by α-helices with interspersing loops. Four of these loops make up the site that binds GTP, and three of them contain the residues that are highly conserved in G-proteins (see Section 3.2).

The [EF-Tu·GTP] complex has not been crystallized in a form suitable for X-ray analysis, but the structure of the [EF-Tu·GDP] complex has been described (la Cour *et al.*, 1985; Jurnak, 1985). The binding site is shown in detail in Fig. 4 (from Nyborg and la Cour, 1989). The placing of the GDP molecule amid the various protein groups suggests the following assignment of roles of binding (Nyborg and la Cour, 1989):

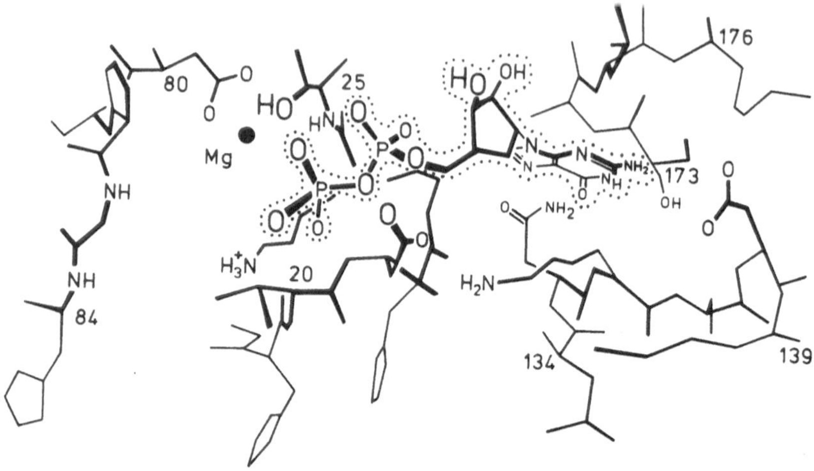

Figure 4. A closeup view of the bound GDP molecule. Redrawn from Nyborg and la Cour (1989). The heteroatoms around the GDP are shown explicitly.

The phosphates. The β-phosphate group appears to be held in place by a combination of three electrostatic factors: the dipole of helix A, the dipoles of the amides of loop aA (i.e., the loop connecting β-strand a and α-helix A), which is $G_{18}HVDHGKT_{25}$, and the positive charge of Lys-24. In addition, a magnesium ion appears to anchor the β-phosphate of GTP to residues Thr-25 and Asp-80. The binding pocket does not have room for a third (γ) phosphate, which hints at the nature of the mechanism that utilizes the chemical energy of GTP hydrolysis for the mechanical switching by EF-Tu referred to earlier. It has been suggested (la Cour *et al.*, 1985) that this switching may involve a displacement of helix B that is passed on to the adjacent domain III. This can be seen by inspecting the position of the β-phosphate (Fig. 4) and envisaging the displacement of helix B (Fig. 3) that would accompany the incorporation of the extra, γ-phosphate group of GTP.

The ribose. The ring is in the 2'-*endo* conformation. The hydroxyl groups are exposed to the solvent, and the ring is not obviously in direct contact with any of the groups of the protein.

The guanine base. The base-binding region of a protein would be expected to be an apolar pocket with hydrogen-bonding groups at the periphery for recognition of the correct nucleotide. This is largely what is found for EF-Tu, although oddly enough, a part of the apolar region is the extended side chain of Lys-136, whose polar end does not interact with the guanine. O(6) of the guanine interacts with the amide NH of Asn-135, and probably with the hydroxyl of Ser-173. Asp-138 probably interacts with the guanine N(1) and/or N(2).

It is clear that, with the availability of this structure on the one hand and systems for mutagenesis and expression on the other, there is a wealth of information to be gleaned that will lead to a better understanding of structure–function relationships at the level of atomic detail. However, we are only at the beginning of this at present. For example, the EF-Tu described here binds GDP some 10^3 times more strongly than the isolated G-domain (prepared by genetic deletion of the rest of the molecule (Parmeggiani *et al.*, 1987). The interaction *between* domains therefore exerts a major influence on interactions *within* one of these domains. This is not just a general binding effect, as is made clear by the fact that the corresponding ratio for GTP is only about 10. This difference is clearly related to the transmission of chemical energy from GTP, as mentioned earlier: if the GTP molecule is the "fuel" that is burnt in the "motor" of the G-domain, then the uncoupling of the "crankshaft" (B-helix?) from the "wheels" of domains II and III will clearly reduce the force needed to run the motor, and thus change the relative potential energies of fresh and spent fuel (GTP and GDP, respectively).

We get into less deep water when we simply change individual residues of EF-Tu. The most interesting and clearest example of this to date has been the replacement of Asp-138. As mentioned earlier, this probably interacts with the guanine's $N(1)$ or $N(2)$ nitrogen. It is tempting to speculate that this may occur by way of the double hydrogen bond associated with the relatively strong interaction between carboxylate anions and guanine in solution. This replacement of Asp-138 by Asn should therefore change the specificity of the EF-Tu, and this mutated EF-Tu does have a greatly reduced affinity for GTP/GDP; instead, it binds and hydrolyses XTP (X = xanthosine) (Hwang and Miller, 1987), which contains a carbonyl group in position 2 and is therefore precisely the substrate that one would expect to be preferred by the mutant, on the basis of the change in hydrogen bonding pattern. Some G-proteins (see Section 3.2, and Dever *et al.*, 1987) have Trp at this position. Tryptophan breaks the hydrogen-bonding pattern and destroys the specificity: these proteins can accept either GTP or ITP (I = inosine) as substrate.

2.2. *ras Protein p21*

The *ras* family of genes (for review, see Barbacid, 1987) has been found in many eukaryotes ranging from yeast to humans. Although their function is still unknown, it is generally accepted that they have an important cellular function related to the growth, proliferation, and differentiation of cells, because (1) many of them are known to be protooncogenes, requiring only a point mutation to turn into so-called oncogenes, which transform the cells containing them into cancer cells, (2) they are not found in prokaryotes, and (3) they show exceptionally high homology between species (cf. Section 3.2).

There are three kinds of *ras* p21 proteins in humans; with 188 or 189 amino acids, they resemble the G-domain of prokaryotic EF-Tu in size. Like other G-proteins, they bind GTP and hydrolyze it to GDP, and they possess the sequence element characteristic of these, as we shall discuss later. The crystal structure of human c-Ha-*ras* p21 was published recently (de Vos *et al.*, 1988), and here we compare the structural features that have emerged with those of EF-Tu. It is especially interesting to do this because we had earlier attempted a knowledge-based prediction of the p21 structure using the known structure of the G-domain of EF-Tu as a starting point (McCormick *et al.*, 1985).

The overall structure of p21 is shown in Figure 5 (de Vos *et al.*, 1988). It is immediately clear that the general pattern of this structure is the same as that of EF-Tu: a twisted β-sheet comprising five parallel strands and a further end strand antiparallel to the rest. Except for the turn connecting the two consecutive anti-parallel strands, these are separated by α-helices, as in EF-Tu, with one exception: the helix before the third of the parallel strands in EF-Tu is replaced in p21 by an extended loop. However, there is a major difference in the order of the β-strands and connecting helices: the antiparallel end strand is not connected to its adjacent stand by a short turn, but to the next stand by a longer loop. This can be seen clearly by tracing the two structures shown in Figs. 2 and 4, and it is shown in a formal way in Fig. 6 (for a more detailed comparison see Jurnak, 1988).

If it emerges that there is a reasonably clear homologous relationship be-tween the two sets of six β-strands, then an uncomfortable evolutionary question is posed. If *ras* p21 protein and EF-Tu are related by divergent evolution from a

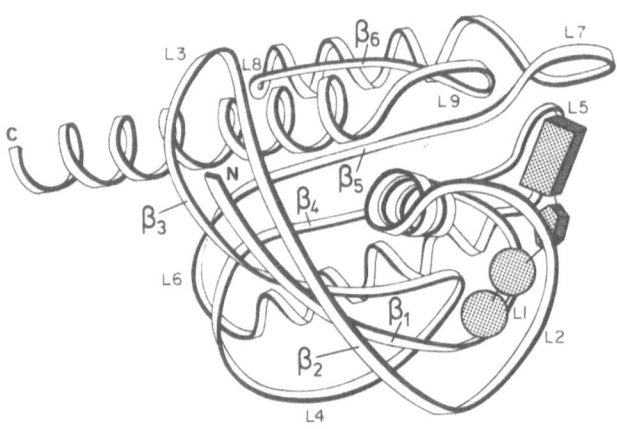

Figure 5. The backbone structure of residues 1–171 of human c-H-*ras* p21 protein. (The C-terminal residues 172–189 are believed to represent a flexible tail, probably interacting with a membrane, and were omitted in this study.) From Vos *et al.* (1988), with permission; we have numbered the β-strands to clearly show the relation to Fig. 6.

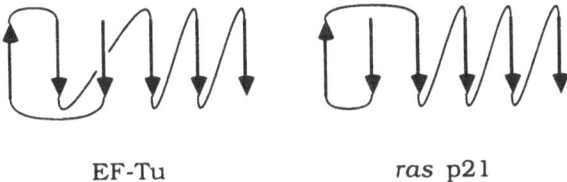

EF-Tu   *ras* p21

Figure 6. Folding pattern of β-strands in the G-domain of EF-Tu and in *ras* p21 protein.

common ancestor, then either a jump in the connectivity pattern occurred at some point, or the ancestor was much smaller and the N-terminal strands (almost half of the molecule/domain!) arose later, independently, and evolved convergently to a similar structure. Alternatively, of course, the similarity of these two proteins could be due entirely to convergent evolution, yet convergence of this degree for proteins with different functions seems unlikely. A possible explanation is that there was an evolutionary jump in connectivity which resulted in a poorly functioning protein whose structure then gradually evolved back to the old one while retaining the new connectivity. But then one must ask why the initial reconnected mutant was not simply wiped out by its more efficient parent. All of these solutions, therefore, are somewhat problematic. Yet the structural data seem clear enough, and both molecules have well-resolved electron density maps for the points in question.

There are some interesting similarities and differences between the GDP-binding sites of EF-Tu and *ras* p21 protein. Common aspects of GDP binding are: (1) Three loops involved in this binding are the three consensus loops common to all G-proteins (see Section 3.2). (2) The guanine base is held by an aspartic acid residue at N(1) and N(2), by a serine plus an asparagine at O(6), and by contact with the apolar part of a lysine side chain. (3) The β-phosphate group is bound by a lysine (part of a conserved loop), by main-chain interactions, and by a metal ion shared with a serine (the equivalent residue is threonine in EF-Tu).

Differences in the GTP binding include the following: (1) The ribose ring appears hydrogen-bonded to an aspartic acid side chain in *ras* p21 protein, and there may also be direct interaction between the furanose oxygen and the charged lysine whose aliphatic chain lies in contact with the G-base. (2) The loop in p21 that corresponds to the missing loop (Fig. 3) of EF-Tu is folded over a β-phosphate. (3) As a consequence of the different connectivity, Asp-57 in p21, which is homologous in sequence to the metal-binding Asp-80 in EF-Tu, now appears at a completely different location, in the third rather than the second β-strand (Jurnak, 1988), and may not have a role in coordinating the Mg^{2+}. (4) The GDP molecule as a whole is much more exposed to the solvent in p21 than in EF-Tu. This may be a reason for the differing pucker of the ribose ring (C3′-*endo*), and for the provision of a contact between the guanine and a Phe ring in p21 that is not made in EF-Tu. (5) The helix that, in EF-Tu, is the putative mediator of mechanical displacement around the phosphates (see Section 2.1) is no longer in contact with the β-phosphate. Nonetheless, there is still no room to accommodate a γ-phosphate, and the role of the GTP site can still be anticipated to be that of a chemomechanical transducer.

This list of differences and similarities cannot yet be taken as final, both because neither structure has been fully refined, and because we shall need further structures in order to obtain a perspective on what differences are to be regarded as significant.

The structure of an important mutant of *ras* p21 protein has been published very recently (de Vos *et al.*, 1989). It contains a Gly → Val substitution at position 12. Biologically, the mutated *ras* gene is the oncogene derived from the unmutated protooncogene; its expression results in the transformation of the host cell to a tumor cell. Biochemically, the mutant has a greatly reduced GTPase activity, suggesting that the protein has a switching function (like many other G-proteins; see Section 3.1.5) and gets stuck in a state that signals "proliferate". Many other mutations have a similar effect.

The structural effect of the mutation G12V is, as expected, very local, but, significantly, it is largely seen in loop 1 of the G-binding region (the loop numbers are shown in Fig. 5). This is summarized in Fig. 7 (de Vos *et al.*, 1989). As expected from the change in function, the two loops affected are in the region close to the phosphate groups of the GDP. However, this does not, as one might have hoped, lead to an immediate explanation of the change in properties, be-

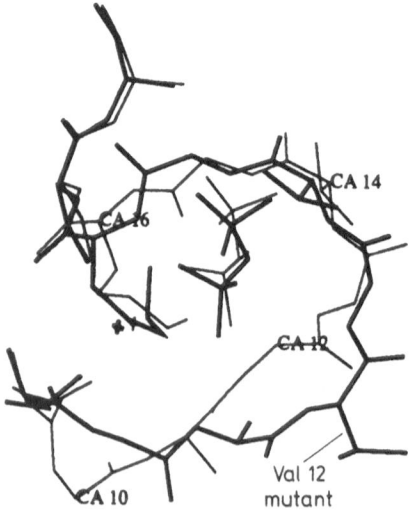

Figure 7. Schematic illustration of the structural changes in *ras* p21 protein accompanying the mutation G12V. The loops largely affected by the mutation are shown in black. From de Vos *et al.* (1989), with permission; thin lines show the wild-type, and thick lines show the mutant type.

cause the mutation G12V strongly reduces the rate of GTP hydrolysis. In EF-Tu, the corresponding residue is a valine in the wild-type (Val-20). The mutation V20G in EF-Tu might therefore be expected to show an increased rate of GTP hydrolysis, since its sense is opposite to that accompanying the mutation in p21. Unfortunately, the mutant EF-Tu V20G shows a reduced rate of GTP hydrolysis (Jacquet and Parmeggiani, 1988), shattering our naïve hopes of a simple, homology-based, rule-of-thumb correlation between structure and function, even for this apparently straightforward example of parallel mutation between two rather similar proteins.

A more detailed comparison between the G-domain structure of EF-Tu and p21 will arise from our protein engineering studies on EF-Tu. These studies, which involve a three-country cooperation between our group and the groups of A. Parmeggiani at Ecole Polytechnique, Paris, and L. Bosch at Leiden University, attempt to relate structure and function using protein engineering methods. This will provide information on how to design the structure of better enzymes or enzyme inhibitors, in keeping with the theme of this volume. In particular, we have been able to isolate the GDP-binding domain of EF-Tu as a separate entity for studies of GDP/GTP binding and GTP hydrolysis (Parmeggiani *et al.*, 1987; Jacquet and Parmeggiani, 1988). Thus, we now have available a prokaryotic version of p21 for further investigation of structure–function predictions.

3. Sequence Homology

3.1. Overall Homology

The existence of sequence homology between individual G-proteins or families of G-proteins has been noticed by many workers. First reports included a survey of regional homology in protooncogene products and elongation factors (Halliday, 1984), a comparison of the nucleotide-binding sites of EF-Tu and p21 (Leberman and Egner, 1984), and an analysis of phosphate-binding sequences (Möller and Amons, 1985). An exhaustive, computer-based study carried out by the group of W. Merrick and published in 1987 (Dever *et al.*) was followed by a recent extension to cover newly discovered sequences (Dever and Merrick, 1989). Two kinds of sequence homology have been seen. The first is a general homology stretching over regions of 100 amino acids or more. It can be expressed as a percentage according to whatever criteria are applied for its quantification: the weight placed on insertions, deletions, conservative and nonconservative substitutions, and so on. It resembles the kind of homology used as evidence for evolutionary kinship, and it may indeed reflect a common origin in many cases, even though the functions represented by the proteins are diverse. The second kind of homology is the rigorous conservation of specific groups, which, not surprisingly, is related to the G-binding proteins' single common function—the binding of G. We will take this up in Section 3.2, after we briefly consider the different families of G-proteins and their functions.

3.1.1. EF-Tu, EF-1α, and IF-2

EF-Tu has already been described, and EF-1α is its eukaryotic analog, which is roughly the same size and to which is attributed the same function in protein synthesis. In addition, it appears to have a connection with the regulation of general levels of protein synthesis and with cellular and/or organismic aging; however, this connection is not yet understood. The initiation factor, although much larger, has a function in the initiation step analogous to that of EF in the elongation step, and a structure model for the G-domain of this protein has been suggested (Cenatiempo *et al.*, 1987).

3.1.2. EF-G and EF-2

These factors (the first prokaryotic, the second eukaryotic) are involved in the second part of the elongation cycle at the ribosome; they are responsible for the events following the peptidyl transfer reaction, and as for EF-Tu, a conformational change following GTP hydrolysis is transformed into mechanical movement: uncharged tRNA is expelled from the ribosome and the mRNA and peptidyl-tRNA move across it in the well-known translocation reaction. Again molecular details of this process are still unclear.

3.1.3. LepA

This is a membrane-bound G-protein, expressed in *Escherichia coli* together with signal protein, and its function is not yet known (March and Inouye, 1985).

3.1.4. ras Proteins

As discussed earlier, these proteins appear to be associated with signaling mechanisms in cell proliferation. It is therefore interesting that they should also be expressed in yeast, in which the cells are not subject to the same critical constraints upon growth as are those which regulate the proliferation of cells in higher organisms. The equivalent of the "oncogenic" transformation in yeast causes a defect in sporulation and a loss of the capacity to accumulate glycogen (D. Gallwitz, personal communication). The protein products of two *ras* genes in yeast regulate adenylate cyclase (Toda *et al.*, 1985), which suggests a connection with the intercellular chemical signaling used by slime-mold cells. There are further *ras*-related genes, such as *rho*1 and 2, *ypt*1 in yeast (Haubruck *et al.*, 1987), and *ral* genes in animal cells (Chardin and Tavitian, 1986).

3.1.5. Eukaryotic Membrane-Bound G-Proteins (For general reviews, see Gilman, 1987; Rawis, 1987)

These are a part of the signal transmission chain that sends messages across the cell membrane, amplifying them at the same time. Briefly, the common feature of these systems is that a signal (a hormone molecule in the case of a hormone receptor, or light in the case of rhodopsin) causes a conformational change in a protein that makes it act like EF-Ts; i.e., it opens the jaws of a G-protein, removes the GDP that it finds there, and inserts a GTP instead. This reaction is catalytic, so many G-protein molecules are charged with GTP by the action of a single hormone molecule or quantum of light. The G-protein stimulates adenylate cyclase (hence its name G_s—in photoexcitation it is called transducin), and this enzyme produces cAMP molecules, enhancing the amplification effect. The G-protein switches itself off by spontaneous hydrolysis of its GTP, reverting to the inactive GDP complex. The many further aspects of these complex mechanisms include the existence of G-proteins with a complementary mode of action: they inhibit instead of simulating (hence the term G_i). A further group of G-proteins of unknown function is termed G_o.

3.1.6. G-Binding Enzymes

These include phosphoenolpyruvate carboxy kinase (PEPCK) and GTP:AMP phosphotransferase (Dever *et al.*, 1987), whose names indicate their respective functions. In both cases, as for the other G-proteins, the product is GDP. In contrast, guanylyltransferase, which is responsible for capping mRNA, cleaves GTP between the α- and β-phosphates, releasing pyrophosphate and generating a covalent Gp–protein complex that then reacts with the mRNA.

3.1.7. Tubulin

This is a structural protein with two 50-kDa forms, α and β (Mandelkow *et al.*, 1985). Microtubules are hollow cylinders consisting of alternating α- and β-tubulin molecules. These molecules bind GTP, and their conformation, in the GTP-complexing state, causes microtubules with a free end to grow spontaneously. Tubulin hydrolyzes its bound GTP to GDP, and the resulting conformation is conducive to dissociation of ends. Thus, as for EF-Tu and G_s and G_i proteins, the GTP hydrolysis has a timeswitch function.

The general homology among the G-binding proteins is high. The following discussion compares primary structures in groups of increasingly different origin and function.

Table I compares EF-Tu sequences from *E. coli, Thermus thermophilus* (Kushiro *et al.*, 1987; Seidler *et al.*, 1987), the mitochondrion of baker's yeast (Nagata *et al.*, 1983), and the chloroplast of *Euglena gracilis* (Montandon and Stutz, 1983). The number of strictly conserved residues, indicated by a star, is 198, or about 50%. This homology is very unevenly distributed, and in many regions, such as the N-terminal stretch or the center of the G-domain, it reaches over 90%, in spite of the very different origins (eubacterium, archaebacterium, mitochondrion, chloroplast).

Table II compares a representative prokaryote and four eukaryotic EF-1α sequences. These are human (Brands *et al.*, 1986), brine shrimp (*Artemia salina*) (Lenstra *et al.*, 1986), and two *Drosophila* genes (Hovemann *et al.*, 1988). Rigorous conservation among the eukaryotes is indicated by a vertical dash and among all sequences by a star. Again, the degree of homology is very high and is clearly highest for the G-domain. A cluster of universally conserved residues is seen around the first and second consensus sequences (see Section 3.2) and less clearly around the third. These correspond to G-binding loops. There is no significant conservation around the fourth G-binding loop seen in the crystal structure of EF-Tu from *E. coli;* interestingly, this loop does not correspond to a consensus sequence (compare Table VI).

As an approximation, it can be assumed that the elongation factors all share a common basic structure which is related to the function of the binding of EF-Tu to tRNA, EF-Ts, and the ribosome. The regions of conservation can therefore point to areas on the surface of the EF-Tu molecule that have potential interest for the location of the binding specificity. These regions are indicated by computer-generated, space-filling models, such as the example shown in Fig. 8.

The regions conserved in all sequences are probably associated with the properties most unaffected by evolution, especially the binding of transfer RNA. (Eukaryotic EF-1α will bind prokaryotic tRNA, and *vice versa*.) The regions that have been conserved only in prokaryotes are more difficult to interpret, as they are bound to contain some evolutionary "noise." To some extent, these regions could relate to specific features of the prokaryotic elongation factors, for instance, binding to EF-Ts.

Interestingly, there is a notable difference in the degree of conservation between the two sides of the EF-Tu molecule, which agrees well with experimental data on the EF-Tu–tRNA interaction.

Table III compares one EF-Tu, one EF-1α (G-domains only), and one *ras* protein (Ha-*ras* in Gibbs *et al.*, 1985). Here the degree of homology is much lower, although it rises substantially when conservative substitutions are allowed. Again, rather more homology is detectable in the regions of the G-loops.

Table IV compares EF-Tu and three eukaryotic signaling proteins: G_s, G_i, and transducin (Dever *et al.*, 1987; Gilman, 1987). The homology is very clear and, again, clusters of higher homology are seen in the region of the consensus sequences (see Section 3.1.5).

Finally, Table V compares the secondary structure patterns predicted for nine EF-Tu and EF-1α G-domains and the actual secondary structure type as seen in the three-dimensional structure of EF-Tu. The predictions were made using the method of Garnier *et al.* (1978). It is clear that the homology among the proteins leads to consistent predictions, but the agreement between the secondary structure predicted and that found is very poor. The now accepted failure of traditional statistical methods in making useful predictions must focus attention on new methods, such as neural networks (Bohr *et al.*, 1988), to see whether they will be capable of better things.

An attempt has recently been made at a reasoned prediction of the secondary and tertiary structures of a family of G-domains from membrane-binding signal proteins (Masters *et al.*, 1986). A comparison with experimental data, when these become available, will be interesting.

3.2. Consensus Sequences

The definitive survey of all G-protein sequences was carried out by Dever *et al.* (1987), and later expanded to include newly discovered sequences (Dever and Merrick, 1989). Three consensus sequence elements were identified. The first is GxxxxGK (rarely, AxxxxGK), which corresponds in the crystal structures of EF-Tu and *ras* p21 protein to a loop binding the β-phosphate GDP. The second is DxxG, conserved throughout, which also corresponds to a phosphate-binding loop. The third is NKxD, which is conserved throughout except in the enzymes, where the N can be G or T and the D can be E (in guanylyltransferases, with unaltered specificity) or W (in the others). This sequence, as discussed earlier, determines the G-specificity, and this is clearly of less importance for the latter two enzymes, which have a greater interest, for example, in phosphorylating phosphoenolpyruvate than in pedantically ensuring that this is done by GTP, and only by GTP. The consensus sequences are compared with the GDP-binding loops in EF-Tu in Table VI.

The three consensus sequences are spaced in a characteristic way: between the first and second consensus sequences, the spacing is either 40–44, 56–75, or 123–170 amino acids, and between the second and third it is 50–65 amino acids, with very few outliers.

This pattern was used to search known sequences for proteins that might have a heretofore unsuspected G-binding function. Three have been identified as fitting the pattern (foot-and-mouth disease virus protein 2C, yeast protein gstl, and *Ureaplasma urealyticum* protein *tetM*), but none of these has been tested.

A remarkable exception to the preceding generalizations is provided by the tubulins, which do not show any of the consensus sequences. An intermediate

Table I. Sequence Comparison of EF-Tu from *Escherichia coli, Thermus thermophilus,* the Mitochondrion of Baker's Yeast, and the Chloroplast of *Euglena gracilis* from Top to Bottom[a,b]

```
                                                   *  *  ***  *  ********
HOM    1  ................................................MSKEKFERTKPHVNVGTIGHVDH   59
EFECT  0  ......................................................AKGEFVRTKPHVNVGTIGHVDH  23
EFTTT  0  ......................................................AKGEFVRTKPHVNIGTIGHVDH  22
EFBYMT 1  MSALLPRLLTRTAFKASGKLLRLSSVISRTFSQTTTSYAAAFDRSKPHVNIGTIGHVDH                     59
EFEGCT 0  .....................................................ARQKFERTKPHINIGTIGHVDH   22

          ********  *          ****  *  *****  *  ***
HOM    60  GKTTLTAAITTVL.AKTYGGAAR....AFDQIDNAPEEKARGITINTSHVEYDTPTRHY   118
EFECT  24  GKTTLTAALTYVAAAENPNVEVKDYGD....IDKAPEERARGITINTAHVEYETAKRHY    77
EFTTT  23  GKTTLTAAITKTLAAK.GGANFLDYA....AIDKAPEERARGITISTAHVEYETAKRHY    77
EFBYMT 60  GKTTLTAAITMALAAT.........................GGGITINTAHVEYETKNRHY   113
EFEGCT 23  GKTTLTAAITMALAAT..............GGGITINTAHVEYETKNRHY             58

          *********  ******  **  *  **  ****  **  *  **  *
HOM    119  AHVDCPGHADYVKNMITGAAQMDGAILVVAATDGPMPQTREHILLGRQVGVPYIIVFLN   177
EFECT  78  SHVDCPGHADYIKNMITGAAQMDGAILVVSAADGPMPQTREHILLARQVGVPYIVVFMN    136
EFTTT  78  SHVDCPGHADYIKNMITGAAQMDGAIIVVAATDGQMPQTREHLLLARQVGVGVQHIVVFVN  136
EFBYMT 114  SHVDCPGHADYIKNMITGAAQMDGAIIVVAATDGQMPQTREHLLLARQVGVGVQHIVVFVN  172
EFEGCT 59  AHVDCPGHADYVKNMITGAAEMDGAILVVSAADGPMPQTKEHILLAKQVGVPNIVVFLN    117

          *  *   ***  *  *  **  *  **  *   ****
HOM    178  KCDMVDDEELLELVEMEVRELLSQYDFPGDDTPIVRGSALKALEG.........DAEW     236
EFECT  137  KVDMVDDPELLDLVEMEVRDLLNQYEFPGDEVPVIRGSALLALEQMHRNPKTRRGENEW    185
EFTTT  137  KVDTIDDPEMLELVEMEMRELLNEYGFPGDGNAPIIMGSALCALEG.....RQPEIDG     195
EFBYMT 173  ......................................RQPEIDG                  224
EFEGCT 118  KCDMVDDSELLELVELEIRETLSNYEFPGDDIPVIPGSALLSVQALTKNPKITKGENKW    176
```

Protein sequence alignment (rows: HOM 1, EFECT, EFTTT, EFBYMT, EFEGCT).

Block 1

```
                *    *  * ***  * *   * **    *  * ***** *** ***  * *   *
HOM   1  237  EAKILELAGFLDSYIPEPERAIDKPFLLPIEDVFSISGRGTVVTGRVERGIIKVGEEVE  295
EFECT  186  VDKIWELLDAIDEYIPTPVRDVDKPFLMPVEDVFTITGRGTVATGRIERGKVKVGDEVE  244
EFTTT  196  EQAIMKLLDAVDEYIPTPERDLNKPFLMPVEDIFSISGRGTVVTGRVERGNLKKGEELE  254
EFBYMT 225  VDKILNLMDQVDSYIPTPTRDTEKDFLMAIEDVLSITGRGTVATGRVERGTIKVGETVE  283
EFEGCT 177                                                               235
```

Block 2

```
          **             ** *     ** ***     ** ****** ** ***** *
HOM   1  296  IVGI.KETQKSTCTGVEMFRKLLDEGRAGENVGVLLRGIKREEIERGQVLAKPGTIKPH  354
EFECT  245  IVGLAPETRRTVVTGVEMHRKTLQEGIAGDNVGVLLRGVSREEVERGTVLAKPGSITPH  302
EFTTT  255  IVGHNSTPLKTTVTGIEMFRKELDSAMAGDNAGVLLRGIRRDQLKRGMVLAKPGTVKAH  313
EFBYMT 284                                                               342
EFEGCT 236  LVG.........TGVEMFQKSLDEALAGDNVGVLLRGIQKNDVERGMVLAKPRTINPH  284
```

Block 3

```
          **    * * *  ****** *   ***      ** ***               *****
HOM   1  355  TKFESEVYILSKDEGGRHTPFFK.GYRPQFYFRTTDVTGTIELPEGV.....EMVMPGD  413
EFECT  303  TKFEASVYVLKKEEGGRHTGFFS.GYRPQFYFRTTDVTGVVQLPPGV.....EMVMPGD  355
EFTTT  314  TKILASLYILSKEEGGRHSGFGE.NYRPQMFIRTADVTVVMRFPKEVEDHS.MQVMPGD  366
EFBYMT 343                                                               399
EFEGCT 285  TKFDSQVYILTKEEGGRHTPFFECGYRPEFYVRTTDVTGKIESFRSDNDNPAQMVMPGD  343
```

Block 4

```
              **    *      * **   * *** **********  *
HOM   1  414  NIKMVVTLIHPIAMDDGLRFAIREGGRTVGAGVVAKVLS  452
EFECT  356  NVTFTVELIKPVGLEEGLRFAIREGGRTVGAGVVTKILE  394
EFTTT  367  NVEMECDLIHPTPLEVGQRFNIREGGRTVGTGLITRIIE  405
EFBYMT 400  RIKMKVELIQPIAIEKGMRFAIREGGRTVGAGVVLSIIQ  438
EFEGCT 344                                            382
```

[a]Rigorous homology is indicated by asterisks and conservations among the signal G-proteins by vertical lines.

[b]The numbering in the EF-Tu sequence in the tables is based on gene sequencing, so the protein sequence numbers in the text are minus 1 compared with the tables.

Table II. Sequence Comparison of EF-1α from Various Organisms and a Representative EF-Tu[a,b]

```
            *  ***   *    *   *   *****  ** *  *
HOM  3
EFECT    1  MSKEKFERTKPHVNVGTIGHVDHGKTTLTAAITTVLAKTYGGAAR..........    45
EFAS1A   1  MGKEK......IHINIVVIGHVDSGKSTTTGHLIYKCGSIDKRTIEKFEKEAQEMGKGSF  54
EFHST    1  MGKEK......THINIVVIGHVDSGKSTTTGHLIYKCCGIDKRTIEKFEKEAAEMGKGSF  54
EFDMT1   1  MGKEK......IHINIVVIGHVDSGKSTTTGHLIYKCCGIDKRTIEKFEKEAQEMGKGSF  54
EFDMT2   1  MGKEK......IHINIVVIGHVDSGKSTTTGHLIYKCCGIDKRTIEKFEKEAQEMGKGSF  54
HOM  4      |||||      ||||||||||||||||||||||||||||  ||||||||||| |||||


            *    *   *****        *         * *** * *****   * *
HOM  3
EFECT   45  .AFDQIDNAPEEKARGITINTSHVEYDTPTRHYAHVDCPGHADYVVKNMITGAAQMDGAI  103
EFAS1A  55  KYAWVLDKLKAERERGITIDIALWKFETAKYYVTIIDAPGHRDFIKNMITGTSQADCAV   113
EFHST   55  KYAWVLDKLKAERERGITIDISLWKFETSKYYVTIIDAPGHRDFIKNMITGTSQADCAV   113
EFDMT1  55  KYAWVLDKLKAERERGITIDIALWKFETAKYYVTIIDAPGHRDFIKNMITGTSQADCAV   113
EFDMT2  55  KYAWVLDKLKAERERGITIDIALWKFETSKYYVTIIDAPGHRDFIKNMITGTSQADCAV   113
HOM  4      |||||||||||||||||| ||| |||| |||||||||||||||||||||||||||||


            **   *       *****  **     **    **  ** *
HOM  3
EFECT  104  LVVAATDGPM.......PQTREHILLGRQVGVPYIIVFLNKCD..MVDDEELLELVEME   153
EFAS1A 114  LIVAAGVGEFEAGISKNGQTREHALLAYTLGVKQLIVGVNKMDSTEPPFSEARFEEIKK  172
EFHST  114  LIVAAGVGEFEAGISKNGQTREHALLAYTLGVKQLIVGVNKMDSTEPPYSQKRYEEIVK  172
EFDMT1 114  QIDAAGTGEFEAGISKNDQTREHALLAFTLGVKQLIVGVNKMDSSEPPYSEARYEEIKK  172
EFDMT2 114  LIDAAGTGEFEAGISKNGQTREHALLAFTLGVKQLIVGVNKMDSTEPPYSEARYEEIKK  172
HOM  4      |||               |||  ||| ||    |||
```

```
HOM 3                       .         .        .     .
                                      **       **    *    *
EFECT  154  VREL.LSQYDFPGDD..T.PIVRGSALKALE............GDAEWEAKILELAGF  195
EFAS1A 173  EVSAYIKKIDYNPAAVAFVPISGWHGDNMLEASDRLPWYKGWNIERKEGKADGKTLLDA  231
EFHST  173  EVSTYIKKIGYNPDTVAFVPISGWNGDNMLEPSANMPWFKGWKVTRKDGNASGTTLLEA  231
EFDMT1 173  EVSSYIKKVGYNPAAVAFVPISGWHGDNMLEPSTNMPWFKGWEVGRKEGNADGKTLVDA  231
EFDMT2 173  EVSSYIKKIGYNPASVAFVPISGWHGDNMLEPSEKMPWFKGWSVERKEGNAEGKCLIDA  231
HOM 4        ||| |||||  |||   || |||||||| |||||||  |  ||  ||   ||    |


HOM 3         **    *   *  .   **   **  * ** *** ****  *  *  *
EFECT  196  LDS  198
EFAS1A 232  LDA  234
EFHST  232  LDC  234
EFDMT1 232  LDA  234
EFDMT2 232  LDA  234
HOM 4        ||
```

aFrom top: *Escherichia coli*, *Artemia salina*, *Homo sapiens*, two *Drosophila melanogaster* cytoplasmic proteins. An asterisk indicates conservation and a vertical dash (below) conservation among eukaryotes only.

bThe numbering in the EF-Tu sequence in the tables is based on gene sequencing, so the protein sequence numbers in the text are minus 1 compared with the tables.

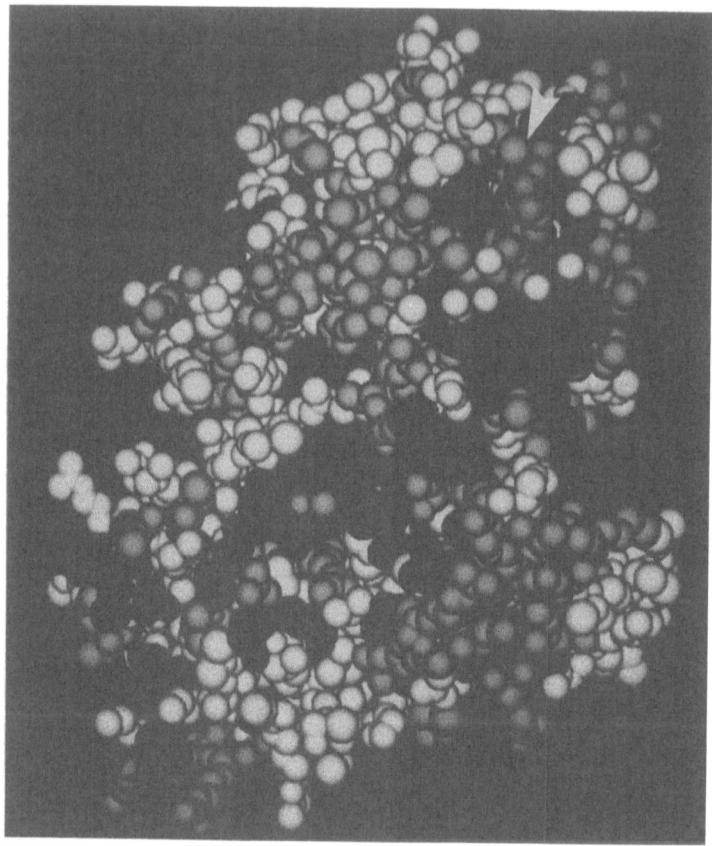

Figure 8. Conservation in three dimensions. The figure shows a space-filling representation of the elongation factor Tu-GDP complex. The shading represents conservation of amino acid residues in elongation factors from different organisms. The homology was determined by an alignment of nine different elongation factors, using the alignment editor *Alias* (S. Thirup and N. Larsen, personal communication). The EF-Tu and EF-1α sequencers are from *E. coli, Thermus thermophilus,* yeast mitochondria, *Euglena gracilis* chloroplast, *Methanococcus vannielii* (Lechner and Böck, 1987), yeast cytoplasm (Schirmaier and Philippsen, 1984), *Artemia salina, Homo sapiens, Mucor racemosus* (Linz *et al*. 1986; other references are already in the text). Residues found unchanged in all sequences are shown with dark gray shading. Residues conserved in prokaryotes only are shown in lighter gray. An arrowhead indicates the position of GDP.

case may be the F-protein of human and simian immunodeficiency viruses HIV and SIV (Guy *et al.,* 1987). These proteins are myristylated at the N terminus (compare *ras* p21 protein, which is palmitylated at the C terminus), and they bind and hydrolyze GTP with roughly the same avidity as v-Ha-*ras* protein. However, they do not possess the second or third consensus sequence, and instead of

Table III. Sequence Comparison of EF-Tu (from *E. coli*), Human EF-1α, and Ha-*ras* Protein p21 from Top to Bottom[a,b]

```
HOM_1
EFECT   1   MSKEKFERTKPHVNVGTIGHVDHGKTTLTAAITTVLAKTYGG.AARAFD..........   48
EFHST   1   MGKEK.....THINIVVIGHVDSGKSTTTGHLIYKCGGIDKR.TIEKFEKEAAEMGKGS   53
TVHUH   0   ........MTEYKLVVVGAGGVGKSALTIQLIQNHFVDEYDPTIED...........     38
HOM_2


HOM_1
EFECT  48   .....QIDNAPEEKARGITINTSHVEYDTPTRHYAHVDCPGHADYVKNMITGAAQMDGA  102
EFHST  54   FKYAWVLDKLKAERERGITIDISLWKFETSKYYVTIIDAPGHRDFIKNMITGTSQADCA  112
TVHUH  38   .........SYRKQVVIDGETCLL............DILDTAGQEEYSA.MRDQYMRTGEG   77
HOM_2


HOM_1
EFECT 103   ILVVAATDGPM.......PQTREHI.LLGRQVGVPYIIVFLNKCD.......MVDDEEL  146
EFHST 113   VLIVAAGVGEFEAGISKNGQTREHA.LLAYTLGVKQLIVGVNKMDSTEPPYSQKRYEEI  170
TVHUH  78   FLCVFAINNTK..SFEDIHQYREQIKRVKDSDDVPMVLVG.NKCD.......LAARTVE  126
HOM_2


HOM_1
EFECT 147   LELVEMEVREL.LSQYDFPGDDTPIVRGSALKALEGDAEWEAKILELAGFLDS        198
EFHST 171   VKEVSTYIKKIGYNPDTVAFVPISGWNGDNMLEPSANMPWFKGWKVTRKDGNA        223
TVHUH 127   SRQAQDLARSYGIPYIETSAKTRQGVEDAFYTLVREIRQHKLRKLNPPDESGP        179
HOM_2
```

[a] Rigorous homology is indicated by asterisks and conservations between EF-1α and Ha-*ras* by vertical lines.
[b] The numbering in the EF-Tu sequence in the tables is based on gene sequencing, so the protein sequence numbers in the text are minus 1 compared with the tables.

Table IV. Sequence Comparison of EF-Tu with the Signal G-Proteins G_I, Transducin, and G_S from Top to Bottom (for Sequence References, see Dever et al., 1987)[a,b]

```
HOM  1                                          *        *  * *
EFECT    0  .................MSKEKFERTKPHVNVGTIGHVDHGKTTLTAA             31
RGRAT    0  .........MGCTVSAEDKAAAERSKMIDKNLREDGEKAAREVKLLLGAGESGKSTIVKQ   52
RGBOAT   0  ......MGAGASAEEKHSRELEKKLKEDAEKDARTVKLLLLGAGESGKSTIVKQ         48
RGBOGA   1  MGCLGNSKTEDQRNEEKGQREANKKIEKQLQKDKQVYRATHRLLLGAGESGKSTIVKQ     59
HOM  2                                                    ||||||||||||


HOM  1                                  .
EFECT   32  ITTVLAKTYGGAARAFDQIDNAPEEKARGITINTSHVEYDT...............        72
RGRAT   53  MKIIHEDGYSEEECRQYRAVVYSNTIQSIMAIVKAMGNLQIDFADPQRADDARQLFALS    11-
RGBOAT  49  MKIIHQDGYSLEECLEFIAIIYGNTLQSILAIVRAMTTLNIQYGDSARQDDARKLMHMA    107
RGBOGA  60  MRILHVNGFNGEGGEEDPQAARSNSSGEKATKVQDIKNNLKEAIETIVAAMSNLVPPVE    118
HOM  2       |   |   |       |                        |


HOM  1                                                           .
EFECT   72  .................................PTRHYAHVDCPGHA                86
RGRAT  112  CAAEEQGML.............PEDLSGVIRRLWADHGVQACFGRSREYQLNDSAAY      155
RGBOAT 108  DTIEE.GTM...........PKEMSDIIQRLWKDSGIQACFDRASEYQLNDSAGY        150
RGBOGA 119  LANPENQFRVDYILSVMNVPDFDFPPEFYEHAKALWEDEGVRACYERSNEYQLIDCAQY    177
HOM  2       |   |                 |
```

```
                                             *
HOM  1
EFECT    87  DYVKNMITGAAQMDGAILVVAATDGPMPQTREHILLGRQV...............  126
RGRAT   156  YLNDLERIAQSDYIPTQQDVLRTRVKTTGIVETHFTFKDLHFKMFDVGGQRSERKKWIH  214
RGBOAT  151  YLSDLERLVTPGYVPTEQDVLRSRVKTTGIIETQFSFKDLNFRMFDVGGQRSERKKWIH  209
RGBOGA  178  FLDKIDVIKQDDYVPSDQDLLRCRVLTSGIFETKFQVDKVNFHMFDVGGQRDERRKWIQ  236
HOM  2


                                                            ****
HOM  1
EFECT   126  .................................................GVPYIIVFLNKCD  139
RGRAT   215  CFEGVTAIIFCVALSAYDLVLAEDEEMNRMHESMKLFDSICNNKWFTDTSIILFLNKKD  273
RGBOAT  210  CFEGVTCIIFIAALSAYDMVLVEDDEVNRMHESLHLFNSICNHRYFATTSIVLFLNKKD  268
RGBOGA  237  CFNDVTAIIFVVASSSYNMVIREDNQTNRLQEALNLFKSIWNNRWLRTISVILFLNKQD  295
HOM  2


HOM  1
EFECT   140  MVDDEELLELVEMEVRELLSQYDFPGDDTPIVRGSALKALEGDAEWEAKILELAGFLDS  198
RGRAT   274  LFEEKITQSPLTICFPEYTGANKYDEAASYIQSKFEDLNKRKDTKEIYTHFTCATDTKN  332
RGBOAT  269  VFSEKIKKAHLSICFPDYNGPNTYEDAGNYIKVQFLELNMRRDVKEIYSHMTCATDTQN  327
RGBOGA  296  LLAEKVLAGKSKIEDYFPEFARYTTPEDATPEPGEDPRVTRAKYFIRDEFLRISTASGD  354
HOM  2
```

[a] Rigorous homology is indicated by asterisks and conservations among the signal G-proteins by vertical lines.

[b] The numbering in the EF-Tu sequence in the tables is based on gene sequencing, so the protein sequence numbers in the text are minus 1 compared with the tables.

Table V. Comparison of Found and Predicted Secondary Structures for Several EF-Tu and EF-1α Proteins[a]

```
found    -----------EEEEEECCCCCCHHHHHHHHHHHHHHHHHHH----,---------,......
pred.1   HHHHHHHHH---EEEEEEEEEE----------EEEEEEE-----,------CCC.......
pred.2   HHHHHHH-----EEEEEEEEEE------------HHHHHHHHH---------H.......
pred.3   EEEHHHHHCCC-EEEEEEEEEE------HHHHHHH-HH,H-----HHHHHHHHH.......
pred.4   HHHHHHH--CC-EEEEEEEEEE------HHHHHH-HHH,--,................
pred.5   ....HHHHH----EEEEEE-------EEEEEEEEEE---EEEEEE-HHHHHHHHHHHHHHHH
pred.6   ....HHHHHHHH-EEEEEEEEE----EEEEEEEEE------HHHHHHHHHHHHHHHHH----
pred.7   ....HHHHHHHHEEEEEEEEEE-----EEEEEEEEE-----HHHHHHHHHHHHHHHHH-----
pred.8   ....HHHHHHHHEEEEEEEEEE-----EEEEEEEEEE------HHHHHHHHHHHHHHHHH---
pred.9   ....HHHHHHHH-EEEEEEEEEE----EEEEEEEEEE-------HHHHHHHHHHHHHHHH----


found    .......-----EEEEEEEEEEEEEEEETTTEEEEECCCCHHHHHHHHHCCCCCCEEEEEEEEEC
pred.1   .......HHHHHHHHH-EEE----------EEEE-EEETHHHHEEEE--HHHHHHHHEEEEE
pred.2   .......HHHHHHHH-EEE-HHHHHHHHHH----EEETHHHHEEEE--HHHHHHH-EEEEE
pred.3   .......HHHHHHHHEEEHHHHHHHHHH----EEETHHHHEEEE--HHHHHHH-EEEEE
pred.4   ............--EEEEEE-HHHHHHHH--EEEEETHHHHEEEEE-HHHHHHHHHHHHEEE
pred.5   HHHHHHHHHHHHHHHHHHHHHHHHHHHHHHEEEEEEEEETHHHEEEE---HHHHHHHHEEEEE
pred.6   HHHHHHHHHHHHHHHHH---HHHHHCCTTEEEEEEEE--HHHHEEEE-----HHHEEEEEEE
pred.7   HHHHHHHHHHHHHHHHH--HHHHHHHHEEEEEEEEE--HHHHEEEE---------EEEE--
pred.8   HHHHHHHHHHHHHHHH-EEEEEHHHHH-TTEEEEEEEE--HHHHEEEE---------EEEE--
pred.9   HHHHHHHHHHHHHHHHH---HHHHH-CTTTEEEEEEEE-HHHHEEEE-----HHHEEEEEEE


found    CCCCC.......CHHHHHHHHHHHHHCCCEEEEEECCCCCCCCCHHHHHHHHHHHHHHHHHHHCC
pred.1   EECCC.......CC----EEEEE--EEEEEEEEEE---HHHHHHHHHHHHHHHHHHHHH--
pred.2   EECCC.......CCHHHHH-------EEEEEEEEEE------HHHHHHHHHHHHHHHHHHHH--
pred.3   E----.......---HHHHHHHHHHHEEEEEEEEEEEE-----HHHHHHHHHHHHHHHHHHHH-
pred.4   EECCC.......CHHHHHHHHHHH--EEEEEEEEE----HHHHHHHHHHHHHHHHHHHH----
pred.5   HHHH.,---.,---EEEEEEEEEE---------------E..EEHHHHHHHHHHHHHHHE
pred.6   ECCCHHHHHHH----HHHHHHHHHHHHHHH-EEEE--HHHHH..HHHHHHHHHHHHH.HHHH
pred.7   -HHHHHHHHH---------HHHHHHHHEEEEEEEEEE-----CCCHHHHHHHHHHHH.HHHH
pred.8   -HHHHHHHHH---------HHHHHHHHEEEEEEEEEE-------TT--HHHHHHHHH.HHHH
pred.9   ECCCHHHHHHH----HHHHHHHHHHHHHEEE-----------,,-----------HH.HHHH


found    CCCCCCCEEEEECHHHHHH.........CCCCCHHHHHH...........HHHHHHHHHH
pred.1   --TTTEEEEEEEHHHHHH.........HHHHHHHHHH...........HHHHHH--CC
pred.2   --TTTEEEEE-HHHHHHHHHHHHCC--CC----HHHHHHH...........HHHHHHHEEE
pred.3   TTT--CEEEEEE---HHH.H.----H..HHHH.HHH.............HHHHHHEEEE
pred.4   --TTTEEEEEEE------HHHH----EEEHHHHHHHH-,...........---EEEEEEE
pred.5   EEEEE-CC--EEEEEEEEEHHHHHH-------,TT.TC............CCEEEEEEE-CC
pred.6   EEE-----EEEEEEEEE--CC-EEEEE--CC.--,.--HHHHHHHHHHHH--HHHHHHHHHH
pred.7   HEEEE----EEEEEEE---CCC--HHHHH--.TT.-----HHHHHHHHHHHH-HHHHEEEE
pred.8   EEEEE---EEEEEEEE---CC-------HHH.HH.HH---EEE----CCCEEE-HHHEEEE
pred.9   HHEEE-CCTEEEEEEEE--CC----------,--.HH---HHHHH-------HHHHHHEE
```

[a](Top line): EF-Tu from *E. coli*, as seen in the crystal structure. (Others, from top): Predicted structures for EF-Tu from *E. coli*, *T. thermophilus*, yeast mitochondrion, *E. gracilis* chloroplast, and *M. vannielii* and for EF-1α from yeast, brine shrimp, human, and fungus (*Mucor*). H = helix; E = extended structure; T = turn; C = coil; − = no clear assignment; . = no residue.

Table VI. Correlation of the GDP-Binding Loops in the X-Ray Structure of EF-Tu with the Consensus Sequences in G-Proteins

Consensus sequence (Merrick *et al.*)	Loop sequence (Aarhus)	Function
G X X X X G K	aA = G_{18} H V D H G K_{24}	Phosphate binding
D X X G	cB = D_{80} C P G_{83}	Phosphate binding
N K X D	eD = N_{135} K C D_{138}	G-selectivity
	fE = S_{173} A L K_{176}	G-binding

GxxxxGK they show G-GLEG(L/M/I), which in view of the universal conservation of GxxxxGK does not seem very impressive.

4. Homology and Functional Similarity

The striking resemblances and correlations observed among the G-proteins suggest at first that a very elegant structure–function relationship is emerging. When examined in more detail, however, the woods give way to a disunity of trees which, viewed individually or in small copses, do not seem to be growing in the orderly direction expected of them.

On the lowest level, molecular structures, we can cite the case already mentioned of *ras* p21 protein and EF-Tu: in the former, a G → V mutation reduces GTPase activity, while in the latter, at the corresponding position, a V → G mutation reduces GTPase activity. This means that even highly homologous protein structures do not necessarily obey a linear law of small perturbations: it would be more true to say that the wild-type structure is *right* and everything departing from this is to a greater or lesser extent *wrong*. Such a statement is more an expression of realism than of defeatism; it implies that we have not found the answer yet and had better go on searching.

On a higher level, the comparison of families, the details seem to fall into place. This applies in particular to the much closer resemblances within the prokaryotic and the eukaryotic elongation factors than between these families; the universality of EF-Tu/1α and its absolutely fundamental role in the cell will make it a good candidate, when more sequences are available, for serious phylogenetic studies embracing all life forms.

On the level of function, we are presented with serious problems in interpretation. It is fascinating to see the way in which the cell uses GTP as the energy source for its "timeswitches," such as EF-Tu, transducin, G_s and G_i proteins, tubulin, and perhaps IF2, EF-G, and others. In this way, the energy supply for these logical devices can be kept independent of, or at least buffered from, fluctuations in the supply of the principal purveyor of energy, ATP. However, it is frustrating to observe that tubulin, which is a timeswitch, does not fit into the sequence pattern of the others, while PEPCK and other enzymes with no switching function fit quite nicely. It is easy to understand the enzymes as having arisen and diverged from the other G-proteins, but the absence of Tubulin from the G-proteinaceous consensus club cannot be explained.

Perhaps there is a unifying principle after all. Evolution is concerned exclusively with the here and now, with making the children slightly better at surviving to reproduce than their parents were. Contrary to general belief, evolution has no memory; it retains what it retains either because the old forms are still well adapted to present function, or because there has been no need to overwrite them yet. Thus, while in a given cell, organism, or species the correlation of one kind of macromolecule's structure with its function is physically meaningful, cross-correlations between molecular or biological species will always smack of archeology. We may be lucky enough, one day, to stumble over some molecular shards that will give us a unifying key to the separate ways in which the diaspora of G-proteins has developed.

ACKNOWLEDGMENTS. We acknowledge Paul Woolley's many suggestions and innovative ideas regarding this manuscript, and we thank Troels la Cour and Jens Nyborg for discussions and use of structural data. We also thank S.-H. Kim and his co-workers for permission to use Figs. 5 and 7. The preparation of the manuscript was greatly expedited by the efficiency of Lisbeth Heilesen. In addition, we are grateful for financial support from the Danish Natural Science Research Council (an FTU project), the Danish Biotechnology Research Programme (via BIOREG), and the BAP Programme (DK58) of the European Economic Community.

References

Barbacid, M., 1987, *ras* gene, *Ann. Rev. Biochem.* **56:**779–827.

Bohr, H., Bohr, J., Brunak, S., Cotterill, R. M. J., Lautrup, B., Nørskov, L., Olesen, O., and Petersen, S. B., 1988, Protein secondary structure and homology by neural networks: the α-helices in rhodopsin, *FEBS Lett.* **241:**223–228.

Brands, J. H. G. M., Maassen, J. A., Van Hemert, F. J., Amons, R., and Möller, W., 1986, The primary structure of the α subunit of human elongation factor 1. Structural aspects of guanine-nucleotide-binding sites, *Eur. J. Biochem.* **155:**167–171.

Cenatiempo, Y., Deville, F., Dondon, J., Grunberg-Manago, M., Sacerdot, C., Hershey, J. W. B.,

Hansen, H. F., Petersen, H. U., Clark, B. F. C., Kjeldgaard, M., la Cour, T. F. M., Mortensen, K. K., and Nyborg, J., 1987, The protein synthesis initiation factor 2 G-domain. Study of a functionally active C-terminal 65-kilodalton fragment of IF2 from *Escherichia coli*, *Biochemistry* 26:5070–5076.

Chardin, P., and Tavitian, A., 1986, The *ral* gene: A new *ras* related gene isolated by the use of a synthetic probe, *EMBO J.* 5:2203–2208.

Clark, B., 1980, The elongation step of protein biosynthesis, *Trends Biochem. Sci.* 5:207–210.

Dever, T. E., Glynias, M. J., and Merrick, W. C., 1987, GTP-binding domain: Three consensus sequence elements with distinct spacing, *Proc. Natl. Acid Sci. USA* 84:1814–1818.

Dever, T. E. and Merrick, W. C., 1989, The GTP-binding domain revisited, in: *The Guanine-Nucleotide Binding Proteins. Common Structural and Functional Properties*, (L. Bosch, B. Kraal, and A. Parmeggiani, eds.), EMBO/NATO/CEC Adv. Res. Workshop, Renesse, The Netherlands, Aug. 1988, Plenum Press, New York, pp. 35–48.

de Vos, A. M., Tong, L., Milburn, M. V., Matias, P. M., Jancarik, J., Noguchi, S., Nishimura, S., Miura, K., Ohtsuka, E., and Kim, S-H., 1988, Three-dimensional structure of an oncogene protein: Catalytic domain of human c-H-*ras* p21, *Science* 239:888–891.

de Vos, A. M., Tong, L., Milburn, M. V., Matias, P. M. and Kim, S-H., 1989, Three-dimensional structure of *ras* p21 proteins, in: *The Guanine-Nucleotide Binding Proteins. Common Structural and Functional Properties*, (L. Bosch, B. Kraal and A. Parmeggiani, eds.) EMBO/NATO/CEC Adv. Res. Workshop, Renesse, The Netherlands, Aug. 1988 Plenum Press, New York.

Garnier, J., Osguthorpe, D. J., and Robson, B., 1978, Analysis of the accuracy and implications of simple methods for predicting the secondary structure of globular proteins, *J. Mol. Biol.* 120:97–120.

Gibbs, J. B., Sigal, I. S., and Scolnick, E. M., 1985, Biochemical properties of normal and oncogenic *ras* p21, *Trends Biochem. Sci.* 10:350–353.

Gilman, A. G., 1987, G proteins: Transducers of receptor-generated signals, *Annu. Rev. Biochem.* 56:615–649.

Guy, B., Kieny, M. P., Riviere, Y., Le Peuch, C., Dott, K., Girard, M., Montagnier, L., and Lecocq, J-P., 1987, HIV F/3' *orf* encodes a phosphorylated GTP-binding protein resembling an oncogene product, *Nature* 330:266–269.

Halliday, K. R., 1984, Regional homology in GTP-binding proto-oncogene products and elongation factors, *J. Cyclic Nucleotide Prot. Phosphoryl, Res.* 9:435–448.

Haubruck, H., Disela, C., Wagner, P., and Gallwitz, D., 1987, The *ras*-related *ypt* protein is a ubiquitous eukaryotic protein: Isolation and sequence analysis of mouse cDNA clones highly homologous to the yeast YPT1 gene, *EMBO J.* 6:4049–4053.

Hovemann, B., Richter, S., Walldorf, U., and Ciepluch, C., 1988, Two genes encode related cytoplasmic elongation factor 1α (EF-1α) in *Drosophila melanogaster* with continuous and stage specific expression, *Nucl. Acids Res.* 16:3175–3194.

Hwang, Y-W., and Miller, D. L., 1987, A mutation that alters the nucleotide specificity of elongation factor Tu, a GTP-regulatory protein, *J. Biol. Chem.* 262:13081–13085.

Jacquet, E., and Parmeggiani, A., 1988, Structure–function relationships in the GTP binding domain of EF-Tu: Mutation of Val20, the residue homologous to position 12 in p21, *EMBO J.* 7:2861–2867.

Jurnak, F., 1985, Structure of the GDP domain of EF-Tu and location of the amino acids homologous to *ras* oncogene proteins, *Science* 230:32–36.

Jurnak, F. A., 1988, The three-dimensional structure of c-H-*ras* p21: Implications for oncogene and G protein studies, *Trends Biochem. Sci.* 13:195–198.

Kushiro, A., Shimizu, M., and Tomita, K-I., 1987, Molecular cloning and sequence determination of the *tuf* gene coding for the elongation factor Tu of *Thermus thermophilus* HB8. *Eur. J. Biochem.* 170:93–98.

la Cour, T. F. M., Nyborg, J., Thirup, S., and Clark, B. F. C., 1985, Structural details of the binding of guanosine diphosphate to elongation factor Tu from *E. coli* as studied by X-ray crystallography, *EMBO J.* **4**:2385–2388.

Leberman, R., and Egner, U., 1984, Homologies in the primary structure of GTP-binding proteins: The nucleotide-binding site of EF-Tu and p21, *EMBO J.* **3**:339–341.

Lechner, K., and Böck, A., 1987, Cloning and nucleotide sequence of the gene for an archaebacterial protein synthesis elongation factor Tu, *Mol. Gen. Genet.* **208**:523–528.

Lenstra, J. A., Vliet, A. V., Arnberg, A. C., Van Hemert, F. J., and Möller, W., 1986, Genes coding for the elongation factor EF-1α in *Artemia, Eur. J. Biochem.* **155**:475–483.

Linz, J. E., Lira, L. M., and Sypherd, P. S., 1986, The primary structure and the functional domains of an elongation factor 1-α from *Mucor racemosus, J. Biol. Chem.* **261**:15022–15029.

Mandelkow, E-M., Hermann, M., and Rühl, U., 1985, Tubulin domains probed by limited proteolysis and subunit-specific antibodies, *J. Mol. Biol.* **185**:311–327.

March, P. E., and Inouye, M., 1985, GTP-binding membrane protein of *Escherichia coli* with sequence homology to initiation factor 2 and elongation factors Tu and G, *Proc. Natl. Acad. Sci. USA* **82**:7500–7504.

Masters, S. B., Stroud, R. M., and Bourne, H. R., 1986, Family of G protein α chains: Amphipathic analysis and predicted structure of functional domains, *Prot. Eng.* **1**:47–54.

McCormick, F., Clark, B. F. C., la Cour, T. F. M., Kjeldgaard, M., Nørskov-Lauritsen, L., and Nyborg, J., 1985, A model for the tertiary structure of p21, the product of the *ras* oncogene, *Science* **230**:78–82.

Montandon, P-E., and Stutz, E., 1983, Nucleotide sequence of a *Euglena gracilis* chloroplast genome region coding for the elongation factor Tu; evidence for a spliced mRNA, *Nucl. Acids Res.* **11**:5877–5892.

Möller, W., and Amons, R., 1985, Phosphate-binding sequences in nucleotide-binding proteins, *FEBS Lett.* **186**:1–7.

Nagata, S., Tsunetsugu-Yokota, Y., Naito, A., and Kaziro, Y., 1983, Molecular cloning and sequence determination of the nuclear gene coding for mitochondrial elongation factor Tu of *Saccharomyces cerevisiae, Proc. Natl. Acad. Sci. USA* **80**:6192–6196.

Nyborg, J., and la Cour, T. F. M., 1989, New structural data on elongation factor-Tu:GDP based on X-ray crystallography, in: *The Guanine-Nucleotide Binding Proteins. Common Structural and Functional Properties*, (L. Bosch, B. Kraal and A. Parmeggiani, eds.), EMBO/NATO/CEC Adv. Res. Workshop, Renesse, The Netherlands, Aug. 1988, Plenum Press, New York, pp. 3–14.

Parmeggiani, A., Swart, G. W. M., Mortensen, K. K., Jensen, M., Clark, B. F. C., Dente, L., and Cortese, R., 1987, Properties of a genetically engineered G domain of elongation factor Tu, *Proc. Natl. Acad. Sci. USA* **84**:3141–3145.

Rawis, R. L., 1987, G-proteins: Research unravels their role in cell communication, *Chem. Eng. News* Dec. 21:26–39.

Schirmaier, F., and Philippsen, P., 1984, Identification of two genes coding for the translation elongation factor EF-1α of *S. cerevisiae, EMBO J.* **3**:3311–3315.

Seidler, L., Peter, M., Meissner, F. and Sprinzl, M., 1987, Sequence and identification of the nucleotide binding site for the elongation factor Tu from *Thermus thermophilus* HB8, *Nucl. Acids Res.* **15**:9263–9277.

Toda, T., Uno, I., Ishikawa, T., Powers, S., Kataoka, T., Broek, D., Cameron, S., Broach, J., Matsumoto, K., and Wigler, M. 1985, In yeast, *ras* proteins are controlling elements of adenylate cyclase, *Cell* **40**:27–36.

V

MODELING AND STRUCTURE PREDICTION IN MACROMOLECULES

Knowledge-Based Protein Modeling and the Design of Novel Molecules

TOM L. BLUNDELL, MARK S. JOHNSON,
JOHN P. OVERINGTON, and ANDREJ ŠALI

1. Introduction

Each year protein crystallographers determine about 30 new three-dimensional protein structures, most of which are published in some form, and many of which are deposited in the Brookhaven Protein Databank (Bernstein *et al.*, 1977). Some of these are novel arrangements of a polypeptide chain, but others represent classes of structure that have been observed earlier in other proteins. Thus, each year we learn about an increasing proportion of the folds available to a polypeptide chain.

We are also, at the same time, gaining a better understanding of the principles of a polypeptide architecture. We now understand that proteins have evolved as hierarchical entities consisting of sequence, secondary structure, motif, domain, globular protomer, and oligomer levels (Richardson, 1981). We know that there are many lower-level structures that are compatible with each structure at a higher level. For example, many differing sequences can form an equivalent helix in a globular protein, and helices, strands, and turns of differing lengths and sequences can assemble as topologically similar motifs, such as nucleotide binding domains, alpha-helical bundles, or Greek keys (Rossmann and Argos, 1976; see Bajaj and Blundell, 1984, for a review).

TOM L. BLUNDELL, MARK S. JOHNSON, and JOHN P. OVERINGTON • Laboratory of Molecular Biology, Department of Crystallography, Birkbeck College, London University, London WC1E 7HX, England. *ANDREJ ŠALI* • J. Stefan Institute, Ljubljana, Yugoslavia; *present address:* Laboratory of Molecular Biology, Department of Crystallography, Birkbeck College, London University, London WC1E 7HX, England.

We can now begin to formulate a powerful set of rules that describe protein construction and architecture at many different levels, for example, the conformations of amino acid side chains at topologically equivalent positions in families of homologous proteins (Summers *et al.*, 1987; Sutcliffe *et al.*, 1987b; McGregor *et al.*, 1987) and the classes and key residues that characterize beta hairpins (Sibanda and Thornton, 1985; Milner-White and Poet, 1986; Efimov, 1986; Edwards *et al.*, 1987). We can consider using these rules, combined with other facts and hypotheses, to indicate which sequences adopt a particular higher-level structure or, in other words, to describe tertiary templates that define all sequences that are consistent with a particular motif, globular domain, or tertiary structure (Ponder and Richards, 1987). We seek to use these facts and rules in a knowledge-based procedure for the modeling and design of proteins.

Knowledge-based modeling can be envisaged as a process concerned with establishing and using rules to generate a model of a protein. One of the most powerful procedures in rules construction is the comparison of related structures, either through an alignment of sequences to identify conserved residues or through a superposition of three-dimensional structures to identify conserved conformations or motifs. Thus, the first step in a knowledge-based modeling procedure is systematic comparison of families of topologically similar structures. This step will lead to establishment of "equivalences" between the structures compared, and to their clustering based on measures of general similarity. The second step involves projection of the results of these comparisons of three-dimensional structures down onto the level of sequence. This step establishes rules relating sequence to structure. These can be expressed as consensus sequences—templates—for topologically equivalenced residues, or as key residues in canonical structures, which are then used to align the sequence of the protein of unknown tertiary structure with the known structures. The third step uses the rules established in the second step to generate a three-dimensional model. The three steps of knowledge-based modeling are shown diagrammatically in Figure 1.

The classical form of knowledge-based modeling is modeling by homology, or comparative modeling. This procedure depends on the knowledge that homologous sequences have similar tertiary structures involving a conserved "framework" of packed helices and strands connected by structurally variable regions that accommodate much of the sequence variation and almost all of the insertions and deletions. The method was first used by Browne *et al.* (1969) to model alpha-lactalbumin on the basis of the known three-dimensional structure of lysozyme. In subsequent years it was used to model insulinlike growth factors and relaxins from the three-dimensional structure of insulin (Bedarkar *et al.*, 1977; Isaacs *et al.*, 1978; Blundell *et al.*, 1978), serine proteinases on the basis of trypsin, chymotrypsin, and elastase (Greer, 1981), renin from the three-dimensional structures of aspartic proteinases (Blundell *et al.*, 1983; Sibanda *et*

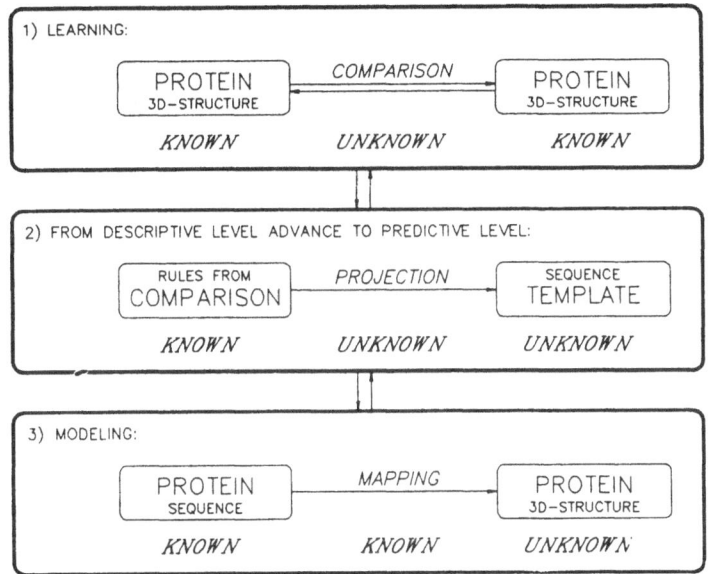

Figure 1. Scheme showing the basic steps of knowledge-based protein modeling. The first uses comparison methods to learn about the structural features of the protein family that contains the protein to be predicted. The second step uses this knowledge to derive the tertiary template and align it with the sequence of the unknown. The third, and last, step uses the constraints imposed on the sequence of the unknown by its alignment with the tertiary template and by general rules of protein structure to map the sequence onto its tertiary structure.

al., 1984), and many other structures. The method has recently been developed into a systematic approach (COMPOSER) in which several homologous structures can be used simultaneously in modeling the unknown (Sutcliffe *et al.*, 1987a,b; Blundell *et al.*, 1988). Rules are used to establish the relative positions of the framework (Sutcliffe *et al.*, 1987a), to select appropriate fragments for variable regions not only from homologous proteins (Greer, 1981; Chothia *et al.*, 1986), but also from other protein structures (Blundell *et al.*, 1988; Sutcliffe, 1988; Sibanda *et al.*, 1989), and to replace side chains (Sutcliffe *et al.*, 1987b; Summers *et al.*, 1987; McGregor *et al.*, 1987). In a parallel development, Jones and Thirup (1986) have shown that modeling into electron density during protein crystallography can also be aided by selecting conformational fragments from a series of other proteins of known three-dimensional structures.

Approaches, such as COMPOSER, that depend on rigid body superposition of three-dimensional structures are restricted to closely related motifs or homologous structures. Chothia and Lesk (1986) showed that for increasingly divergent structures, the number of topologically equivalent residues obtained by superposition decreases and the root mean square difference increases. This is due mainly to small, but cumulative relative translations and rotations of the packed secondary structural elements. This also affects the core residues (Hubbard and Blundell, 1987) and results in an insufficient framework for modeling. Clearly, a more flexible approach to defining topological equivalence is required.

The problem of defining topological equivalence was addressed more than a decade ago by Rossmann, Matthews, and their colleagues (see Matthews and Rossmann, 1985, for a review), who compared local main-chain direction, conformation, and position to establish topological equivalence. An alternative approach is to simplify the structure to a series of vectors representing the axes of the helices and strands, which are then compared (Murthy, 1984; Richards and Kundrot, 1988). In our approach, we compare properties and relations at each level in the hierarchy of protein structure and derive weight matrices from which the optimal alignment can be deduced using the dynamic programming approach of Needleman and Wunsch (1970). This approach works well for related structures that have little or no significant sequence identity (Šali and Blundell, 1990).

Once the topologically equivalent residues have been defined by comparing properties and relations, templates and key residues can be derived and used to align the sequence of the unknown. However, new approaches are required for the third step in the procedure, whereby a model is generated. The fact that properties and relations have been equivalenced indicates that internal coordinates should be used. In many ways the problem is closely related to that of reconstructing a model from upper and lower bounds on distances obtained from two-dimensional nuclear magnetic resonance (NMR) experiments. Thus, distance geometry (Crippen, 1977; Havel and Wuthrich, 1985) or dihedral angle optimization techniques (Braun and Go, 1985) can be adopted.

In this chapter we discuss the various approaches available in each step of the modeling procedure as they are being developed in our laboratory and then briefly describe applications to protein design.

2. Learning by Comparison of Structures

At Birkbeck, we have been concerned with two aspects of the comparison of structures. The first of these is designed to obtain a simultaneous rigid-body superposition of a family of protein structures. The second compares properties and relations.

2.1. Superposition of Structures

In order to gain as much information as possible from a family of protein structures, it is advantageous to obtain a multiple superposition that is unbiased by the order in which the proteins are compared. Thus, in the program MNYFIT (Sutcliffe *et al.*, 1987a), the alpha-carbon positions of a set of proteins are compared by least-squares fitting to an average structure or "framework." In this procedure one of the structures is chosen at random as the first approximation to the framework, and all other structures are fitted pairwise, using unit weights. A new framework is then chosen from the average of the fitted structures, using weights that depend on the estimated precision in the molecule (a function of the resolution of the X-ray analysis) and the distance of the atoms from the previous framework. An algorithm by McLachlan (1982) is used for the least-squares-fitting step. The procedure is continued iteratively until stability is obtained in both the number of equivalences and the root mean square distances over the topologically equivalent positions.

This procedure works efficiently for families of closely related protein structures such as the mammalian serine proteinases (Overington *et al.*, 1988), the vertebrate globins (Sutcliffe *et al.*, 1987a), and the constant domains of the immunoglobulins (Sutcliffe *et al.*, 1987a). The result of such a multiple super-position for the mammalian serine proteinases is shown in Fig. 2. With more divergent families the number of topological equivalent residues common to the complete set decreases rapidly, so that the framework is not useful as the basis for modeling.

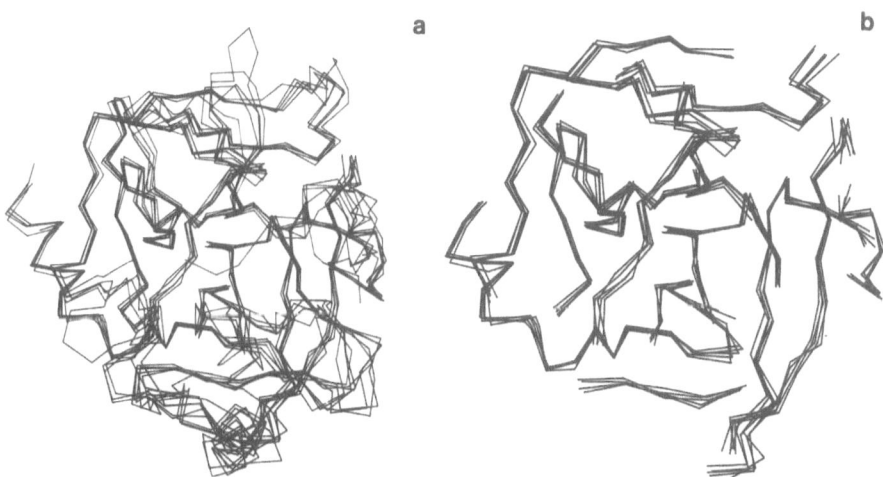

Figure 2. (a) Five mammalian serine proteinases (gamma-chymotrypsin, elastase, kallikrien, trypsin, and rat mast-cell protease type II) optimally superposed using the program MNYFIT; initial equivalences were the three residues in the catalytic triad. Coordinates were taken from the Brookhaven Protein Databank (Bernstein *et al.*, 1977); datasets used were 2GCH, 3EST, 2PKA, 2PTN, and 3RP2. (b) Regions determined to be in the structurally conserved core from the results of the previous fitting.

2.2. Comparison of Properties and Relations

In order to overcome the problems encountered in global rigid body super-position procedures, we compare proteins for features that indicate a common fold. These features may exist at any level in the hierarchy of protein structure— residue, secondary structure, supersecondary structure, motif, domain, or globular protomer. At each level, the features compared could be properties or relationships. Residue properties include sequence identity, hydrophobicity, size of side chain, and so on, as discussed by Argos (1987). When the three-dimen-sional structure is included in the comparison, the residue properties include local conformation, orientation of a side chain and main chain relative to the center of mass of the globular structure, accessibility, main chain dihedral an-gles, and so on, as shown in Table I. Equivalent properties at higher levels include the nature of the secondary structure element i, its accessibility, the orientation of the vector defining a helix or strand compared to the center of mass, and the local dihedral angle formed by secondary structure elements $i - 1$, i, $i + 1$. Comparisons of all such properties are included in a residue-by-residue

Table I. Protein Features that Can Be Used in Comparing Proteins[a]

Residues	Segments
Properties	Properties
Identity	Secondary structure type
Physical properties	Amphipathicity
Local conformation	Improper dihedral angle
Distance from gravity center	Distance from gravity center
Side-chain orientation	Orientation relative to gravity center
Main-chain orientation	Side-chain accessibility
Side-chain accessibility	Main-chain accessibility
Main-chain accessibility	Position in space
Position in space	Global orientation
Global direction in space	
Main-chain dihedral angles	
Relations	Relations
Hydrogen bond	Distances to one or more nearest
Distance to one or more	neighbors
nearest neighbors	Relative orientation of two or more
Disulfide bond	segments
Ionic bond	
Hydrophobic cluster	

[a]Various features are represented by rows; different levels of protein organization by columns. The table can be easily expanded to the right by adding features at higher levels of protein structure. The term "property" is used here for all protein features that imply comparison of only one element from each protein. Conversely, the term "relationship" is used for a feature that implies comparison of at least two elements from each protein.

weight matrix (Šali and Blundell, 1990). Optimal alignment is then obtained using the dynamic programming approach of Needleman and Wunsch (1970).

We also compare relations between different elements at each level of the hierarchy. At the residue level, the hydrogen bonding pattern represents a highly conserved feature of a protein fold. For example, the two lobes of aspartic proteinases, each of about 160 amino acid residues, appear to have evolved by gene duplication. Even though the sequence identity is not significant between the two lobes, about 43 residues are topologically equivalent by pairwise super-position. In fact, when the general arrangement of the beta-strands and their hydrogen bonding arrangements are compared, roughly twice as many residues appear to be equivalent (Blundell *et al.*, 1985). Hence, we include in our comparisons a consideration of hydrogen bond and local packing relationships. However, as such relationships affect more than one element in the sequence, this makes the dynamic programming approach computationally difficult. Instead, a simulated annealing technique is used to provide initial equivalences of relationships that are then directly introduced into the residue by residue weight matrix. Using this approach, 122 residues are found to be equivalent between the two lobes of the aspartic proteinase endothiapepsin. This method also works well for the globins and gives virtually the same set of topological equivalences—and therefore identical alignment—that has been obtained by careful analysis of the packing relationships in this family (Lesk and Chothia, 1980; Bashford *et al.*, 1988).

2.3. Clustering and Tree Construction

Phylogenetic relationships can be derived from distances that measure the dissimilarity between structures, and these can be obtained from any of the quantities used in the comparison procedures described earlier (Johnson *et al.*, 1990a,b). For multialigned structures, the root mean square distances for the framework regions can be used. This is not useful for divergent families, and we have calculated pairwise distance scores from a function that combines information from the root mean square distances and the number of topologically equivalent atoms obtained from pairwise superpositions (Johnson *et al.*, 1990a), as suggested by the work of Chothia and Lesk (1986) and Hubbard and Blundell (1987). The topologies and branch lengths of the phylogenetic trees were then constructed using a program KITSCH from the Phylogeny Inference Package (PHYLIP) written by Felsenstein (1985). Figure 3 shows an example of the optimal tree determined using global optimization and excluding negative branch lengths. Although this procedure uses no information from the sequence, it gives a tree that is generally topologically equivalent with the sequence phylogenies based on a residue type. Figure 3 compares the two types of trees; a more general discussion of trees for six homologous sets of proteins (immunoglobulin do-

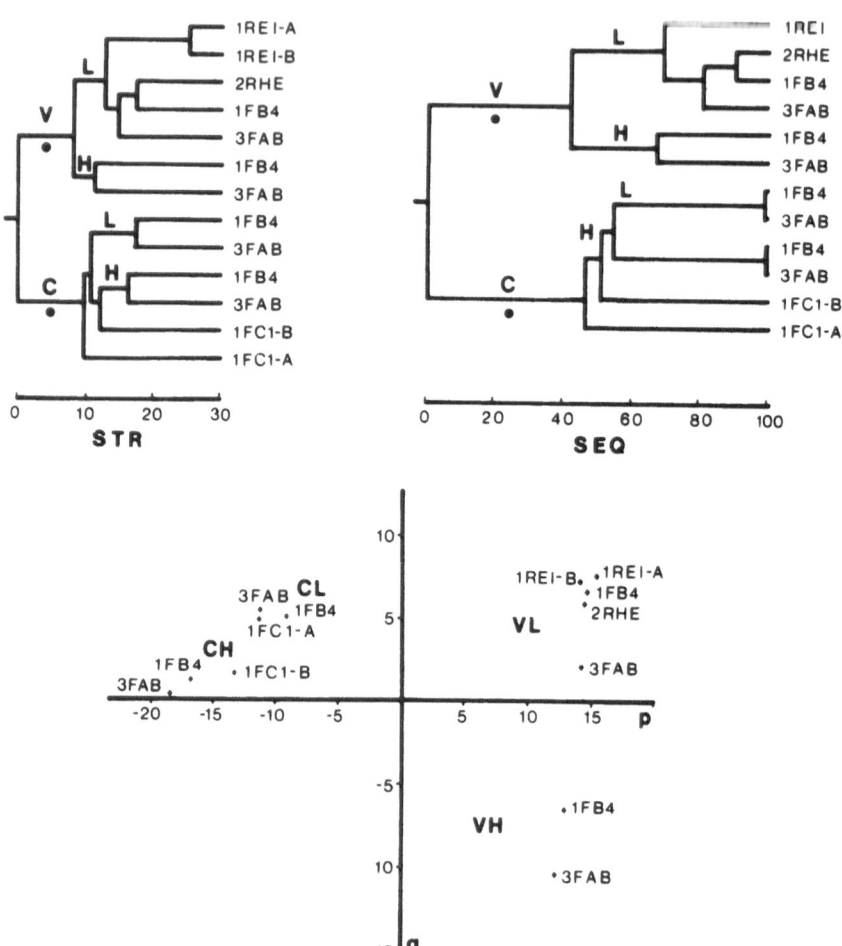

Figure 3. (Upper) Cladograms determined for various immunoglobulin fragments and derived from structural distances (STR) and sequence-based distances (SEQ). The separations between constant (C) and variable regions (V) of the light (L) and heavy (H) chains are indicated. Solid dots indicate branches whose lengths may change depending on the value of an outlier used to construct branches near the root. (Lower) Multidimensional scaling of the structural distances. The variable-domain/light-chain (VL), variable-heavy (VH), constant-light (CL), and constant-heavy (CH) clusters are indicated. The p and q axes are determined from the two largest eigenvalues and the corresponding eigenvectors of the cross-correlation matrix of distances derived from the structural data. This projection accounts for 72% of the total variance for this dataset. The fragments are designated by the following Brookhaven codes (Bernstein *et al.*, 1977): the chains of the human Bence–Jones dimer Rei—1REI-A and 1REI-B; the human Bence–Jones monomer Rhe—1RHE; the human Fab fragment Kol—1FB4; the human Fab fragment New—3FAB; the two domains of 1FC1—1FC1-A and 1FC1-B.

mains, globins, cytochromes c, serine proteinases, eye lens gamma-crystallins, and dinucleotide binding domains) is described in Johnson *et al.* (1990a). For more divergent families, distances can be obtained from any of the properties or relations used to compare proteins (Šali and Blundell, 1990). Trees based on different properties and relations appear to have very similar topologies (Johnson *et al.*, 1990b).

Multidimensional scaling (Crippen, 1977) can also be used to perform cluster analysis independently of the tree constructing algorithms. With this technique, columns of the cross-correlation matrix of the distances can be considered as points in a higher dimensional space. These coordinates completely describe the relationships among the data. A projection of these points to two dimensions is easily calculated such that the plane depicted is that which displays the maximum variance among the points with only a modest loss of information. Figure 3 also shows the multidimensional scaling analysis for the same set of structural distances between the immunoglobulin fragments that were used for the tree construction.

3. Rules for Tertiary Templates and Key Residues, and Their Comparison with a Sequence of an Unknown

We describe below how comparisons of three-dimensional protein structures can indicate features of sequence that are important for the adoption of a particular conformation. We discuss the derivation of templates that define variation of sequences consistent with a particular tertiary structure and show how these can be used to identify structures that may be useful in modeling.

3.1. Consensus Sequences for Framework Regions

The comparison of three-dimensional structures by multiple superposition described in Section 2.1. leads to a framework, which defines the topologically equivalent residues that are conserved in a family or subfamily of proteins (Sutcliffe *et al.*, 1987a). The consensus sequences for the frameworks can be used to align the sequence of an unknown tertiary structure using a procedure such as that of Taylor (1986). The consensus sequences emphasize those features of the framework on which there are major three-dimensional structural constraints due to conformation or packing. However, by definition they ignore the regions outside the framework which are variable but which can nevertheless contain useful information concerning the extent and the positions of the framework. A comparison of tertiary structures using properties and relations (Section 2.2) can provide an extension of the region of topological equivalence used to derive the consensus sequence template.

3.2. Key Residues from Comparisons of Properties and Relations

The comparison of structures, particularly through comparison of properties and relations, can give clues to the features that are critical for the adoption of a particular three-dimensional structure. For example, we may identify glycine, aspartic acid, and asparagine residues that have positive PHI angles that are not easily adopted by other residues. Alternatively, we may find residues that are conserved, such as hydrophobic aromatic or aliphatic carbon side chains, because they are buried without hydrogen bonding partners. For each position in the alignment, a score that indicates the conservation of properties and relations in the set compared can be used to weight the positions during alignment with the unknown. This provides a general and automatic procedure for identifying key residues in variable regions of homologous proteins, a problem first addressed by Chothia and Lesk (1986) for the immunoglobulin variable regions.

3.3. Characteristic Sequences from Features of Tertiary Structure

In many cases there will be only one experimentally defined three-dimensional structure for a particular class of motif, domain, or globular protein. The prediction of sequences that are compatible with this structure—the production of a tertiary template—is central to our understanding of protein diversity and evolution.

Ponder and Richards (1987) have adopted a systematic approach in which a library of amino acid side-chain rotomers is generated and used for testing for combinations that are consistent with the tertiary structure given a variation in the main chain of 0.5 Å. This is a useful approach, but it is limited by the fact that the diversity in real proteins of sequence identity less than 50% involves relative shifts of main-chain residues much greater than 0.5 Å (Clothia and Lesk, 1986). However, identification of the possible sequence variation using this method with an even greater variation of main-chain positions soon becomes computationally infeasible.

An alternative approach is to learn from comparisons of many families of proteins, which may reveal general features of tertiary structure that impose strong constraints on the variation available to the amino acid sequence. This has already been done for many protein families, and some generalizations can be made. For example, it is known that glycines are most conserved in evolution (Bajaj and Blundell, 1984; Blundell et al., 1986) where they adopt positive PHI angles or where packing restrictions are incompatible with the existence of a side chain. It is also well established that residues that are inaccessible to solvent are less variable in evolution. This has been quantified by Hubbard and Blundell (1987), who compared the percentage identity and root mean square distances in several protein families, not only for all topologically equivalent residues, but also for those with side chains inaccessible to solvent. Buried polar, hydrogen-bonded residues tend to be even more strongly conserved in highly divergent families (Bajaj and Blundell, 1984; Blundell *et al.*, 1986). Thus, in the cellular and retroviral aspartic proteinases, which have very low sequence identities, only a buried, hydrogen-bonded threonine is invariant (Pearl and Blundell, 1984) apart from the catalytic aspartates and two glycines. Similarly, in the Greek key motifs of the eye lens crystallins only a buried hydrogen-bonded serine is conserved apart from an invariant glycine. Such broad comparisons can assist in formulating rules that restrict the number of sequences that we need consider for a particular structure.

We use such rules in several ways in selecting structures for modeling. We have used them to construct tertiary templates for a globular domain, for example, in the identification of the crystallin Greek key motif in a bacterial surface protein (Wistow *et al.*, 1985). We are also using them to identify key residues in structurally variable regions, in the following manner. First we select a series of fragments from the data base of protein structures that are geometrically compatible with the framework. These are then clustered, and for each cluster—often containing only one example—key residues are selected on the basis of low solvent accessibility, strong hydrogen bonding, unusual torsion angles, and so forth. These approaches to identifying key residues and their use in selecting conformers in variable regions are being encoded in COMPOSER (Sutcliffe, 1988; Blundell *et al.*, 1988).

4. Generation of a Model from a Sequence of an Unknown Using Rules from Comparison of Structures

In the previous sections we have shown that comparing protein three-dimensional structures can define topological equivalence and phylogenies for members of homologous or analogous families. The equivalenced structures provide a basis for deriving rules for the selection and alignment of sequences that adopt the family fold. We will now consider the construction of the protein model using the knowledge derived from these sequence and structural comparisons.

4.1. Construction of the Model by Assembly of Rigid Groups in Three Dimensions

We first consider the construction of a model by a superposition of rigid three-dimensional structures. We assume initially that there are several homologous or analogous structures from which a framework can be generated. The first stage is to select the structures that will be most useful in the modeling exercise.

We have developed an automatic procedure for this which depends on combining the phylogenetic tree from sequence—including that of the unknown structure—with that based on the three-dimensional structures alone (Johnson *et al.*, 1990a,b). This enables selection of the set of structures that are clustered around or near the sequence of the unknown. These structures are then used to produce a framework for the unknown (see Fig. 2b) in which contribution of each structure is weighted according to its percentage sequence identity to the unknown (Sutcliffe *et al.*, 1987a). The framework so derived is an average structure, and it is endowed with real geometry by least-squares fitting the fragments from the homologous structures for each section of the framework.

The next task is to select the structurally variable regions. This is achieved by first using a geometrical filter in a similar way to that of Jones and Thirup (1986), with a three-residue overlap on each end of the fragment. Each fragment selected is then least-squares-fitted to the framework, and the fragments so fitted are clustered using multidimensional scaling analysis or tree construction (see Section 2.3). Key residues are identified using procedures discussed in Section 3.3, and the fragments are ranked. The top-ranking fragment is then tested for overlap with other parts of the model structure. If it is rejected on these grounds, the next-ranking fragment is selected. The optimal fragment is then melded onto the framework.

Alternative procedures may, in fact, need to be adopted. First, it may be best to extend the framework using differing subsets of homologous structures at each variable region. Second, the rules developed by Thornton and her colleagues (Sibanda and Thornton, 1985; Edwards *et al.*, 1987; Wilmot and Thornton, 1988; Sibanda *et al.*, 1989) may be used to select a loop not recognized by the key residues procedure. Third, regions within the structurally conserved core of the known set may require insertions or deletions. In certain cases in which the equivalent fragments are of the same length, but the key residues are changed, an alternative conformer may be required. In these cases, a further definition of the region to be replaced is required. In general, insertions, deletions, and main-chain distortions are made locally and, where possible, in regions of irregular secondary structure.

The third step is to replace side chains. This is achieved using a set of rules derived from an analysis of sequence variation at topologically equivalent positions in homologous families (Sutcliffe *et al.*, 1987b). The 1200 rules include one for each of the 20 by 20 amino acid replacements in each of alpha-helical, beta-sheet, and irregular regions. When there is no useful prediction, the most probable conformation is chosen, and when there is more than one prediction, the conformation closest to the median of the predictions is chosen.

This procedure for modeling is very successful when the known structures cluster around that to be predicted and when the percentage sequence identity is high (>40%). For example, in a model building of a porcine kallikrein from four other structurally known mammalian serine proteinases, the root mean square difference between the model and the real structure is 0.64 Å for the 150 residues defined to be in the framework by spatial superposition. Where the structure to be predicted lies outside the cluster and the sequence identity is <40%, a procedure is required to introduce translations and rotations of the elements of the framework relative to each other. For this, we are currently exploring algorithms that relate distances between elements of secondary structure to the volumes of the side chains within contact regions (Lesk and Chothia, 1980; J. Overington, unpublished results). For closely related structures, the structurally variable regions tend to be small and the predictions are reasonably good, but for divergent structures, long insertions can lead to poorly modeled conformations. Similarly, 80% of the side-chain conformations are correctly predicted for closely homologous structures, but this decreases for structures that are less similar.

4.2. Construction of a Model Using Optimization Techniques

An alternative approach for modeling divergent protein structures is to adopt distance geometry or optimization techniques such as those used for derivation of protein structures from the two-dimensional NMR data (Braun and Go, 1985; Havel and Wuthrich, 1985). First, protein structures and fragments that are homologous or analogous to the sequence of the unknown are selected using the methodology of Johnson *et al.* (1990a). Second, these known structures and fragments are aligned using the comparison method that takes properties and relationships into account (Šali and Blundell, 1990). Third, this alignment is used to derive the tertiary template, and the tertiary template is then aligned with the sequence of the unknown. From the alignment of the unknown with the tertiary template and from the general rules of protein structure, the set of constraints on the structure of the unknown is obtained. For example, hydrogen bonds, secondary structure, solvent accessibilities, and so on, which are conserved in the alignment of the known structures, constrain the degrees of freedom available to the structure of the unknown. Finally, these constraints are used as the input to the optimization program (Šali and Blundell, unpublished results), similar to the variable target function minimization procedure of Braun and Go (1985), which calculates the structure of the unknown, minimizing the violations of the constraints.

4.3. Refinement of the Model

All knowledge-based procedures require simulation of the solvent, energy minimization, and molecular dynamics simulations to optimize the structure and to provide a useful model of the time-and-space-averaged structures determined by X-ray analysis and two-dimensional NMR.

5. Applications to Design of Novel Molecules

Knowledge-based modeling has applications both to receptor-based drug design and to protein engineering (Fig. 4).

5.1. Receptor-Based Drug Design

Although the sequences of many receptors have recently been determined, the three-dimensional structures are known for very few of pharmaceutical interest. In some cases the structure of the receptor from another species or an orthologous protein structure may have been determined by X-ray analysis. For example, when the sequence of human renin was determined in 1984, no three-dimensional structures of renins had been determined. This has taken 4 years; only the structures for homologous fungal aspartic proteinases were accurately analysed by X-ray analysis. Modeling by homology produced rough models for mouse renin (Blundell *et al.*, 1983) and for human renin (Sibanda *et al.*, 1984). These have been extended using experimentally determined structures of aspartic proteinases complexed with human renin inhibitors to give a model of the human renin—human angiotensinogen (fragment) transition state complex (Foundling *et al.*, 1987; Blundell *et al.*, 1987). These models have been used by several pharmaceutical companies as a receptor-based contribution to their design of orally active renin inhibitors for the treatment of hypertension. The model is probably accurate to approximately 0.5 Å (comparison of alpha-carbons) close to the active site, but will have errors in excess of 1.5 Å in the peripheral loops.

5.2. Site-Directed Mutagenesis

Protein engineering using site-directed mutagenesis involves the introduction of insertions, deletions, and replacements in a protein with retention of the three-dimensional structure but modification of catalytic activity, stability to high-temperature or nonaqueous solvents, or other properties in a predictable fashion. The knowledge-based procedures developed for modeling local insertions and deletions and side-chain replacements (Section 4.1) provide a useful starting point, although energy minimization and molecular dynamics procedures in a simulated aqueous environment will be needed to explore local conformations.

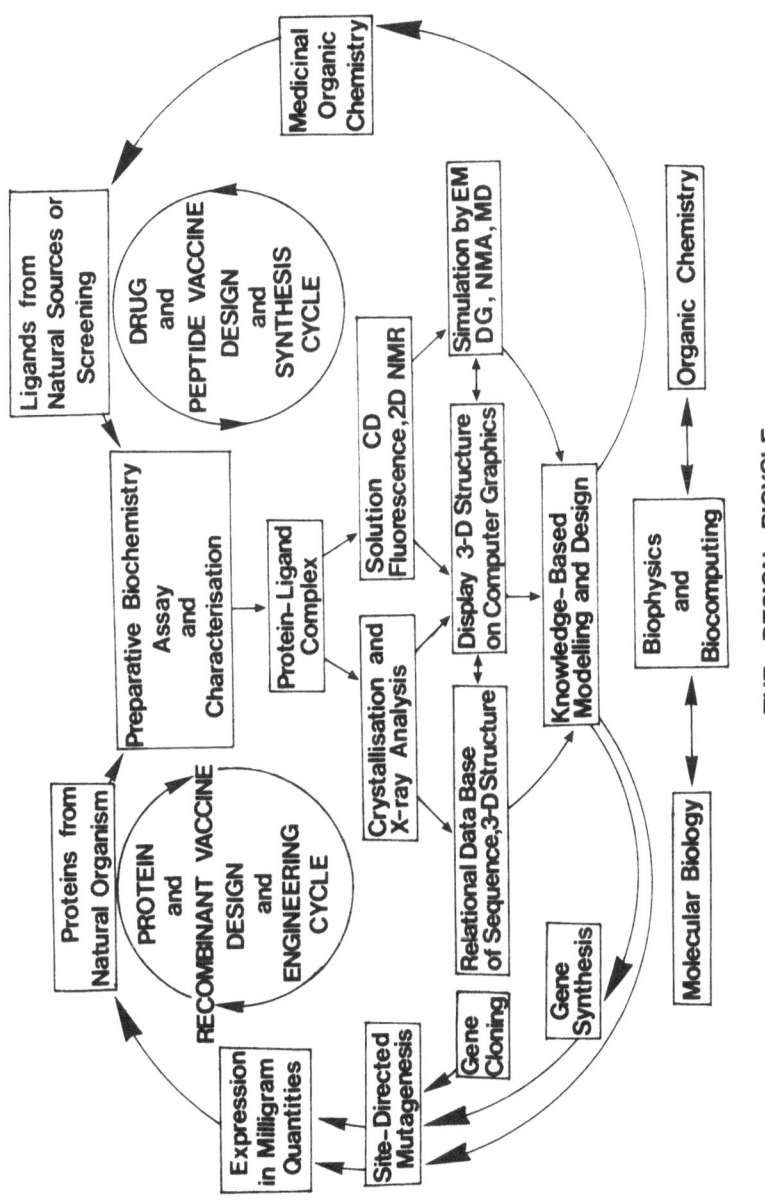

Figure 4. Bi-cycle illustrating the common structural analysis and design steps in the engineering of proteins, vaccines, and drugs.

5.3. Chimeric Molecules

Let us consider the design of a chimeric molecule that comprises a tissue plasminogen activator (tPA) serine proteinase domain linked to the COOH-terminus of an Fab fragment of a monoclonal antibody with fibrin specificity. The modeling procedures are first used to model the serine proteinase and the immunoglobulin domain on the basis of homologous structures (Blundell *et al.*, 1988; Harris, 1987; Overington *et al.*, 1988). Second, the relative disposition of the two fragments is chosen interactively, using computer graphics. A linker region with overlaps in both tPA and Fab domains is then selected from the data base of polypeptide fragments in which the linking residues are small and hydrophilic if the link needs to be flexible. Finally, the contiguous surfaces of the two linked domains are mutated using the side-chain replacement algorithm so that they are compatible with each other and the solvent.

5.4. Ab Initio Protein Design

Analogous approaches can be used in designing novel proteins. Although this area is presently in its infancy, attempts have been made to design alpha-helical bundles, beta-barrels, and other commonly occurring canonical structures. We have used knowledge-based techniques, including COMPOSER, to design a symmetrical, two-Greek-key protein called CRYSTANOVA (based on the stable eye lens crystallins) that is designed to bind copper in a similar way to superoxide dismutase (Hubbard, 1988). Although such projects are currently academic, in the future protein engineers may be requested to design proteins, for example, for rare metal ion scavenging or even biochips where no useful parallel is known to exist in Nature.

ACKNOWLEDGMENTS. We are grateful to all our colleagues in the modeling group for very useful discussion. We thank the American Cancer Society for a fellowship (M.S.J.). J.P.O. is supported by a SERC and CASE studentship with Pfizer. A.S. is funded by the ORS Award Scheme and the Research Council of Slovenia.

References

Argos, P., 1987, A sensitive procedure to compare amino acid sequences, *J. Mol. Biol.* **193**:385–396.

Bajaj, M., and Blundell, T. L., 1984, Evolution and the tertiary structure of proteins, *Annu. Rev. Biophys. Bioeng.* **13**:453–492.

Bashford, D., Chothia, C., and Lesk, A. M., 1987, Determinants of a protein fold: unique features of the globin amino acid sequence, **196**:199–216.

Bedarkar, S., Turnell, W. G., Blundell, T. L., and Schwabe, C., 1977, Relaxin has conformational homology with insulin, *Nature* 270:449–451.

Bernstein, F. C., Koetzle, T. F., Williams, G. J. B., Meyer, E. F., Brice, M. D., Rodgers, J. R., Kennard, O., Shimanovichi, T., and Tasumi, M., 1977, The Protein Data Bank: A computer-based archival file for macromolecular structures, *J. Mol. Biol.* 112:535–542.

Blundell, T. L., Bedarker, S., Rindernecht, E., and Humbel, R. E., 1978, Insulin-like growth factor: A model for tertiary structure accounting for immunoreactivity and receptor binding, *Proc. Natl. Acad. Sci. USA* 75:180–184.

Blundell, T. L., Sibanda, B. L., and Pearl, L., 1983, Three dimensional structure, specificity and catalytic mechanism of renin, *Nature* 304:237–275.

Blundell, T. L., Jenkins, J., Pearl, L., and Sewell, T., 1985, The high resolution structure of endothiapepsin, in: *Aspartic Proteinases and Their Inhibitors* (V. Kostka, ed.), pp. 151–161, Walter de Gruyter, Berlin.

Blundell, T. L., Barlow, D., Sibanda, B. L., Thornton, J. M., Taylor, W., Tickle, I. J., Sternberg, M. J. E., Pitts, J. E., Haneef, I., and Hemmings, A. M., 1986, Three-dimensional structural aspects of the design of new protein molecules, *Phil. Trans. Roy. Soc. Lond.* 317:333–344.

Blundell, T. L., Cooper, J., and Foundling, S. I., 1987, On the rational design of renin inhibitors: X-ray studies of aspartic proteinases complexed with transition-state analogues, *Biochemistry* 26:5585–5590.

Blundell, T. L., Carney, D., Gardner, S., Hayes, F., Howlin, B., Hubbard, T., Overington, J., Singh, D. A., Sibanda, B. L., and Sutcliffe, M., 1988, Knowledge-based protein modelling and design, *Eur. J. Biochem.* 172:513–520.

Braun, W., and Go, N., 1985, Calculation of protein conformations by proton–proton distance constraints: A new efficient algorithm, *J. Mol. Biol.* 186:611–626.

Browne, W. J., North, A. C. T., Phillips, D. C., Brew, K., Vanaman, T. C., and Hill, R. L., 1969, A possible three-dimensional structure of bovine alpha-lactalbumin based on that of hen's egg white lysozyme, *J. Mol. Biol.* 42:65–86.

Chothia, C., and Lesk, A. M., 1986, The relation between the divergence of sequence and structure in proteins, *EMBO J.* 5:823–826.

Chothia, C., Lesk, A. M., Levitt, M., Amit, A. G., Mariuzza, R. A., Phillips, S. E. V., and Poljak, R. J., 1986, The predicted structure of immunoglobulin D1.3 and its comparison with the crystal structure, *Science* 233:755–758.

Crippen, G. M., 1977, A novel approach to calculation of conformation: Distance geometry, *J. Comp. Physiol.* 24:96–107.

Edwards, M. S., Sternberg, M. J. E., and Thornton, J. M., 1987, Structural and sequence patterns in the loops of beta-alpha-beta units, *Prot. Eng.* 1:173–181.

Efimov, A. V., 1986, The conformation of beta-turns, *Mol. Biol. (Moscow)* 20:2250–2260 (in Russian).

Felsenstein, J., 1985, Confidence limits on phylogenies: an approach using the bootstrap, *Evolution* 39:783–791.

Foundling, S. I., Cooper, S. I., Watson, J., Cleasby, F. E., Pearl, L. H., Sibanda, B. L., Hemmings, A., Wood, S. P., Blundell, T. L., Valler, T. L., Norey, Kay, J., Boger, J., Dunn, B. M., Leckie, B. J., Jones, D. M., Atrash, B., Hallett, A., and Szelke, M., 1987, High resolution X-ray analyses of renin inhibitor–aspartic proteinase complexes, *Nature* 327:349–352.

Greer, J., 1981, Comparative model-building of the mammalian serine-proteases, *J. Mol. Biol.* 153:1027–1042.

Harris, T. J. R., 1987, Second-generation plasminogen activators, *Prot. Eng.* 1:449–450.

Havel, T., and Wuthrich, K., 1985, An evaluation of the combined use of NMR and distance geometry for the determination of protein conformations in solution, *J. Mol. Biol.* 182:281–294.

Hubbard, T. J. P., 1988, The design, expression and characterization of a novel protein, Ph.D. thesis, University of London.

Hubbard, T. J. P., and Blundell, T. L., 1987, Comparison of solvent inaccessible cores of homologous proteins: Definitions useful for protein modelling, *Prot. Eng.* **1**:159–171.

Isaacs, N., James, R., Niall, H., Bryant-Green, G., Wood, S., Dodson, G. G., Evans, A., and North, A. C. T., 1978, Relaxin and its structural relationship to insulin, *Nature* **271**:278–281.

Johnson, M. S., Sutcliffe, M. J., and Blundell, T. L., 1990a, Molecular anatomy: Phyletic relationships derived from three-dimensional structures of proteins, *J. Mol. Evol.* **30**:43–59.

Johnson, M. S., Šali, A., and Blundell, T. L., 1990b, Phylogenetic relationships from three-dimensional protein structures, *Meth. Enzymol.* **183**:670–690.

Jones, T. H., and Thirup, S., 1986, Using known substructures in protein model building and crystallography, *EMBO J.* **5**:819–822.

Lesk, A. M., and Chothia, C., 1980, How different amino acid sequences determine similar protein structures: The structure and evolutionary dynamics of the globins, *J. Mol. Biol.* **136**:225–270.

Matthews, B. W., and Rossmann, M. G., 1985, Comparison of protein structures, *Meth. Enzymol.* **115**:397–420.

McGregor, M. J., Islam, S. A., and Sternberg, M. J. E., 1987, Analysis of the relationship between side-chain conformation and secondary structure in globular proteins, *J. Mol. Biol.* **198**:295–310.

McLachlan, A. D., 1982, Rapid comparison of protein structures, *Acta Cryst.* **A38**:871–873.

Milner-White, J., and Poet, R., 1986, Four classes of beta-hairpins in proteins, *Biochem. J.* **240**:289–292.

Murthy, M. R. N., 1984, A fast method of comparing protein structures, *FEBS Lett.* **168**:97–102.

Needleman, S. B., and Wunsch, C. D., 1970, A general method applicable to the search for similarities in the amino acid sequence of two proteins, *J. Mol. Biol.* **48**:443–453.

Overington, J. P., Sutcliffe, M. J., Watson, F., James, K., Campbell, S., and Blundell, T. L., 1988, The knowledge-based modelling of tissue-type plasminogen activator and its interactions with its inhibitor endothelial cell plasminogen activator inhibitor, *Proceedings of the 8th International Biotechnology Symposium*, Vol. 1, Paris, pp. 279–304.

Pearl, L., and Blundell, T. L., 1984, The active site of aspartic proteinases, *FEBS Lett.* **174**:96–101.

Ponder, J. W., and Richards, F. M., 1987, Tertiary templates for proteins: Use of packing criteria in the enumeration of allowed sequences for different structural classes, *J. Mol. Biol.* **193**:775–791.

Richards, F. M., and Kundrot, C. E., 1988, Identification of structural motifs from protein coordinate data: Secondary structure and first-level supersecondary structure, *Proteins* **3**:71–84.

Richardson, J. S., 1981, The anatomy and taxonomy of protein structure, *Adv. Prot. Chem.* **64**:167–339.

Rossmann, G. M., and Argos, P., 1976, Exploring structural homology of proteins, *J. Mol. Biol.* **105**:75–95.

Šali, A., and Blundell, T. L., 1990, The definition of general topological equivalence in protein structures, *J. Mol. Biol.* **212**:403–428.

Sibanda, B. L., and Thornton, J. M., 1985, Beta-hairpin families in globular proteins, *Nature* **316**:170–316.

Sibanda, B. L., Blundell, T. L., Hobart, P. M., Fogliano, M., Bindra, J. S., Dominiy, B. W., and Chirgwin, J. M., 1984, Computer graphics modelling of human renin: Specificity, catalytic activity and intron–exon junctions, *FEBS Lett.* **174**:102–111.

Sibanda, B. L., Blundell, T. L., and Thornton, J. M., 1989, The conformation of beta-hairpins in protein structures: A systematic classification with applications to modelling by homology, electron density fitting and protein engineering, *J. Mol. Biol.* **206**:759–777.

Summers, N. L., Carson, W. D., and Karplus, M., 1987, Analysis of side chain orientations in homologous proteins, *J. Mol. Biol.* **196**:175–198.

Sutcliffe, M. J., 1988, An automated approach to the systematic model building of homologous proteins, Ph.D. thesis, University of London.

Sutcliffe, M. J., Haneef, I., Carney, D., and Blundell, T. L., 1987a, Knowledge-based modelling of homologous proteins; Part I: Three-dimensional frameworks derived from simultaneous superposition of multiple structures, *Prot. Eng.* **1**:377–384.

Sutcliffe, M. J., Hayes, F. R. F., and Blundell, T. L., 1987b, Knowledge based modeling of homologous proteins; Part II: Rules for the conformation of substituted sidechains, *Prot. Eng.* **1**:385–392.

Taylor, W. R., 1986, Identification of protein sequence homology by consensus template alignment, *J. Mol. Biol.* **188**:233–258.

Wilmot, C. M., and Thornton, J. M., 1988, Analysis and prediction of the different types of beta-turn in proteins, *J. Mol. Biol.* **203**:221–232.

Wistow, G., Summers, L., and Blundell, T. L., 1985, *Myxococcus xanthus* spore coat protein S may have a similar structure to vertebrate lens beta/gamma crystallins, *Nature* **316**:771–773.

Computer Simulation Methods as Applied to Site-Specific Mutations

PETER KOLLMAN

1. Introduction

In parallel with the revolution in molecular biology, which has enabled the production and characterization of enzymes and their site-specific mutants in large quantities, there has been a revolution in computer technology. This has led to powerful new tools for computer graphics visualization of complex molecules, data base analysis, artificial intelligence approaches to analyze the structural principles in complex molecular systems, and numerical simulations of the properties of complex molecules. This chapter focuses on numerical simulation methods, which include quantum mechanical solutions of molecular energy and molecular mechanical and dynamical simulation.

It is clear that one cannot currently predict a protein structure from the amino acid sequence: there are too many local energy minima that must be explored in the conformational study of a molecule with the large number of degrees of freedom of a protein. Not only is it difficult to calculate the energy for each of the reasonable candidate structures, but the accuracy of the energy calculation is too imprecise to correctly assign the lowest energy structures. The approaches that have led to the most progress in predicting the three-dimensional protein structure include combining secondary structure predictions with the docking of secondary structure elements to predict tertiary structure (Cohen and Kuntz, 1987). If such approaches ever become accurate enough to suggest a nativelike structure among a small ($\sim$10) number of candidate structures, energy

PETER KOLLMAN • Department of Pharmaceutical Chemistry, School of Pharmacy, University of California at San Francisco, San Francisco, California 94143.

calculations will, hopefully, be able by then to accurately calculate the correct lowest free-energy structure. We are still a considerable distance from this goal.

A more limited and realistic goal for computer-based methods is to predict the results of making *small perturbations* in a known protein structure. Given the extensive X-ray studies on subtilisin (Bott *et al.*, 1987) and T4 lyzozyme (Alber *et al.*, 1987), where single amino acid mutations lead to local and small structural changes, we would hope that the theoretical methods can indeed by predictive of these small changes. Even this is a challenging problem for numerical simulation methods, for reasons we detail below. In addition to wishing for the ability to predict structural changes upon mutation, other properties of interest include changes in ligand binding, enzyme catalytic rate, and protein stability upon mutation. These latter three properties have recently become approachable by free-energy perturbation methods; we also detail some applications of this in this chapter.

2. Methodology

The choice of the method of determining the energy of the molecular assembly as a function of atomic position is at the heart of simulations on molecules. Approaches include those that solve the quantum mechanical Schroedinger equation and those that use an empirical energy to express and analyze this equation using classical mechanics. It is obvious that the first approach is more "correct," since it has been demonstrated that solutions of the Schroedinger equation can lead to calculated molecular properties in agreement with experiment for small molecules. However, it is completely impractical to use quantum mechanical methods to calculate the energy and analytical derivatives of the energy for all of the coordinates of a system as large as a protein. Fortunately, it has been demonstrated that empirical energy equations and classical mechanical methods are surprisingly accurate at representing the conformations and noncovalent interactions for chemical systems in which the chemical bonding does not change during the process of interest (Kollman, 1985). Thus, in cases in which chemical bonding is changing (e.g., enzyme catalysis) one can use hybrid approaches, analyzing the part of the system in which the reaction occurs with quantum mechanics and the remainder of the system with (classical) molecular mechanics.

Such hybrid approaches have been developed by Warshel *et al.* (Warshel and Levitt, 1976; Warshel and Weiss, 1980), using MINDO3 and Empirical Valence Bond quantum mechanical methods, by Bash *et al.* (1987a), using semiempirical (MNDO and AM1) quantum mechanical methods, and by Singh and Kollman, using *ab initio* quantum mechanical methods (Singh and Kollman, 1986).

When *ab initio* quantum mechanics are coupled to molecular mechanics, there is a considerable imbalance in the time it takes to calculate the quantum mechanical energy and analytical derivatives of this energy with respect to coordinates and the time to calculate the corresponding molecular mechanical terms for the rest of the system. When one uses semiempirical quantum mechanics, as Warshel and Bash *et al.* have done, the quantum mechanical calculations can be done quickly enough to allow molecular dynamic trajectories with this hybrid approach. The reliability of the empirical or semiempirical quantum mechanical approach is generally lower, and the approach is less transferable from molecule to molecule, although Warshel calibrates his energies empirically on the corresponding reaction in solution and the MNDO/AM1 methods have been extensively tested against much empirical data.

The molecular mechanics/dynamics approaches (see McCammon and Harvey, 1987, for a review) use a simple energy function such as described in Eq. (1):

$$
\begin{aligned}
E_{total} = \sum_{bonds} K_r(r - r_{eq})^2 + \sum_{angles} K_\theta(\theta - \theta_{eq})^2 + \\
\sum_{dihedrals} \frac{V_n}{2} [1 + \cos(n\phi - \gamma)] + \\
\sum_{i<j} \left[\frac{A_{ij}}{R_{ij}^{12}} - \frac{B_{ij}}{R_{ij}^6} + \frac{q_i q_j}{\epsilon R_{ij}} \right] + \sum_{H\text{-bonds}} \left[\frac{C_{ij}}{R_{ij}^{12}} - \frac{D_{ij}}{R_{ij}^{10}} \right]
\end{aligned}
\tag{1}
$$

The various terms in this equation have been discussed in detail elsewhere; they involve simple harmonic terms for bond stretching and bond bending, a Fourier series for torsional rotations around bonds, and a nonbonded interaction between atoms separated by three or more bonds. These nonbonded interactions include electrostatic (R^{-1}) and van der Waals (R^{-6} and R^{-12}) terms. The energy function has the advantage that it and its analytical derivatives are simple and rapidly evaluated. More elaborate versions of Eq. (1) exist in the literature, but there is no consensus about whether the extra computation involved is worth the extra of deriving parameters for the more elaborate function. Molecular mechanics takes Eq. (1) and calculates $- \nabla E_{total}$ and then uses energy minimization methods to move to the point in conformation space where $\nabla E_{total} = 0$. Molecular dynamics involves setting $- \nabla E_{total} = F = m \frac{d^2 \vec{r}}{dt^2}$. Given initial velocities and forces, molecular dynamics leads to a trajectory in phase space for the molecular system.

What are the major problems in using Eq. (1)? The largest energy term in Eq. (1) is the electrostatic one, although its magnitude depends on ϵ, the dielectric constant used in the simulation. For computational efficiency, one typically truncates the atom–atom nonbonded interactions in Eq. (1) at a cutoff distance of 8–12 Å. This truncation can have a large effect on the truncated energy, particularly if the number of charges in the system is large and ϵ is small.

These energy effects are usually much larger than more subtle differences in Eq. (1). In addition, particularly using molecular mechanics (which moves to the nearest local minima), and also molecular dynamics (which scans a limited area of conformation space), only a limited number of conformations can be studied in applying Eq. (1) to large molecules.

The most exciting recent uses of molecular dynamics (van Gunsteren, 1988) are in (1) x-ray refinement, (2) the generation of structures consistent with NMR–NOE data, and (3) free-energy perturbation applications. In the latter case, one calculates the free-energy difference between related systems A and B by the equation (Singh et al., 1987)

$$\Delta G = -RT < \exp(-\Delta H / RT) >_A$$

where $\Delta H = H_B - H_A$, the difference in the Hamiltonians between systems A and B, and $< >_A$ designates an ensemble average of configurations characteristic of system A. In almost all practical applications, one must create hybrid nonphysical systems that lie between A and B and calculate the free-energy difference between each of these, summing up these free-energy differences to determine the total free-energy difference.

The key to calculations of ΔG include determining the ensemble average $< >_A$, which can be done using either Monte Carlo (Jorgensen and Ravimohan, 1985) or molecular dynamics (Singh et al., 1987) techniques. Molecular dynamics approaches allow more general macromolecular simulations and thus are usually the method of choice. The key limitation in applying the free-energy method is the sampling problem, or accurately determining a converged $< >_A$. This does not appear to be a limitation in simulations to determine solvation free energies, but it is the key difficulty in macromolecular simulations. This sampling problem is nothing more than a restatement of the local minimum problem for the system. Such macromolecule simulations involve some drift away from the X-ray structure, since the potential function/representation of environment is not perfect. Because of this drift, more extensive sampling may actually lead to less *correct* sampling near the experimental structure. This is the most serious problem with the free-energy approach. Nonetheless, many different applications have led to encouraging agreement with experimental observations, as we describe in Section 3.

3. Applications of Molecular Dynamics/Free-Energy Calculations to Site-Specific Mutagenesis

3.1. Structure Correlation with X-Ray Observations

Relatively few studies have been done in this area. Alagona et al. (1984) have used molecular mechanics to predict that a His 95 → Gln mutant will place the Gln in a position that is qualitatively different from the His. No data are available on this. Shih *et al.* (1985) have used molecular mechanics to compare the structure of a Gly → Asp mutant in hemagglutinin, with good correspondence to the X-ray structure.

The extensive X-ray structures of T4 lysozyme and subtilisin (Bott *et al.*, 1987; Alber *et al.*, 1987) make them very useful data bases for critical evaluation of simulation methods and force fields. We see extensive and critical comparison of simulated *structures* vs. X-ray structures on these two systems as important research for the next few years.

3.2. Free Energies of Solvation and Ligand Binding

Bash *et al.* (1987b,c) have calculated the solvation free energies of many of the amino acid side chains, and are generally in good agreement (within 0.5–1.0 kcal/mole) with the results of Wolfenden's experiments (Wolfenden, 1983). By calculating the difference in free energy of solvation of a ligand and the difference in free energy of interaction of the ligand to a protein, one can predict net free-energy differences in binding of ligands to proteins. Bash *et al.* (1987c) did this for two inhibitors of thermolysin and calculated the binding free-energy difference in good agreement with experiment.

Merz (Merz and Kollman, submitted) has *predicted* the binding free energy of another analog. Merz has also calculated the free-energy differences for carbonic anhydrase inhibitors (Merz *et al.*, manuscript in preparation), which are in generally good agreement with experiment. Caldwell has calculated the relative K_M values for α-lytic protease substrates with varying P_1 residues, in qualitative agreement with experiment (Caldwell and Kollman, submitted). Finally, Gough has calculated the free-energy difference in binding of two penicillopepsin inhibitors (difluorostatine vs. difluorostatone), in rather good agreement with experiment (Gough and Kollman, submitted).

3.3. Free Energies of Binding and Catalysis of Site-Specific Mutants

Warshel *et al.* (Hwang and Warshel, 1987) and Rao *et al.* (1987) have calculated the change in free energies of binding and catalysis by mutating Asn 155 → Ala in subtilisin. Both found rather good agreement with subsequent experiments, despite using rather different methods. Warshel and Sussman, in their studies of trypsin (1986) and subtilisin, used a combination of the Empirical Valence Bond quantum mechanical method with molecular dynamics to calculate the free energies for a proposed catalytic pathway and then calculated the free-energy difference for mutating the Asn → Ala at various steps along this pathway. Rao *et al.* (1987) assumed a tetrahedral model for the transition state of acylation and then used only molecular dynamics to calculate the $\Delta\Delta G_{cat}$ due to the mutation. The fact that both calculations agreed with experiment is encouraging and suggests that the transition state is indeed close to a tetrahedral structure.

Other such studies include those by Warshel and Sussman (1986) and Seibel (Seibel and Kollman, manuscript in preparation) on the Gly 216 and Gly 226 → Ala mutants in trypsin. In general, good agreement with the experimental observations have been achieved, as well as a clearer understanding of why the experiments turned out as they did. Hwang and Warshel (1988) have also studied the "ion-pair" reversal in aspartate aminotransferase and have shown that in general such mutations will not be favorable for binding and catalysis.

3.4. Free Energies of Protein Stability

McCammon *et al.* (Wong and McCammon, 1987) Bash *et al.* (1987b), and Dang *et al.* (1989) have studied the free energies of protein stability upon site specific mutation. Such calculations have a much weaker basis than binding calculations, because one does not have a structure for the denatured state and must thus make some assumption about the properties of this state. Dang *et al.* used an isolated tripeptide in solution to represent residues 156–158 of T4 lyzozyme in a study of the Thr 157 → Val mutation. The calculated $\Delta\Delta G$ upon denaturation was in good agreement with the experimental destabilization of 1.7 kcal/mole due to this mutation.

4. Summary and Critique

It is clear that one can use theoretical methods in useful ways in studying protein structure and site-specific mutation. To date, the most exciting applications of numerical simulation methods to proteins have been in X-ray/nuclear magnetic resonance refinement and in free-energy perturbation calculations. The first two applications use experimental data to constrain the calculations; the latter does not. Nonetheless, the most critical problems still remain in the theoretical methodology: (1) inaccuracies in the empirical energy functions and long-range cutoff and electrostatics, and (2) the limited sampling possible in a computationally feasible simulation of ∼100 pisec. Nonetheless, the agreement with experiment found in many simulations and the progress over techniques of 5 years ago are impressive and encouraging. It is likely that the usefulness of computer simulation in understanding and predicting the results of site-specific mutations will increase in the future.

ACKNOWLEDGMENTS. We acknowledge the National Institutes of Health (NIH-GM-29072), the National Science Foundation (NSF-DMB-87-14775), and

DARPA (under Contract N-00014-86-K-0757 administered by the Office of Naval Research, R. Langridge, P.I.) for research support.

References

Alagona, G., Desmeules, P., Ghio, C., and Kollman, P. A., 1984, Quantum mechanical and molecular mechanical studies on a model for the dihydroxyacetone phosphate–glyceraldehyde phosphate isomerization catalyzed by triose phosphate isomerase (TIM), *J. Am. Chem. Soc.* **106**(12):3623–3632.

Alber, T., Sun, D. P., Wilson, K., Wozniak, J. A., Cook, S. P., and Matthews, B. W., 1987, Contributions of hydrogen bonds of Thr 157 to the thermodynamic stability of phase T4 lysozyme, *Nature* **330**(6143):41–46.

Bash, P. A., Field, M. J., and Karplus, M., 1987a, Free energy perturbation method for chemical reactions in the condensed phase: A dynamic approach based on a combined quantum and molecular mechanics potential, *J. Am. Chem. Soc.* **109**(26):8092–8094.

Bash, P. A., Singh, U. C., Brown, F. K., Langridge, R., and Kollman, P. A., 1987b, Calculation of the relative change in binding free energy of a protein–inhibitor complex, *Science* **235**(4788):574–576.

Bash, P. A., Singh, U. C., Langridge, R., and Kollman, P. A., 1987c, Free energy calculations by computer simulation, *Science* **236**(4801):564–568.

Bott, R., Ultsch, M., Wells, J., Powers, D., Burdick, D., Struble, M., Burnier, J., Estell, D., Miller, J., Graycar, T., Adams, R., and Power, S., 1987, Importance of conformational variability in protein engineering of subtilisin, *ACS Symp. Ser.* **334**:139–142.

Cohen, F. E., and Kuntz, I. D., 1987, Prediction of the three-dimensional structure of human growth hormone, *Proteins* **2**(2):162–166.

Dang, L., Merz, K., and Kollman, P., 1989, *J. Amer. Chem. Soc.* **111**:8507.

Hwang, J. K., and Warshel, A., 1987, Semiquantitative calculations of catalytic free energies in genetically modified enzymes, *Biochemistry* **26**(10):2669–2673.

Hwang, J. K., and Warshel, A., 1988, Why ion pair reversal by protein engineering is unlikely to succeed, *Nature* **334**(6179):270–272.

Jorgensen, W. L., and Ravimohan, C., 1985, Monte Carlo simulation of differences in free energies of hydration, *J. Chem. Phys.* **83**(6):3050–3054.

Kollman, P., 1985, Theory of complex molecular interactions: Computer graphics, distance geometry, molecular mechanics, and quantum mechanics, *Acc. Chem. Res.* **18**:105–111.

McCammon, J. A., and Harvey, S. C., 1987, *Dynamics of Proteins and Nucleic Acids*, p. 220, Cambridge University Press, Cambridge.

Rao, S. N., Singh, U. C., Bash, P. A., and Kollman, P. A., 1987, Free energy perturbation calculations on binding and catalysis after mutating Asn 155 in subtilisin. *Nature* **328**(6130):551–554.

Shih, H. H., Brady, J., and Karplus, M., 1985, Structure of proteins with single-site mutations: A minimum perturbation approach, *Proc. Natl. Acad. Sci. USA* **82**(6):1697–1700.

Singh, U. C., and Kollman, P. A., 1986, A combined abinitid quantum-mechanical and molecular mechanical method for carrying out simulations on complex molecular-systems—Applications to the CH_3Cl + Cl^- exchange-reaction and gas-phase protonation of polyethers, *J. Comp. Chem.* **7**(6):718–730.

Singh, U. C., Brown, F. K., Bash, P. A., and Kollman, P. A., 1987, An approach to the application of free energy perturbation methods using molecular dynamics: Applications to the transformations of $CH_3OH \rightarrow CH_3CH_3, H_3O^+ \rightarrow NH_4^+$, glycine $\rightarrow$ alanine, and alanine $\rightarrow$ phenylalanine in aqueous solution and to $H_3O^+(H_2O)_3 \rightarrow NH_4^+(H_2O)_3$ in the gas phase, *J. Am. Chem. Soc.* **109**(6):1607–1614.

van Gunsteren, W. F., 1988, The role of computer simulation techniques in protein engineering, *Protein Eng.* **2**(1):5–13.

Warshel, A., and Levitt, M., 1976, Theoretical studies of enzymic reactions: Dielectric, electrostatic and steric stabilization of the carbonium ion in the reaction of lysozyme, *J. Mol. Biol.* **103**(2):227–249.

Warshel, A., and Sussman, F., 1986, Toward computer-sided site-directed mutagenesis of enzymes, *Proc. Natl. Acad. Sci. USA* **83**(11):3806–3810.

Warshel, A., and Weiss, R. M., 1980, An empirical valence bond approach for comparing reactions in solutions and in enzymes, *J. Am. Chem. Soc.* **102**(20):6218–6226.

Wolfenden, R., 1983, Waterlogged molecules, *Science* **222**(4628):1087–1093.

Wong, C. F., and McCammon, J. A., 1987, Thermodynamics of enzyme folding and activity: Theory and experiment, *Springer Ser. Biophys.* **1**:51–55.

Dihydrofolate Reductase
A Paradigm for Drug Design

STEPHEN J. BENKOVIC and CARSTON R. WAGNER

1. Introduction

Dihydrofolate reductase (5,6,7,8-tetrahydrofolate:NADP$^+$ oxidoreductase) catalyzes the NADPH-dependent reduction of 7,8-dihydrofolate (H_2F) to 5,6,7,8-tetrahydrofolate (H_4F). This enzyme is necessary for maintaining intracellular pools of H_4F and its derivatives, which are essential cofactors in the one-carbon transfer reactions utilized in the biosynthesis of purines, thymidylate, and several amino acids. It is also the target enzyme for antifolate drugs such as the antineoplastic drug methotrexate (MTX) and the antibacterial drug trimethoprim (Scheme 1). Because of its biological and pharmacological importance, dihydrofolate reductase (DHFR) has been the subject of intensive structural and kinetic studies (Blakley, 1985). The structures of the *Escherichia coli* and the *Lactobacillus casei* enzymes have been determined to 1.7 Å for several binary and ternary complexes containing MTX and/or NADP$^+$ (Bolin *et al.*, 1982; Filman *et al.*, 1982; Matthews *et al.*, 1985). The primary sequences of DHFR for eight bacterial and vertebrate sources are also available for comparison (Blakley, 1985). In addition, a complete kinetic scheme for wild-type *E. coli* DHFR has been derived from pre-steady-state and steady-state kinetics (Fierke *et al.*, 1987).

The availability of this quantitative information makes this protein the enzyme of choice for the study of a number of fundamental issues underlying biological catalysis: (1) the relationship between ligand binding (thermodynamics) and catalytic turnover (kinetics); (2) the identity and quantitative contribution of individual residues at the active site of an enzyme to both ligand binding

STEPHEN J. BENKOVIC and CARSTON R. WAGNER • Department of Chemistry, Pennsylvania State University, University Park, Pennsylvania 16802.

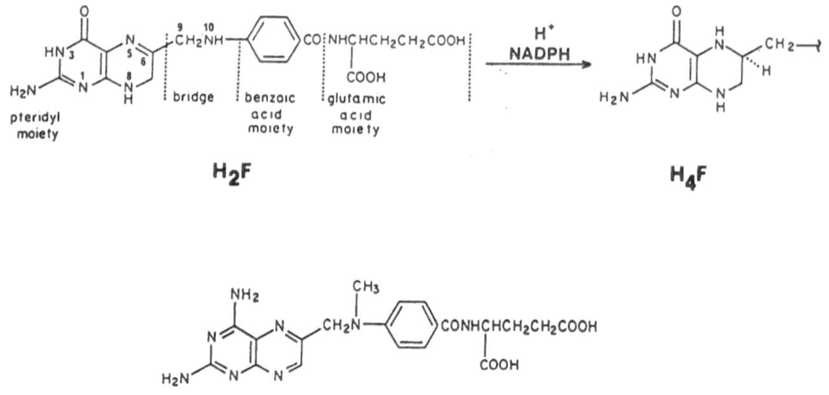

$$H_2F \qquad\qquad\qquad\qquad H_4F$$

MTX (methotrexate)

Scheme 1

and chemical catalysis; (3) the role of conserved vs. nonconserved residues in key interactions for both binding and turnover.

2. Structure and Kinetics of Dihydrofolate Reductase from E. coli

2.1. Structure

The X-ray crystallographic structure of the ternary complex between DHFR·MTX·NADP$^+$ has been obtained for the *L. casei* enzyme (Filman *et al.*, 1982) and can be used to model a similar ternary complex for the *E. coli* enzyme for which only the binary DHFR·MTX complex is available (Bolin *et al.*, 1982). A closeup of the active site of this complex is shown in Fig. 1. The active site is a pronounced cavity some 15 Å deep cutting across one face of the enzyme which is lined by hydrophobic side chains, indicating the importance of hydrophobic and van der Waals interactions in the binding of substrates and inhibitors. The only polar side chain in this cavity is Asp-27, which has been shown by mutagenesis experiments (Howell *et al.*, 1986) to participate in the protonation of N5 during reduction of the C6-N5 imine.

The groups highlighted in Fig. 1 are conserved residues that have been altered by site-specific mutagenesis. Phe-31 and Leu-54 make hydrophobic contacts with the *p*-aminobenzoyl and pteridine moieties of H$_2$F, respectively; Thr-113 is hydrogen-bonded through a water molecule to Asp-27, and Arg-44 forms a salt bridge to a phosphoryl residue of NADP$^+$. These particular amino acids are the focus of this chapter.

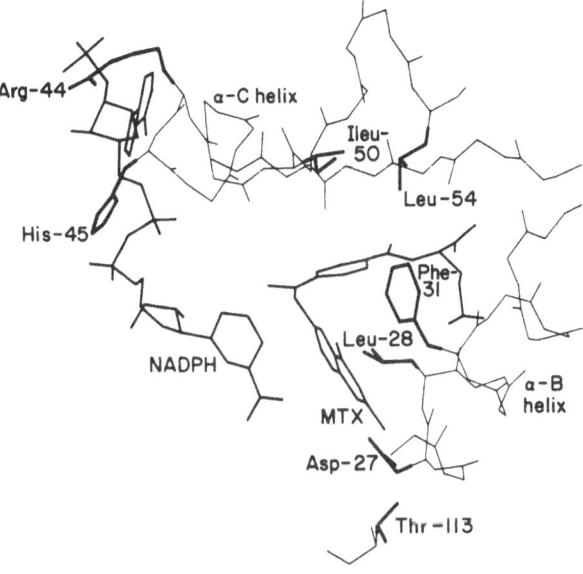

Figure 1. Active site for simulated *E. coli* DHFR·MTX·NADPH.

2.2. Kinetics

The overall kinetic scheme for dihydrofolate reductase at low pH where the enzyme is maximally active is shown in Scheme 2 (Fierke *et al.*, 1987). The association and dissociation rate constants of the binary complexes and all of the ternary complexes except E·H$_2$F·NADPH have been measured directly by a combination of stopped-flow absorbance or fluorescent measurements. The association rate constants for H$_2$F to E·NADPH or NADPH to E·H$_2$F to form the reactive ternary complex, E·H$_2$F·NADPH, were measured in single turnover experiments under conditions in which formation of the ternary complex was slower than hydride transfer. The dissociation rate constants of H$_2$F and NADPH from this complex were more difficult to estimate directly, so a thionicotinamide (TNADPH) was employed as an NADPH surrogate. The value for k_{off} for NADPH from E·NADPH·H$_2$F was then chosen to balance the equilibrium binding within the closed system. Finally, the pH dependence of the reaction was rationalized by the same scheme by introducing a single pK_a of 6.5 for the E·NADPH·H$_2$F ternary complex associated with the Asp-27. Since there is little

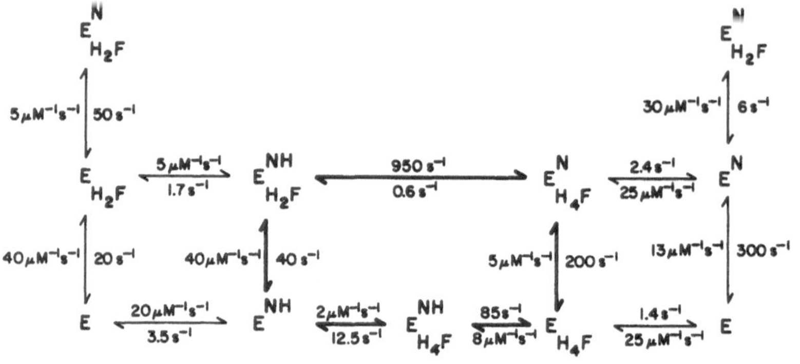

$$E{\cdot}\overset{N}{\underset{H_2F}{}} \qquad\qquad\qquad E{\cdot}\overset{N}{\underset{H_2F}{}}$$

$5\,\mu M^{-1}s^{-1}$ | $50\,s^{-1}$ $\qquad\qquad$ $30\,\mu M^{-1}s^{-1}$ | $6\,s^{-1}$

$$E{\cdot}\underset{H_2F}{} \xrightleftharpoons[1.7\,s^{-1}]{5\,\mu M^{-1}s^{-1}} E{\cdot}\overset{NH}{\underset{H_2F}{}} \xrightleftharpoons[0.6\,s^{-1}]{950\,s^{-1}} E{\cdot}\overset{N}{\underset{H_4F}{}} \xrightleftharpoons[25\,\mu M^{-1}s^{-1}]{2.4\,s^{-1}} E{\cdot}\overset{N}{}$$

$40\,\mu M^{-1}s^{-1}$ | $20\,s^{-1}$ $\quad$ $40\,\mu M^{-1}s^{-1}$ | $40\,s^{-1}$ $\quad$ $5\,\mu M^{-1}s^{-1}$ | $200\,s^{-1}$ $\quad$ $13\,\mu M^{-1}s^{-1}$ | $300\,s^{-1}$

$$E \xrightleftharpoons[3.5\,s^{-1}]{20\,\mu M^{-1}s^{-1}} E{\cdot}\overset{NH}{} \xrightleftharpoons[12.5\,s^{-1}]{2\,\mu M^{-1}s^{-1}} E{\cdot}\overset{NH}{\underset{H_4F}{}} \xrightleftharpoons[8\,\mu M^{-1}s^{-1}]{85\,s^{-1}} E{\cdot}\underset{H_4F}{} \xrightleftharpoons[25\,\mu M^{-1}s^{-1}]{1.4\,s^{-1}} E$$

Scheme 2

or no pH dependence on substrate binding, the pK_as for the free enzyme and enzyme-substrate binary complexes were also estimated to be 6.5.

Scheme 2 correctly predicts the full time course kinetics and the steady-state parameters at a variety of substrate concentration and pHs. Note that at pH 7 the steady state is rate-limited by the dissociation of H_4F from the enzyme from the E·NADPH·H_4F complex. Consequently, the promotion of product release from the previous turnover by cofactor binding for subsequent turnover causes the catalytic cycles to be interlocked (Morrison and Stone, 1988; Fierke et al., 1987; Taira et al., 1987).

3. Mutagenesis Studies

3.1. Ligand Binding and Catalytic Turnover

The effects of various mutations at the Phe-31, Leu-54, and Thr-113 positions are summarized in Table I. The mutations can be divided into two groups: Phe-31 and Leu-54, involving hydrophobic residues (Mayer et al., 1986; Murphy and Benkovic, 1989), and Thr-113, associated with hydrogen bonding (Chen et al., 1987). Replacement of the hydrophobic residues with groups that are smaller in steric size results in small changes in V_{max}, but larger increases in the dissociation of the substrates and the MTX inhibitor. The changes in V_{max}, however, mask larger alterations in the off rate for tetrahydrofolate, as well as larger decreases in hydride transfer. The increase in V_{max} for the Val-31 mutant reflects the increase in k_{off}. However, in the case of the Gly-54 mutant the rate of hydride transfer has decreased to the extent that it now determines V_{max}. Overall the structure tolerates the mutational changes well, since the double mutation exhibits binding properties expected from the combination of the two individual

Table I. Effects of Mutations on Kinetic and Thermodynamic Parameters Characteristic
of *E. coli* DHFR

	$k_H{}^a$ (sec^{-1})	$k_{off}{}^a(H_4F)$ (sec^{-1})	$V_M{}^a$ (sec^{-1})	$K_D(H_4F)$ (μM)	$K_D(H_2F)$ (μM)	$K_D(MTX)$ (nM)
W.T.	950	12	12	0.06	0.21	0.02
(Phe-31, Leu-54, Thr-113)						
Val-31	400	>20	26	0.3	6.6	3.2
Gly-54	14	>300	14	>15	140	100
Gly-54, Val-31	0.9	>300	0.9	—	2000	900
Val-113	120	60	32	4.0	30	0.5

^{a}pH independent values in MTEN buffer, 25°C.

mutations at positions 31 and 54, respectively (Mayer *et al.*, 1986). Finally, the second grouping represented by Val-113 replacing the Thr indicates that the presence of the hydroxyl function is not an absolute requirement for proton transfer to N5, but that its replacement leads to a mutant that does not bind the substrate, H_2F, as well.

Two questions may be asked: (1) does a single mutation affect all of the steps shown in Scheme 1 or is the effect localized to a particular step; (2) is there, in addition to the obvious qualitative relationship between binding weakened H_2F (K_D) and decreased catalysis (k_H) manifested in Table I, a quantitative linear free-energy relationship?

A complete kinetic analysis of turnover by the Val-113 mutant (Scheme 3) shows that this mutation affects all steps in the pathway (Fierke and Benkovic, 1988). However, key features remain: (1) the rate constant for steady-state turnover at neutral pH is partially or completely limited by H_4F release; (2) successive catalytic cycles are interlocked; and (3) the internal equilibrium constant for hydride transfer (K_{int} = 513) reflects the favorable overall equilibrium [K_{ov} (pH 7) = 10^4].

From the complete kinetic schemes for wild-type and Val-113 DHFR it is possible to construct a reaction coordinate diagram for turnover at defined substrate concentrations. We have chosen those approximating the *E. coli* cell, namely, 1.0 mM NADPH, 1.5 mM NADP (Lilius *et al.*, 1979), 0.3 μM H_2F, 13 μM H_4F (D. Duch, personal communication), 0.1 M NaCl, pH 7.0, 25°C. The two diagrams (Fig. 2) aligned at the binary E·NADPH complex clearly illustrate that under these conditions the mutation causes two major effects: (1) the binding of H_2F and H_4F is decreased, and (2) the internal equilibrium constant for hydride transfer is unaffected, but the transition state is destabilized.

In terms of the evolution of catalytic behavior (Albery and Knowles, 1976), hypothetical improvements in enzyme efficiency may be achieved by three

Scheme III

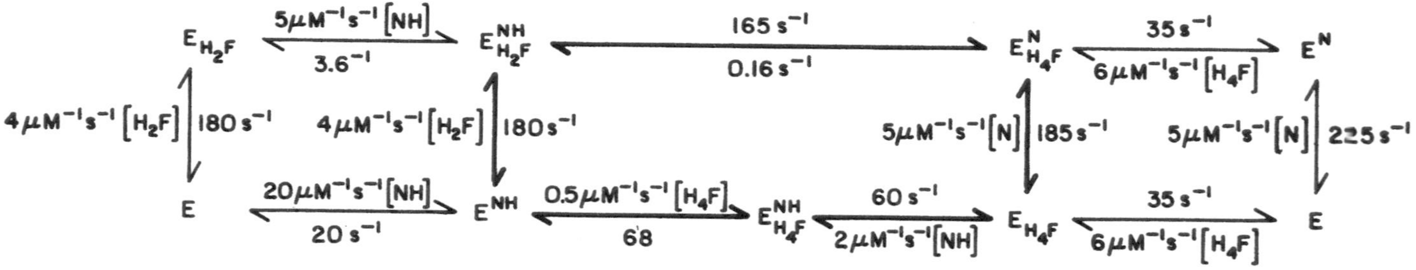

Scheme 3

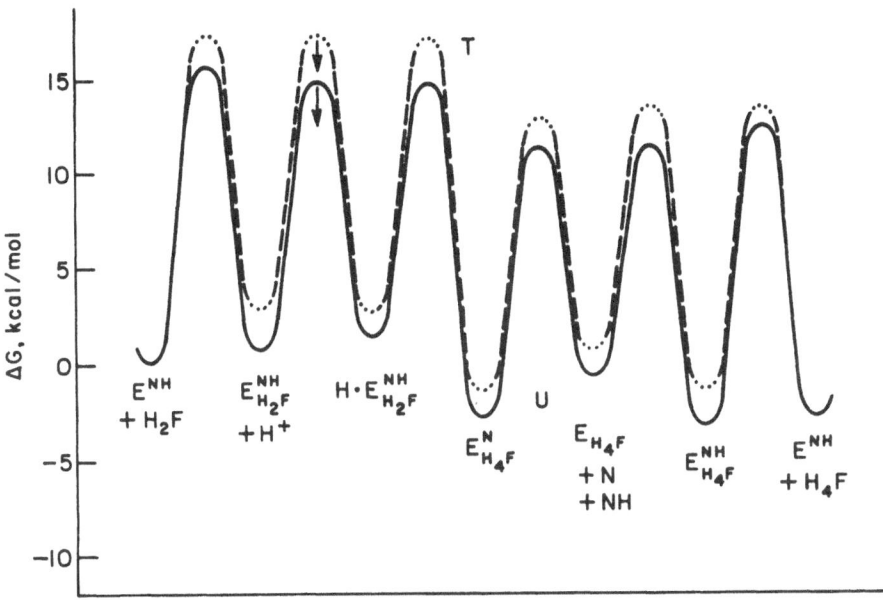

Figure 2. Reaction coordinate diagram for turnover of wild-type (---) and Val-113 (····) DHFR at defined substrate concentrations (see text).

means: uniform binding in which the free energies of internal ground and transition states change equally; differential binding in which the free energies of the internal intermediates and their associated transition states vary relative to one another in a linear fashion; and fine turning of the free energy of a specific transition state. The Thr-113 to Val DHFR mutation is manifest in both uniform free-energy changes for the intermediates $E^{NH}_{H_2F}$ and $E^{N}_{H_4F}$ and the intervening transition state, as well as a specific effect on the transition state in hydride transfer. Differential effects are exhibited by the remaining intermediates. Thus all effects are manifested.

Although the data in Table I can be fitted to a linear free-energy correlation, i.e., a plot of log k_H vs. K_D (H_2F) with the slope term $\beta = 0.95$, the correlation is misleading (Benkovic *et al.*, 1988). If one views the β parameter as an interaction coefficient, then there is little reason that the coefficient should remain constant for the various residues that are being altered. Specifically within the hydrophobic grouping, the Phe-31 is primarily an aromatic–aromatic interaction, whereas the Leu-54 is an alkyl–aromatic interaction that is further combined in a double mutant. It is highly unlikely that the value of β should be the same for all these positions. However, one might hope that a series of mutations at a single site would indeed be linearly correlated.

A series of mutations have been made at Leu-54 where this residue has been changed to Leu, Asn, and Ile (Murphy and Benkovic, 1989). Table II shows the effect of these mutations on key stages in DHF turnover. Note that the binding constants for H_2F vary by factors $\leq 10^3$, but k_H does not. Thus there is no correlation between the substrate binding affinity and the rate of hydride transfer. Removal of the Leu-54 residue results in a constant loss of 2.0 kcal/mole, reducing k_H by a factor of approximately 30-fold regardless of the mutant. In particular, the inability to substitute leucine with isoleucine (and to a lesser extent asparagine and glycine) implies that this region of the active site is responsible for the unique interaction with the folate in order to gain selective transition state stabilization. This mandate may not be met by side chains of similar volume, with or without different polarity, as far as transition-state stabilization is concerned, but it is relaxed in ground-state binding of ligand, as manifested in the greater changes in K_D for H_2F. When these data are added to our earlier linear free-energy relationship, it is apparent that there is no global relationship that rationalizes the behavior of all the mutants.

Previous calculations have indicated that hydride transfer is extremely sensitive to deviations from the colinearity of the atoms involved as well as the distance between them (Wu and Houk, 1987). The transition state for hydride transfer from either methylamine or 1,4-dihydropyridine to the methyliminium cation favors a syn transition state with a $-C{\cdots}H{\cdots}C$ = bond angle (150–160°) and an optimal bond distance $(C{\cdots}C)$ of 2.6 Å. Although the data are still insufficient, the role of conserved residues thus may be to increase, in an incremental manner, the efficiency of hydride transfer by eliminating unproductive conformations of H_2F by precisely positioning dihydrofolate for reaction.

Amino acid substitutions, whether they are conservative deletions or replacements, will lead to conformational changes that may be localized or subtly propagated through the protein. In estimating the value of a side-chain contribution to the free energy of binding, the most evident comparison is its deletion to a glycine. However, because of its reduced volume the active site will certainly be solvated differently. Thus, in general, ΔG_{rel} may not be equated to the binding energy of the side chain in question, although this term may be the predominant contribution (Fersht, 1987). With that caveat one can still assess the apparent

Table II. Effect of Mutations on Key Steps in DHFR Turnover

	k_H(sec^{-1})	K_D (H_2F) (μM)	pK_a
Leu-54	950 ± 50	0.2 ± 0.03	6.5 ± 0.1
Ile-54	31 ± 9	1.9 ± 0.3	6.5 ± 0.1
Gly-54	28 ± 4	350 ± 50	5.7 ± 0.1
Asn-54	42 ± 8	75 ± 15	5.6 ± 0.2

binding energy and the relative contribution of a given residue to the active site ensemble. A partial listing of these values can be found in Table III for mutations grouped according to whether the interaction is ionic, hydrophobic, or hydrogen bonding. It is apparent that the major contributions are made through hydrophobic interactions with values as high as 4.0 kcal/mole. Similar interactions have been seen for the binding of MTX, which far outweigh the strength of the salt-bridge formed between N1 and N2 of MTX and Asp-27, which contributes ≈ 1.8 kcal/mole to the drug's binding affinity (Howell *et al.*, 1986). Thus, hydrophobic interactions are at least as important in determining the tight binding of MTX as those of hydrogen bonds or salt-bridges (Taira *et al.*, 1987).

3.2. Long-Range Effects

As noted earlier, the kinetic scheme for turnover of the wild-type enzyme at saturating NADPH features, as the favored kinetic pathway, NADPH-promoted dissociation of tetrahydrofolate. The synergistic effects of NADPH and H_4F are complementary: the presence of NADPH (but not $NADP^+$) increases k_{off} for H_4F by ninefold from free enzyme, and the presence of H_4F increases k_{off} for NADPH by 24-fold (Table IV). The Ile-54 mutant behaves very differently; binding of NADPH assists H_4F dissociation by only a factor of 1.5, in marked contrast to the wild-type enzyme. The differential effect of reduced vs. oxidized cofactor on H_4F dissociation is also lost. Moreover, H_4F has little effect on the rate constant for NADPHdissociation: $k_{off} = 15-20$ sec^{-1} from E·NADPH or E·NADPH·H_4F. The conclusion to be reached from these data is that the substrate synergism has been structurally disconnected by this mutation. Examina-

Table III. Evaluation of Various Types of Interactions through Comparison of Mutant Enzymes[a]

Mutations	Change	ΔG_{rel} (kcal/mole)
His-45-Gln	Salt-bridge	1.1
Arg-44-Leu		1.4
Ile-14-Ala	Hydrophobic	0.9
Tyr-100-Ile		2.1
Phe-31-Tyr	Hydrophobic	1.5
Phe-31-Val		1.9
Leu-54-Gly		3.9
Val-31, Gly-54		5.4
Thr-113-Val	H-bond (2)	2.4

[a]Relative to K_D (NADPH) or K_D (H_2F).

Table IV. Ligand Dissociation Rates

	E_{H_2F}	E^{NH}	$E^{NH}_{(H_4F)}$	E_{H_4F}	$E^{(N)}_{H_4F}$	$E^{(NH)}_{H_4F}$
			$(k_{off}\ sec^{-1})$			
Leu-54	22 ± 5	3.6 ± 0.5	85 ± 10	1.4 ± 0.2	$2.5 \pm .02$	12.3 ± 2
Ile-54	300 ± 10	15 ± 2	21 ± 1	60 ± 3	100 ± 2	90 ± 10
Gly-54	$>>300$	—	—	>300	—	—
Asn-54	$>>300$	—	—	>300	—	—

tion of the X-ray crystallographic structures, particularly those of E. coli·MTX and L. casei·MTX·NADP$^+$ enzyme complexes, reveals that Leu-54 is at the C-terminal of the C α-helix whose N-terminal end has residues that interact with the adenine moiety of the NADPH, suggesting that this substructure might be transmitting this effect.

The subtle long-range effects transmitted through helical or loop-type structures within this enzyme are dramatically revealed by the mutation of Arg-44 to Leu-44 (Adams, 1987, unpublished results). This change effectively deletes the ionic interaction of this side chain with the 2'-phosphate and pyrophosphate moiety of the nicotinamide cofactor in the binary complex as well as two neighboring residues in the free enzyme, e.g., Pro-66 and Ser-63. Table V shows the effect of this mutation in the NADPH binding site on key steps in DHFR turnover. Particularly striking is the effect of the Arg-44 to Leu-44 mutation on the pK_a of the free, binary, and ternary enzyme complexes, which is increased approximately 2 pK_a units to 8.4. The distance from Arg-44 to Asp-27 is approximately 25', suggesting that a subtle movement of the peptide chain is responsible for pK_a change, rather than an environmental effect. If the effect were purely electrostatic, a simple coulombic calculation would require a dielectric constant of 5. This is far less than the range 50–100 reported for the influence of mutations on the pK_a of His-64 of subtilisin (Russel and Fersht, 1987).

The leucine-44 mutant is also characterized by an accelerating effect of NADPH on H_4F dissociation and a reduced rate for hydride transfer. The increased rate of H_4F dissociation for this mutation suggests a more complex effect than that exerted by the C α-helix alone. The reduced rate of hydride transfer in

Table V. Effect of Mutations on Key Steps in DHFR Turnover

	k_H (sec^{-1})	K_D (NADPH) (μM)	K_D (H$_2$F) (μM)	pK_a
W.T.	950	0.33	0.22	6.5
R44L	45	3.46 ± 0.21	0.26 ± 0.02	8.4
H45Q	340	2.0 ± 0.2	0.33 ± 0.05	7.5

this mutant also observed to a lesser extent with the Glu-45 mutant is probably a result of the increase in the pK_a of Asp-27, resulting in a reduced concentration of reactive ternary complex protonated at N5.

3.3. Nonconserved Residues

The effect of nonconserved residues on ligand binding and enzyme turnover is readily assessed by a comparison of the *E. coli* and *L. casei* DHFRs. Despite only 30% primary sequence homology, the tertiary structure of the two enzymes may be overlaid with a root mean square deviation of only 1.8 Å. Using MTX as a center for alignment, one finds an ensemble of 40 amino acids within 5´ of the drug. A similar overlay was constructed for the NADPH binding site after positioning of the NADPH in the *E. coli* site by analogy to its known position in the *L. casei*. In this case an active-site ensemble of 58 amino acids was found 7 Å from the cofactor. The active-site equivalence was computed to be 55% and 72%, respectively, assuming equivalent amino acids were either identical or had only backbone interactions with either the substrate or cofactor. The inclusion of the van der Waals envelope for the side chains of the amino acids showed a remarkable spatial congruence between the two active-site surfaces (Benkovic *et al.*, 1988). Thus, the many changes—some 22 amino acids in the case of the H_2F binding site—are compensatory, so that the same active-site surface is maintained.

Elucidation of the kinetic sequences for the *L. casei* DHFR using the stopped-flow methods described earlier revealed the same features reported above for the *E. coli* enzyme (Andrews *et al.*, 1989). Most important, when the

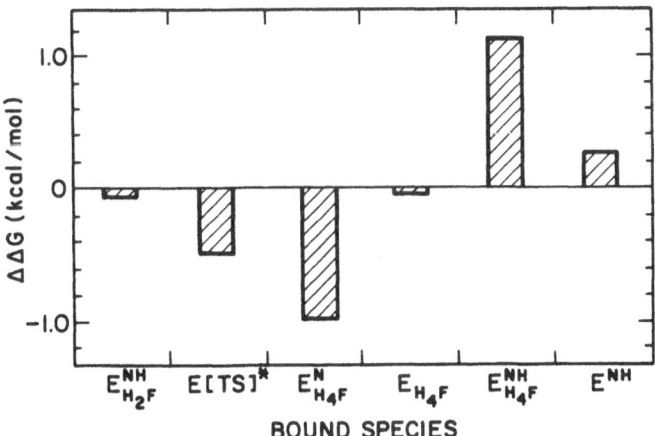

Figure 3. Free-energy differences for *E. coli* vs. *L. casei* enzymes for species indicated.

individual rate constants are expressed in terms of free-energy differences for each step (*E. coli* vs. *L. casei*) they are within 1 kcal/mole. These free-energy differences for the two enzymes are depicted for the bound species in the bar graph shown in Fig. 3. Thus, an active site with multiple amino acid changes can result in a similar kinetic sequence and catalytic efficiency, provided the dimensions and surface of the active site are retained despite these changes. Since single mutations can result in a marked loss in catalytic activity, i.e., Leu-54 to Ile-54, where it is reasonable to presume that some of these changes must be compensatory, others are probably neutral.

4. Conclusions

One may hypothesize that the active-site residues act as an ensemble to create a unique surface with the observed novel kinetic properties. The answer to the theoretical question "why this surface and not others?" may lie not only in the need for catalytic efficiency within the cell, but also in the stereochemical requirements for binding those substrates and curtailing inhibition by others. Other combinations of amino acids leading to the same shape may not provide a stable folded protein. One may also speculate that strictly conserved residues that arose during evolution can no longer be changed because no compensatory mutation exists that satisfies all of these requirements within the cell.

References

Albery, W. J., and Knowles, J. R., 1976, Evolution of enzyme function and the development of catalytic efficiency, *Biochemistry* **15**:5631–5640.

Andrews, J., Fierke, C. A., Birdsall, B., Feeney, J., Roberts, G. C. K., and Benkovic, 1989, A kinetic study of wild-type and mutant dihydrofolate reductases from *Lactobacillus casei Biochemistry* **28**:5743–5750.

Benkovic, S. J., Fierke, C. A., and Naylor, A. M., 1988, Mechanism of oxygen activation by pteridine-dependent monooxygenases, *Acc. Chem. Res.* **21**:101–107.

Blakley, R. L., 1985, Dihydrofolate reductase, in: *Folates and Pterins* (R. L. Blakley and S. J. Benkovic, eds.), pp. 191–253, Wiley, New York.

Bolin, J. T., Filman, D. J., Matthews, D. A., Hamlin, R. C., and Kraut, J., 1982, Crystal structures of *Escherichia coli* and *Lactobacillus casei* dihydrofolate reductase refined at 1.7 Å resolution, *J. Biol. Chem.* **257**:13650–13663.

Chen, J-T., Taira, K., Tu, C-P. D., and Benkovic, S. J., 1987, Probing the functional role of phenylalanine-31 of *Escherichia coli* dihydrofolate reductase by site-directed mutagenesis, *Biochemistry* **26**:4093–4100.

Fersht, A. R., 1987, The hydrogen bond in molecular recognition, *Trends Biochem. Sci.* **12**: 301–304.

Fierke, C. A., and Benkovic, S. J., 1989, Probing the functional role of threonine-113 of *Escherichia coli* dihydrofolate reductase for its effect on turnover efficiency, catalysis and binding, *Biochemistry* **28**:478–486.

Fierke, C. A., Johnson, K. A., and Benkovic, S. J., 1987, Construction and evaluation of the kinetic scheme associated with dihydrofolate reductase from *Escherichia coli, Biochemistry* **26**:381–391.

Filman, D. J., Bolin, J. T., Matthews, D. A., and Kraut, J., 1982, Crystal structures of *Escherichia coli* and *Lactobacillus casei* dihydrofolate reductase refined at 1.7 Å resolution. Environment of bound NADPH and implications for catalysis, *J. Biol. Chem.* **257**:13663–13672.

Howell, E. E., Villafranca, J. E., Warren, M. S., Oatley, S. J., and Kraut, J., 1986, Functional role of aspartic acid-27 in dihydrofolate reductase revealed by mutagenesis, *Science* **231**:1123–1128.

Lilius, E-M., Multanen, V-M., and Toivonen, V., 1979, Quantitative extraction and estimation of intracellular nicotinamide nucleotides of *Escherichia coli, Anal. Biochem.* **99**:22–27.

Matthews, D. A., Bolin, T. J., Burridge, J. M., Filman, D. J., Volz, K. W., Kaufman, B. T., Beddell, C. R., Champness, J. N., Stammers, D. K., and Kraut, J., 1985, Refined crystal structures of *Escherichia coli* and chicken liver dihydrofolate reductase containing bound trimethoprim, *J. Biol. Chem.* **260**:381–391.

Mayer, R. J., Chen, J-T., Taira, K., Fierke, C. A., and Benkovic, S. J., 1986, Importance of a hydrophobic residue in binding and catalysis by dihydrofolate reductase, *Proc. Natl. Acad. Sci. USA* **83**:7718–7720.

Morrison, J. F., and Stone, S. R., 1988, Mechanism of the reaction catalyzed by dihydrofolate reductase from *Escherichia coli:* pH and deuterium isotope effects with NADPH as the variable substrate, *Biochemistry* **27**:5499–5506.

Murphy, D. J., and Benkovic, S. J., 1989, Studies of hydrophobic interactions via mutants of *E. coli* dihydrofolate reductase: Separation of binding and catalysis, *Biochemistry* **28**:3025–3031.

Russell, A. J., and Fersht, A. R., 1987, Rational modification of enzyme catalysis by engineering surface charge, *Nature* **328**:496–500.

Taira, K., and Benkovic, S. J., 1988, Evaluation of the importance of hydrophobic interactions in drug binding to dihydrofolate reductase, *J. Med. Chem.* **31**:129–137.

Taira, K., Fierke, C. A., Chen, J. T., Johnson, K. A., and Benkovic, S. J., 1987, On interpreting the inhibition of catalysis dihydrosolute reductase, *Tr. Biochem. Sci.* **12**:275–278.

Wu, Y., andHouk, K. N., 1987, Theoretical transition structures for hydride transfer to methyleneiminium ion from methylamine and dihydropyridine. On the nonlinearity of hydride transfers, *J. Am. Chem. Soc.* **109**:2226–2227.

VI

PROTEIN ENGINEERING AND ENZYME DESIGN

13

Chemical Approaches to Protein Engineering

ROBIN E. OFFORD

1. Introduction

Other contributions to this Volume provide an excellent account of the current stage of protein engineering using the technology of recombinant DNA. This approach has led to dramatic advances in the last 5 years or so. For the moment, it is limited to the same kinds of changes, as is the natural process of protein evolution, which it mimics. The same L-amino acid residues are involved, and once again all must be linked by the peptide bond. Chemical methods presently have something additional to offer, in that they permit the site-specific introduction into the covalent structures of proteins of noncoded amino acids (L or D), the insertion of molecules that are not, strictly speaking, amino acids, or even the insertion of molecules that are quite definitely not amino acids, and that are linked by something other than the peptide bond.

As long as recombinant methods are limited to coded exchanges (and it is easy to see how that could be overcome), there is a tendency to reserve chemical methods for the introduction of noncoded structures and to reserve all other problems for recombinant methods. This is often, but not always, a wise course. The production of insulin of human sequence, widely regarded as the first true test of recombinant technology at the pharmaceutical scale, did indeed succeed brilliantly by that means. However, the semisynthetic material that results from the conversion by protein–chemical means of pig insulin into the human structure has, at the very least, been available in substantial quantities from the time the recombinant form was first produced: some would claim that the semi-

This chapter was submitted in Autumn, 1988.

ROBIN E. OFFORD • Department of Medical Biochemistry, University of Geneva, Geneva 4, Switzerland.

synthetic material was the first to be available in any notable quantity. Without entering into that discussion, it can be said without controversy that serious development of the recombinant and semisynthetic techniques began at about the same time, and that the chemical route was found to be exceptionally convenient to use. Naturally, there are many other cases in which recombinant methods are greatly preferred even by groups familiar with the chemical techniques (e.g., from our own laboratory—Varley *et al.*, 1988). However, the really interesting fact is that the advanced state of progress in both the recombinant and chemical approaches should permit developments that would be denied to either if used alone. Concrete examples of the efficiency of the combined approach are now appearing in the literature. These will be mentioned below, together with a clear indication of how much more could be achieved in the future.

2. Types of Chemical Approach

Variant protein and polypeptide structures can be made by total synthesis, by specific modification of protein side chains, or by semisynthesis.

2.1. Total Synthesis

Total synthesis involves construction of molecules that are orders of magnitude larger than those normally attempted by the synthetic organic chemist. The problem posed by the sheer complexity of the synthesis is mitigated to some extent by the fact that the synthetic chemist, like the ribosome, can work in preformed amino acid units. However, the tendency of even protected side chains to participate in undesirable reactions, and the stringent requirement for an effective 100% yield in the formation of each peptide bond, has long delayed the hoped-for moment when any given protein can be synthesized at will. The introduction of automated peptide synthesizers has meant that virtually any 30-residue peptide, for example, which is made up from coded amino acids can be synthesized, and many noncoded residues can be incorporated without too much trouble. As we go from 30 to 100 residues, for example, a progressively smaller number of laboratories are found that have the expertise needed to produce molecules in the necessary yields and state of purity for the product to be of any practical use. Even in this range, cases involving the introduction of noncoded amino acids constitute a relatively minor, although increasing, proportion of the work successfully carried out. Beyond 100 residues there is still relatively little to report. The present limitations will undoubtedly recede: it is only necessary to contrast the present situation with that 10 years ago to feel extremely optimistic. Nonetheless, certain sequences will be more difficult than others and there will always be the problem, which aggravates exponentially with chain length, that the yield of a peptide of n residues, each coupled onto the growing chain with an efficiency of $r\%$, will be $(r/100)^{n+1} \times 100\%$.

The current state of this technique is best appreciated by reading in some detail the most recent volumes from the European and the American peptide symposia. The situation is already good, and can only get better, but total synthesis does not hold the answer to all possible problems.

2.2. Side-Chain Modification

Side-chain modification normally involves conversion of the natural sequence into a modified form by the use of reagents that are more or less specific for a given type of side chain. This approach has a long and successful history and has been reviewed many times (i.e., Means and Feeney, 1971). It is often successful in converting all side chains of a given type that are present in the target protein. Its success rate is much lower in converting just one chosen site, while leaving the same amino acid residue untouched in all the other sites in the protein in which it occurs. The classic experiment of this kind remains the conversion of a serine protease to a cysteine protease (Neet and Koshland, 1966; Polgar and Bender, 1966), but the impressive experiment carried out more recently by Kaiser and Lawrence (1984) reminds us of what might still be done. Kaiser and Lawrence reacted appropriate analogs of coenzymes with amino acid residues in or near the catalytic sites of hydrolases. Provided that the attachment of the coenzyme does not block the entry of substrates to their binding site, the possibility exists of creating a new enzymic activity. Kaiser and Lawrence regard it as particularly important that the high-resolution crystal structure of the enzyme should be known in advance, in order to facilitate rational planning of the modification structure. Flavins were chosen for the first series of derivatives, since these already have some intrinsic catalytic activity and are relatively tolerant of the circumstances under which they act. Carefully chosen alkylating derivatives of flavins were allowed to react with the catalytic residue. Cys^{25}, of the proteolytic enzyme papain. In the most favourable case, the resulting flavopapain was shown to be quite an efficient oxidoreductase against certain dihydronicotinamide substrates. Its best k_{cat}/Km value (5.7×10^5 $M^{-1}sec^{-1}$) is about 600 times better than the corresponding model (nonenzymic) reaction and is comparable to the k_{cat}/Km value for a modestly efficient natural enzyme, such as glucose oxidase.

Wilson and Whitesides (1978) approached the problem in a way which, although it did not strictly speaking involve covalent modification, could easily be made to do so. Once again, a catalyst of low molecular weight was linked to an agent that would, in turn, link it to a protein. Instead of the alkylating agents of Kaiser and Lawrence, the linker was biotin, and the catalyst was an organic complex of rhodium. This derivative, as would be expected, bound well to avidin, which, as an asymmetrical surface, conferred a degree of stereoselectivity on the catalyst. Naturally enough, in the case in question this approach destroyed the natural specificity of the protein, but in general, it is possible to imagine how the specificity of a binding site could be either partially conserved or modulated in some useful way while still anchoring a catalytic complex.

Noncovalent binding of the modified catalyst could easily be reinforced in favorable cases by covalent binding, and thus the idea of Wilson and Whitesides, in principle at least, extends the range of site-specific modifications of side chains beyond those that participate directly in catalytic sites.

These promising general approaches have been little exploited and are by no means superseded by recombinant technology. They are worthy of further attention and development. Our ability to manipulate protein structures by chemical techniques grows year by year, and perhaps this area should be looked at again.

3. Semisynthesis

The more detailed part of this chapter will be restricted to the approach to protein modification known as "semisynthesis." I wrote an extensive review of this technique 9 years ago (Offord, 1980), and a brief one more recently (Offord, 1987), and other authors have also reviewed it (e.g., Sheppard, 1980; Chaiken, 1981). The present review will seek to set more recent progress (mainly over the last 2 years) in the context of the earlier work and will stress the way in which the possibilities for a fruitful integration of recombinant and semisynthetic work are rapidly improving.

Semisynthesis is defined as the use of fragments of natural (or recombinant) proteins as ready-made intermediates in the chemical synthesis of proteins. It can involve either noncovalent or covalent association of the biologically synthesized portions with the chemically synthesized ones. If the association is to be covalent, this may be by means of disulfide bridges, or through the peptide bond, or by nonnatural types of chemical linkage.

3.1. Noncovalent Semisynthesis

The ability of fragments of certain proteins to form functioning noncovalent complexes has been exploited for a long time. In particular, work has appeared on pancreatic ribonuclease (e.g., Richards, 1958); *Staphylococcus aureus* nuclease (e.g., Taniuchi *et al.*, 1967); cytochrome c (Harris and Offord, 1977); thioredoxin (Slaby and Holmgren, 1975); somatotropin (Li *et al.*, 1978); phospholipase A_2 (Kihara *et al.*, 1981); myoglobin (Hagenmeier *et al.*, 1978); and IgA fragment Fv (Gavish *et al.*, 1978).

A recent example shows the continuing relevance of this approach. Knoblauch *et al.* (1988) report extensive ^{15}N nuclear magnetic resonance (NMR) studies of the active-site region of pancreatic ribonuclease by the use of the N-terminal pentadecapeptide of this molecule made by the solid-phase method using ^{15}N amino acids. In particular, peptides were prepared with ^{15}N in the ϵ position of Lys[7], the ω position of Gln[11], and the π, τ position of His[12]. The appropriate peptides were combined noncovalently with the natural fragment of ribonuclease A, 21–124. Residues 16–20 of the enzyme are not important, either for activity or for stabilization of the noncovalent complex. The various ^{15}N complexes were studied at a range of pH values, in both the presence and absence of the nucleotides 2'-CMP, 3'-CMP, and 5'-AMP. The NMR data from these complexes were compared both with spectra of the free ^{15}N-substituted peptides and with proton NMR spectra of the native protein. The authors were able to conclude from chemical shifts, [15]N-^{1}H couplings, and the ^{15}N line broadening due to proton exchange that the side chains of Lys[7] and Glu[11] do not contribute to the stabilization of the noncovalent complex. His[12] was shown to form a stable hydrogen bond between its Nτ atom and the phosphate group of 2'-CMP, and it appears that the carbonyl oxygen of the Glu[11] side chain also interacts in some way with this phosphate group. Among other results, the authors have noted a slight inflection in the Hill plots of the imidazole resonances; the simplest explanation for this seems to be the partial formation of an H-bond between the active-site histidines.

3.2. Covalent Semisynthesis: Disulfide Bridges

The method of linking a chemically synthesized fragment to a biosynthetic one by correctly positioned disulfide bridges has been well known since the beginning of synthetic studies on proteins. It has been used particularly intensively with insulin (e.g., Zahn *et al.*, 1981). A notable recent example is the semisynthetic superactive analog [Asp[B10]]insulin (Schwartz *et al.*, 1987), which has a binding affinity for the receptor that is about five times higher than that of the natural form, and which shows a corresponding increase in biological activity. The analog was made by combining the natural A-chains (from porcine insulin) with synthetic B-chains.

The approach has also proved valuable in connection with antibody problems (Rosenberg and Terry, 1977; Kubiak *et al.*, 1987) and in studies on the third domain of ovomucoid protease inhibitor (Wieczorek *et al.*, 1987). For instance, Kubiak *et al.* (1987) used the solid-phase strategy to synthesize a peptide corresponding to the variable region 16–68 of the heavy chain of the mouse myeloma IgG M603. They also synthesized another fragment corresponding to regions 27–68. In separate experiments, these synthetic chains were associated with the M603 light chain by spontaneous formation of disulfide bridges, under carefully chosen and optimized conditions. The parent IgG binds phosphocholine, and the successfully recombined semisynthetic product was isolated from the reaction mixture by affinity chromatography on a phosphocholine–Sepharose column. With this work Kubiak *et al.* have already shown that the disulfide bridge in the heavy-chain variable region, which was once believed to be essential to proper antibody function, may not be critical. Above all, the way is now clear for them to construct a series of analogs designed to study the antigen-binding pocket of the M603 immunoglobin.

3.3. Covalent Semisynthesis: The Peptide Bond

Provided that appropriate sites of specific cleavage can be found, or introduced by recombinant methods, biologically synthesized proteins form a ready source of peptide fragments to which single synthetic residues or whole synthetic sequences can be coupled. The condition that one must have, or must be able to make, a suitable fragment is satisfied more frequently that might be thought. The knowledge of protease specificity that is accumulated as a by-product of protein sequencing is of immense value here. The work of Chang *et al.* (1985) demonstrates how such information can be built upon to produce fragmentation schemes specifically tailored for semisyntheses: they showed that α-thrombin cleaves selectively at the junctions between the variable and joining regions of the antibody light chain, as well as at the junction between the joining and constant region. These authors conclude that such a specificity should greatly assist in the production of semisynthetic chimerae of antibodies.

3.3.1. Fragment Condensation and Stepwise Semisynthesis

The two major categories of such semisyntheses relate to the fragmentation methods used: the parent protein is cleaved either by methods derived from the Edman-type stepwise degradation approach to protein sequencing, or by methods derived from Sanger's original approach of fragmentation by endoproteolytic agents (at first mainly by proteases such as trypsin, later also by specific chemical reagents like CNBr). The synthetic component can then be coupled back either by stepwise methods, as in Merrifield-type peptide synthesis, or by the more traditional fragment-condensation method.

A previous review (Offord, 1987) referred to the wide range of fragment-condensation work carried out by different laboratories on insulin, proinsulin, ferredoxin, the nucleases, cytochrome c, myoglobin, various protein inhibitors, and phospholipase A_2. It was already apparent from these successful examples that the key problem was the same as that of classical fragment-condensation synthesis of peptides: to ensure that the fragments coupled only through the reaction of the α-COOH of one partner with the α-NH_2 group of the other, and that the side-chain -NH_2 and -COOH groups did not participate in such reactions.

Originally a rather complex problem in selective protection of side chains, this requirement can often be easily satisfied by the use of endoproteolytic enzymes made to work in reverse, that is, by a sort of protease ligation (Laskowski, 1978; Inouye *et al.*, 1979; Markussen, 1980). This approach is not to be confused with the interesting proposal made by Glass and Pelzig (1974) that enzyme-labile protecting groups should be introduced into peptide synthesis.

More recent applications of ligation by proteases are discussed in Section 5.2, including use of the method for the activation of the C-terminus for subsequent chemical coupling, and for the site-specific introduction of linkages other than the peptide bond.

The stepwise approach was also reviewed previously (Offord, 1987). It has been fruitfully applied to the removal and replacement of residues at or near either the N and C termini of the protein chain itself or the N and C termini of fragments destined to be subsequently condensed back to a complete protein sequence. This has principally been done in work on insulin, ferredoxin, phospholipase, and glucagon (for references to the earlier work, see the previous reviews). Here again, it is essential to ensure that reaction takes place only between the relevant α-NH$_2$ and α-COOH groups. Ligation by exoproteases has proved its value in this context too, since at least one, carboxypeptidase Y, is readily capable of synthesis as well as degradation (e.g., Widmer and Johansen, 1979).

3.3.2. Ligation by Proteases

Since ligation by proteases is so important, it is probably worthwhile to say something about it in general before reviewing recent advances obtained by applying this and other techniques to semisynthetic problems. Briefly, the hydrolysis of a peptide bond

$$—NH—CH(R_1)—CO—NH—CH(R_2)—CO—$$
$$\downarrow H_2O$$
$$—NII—CII(R_1)—COOH + H_2N—CH(R_2)—CO—$$

is, like all other processes, reversible in theory. From an energetic point of view, it is largely the free energy of ionization of the nascent -COOH and -NH$_2$ groups that drives the process in the direction of the hydrolysis, together, of course, with the fact that the concentration of water is usually so high. It had long been known that if the product of the reverse reaction could be trapped in some way (usually by choosing instances in which the product of synthesis was insoluble while the starting materials were not), synthesis could be readily and rapidly achieved. In practice, however, trapping by insolubility is less easy to arrange when seeking to couple two large natural fragments. It has often been remarked that it should be possible to trap these products by such techniques as immunoaffinity, or even absorption to an ion exchanger, but this does not seem to have been realized in general practice. The best example of trapping that uses biological specificity is the work of Laskowski *et al.* (e.g., Sealock, and Laskowski, 1969) and of Tschesche (e.g., Tschesche and Kupfer, 1976) on protein inhibitors of proteases.

The active sites of such inhibitors consist of a bond susceptible to the cleavage by the enzyme to be inhibited; the enzyme has a high affinity for the inhibitor and is thus arrested in the act of cleaving the susceptible bond. The protease is equally able to complex with a molecule of inhibitor in which the susceptible bond has already been cleaved. This phenomenon has long been recognized (Sealock and Laskowski, 1969) as constituting a trap in the sense defined above. The inhibitor can be represented by

$$\downarrow$$
$$A\text{---}X\text{---}Y\text{---}B$$
$$|\underline{\hspace{3cm}}|$$

in which the susceptible bond is between X and Y. A form in which X, say, has been removed by a carboxypeptidase after cleavage between X and Y can be represented by

$$A \qquad Y\text{---}B$$
$$|\underline{\hspace{3cm}}|$$

In contrast to the form in which the bond between X and Y is cleaved without the loss of either residue, the new derivative binds only weakly to its target protease. However, in the presence of the target protease and another appropriate amino acid, Z, the system constitutes a trap for the (normally very weak) synthetic action of carboxypeptidase, provided that

$$A\text{---}Z \qquad Y\text{---}B$$
$$|\underline{\hspace{3cm}}|$$

is strongly bound to the target enzyme.

It is the peculiarity of these inhibitors that, when the enzyme–inhibitor complex is dissociated by acid or by denaturation, the broken bond re-forms

$$A\text{---}Z\text{---}Y\text{---}B$$
$$|\underline{\hspace{3cm}}|$$

thus permitting isolation of the new analog.

Leary and Laskowski (1973) used a trypsin inhibitor as the starting material, removing X (arginine) with carboxypeptidase B and then replacing it with Z (tryptophan) by trapping the synthetic action of carboxypeptidase A. Since trypsin requires a basic amino acid in the position of X before it will bind to the inhibitor, they used another enzyme, chymotrypsin, to trap the desired synthetic reaction. They reasoned, correctly, that placing a hydrophobic residue in the position of X would create a hitherto unknown inhibitor analog that would have a good specificity for chymotrypsin. The latter enzyme, in fact, was persuaded to bring its own specific inhibitor into existence. This work was probably the first example of the rational creation of a strong, novel biological activity based on prior knowledge of structure–function relationships in macromolecules. Although more than 15 years old, it is reviewed here both because it remains the best example of the trapping concept and because its implications for current work are not always remembered.

The most recent example of this sort of trapping is provided by Tschesche *et al.* (1987), using aprotinin, the Kunitz trypsin inhibitor from bovine mast cells. The sequence -X-Y-, using the preceding notation, was -Lys^{15}-Ala^{16}-. By strategies analogous to that described above, together in some cases with more chemically based routes involving coupling reagents and reversible protection, these authors replaced the -$(CH_2)_4$-NH_2 side chain of Lys^{15} with those of a variety of coded and noncoded residues: -H, -CH_3, -CH_2CH_3, -$CH_2CH_2CH_3$, -$CH_2CH_2CH_2CH_3$,-$CH(CH_3)_2$, -$C(CH_3)_3$, -$CH_2CH(CH_3)_2$, -$CH_2C(CH_3)_3$, and -$CH_2CH(CH_3)CH_2CH_3$.

The analogs have a range of specificities quite different from that of the parent molecule, and the practical and theoretical implications are discussed by the authors (Tschesche *et al.*, 1987). When the analogs in question were accessible to recombinant methods, their semisynthesis was no more complicated than the DNA-based operations would have been.

Although useful in those relatively limited circumstances when the products could be trapped, reverse proteolysis was not generally valuable for the production of analogs of macromolecular size until about 10 years ago, when Laskowski (1978) reported the advantages of adding a neutral, polar organic compound to the aqueous medium in which the enzyme was called upon to act. Among such compounds, usually called cosolvents, butane-1,4-diol, glycerol, and dimethyl-acetamide have been found to be particularly useful [a more extensive list, together with a detailed theoretical description of ligation by proteases, is given by Kullmann (1987)]. These cosolvents are used in concentrations between 50% and 95%, often without either denaturing or precipitating the proteases. Many proteins and protein fragments are also soluble at quite low concentrations of water. The beneficial effect of the cosolvent goes beyond the mere dilution of the water, although this factor can contribute to improving the yield, particularly when the cosolvent concentration is high. The principal advantage comes from the changes that cosolvents produce in the dielectric constant of the medium, due to the effect of changes in the dielectric constant on the pK values of ionization of amino and carboxyl groups. The optimum pH for synthesis can be shown to be midway between the pK values for the carboxyl and amino ionizations, but even at this pH optimum, the size of the yield depends on the distance between the two pK values: the further apart they are, the lower the yield. We now know many cosolvent systems that move the pK values close enough together to allow very good overall yields.

The advantages mentioned are not restricted to the reformation of peptide bonds. Let us consider those proteases that catalyze hydrolysis by a mechanism involving an acyl-enzyme intermediate:

$$\text{-CO-NH-CH}(R_1)\text{-CO-NH-CH}(R_2)\text{-CO-NH- + H-enzyme}$$
$$\downarrow$$
$$\text{-CO-NH-CH}(R_1)\text{-CO-enzyme} + \text{H}_2\text{N-CH}(R_2)\text{-CO-NH-}$$

Normally, the hydrolysis is then completed by attack on the acyl-enzyme intermediate by OH^-:

$$\text{-CO-NH-CH}(R_1)\text{-CO-enzyme} + OH^-$$
$$\downarrow$$
$$\text{-CO-NH-CH}(R_1)\text{-COOH} + \text{H-enzyme}$$

However, other, better nucleophiles than OH^- exist, and these can often successfully compete in the attack on the intermediate. For instance, if the nucleophile contains the group $-NH_2$,

$$\text{-CO-NH-CH}(R_1)\text{-CO-enzyme} + \text{H}_2\text{N-X}$$
$$\downarrow$$
$$\text{-CO-NH-CO}(R_2)\text{-CO-NH-X} + \text{H-enzyme}$$

We have in a sense already considered this point, since H_2N-X can be a peptide or amino acid derivative, with consequent formation of a true peptide bond. We are not restricted to type of attack, however. In our laboratory, Rose and coworkers (1987) have even succeeded in using an amino acid active ester as the attacking nucleophile:

$$\text{—CO—NH—CH}(R_1)\text{—CO—NH—CH}(R_2)\text{—CO—NH—}$$
$$\downarrow \quad \text{H}_2\text{N—CH}(CH_3)\text{—CO—O—C}_6\text{H}_4\text{Cl}_2$$
$$\text{—CO—NH—CH}(R_1)\text{—CO—NH—CH}(CH_3)\text{—CO—O—C}_6\text{H}_4\text{Cl}_2 + \text{H}_2\text{N—CH}(R_2)\text{—CO—NH}$$

In the case illustrated here, the left-hand fragment has in effect been extended by one residue (an alanine) carrying a dichlorophenyl ester on its α-COOH group. In practice, the dichlorophenyl ester was chosen (Rose *et al.*, 1987, 1988a) because, even with its α-HN$_2$ group free, it neither polymerized nor acylated sidechain amino groups under the pH conditions (an apparent reading of pH $= 5$) that was optimal for the reaction. This particular amino acid ester was, equally, resistant to hydrolysis at this pH.

This reaction has specifically activated the carbonyl group of the peptide. This particular activated form is stable enough to permit it to be isolated, and fragment-condensation couplings are now possible in the usual way. A particularly striking instance of the successful use of this reaction is given in Section 4.3 for cytochrome c. In that instance the fact that the activated peptide tends to form a noncovalent complex with the nonactivated one certainly helped their reaction: it is too early to say whether this is a general requirement.

Other kinds of specific activation of the C-terminus have been reported: for instance, the conversion of the peptide to various kinds of hydrazide derivatives (e.g., Canova-Davies and Carpenter, 1980; Yagisawa, 1981; Jones and Offord, 1982). The advantage of such an activation is that it is not obligatory to elongate the peptide by one residue and there is usually no danger during the activation of reactions involving side-chain amino groups. On the other hand, hydrazo derivatives of peptides are only truly activated for coupling to form peptide bonds after oxidation (e.g., the so-called "azide coupling" of conventional peptide synthesis). The overall yield of an azide coupling is never very high, and other side chains suffer oxidative damage. Thus, if at all possible, the active-ester approach remains the method of choice when peptide bonds are to be formed. On the other hand, the hydrazide approach is excellent if the subsequent coupling is aimed at other than purely peptide-bond formation; this and kindred considerations will be discussed in more detail below.

4. Recent Advances in Knowledge Achieved with Semisynthetic Proteins

4.1. Insulin

Insulin, aside from its medical significance, has long been a particularly rewarding subject for semisynthetic study. The molecule is quite small and easily withstands the solvent and deprotection conditions necessitated even by relatively aggressive semisynthetic schemes. Many of the more important residues from a structure–function point of view lie outside the cycle constituted by the segments of the *A* and *B* chains that lie between the two interchain disulfide bridges and by the two bridges themselves. This means that significant regions of the molecule can be modified without the need to destroy and then re-form the disulfide bridges, which is an immense advantage.

The present account will be restricted to analogs designed to further investigate the various natural mutants of insulin that have been identified by family studies, and analyses used to study insulin degradation by biological systems. A description of the many papers that deal with the clinical and pharmacokinetic properties of semisynthetic human insulin is beyond the scope of this review. The fascinating combination of chemical and recombinant methods that are producing analogs of human insulin with new and beneficial properties is the subject of Chapter 19 of this volume and will not be discussed further here. It is sufficient to say that this work represents the most significant example to date of the combination of chemical and recombinant methods, a trend that the present chapter has been written to promote.

4.1.1. Natural Mutant Insulins and Their Artificial Analogs

The first three mutant insulins discovered in family studies were insulin Chicago (Phe^{B25} → Leu^{B25}), insulin Los Angeles (Phe^{B24} → Ser^{B24}), and insulin Wakayama (Val^{A3} → Leu^{A3}). The three lie outside the disulfide-bridge cycle mentioned earlier, which has greatly facilitated their semisynthesis.

It is necessary to be able to produce such analogs in the laboratory by one means or another, both to serve as authentic standards to confirm the structures proposed for the natural mutants and also because further studies require larger quantities than can be obtained from patients. These analogs could have been made by recombinant methods, but semisynthesis has generally proven more convenient. Once achieved, the semisyntheses have opened the way for the preparation of related analogs that have either not yet been identified in nature (e.g., [Ser^{B25}]insulin, Shoelson *et al.*, 1983) or are presently inaccessible by any biosynthetic route (e.g., the remarkable series of analogs described by Nakagawa and Tager, 1986).

The necessary quantities of the [Leu^{B25}] and [Ser^{B24}] mutant forms were obtained semisynthetically by trypsin-catalyzed condensation of des-octapeptide-(B22-B30)-insulin and an appropriate synthetic octapeptide-(B22-B30) (Shoelson *et al.*, 1983; Haneda *et al.*, 1985). The [Leu^{A3}] form has been made semisynthetically by a combined stepwise and fragment-condensation process (Kobayashi *et al.*, 1986). The α-amino group of residue B-1 was reversibly protected and potential problems with the ε-amino group of residue B29 were avoided by tryptic removal of the sequence B22-B30. Three cycles of Edman degradation of the *A*-chain were then carried out, and the required modification was introduced by chemical coupling of the appropriate tripeptide. The octapeptide B23-B30 was then replaced by an enzyme-catalyzed coupling. Although all were successful, some of these schemes involved what seemed to the present reviewer to be unnecessary protection and deprotection steps; for example, the scheme of Kobayashi *et al.* (1986), just described, seems to owe its complexity to the feeling that no simpler method existed for avoiding reaction of Edman's reagent with the α-amino group of residue B1 and ε-amino group of residue B29. Either these authors were unaware of previous work (reviewed by Offord, 1980) that permits the selective protection of these sites by groups that withstand the conditions of the Edman reaction, or they found defects in these earlier procedures which they do not discuss.

All the semisynthetic mutant forms have been tested as to their receptor binding and negative cooperativity (see the literature cited earlier). The [Leu^{B25}]

and [Ser^{B24}] analogs have only a few percent of normal receptor binding and affinity and show decreased negative cooperativity. The [Leu^{A3}] analog has even lower values for receptor binding and biological activity but, at higher concentrations at least, it shows nearly normal negative cooperativity. This represents one more piece of evidence concerning the location on the molecule of sites responsible for negative cooperativity. The results of Shoelson *et al.* (1984) and Haneda *et al.* (1985), taken together, show that the [Ser^{B24}] mutant is more slowly degraded and more slowly cleared from circulation than the natural form of the hormone.

As stated earlier, the same semisynthetic routes gave access to other analogs. Thus, Haneda *et al.* (1985) studied semisynthetic [Ser^{B25}]insulin, and showed that it too has a diminished rate of degradation. Kobayashi *et al.* (1984) had already shown that this latter derivative had poor biological activity and reduced negative cooperativity. Kobayashi *et al.* (1982) established that the [D-Phe^{B24}]insulin is superactive. In 1986, Kobayashi *et al.* made and characterized a further analog at position A3, [Ala^{A3}]insulin. Nakagawa and Tager (1986) have used semisynthesis to make a long series of analogs, modified in the crucial region around residue B-25. The routes used follow the same principles used by the authors in preparing the [Leu^{B25}] mutant mentioned earlier. Their analogs fall into three classes. The first class consists of those in which the carboxyl-terminal 5, 6, or 7 residues of the B-chain have been removed and replaced with an amide group. The second class consists of analogs in which Phe^{B25} has been replaced by unnatural aromatic amino acids ([L-3(2'-3(1'-naphthyl)-alanine^{B25}]insulin; [L-naphthyl)alanine^{B25}]insulin; [L-2-amino-4-phenylbutyryl^{B25}]insulin), or by natural L-amino acids. The third class consists of residues in which residues B26–B30 have been deleted, and in which Phe^{B25} is replaced by various α-carboxyamido amino acids (including the unnatural aromatic ones already mentioned). As could be expected, this remarkable series of analogs has permitted new insights into the structure–function relationships in this region. The paper should be read in full: it must suffice here to say that the authors are able to draw detailed conclusions regarding the steric role of the C-terminal residues in directing the receptor–hormone interaction, and regarding the cooperation between the side chain of residue B25 and other regions of the molecule that lie close to it.

4.1.2. Studies on Insulin Degradation

The destruction of a message after it has arrived at its destination and has been acted upon is sometimes no less important than the arrival itself. In order to study the biological degradation of insulin and its possible aberrations, we prepared a number of isotopically labeled insulins. These were, in addition to our original [(^{3}H)Phe^{B1}]insulin (Halban and Offord, 1975), [(^{3}H)-Gly]^{A1}]insulin (Davies and Offord, 1985), [(^{3}H)Ala^{B30}]insulin (Davies *et al.*, 1987), [(^{18}O)Lys^{B29}]insulin

(Rose *et al.*, 1984; Savoy *et al.*, 1988), and [(^{2}H)Gly^{A1}]insulin (Savoy *et al.*, 1988). All of these analogs had the porcine sequence; the ^{18}O-labeled material was also made with the human sequence. Scheme 1 represents the most complex of the purely chemical semisyntheses of these analogs, while Scheme 2 illustrates the contribution of protease ligation to such work.

It will be seen from Scheme 1 that the target -NH$_2$ group for this particular semisynthesis, that of Gly^{A1}, was first reversibly protected. The N$^{\alpha A1}$-Boc derivative (II) was not the only product of the first reaction, but it could easily be separated from others (some of which could be used in other semisyntheses) by ion-exchange chromatography. The other two -NH$_2$ groups were then reversibly protected, this time with an acid-stable group, to give III.

Anhydrous acid treatment then gave IV, in which the target amino group alone was exposed, and the N-terminal residue of the A-chain alone could then be removed by the Edman reaction to give V. As already mentioned, the protec-

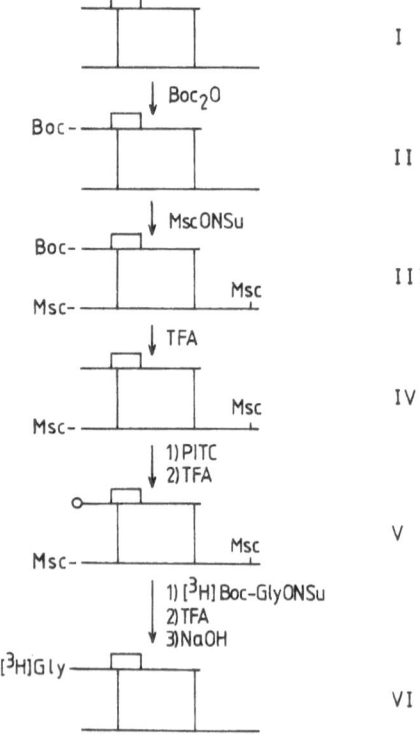

Scheme 1. Abbreviations used are as follows: Boc, *t* butyloxycarbonyl; Boc$_2$O, di-t-butyldicarbonate; Msc, methylsulfonylethyloxycarbonyl; ONSu, hydroxysuccinimide ester; PITC, phenyl isothiocyanate; TFA, trifluoroacetic acid.

tion of the other two amino groups is acid stable, and so V, also, has only the desired -NH$_2$ group in the free state. The missing glycine residue was replaced by the tritiated form in a conventional chemical coupling, and all the protecting groups were then removed to give VI, the wanted product. This procedure, which was also used to give the deuterated analog, is typical of such operations in that it involves the development of appropriate conditions for differential reversible protection, followed by cleavage, resynthesis, and deprotection steps. As always, it was necessary to show by appropriate controls that the biological activity of the product had survived, and as is surprisingly often the case, VI showed no signs of damage.

The enzymic scheme

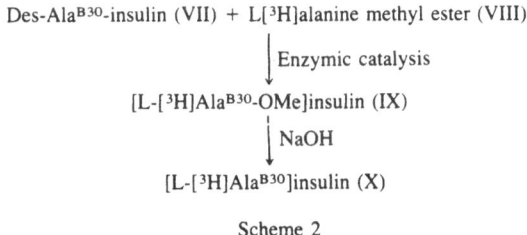

Scheme 2

is apparently simpler. The only problem was to get around the difficulties posed by the requirement that the incoming -NH$_2$ compound (in this case VIII) must normally be present in quite high concentrations. This is clearly not feasible when using compounds labeled at high specific radioactivity, not only because of the cost, but because the release of radiochemical energy in the solution prevents smooth running of the reaction. The problem was overcome by having what would by normal standards be a very low concentration of VII (30 mM) and accepting a relatively low yield of IX as compared to the quantity of VII that was used. This posed no problem, however, since VII is easy to make and IX could be readily isolated from unchanged VII before deprotection to X.

The production of the ^{18}O analogs was much simpler. The reaction analogous to the formation of IX was carried out in H$_2$ ^{18}O, using the ester of natural alanine or threonine, depending on whether the porcine or the human sequence was required. Enough cosolvent was used to bring the water concentration down to less than 10%. The reaction was over in less than 10 min. Since the enzymic catalysis is a reversible process, there was a progressive exchange of ^{18}O into the $>$C=O of the peptide bond between Lys^{B29} and residue B-30. As soon as the enzyme was inactivated, the ^{18}O was no longer exchangeable and the ^{18}O derivative could be isolated. To obtain the highest possible abundance of ^{18}O in the product, special precautions were taken to remove protein-bound water before the reaction components were brought together.

All of these exchanges involve coded amino acids. While it is difficult to imagine anything easier (once one knows how to do it) than the ^{18}U synthesis, Scheme 1 is relatively complicated. Could any of these analogs have been made by recombinant methods? This is conceptually possible, but it would be hard to obtain the purity, and the quantities, required for clinical research. Note that it is important that each label be confined to a single, known site in the molecule.

Taken as a mixture, even two of the tritiated derivatives permit double-labeling experiments in principle. Since the labeling of the amino acids is not lost on strong acid treatment, the products recovered from such an experiment have only to be hydrolyzed and the radioactive amino acids separated for it to be possible to assess the relative proportion of the two labeled forms in the original mixture.

Jones *et al.* (1987) used chemical methods analogous to those mentioned earlier to prepare [(^{3}H)Phe^{1}]human proinsulin from the biosynthetic human material.

How have these analogs been used in practice? The approach was twofold: radiochemical and mass-spectrometric. Since the ^{3}H analogs were made at reasonably high specific activity, it is possible to use them from digests with the putative "insulin-specific proteinase" (Duckworth and Kitabchi, 1981), for example, in which the insulin concentration, while higher than optimal, was not dramatically unphysiological. Iodinated derivatives could have been used at higher specific activities, and thus at lower concentration. In contrast to iodination, however, the isotopic substitutions involved in the present series of analogs should under no circumstances have altered the specificity of proteolytic degradation.

Since the fragments were labeled in a single position, the patterns of radioactivity released were relatively simple and, of course, there was no interference with the analyses by unlabeled impurities. Simple observations of the paper electrophoretic mobilities of the tritiated fragment permitted us, from established empirical relationships between a peptide's electrophoretic mobility, mass, and charge (Offord, 1966), to deduce something of the nature of the fragment. Such ideas could be considerably refined if note was taken of the effect (or lack of effect) of various agents of known specificity on the mobility of the tritium label. If a proposed structure for a fragment included a trypsin-labile site, for instance, it was simple to check the hypothesis by seeing if the mobility of the label changed after treatment with trypsin. If arginine, for example, was proposed as a C-terminal residue, digestion by carboxypeptidase B ought to change the mobility in a predictable way. In addition, performate oxidation was valuable in dealing with fragments thought to have disulfide bridges. Thus, it was already possible to have reasonably precise working hypotheses regarding the structures of the various fragments (Muir *et al.*, 1986; Davies *et al.*, 1988). On the basis of these hypotheses, in which the site of cleavage could normally be assigned to within one or two residues along the A- and B-chains, it was worthwhile to make, as authentic standards, fragments with the expected sequences and fragments differing from the expected sequences by one, two, or a few residues. These standards were obtained either by digestion of insulin by proteases of known specificity followed by isolation and characterization of the wanted fragment, or occasionally, if no known protease would produce the fragment in question, by controlled and characterized semisynthetic transformation of a more readily obtained peptide fragment. If these data (Muir *et al.*, 1986; Davies *et al.*, 1986, 1988) were taken together with evidence obtained by fast-atom bombardment mass spectrometry of the natural hormone and its stable-isotope analogs (Savoy *et al.*, 1988), a fairly complete view could be formed of the cleavages.

The major component representing the N-terminal part of the hormone is an H-shaped fragment consisting of the segment Gly^{A1}–Leu^{A13} joined by the expected disulfide bridge to the segment Phe^{B1}–Ser^{B9}. Apparently less important cleavages also occur one residue to the right of those producing the major fragment, i.e., between His^{B10} and Leu^{B11} as well as between Ser^{B9} and His^{B10}, and between Tyr^{A14} and Glu^{A15} as well as between Leu^{A13} and Tyr^{A14}. These paired cleavages represent a characteristic that we have noted with regard to the action of the enzyme both at other sites in the insulin molecule and during the degradation of other peptide substrates (unpublished work). Preliminary indications are that this is an intrinsic property of the enzyme and is not due to a contaminating carboxypeptidase. A further, unpaired cleavage, between Asn^{B3} and Glu^{B4}, appears progressively in long-term digests. Other sites at which cleavage of insulin takes place are between residues Glu^{B13} and Ala^{B14}, Phe^{B24} and Phe^{B25}, and Phe^{B25} and Phe^{B26}. Because of the complexity of kinetic analysis of such a multihit process, we cannot yet say which of these last three sites are major and which are minor. We also need to see if any of the fragments might have an intrinsic, secondary biological activity, and if any are to be found in other systems, for example, in the circulation after either exogenous or endogenous insulin has been consumed.

It seems far-fetched to imagine that any of the fragments might have their own biological activities, but a check on this point is nonetheless essential. We find that the same enzyme liberates from glucagon, quite cleanly, a fragment already known to be 100 times more active against the calcium–magnesium ATPase than is glucagon itself (Rose *et al.*, 1988c).

4.2. Other Hormones

In a logical continuation of their work, Tager's group (Cara *et al.*, 1988) have used semisynthetic insulins as probes in the study of the receptors for Type 1 insulinlike growth factor. Their results suggest that qualitatively similar mechanisms may be responsible for the binding of ligand both to insulin receptors and to receptors for the growth factor, although (not surprisingly) many quantitative aspects of these phenomena show as yet unexplained differences between the two systems studied.

In an operation that has clear implications for all those concerned with pharmaceutically active polypeptides, Mahrenholz *et al.* (1987) replaced His-1 of glucagon with its D-analog. The resulting analog is a full agonist of the native hormone in the adenylate cyclase assay, but the affinity of binding to the receptor has fallen to approximately 1% of the native value. These authors conclude that the imidazole side chain of the derivative is still able to form the correct hydrogen bonds with the receptor and can thus bring about full signal transduction, but that the rest of the polypeptide backbone then finds itself in a less than optimal position for binding.

In common with several of the semisynthetic schemes reviewed below, the parent hormone was N^ϵ-protected by acetimidylation:

$$CH_3\text{-}C(=NH_2{}^+)\text{-}O\text{-}CH_3 + R\text{-}NH_2$$
$$\downarrow$$
$$CH_3\text{-}C(=NH_2{}^+)\text{-}NH\text{-}R + CH_3OH$$

This reaction was originally introduced into semisynthetic work in our laboratory, not so much because Hunter and Ludwig (1962) had suggested that it was unusually specific for the ϵ-amino group as compared to the α-amino group, but because the N^ϵ-acetimidyl lysine retains its positive charge. Thus the protection, while depriving the lysine side chain of unwanted chemical reactivity and preventing it from constituting a site for tryptic cleavage, does not perturb the solubility or noncovalent binding properties of the macromolecule to any great extent. Unless there is a lysine residue in the peptide or protein in question that plays a particular role in the mechanism of action, biological activity will normally not be significantly altered. [This question is reviewed, with examples, by Offord, (1980).] The N-acetimidyl protection is reversible, and although the ammoniacal reagent in question (Ludwig and Byrne, 1962) is rather harsh, many proteins seem to withstand it. Nonetheless, several workers do not deprotect their analogs before using them in structure–function studies once they are able to establish that the parent macromolecule retains full biological activity when acetimidylated.

This system of protection has proved its value, but in addition, the supposed tendency for acetimidylation to leave the α-NH$_2$ group untouched seems to have been abundantly confirmed in recent years and constitutes a further reason for adopting it. Some care is needed when putting on the protection in order to avoid unwanted side reactions, particularly the formation of cross-links. The article by Wallace and Harris (1984) contains a good, generally applicable method.

Another hormonal system that has recently received attention from a semisynthetic point of view is the oxytocin/bovine neurophysin I biosynthetic precursor (Ando *et al.*, 1987, 1988). These authors made and studied their analogs in the N^ϵ-acetimidyl form. First they made $N^{\alpha 1}$-acetyl[$N^{\epsilon 30,71}$-diacetimidyl, Ala^2]-des-His^{106}-precursor, in which they eliminated two structural elements (the amino terminal α-NH$_2$ group and the side chain of Tyr^2) that were thought to stabilize the contact between the oxytocin and neurophysin domains. In addition, these authors produced [$N^{\epsilon 30,71}$-diacetimidyl, D-Pro^7, D-Leu^8]-des-His^{106}-precursor. The natural sequence is -Pro^7-Leu^8-, and lies in the spacer (and thus, the processing) region between the two domains. In a lengthy series of experiments these authors established a number of important propositions concerning the precursor, its maturation, and the effect on enzymic processing of self-aggregation. They also conclude that the semisynthetic methodology is generally applicable in studies on the processing of hormone precursors.

4.3. Cytochrome c

Cytochrome c has long been the object of semisynthetic studies by a number of groups. The present review covers work on this protein published within the last 2 years. Previous reviews (e.g., Offord, 1987) have dealt with earlier work by various laboratories that have been active in semisynthetic studies on cytochrome c.

In our own laboratory, Wallace and Proudfoot (1987a,b) and Wallace (1987) have studied the preparation of a cytochrome c analog from which the so-called "bottom" or "Ω" loop has been excised. The importance of the "Ω"-loop (residues 40–55) lies in the fact that it constitutes one of the principal differences between mitochondrial cytochrome c and the bacterial cytochrome c_{555} it may explain why the properties of the two cytochromes are so different. The loop, a relatively self-contained structure, has its beginning and its end very close together in three dimensions. This proved an ideal challenge for the enzyme-directed, active-ester method of Rose *et al.* (1987, 1988b), which was described toward the end of Section 3.3.2. It proved a remarkably efficient process to ligate by this means the appropriate truncated fragments of cytochrome c, namely (1-38) and (56-104), bridging across the space between the beginning and the end of the Ω-loop with peptide bonds to and from a single alanine residue. The resulting derivative, [Ala^{39}]des-(41-55)-cytochrome c, permitted Wallace (1987) to reach a definitive conclusion about the role of the Ω-loop.

Proudfoot *et al.* (1986) have prepared a semisynthetic analog of fully acetimidylated cytochrome c in which the peptide chain is nicked between Gly^{37} and Arg^{38}. The two fragments making up this analog form a noncovalent complex with a cytochrome c-like structure. This complex is, in a sense, that of Harris and Offord (1977), but the nick between residues 38 and 39 has been displaced to the left by one peptide bond. In contrast to that earlier complex, however, the new one has a nearly normal redox potential and could more fully restore succinate oxidation to mitochondria depleted of cytochrome c. Study of the new complex, and comparison of the results obtained with those from other analogs, led Proudfoot *et al.* (1986) to propose an explanation for the low biological activity of the original complex and for the evolutionary invariance of Arg[38]. These conclusions were further refined by Proudfoot and Wallace (1987) in the light of the examination of analogs of the new complex in which the evolutionarily invariant Arg[38] was replaced by glutamic acid, lysine, and glutamine. They also studied substitutions of Lys[39], in which some small degree of evolutionary change is permitted, by tyrosine, threonine, and glutamic acid.

Wallace and Corthésy (1986), also in our laboratory, prepared three semisynthetic analogs of [Homoserine[65]]-N^{ϵ}acetimidyl-cytochrome c, in which the substitutions, all at position 66, were by norvaline, glutamine, and lysine. As a control synthesis the native residue, glutamic acid, was removed and replaced. Comparison of the physical and biochemical properties of the analogs indicated two previously unsuspected roles for Glu[66]. First, this residue contributes significantly to the stabilization of the active configuration of the protein, probably by salt-bridge formation. Second, it appears to participate in the redox-state-dependent ATP binding site of cytochrome c. These authors also draw general conclusions from the data about the way in which disposition of surface charge on the molecule influences both its redox potential and the kinetics of electron transfer.

Corradin *et al.* (1986) and Baumhueter *et al.* (1987) used analogs of the above type to identify multiple T-cell antigenic determinants within a defined portion of the cytochrome c molecule. They conclude, among other things, that whether or not a protein has segregated hydrophobic and hydrophilic structures is a useful criterion for the probability of its being recognized by T cells.

Collawn *et al.* (1988) have used monoclonal antibodies as probes of conformational changes in certain of the analogs mentioned in this section. The analogs in question involve substitutions in or near epitopes for four of the monoclonal antibodies used. Not only do such changes alter antibody binding, but it is found that the longer-range perturbations of structure brought about by these substitutions can be detected by the antibodies for epitopes distant from the sites of substitution. It is notable that the antibodies appear to be more sensitive probes of conformational change than circular dichroism.

Wallace *et al.* (1986) made semisynthetic chimerae of cytochrome c in which the N-terminal portion came from one species and the C-terminal portion from another. Some of the C-terminal portions were totally synthetic. The species involved ranged from horse, via tunafish, to yeasts. The chimerae in which both species were vertebrates all exhibited biological activity indistinguishable from that of the parent molecules, providing no support for the "covarion" hypothesis of protein evolution. Study of some vertebrate–yeast chimerae led to the narrowing down of the number of sites likely to be responsible for the functional differences between yeast and vertebrate cytochrome c.

Wallace and Proudfoot (1987b) report four novel two-fragment complexes. Two, (1–55) • (56–194) and (1–50) • (51–104), are contiguous, while one, (1–65) • (56–104), overlaps. The fourth complex, (1–50) • (56–104), has a gap. It was found that if the logarithm of the relative biological activity in the succinate assay was plotted for each complex against its redox potential, a straight line resulted on which the natural cytochrome c and several other derivatives also lay. Wallace and Proudfoot use this linearity to relate electron-transfer rates to thermodynamic factors. They conclude that it is possible to decide whether a particular residue participates directly in electron transfer simply by seeing whether analogs involving substitutions at the site in question do, or do not, lie on their linear plot. Cytochrome c lends itself particularly well to semisynthesis, and it is to be expected that continued work along the lines cited here will lead to further insights into structure–function relationships in this protein.

4.4. Hemoglobins

Hemoglobins are beginning to receive more attention through the semisynthetic method. Hefta *et al.* (1988) have, using a stepwise strategy, removed the N-terminal valine from isolated α chains of human hemoglobin, and then reconstituted the α_2, β_2 tetramer either with the truncated form of the α chain or with a semisynthetic $[(1-{}^{13}C)Gly^1]$analog of the α chain. In order to be able to couple the ${}^{13}C$-enriched glycine exclusively onto the α-NH_2 of the des-Val^1-chain, it was necessary to protect all the ϵ-amino groups by acetimidylation, and to protect the side-chain -SH groups with *p*-hydroxymercurybenzoate. The -SH protection was removed before the tetramers were performed. Unlike many semisyntheses, the scheme in question included removal of the acetimidyl groups, with concentrated ammonium hydroxide/acetic acid 15:1(v/v), pH 11.5. This was done after formation of the tetramer, and without apparent damage to the product. The heme ring was removed during this synthesis and replaced at the end.

The derivatives thus obtained were studied by spectroscopy and other physical methods, and the affinity for oxygen determined in the presence and absence of chloride ions and 2,3-bisphosphoglycerate. The pK value of the α-NH_2 group was determined directly by titration of the ${}^{13}C$ NMR signal. The value obtained was consistent with the previously proposed role of this group in the alkaline Bohr effect, but contrary to what might have been expected from X-ray crystallographic studies. the pK value was inconsistent with the presence of a salt-bridge from the α-NH_2 of the chain to the carboxyl group of Arg$^{\alpha 141}$ when the α-chain has a bound O_2 molecule.

Seetharam and Seetharama Acharya (1986), having discovered the *S. aureus* V8 protease cleaved the α chain of hemoglobin S exclusively at the linkage

-Glu^{30}-Arg^{31}-, sought to drive the reaction in the reverse sense. In the paper cited above they used *n*-propanol and glycerol as organic cosolvents, and successfully coupled fragment 1–30 of the α-chain to fragment 31–47.

More recently, Acharya *et al.* (1987) have also used the fact that fragments 1–30 and 31–131, equivalent to the whole chain, associate noncovalently and give a measure of steric assistance of the type mentioned earlier to the coupling. The efficiency of the coupling between these two fragments in the presence of 30% *n*-propanol is about 55%. This is sufficient for preparing usable quantities of many wanted derivatives of hemoglobin. It even proved possible to use the enzyme to exchange synthetic fragment 1–30 with the natural sequence 1–30 of the α-chain without having to first fragment the α-chain in a separate operation and isolate the natural fragments beforehand. The authors conclude that this method will permit the preparation, starting with the many naturally occurring mutant α-chains having a single mutation, of forms carrying two changes relative to the natural sequence.

4.5. Phospholipase A$_2$

Studies on phospholipase A$_2$ have probably benefited more from semi-synthesis than studies on any other enzyme. A recent review (Offord, 1987) refers to earlier work by the laboratory in question, that of Slotboom and de Haas. Most recently, van Binsbergen *et al.* (1988) have inserted a number of different coded and noncoded (D and L) amino acids in positions 5, 6, and 9 of the acetimidylated enzyme. Notable among these derivatives is the [N^ϵ-palmitoyl-Lys^6) analog. It had been decided, on the basis of known structure–function relationships, to make this region of the molecule selectively more hydrophobic. As hoped, the analog proved to be superactive. The V_{max} against a micellular lipid substrate is more than 270% of the control value, the k_{cat}/Km for a monomeric lipid substrate is more than 130% of the control value, and the K_D for a micellular lipid substrate is 0.02 mM (control value 0.16 mM).

5. Linkages Other than by Peptide Bonds

It could be desirable on occasion to construct protein analogs or chimerae between proteins and other molecules in which covalent linkages other than peptide bonds between L-amino acids were used at specific sites. This could be for any of the usual reasons encountered in structure–function studies, or because one wished to protect a portion of the molecule from hydrolysis. Conversely, it might actually be desirable to introduce a weak link at a particular point. Finally, the formation of a nonpeptide linkage might simply be more convenient to carry out as a site-specific operation than the formation of a peptide bond. If the chemistry involved was sufficiently different from that of peptide-bond formation, it might not be necessary to protect any of the protein side chains that normally interfere.

For the moment, only chemical methods can give access to such derivatives. As far as semisynthesis is concerned, it is appropriate when contemplating the replacement of peptide bond by another linkage to continue to think in terms of the α-COOH and α-NH$_2$ group of the natural fragments involved. This is not obligatory, but such an option could be advantageous both in terms of remaining close to the natural structure and because, since there is only one a-COOH and one a-NH$_2$ group per natural fragment, the proposed substitution will inevitably be site-specific.

5.1. Linkages at the Amino Terminus

Two methods present themselves for preparing the amino terminus of a peptide for such substitutions. The first is applicable if the amino-terminal residue is cither serine or threonine, or if its conversion to either of these residues by site-specific recombinant methods can be tolerated from a structure–function point of view. If so, the N-terminal residue will constitute a unique site on the polypeptide in which a free -NH$_2$ group is carried on a carbon atom adjacent to one that carries a free -OH group. It is well known that such structures are extremely labile to periodate oxidation. In fact, the reaction is several orders of magnitude faster than the more familiar periodate oxidation of -CHOH- CHOH- in, say, sugars. So sensitive are the 1-amino, 2-hydroxy compounds to periodate oxidation that it is possible to oxidize an N-terminal serine or threonine, e.g.,

$$H_2N\text{-}CH(CH_2OH)\text{-}CO\text{-}NH\text{-}(\text{peptide chain})$$
$$\downarrow$$
$$HCO\text{-}CO\text{-}NH\text{-}(\text{peptide chain})$$

in the presence of a large protective excess of ethylene glycol. This permits the site-specific formation of aldehyde groups on proteins (K. Rose, unpublished work; Offord and Rose, 1986a,b) without, for example, there being any significant risk of attacking other sensitive side chains or any carbohydrate moiety that might be present.

Once present, the aldehyde constitutes a group with unique reactivity, and in a unique site. Many different kinds of nucleophile, particularly amines, hydrazines, and hydroxylamines, will react with it quite specifically. The adducts have a wide range of stability, which tends to increase as we move along the list of nucleophiles just mentioned. If stability needs to be increased further, this can often be done by reduction within such reagents as cyanoborohydride (e.g., the reduction of a Schiff base).

$$R\text{-}NH_2 + HCO\text{-}CO\text{-}NH\text{-}(\text{peptide chain})$$
$$\downarrow$$
$$R\text{-}NH = CH\text{-}CO\text{-}NH\text{-}(\text{peptide chain})$$
$$\downarrow$$
$$R\text{-}NH\text{-}CH_2\text{-}CO\text{-}NH\text{-}(\text{peptide chain})$$

A second method of introducing a carboxyl group at the N-terminus is metal-catalyzed transfer of the $-NH_2$ to glyoxylate. This reaction:

$$H_2H\text{-}CH(R)\text{-}CO\text{-}NH\text{-}(\text{peptide chain}) + HCO\text{-}COOH$$
$$\downarrow$$
$$HCO\text{-}CH(R)\text{-}CO\text{-}NH\text{-}(\text{peptide chain}) + \text{glycine}$$

is thought to be confined to the α-amino group by the steric constraints on the metal-ion catalysis. Its use in general protein modification has been reviewed by Dixon and Fields (1972), and in semisynthesis by Offord (1980). It is probably fair to say that this approach is worth considering, but with some caution. When it works, it works well.

5.2. Linkages at the Carboxyl Terminus

It is even harder to find chemical reactions that discriminate between α- and side-chain carboxyl groups. By far the best solution is to use the specificity of proteases. The ability of proteases to couple molecules to the α-carboxyl group by nonnatural linkages has been known for many years. References to the early history of the enzymic formation of amino acid anilides and kindred derivatives, largely associated with the name of J. S. Fruton, are given by Kullmann (1987). As with true peptide synthesis through the peptide bond, the original approach involved trapping the wanted product, normally by exploiting its insolubility relative to the starting materials. Once again, this is less likely to be possible for macromolecules but, freed from the limitation to the peptide bond, other possibilities present themselves. For instance, it is possible to attack the acyl-enzyme intermediate (Section 3.3.2) by $-NH_2$ compounds other than amino acid derivatives or peptides. Such compounds can be much stronger nucleophiles, with the result that the equilibrium lies more strongly in the direction of synthesis, even if no trapping strategy is adopted. It is even possible on occasion to dispense with the organic cosolvents discussed in Section 3.2.2. Hydrazine and hydrazides constitute a good example of such nucleophiles. Canova-Davies and Carpenter (1980) carried out the enzyme-catalyzed conversion of insulin into a derivative in which the last eight residues of the B-chain had been removed and replaced, with an efficiency of 30%, by a hydrazo group. Yagisawa (1981) carried out a detailed study of the optimum conditions for the formation of esters, hydrazides, t-butyloxycarbonyl-, and carbobenzoxyhydrazides of model peptides using trypsin caralysis. Jones and Offord (1982) extended this work to the forma-tion of hydrazides, *t*-butyloxycarbonylhydrazides, and phenylhydrazides of natu-ral fragments of proteins, using trypsin, chymotrypsin, elastase, and subtilisin. Carboxypeptidase Y is also capable of carrying out such substitutions (Widmer *et al.*, 1981).

The enzyme-directed formation of peptide active esters mentioned in Sec-tion 3.2.2 lends itself in principle to the formation of nonpeptide links. The peptide active ester could even couple to D-amino acids, thus going beyond the specificity limitation to L- or optically inactive structures characteristic of most enzyme-based methods.

We (Offord and Rose, 1986a,b; Offord *et al.*, 1986) wished to be able to conjugate molecules of therapeutic and diagnostic importance to proteins in a site-specific manner. We decided to fix, at the C-terminus of the target polypep-tide chain, aldehyde, aromatic-NH_2, and hydrazide groups. These groups have useful reactivities not otherwise found in proteins. It has already been explained how the hydrazide group can be placed at the C-terminus of a polypeptide, and we also found it possible to use proteolysis working in reverse to fix either acetal-protected aldehydes or aromatic amines in this position. More recently (Rose *et al.*, 1988b; Offord and Rose, 1987), we have used carbonyldihydrazide and 1,3 diamino propan-2-ol as nucleophiles to produce, respectively,

$$(\text{peptide chain})\text{-NH-NH-CO-NH-NH}_2$$

and

$$(\text{peptide chain})\text{-NH-CH}_2\text{-CH(OH)-CH}_2\text{NH}_2$$

As already explained, the substituent in the second derivative will be almost uniquely sensitive to periodate oxidation. Under conditions that carry little or no risk of oxidizing sensitive side chains or possible carbohydrate groups, this derivative can be readily converted to

$$(\text{peptide chain})\text{-NH-CH}_2\text{-CHO}$$

The hydrazide substituent in the first derivative is immediately ready for further reaction. K. Rose (unpublished) has drawn attention to the great advantages offered by such compounds as carbonyldihydrazide and 1,3-diaminopropan-2-ol in that they are symmetrical molecules, each carrying two groups capable of participating in the desired reaction. The effective molar concentration of these components is thus always twice the value that one would first suppose. This can have a beneficial effect on the efficiency of protein substitution.

It is extremely simple to make aldehyde- or hydrazide-seeking derivatives of virtually any molecule one wishes to conjugate to the polypeptide chain, and reaction is usually efficient, even under mild conditions. Once again, any tenden-cy for the groups to fall off can be abolished by reduction, if required.

We believe that this approach can make a significant contribution, particularly to the site-specific conjugation of diagnostically and therapeutically useful compounds to target-seeking proteins. This will be particularly true if it is possible, when necessary, to use recombinant methods to modify the sequence in a way that will facilitate the use of the reaction schemes shown here.

6. Conclusions

We hope that this account will give some idea of what is possible, what obstacles remain to be overcome, and above all, of the promising future for a beneficial combination of these techniques with those based on recombinant DNA.

ACKNOWLEDGMENTS. I thank my many co-workers, past and present, for all their intellectual and practical contribution to the work of our laboratory. In addition to the great debt that I owe them in this respect, J. G. Davies, K. Rose, and C. J. A. Wallace have helped with useful discussions on their more recent work. I thank the Fonds National Suisse de la Récherche Scientifique for having made a major financial contribution to our recent work described in this review.

References

Acharya, A. S., Cho, Y. J., and Iyer, K. S., 1987, Staphylococcus aureus V8-protease catalyzed segment exchange reaction of alpha-chain of hemoglobin S: A semisynthetic approach for the preparation of variants of alpha-chain, *Prog. Clin. Biol. Res.* **240**:3–19.

Ando, S., McPhie, P., and Chaiken, I. M., 1987, Sequence redesign and the assembly mechanism of the oxytocin/bovine neurophysin I biosynthetic precursor, *J. Biol. Chem.* **262**(27):12962–12969.

Ando, S., Murthy, S. M., Eipper, B. A., and Chaiken, I. M., 1988, Effect of neurophysin on enzymatic maturation of oxytocin from its precursor, *J. Biol. Chem.* **263**(2):769–775.

Baumhueter, S., Wallace, C. J. A., Proudfoot, A. E. I., Bron, C., and Corradin, G., 1987, Murine T-cell antigenic determinants identified within a limited region of the horse cytochrome *c* molecule, *Eur. J. Immunol.* **17**:651–656.

Canova-Davies, E., and Carpenter, F. H. C., 1980, Specific activation of the arginine carboxyl group of the B-chain of bovine des-octapeptide-(B23–30)-insulin, in: *Insulin* (D. W. Brandenburg, and A. Willmer, eds.), pp. 107–115, Walter de Gruyter, Berlin, New York.

Cara, J. F., Nakagawa, S. H., and Tager, H. S., 1988, Structural determinants of ligand recognition by Type I insulin-like growth factor receptors: Use of semisynthetic insulin analog probes, *Endocrinology* **122**(6):2881–2887.

Chaiken, I. M., 1981, Semisynthetic peptides and proteins, *Crit. Rev. Biochem.* **11**:255–301.

Chang, J. Y., Alkan, S. S., Hilschmann, N., and Braun, D. G., 1985, Thrombin specificity. Selective cleavage of antibody light chains at the joints of variable with joining regions and joining with constant regions, *Eur. J. Biochem.* **151**:225–230.

Collawn, J. F., Wallace, C. J. A., Proudfoot, A. E. I., and Paterson, Y., 1988, Monoclonal antibodies as probes of conformational changes in protein-engineered cytochrome *c*, *J. Biol. Chem.* **263**:8625-8634.

Corradin, G., Wallace, C. J. A., Proudfoot, A. E. I., Verdini, A. S., and Baumhueter, S., 1986, Use of natural and synthetic peptides to determine the fine specificity of cytochrome *c* specific T cell clones, in: *Protides of the Biological Fluids, XXXIV* (H. Peeters, ed.), pp. 145-147, Pergamon, Oxford.

Davies, J. G., and Offord, R. E., 1985, The preparation of tritiated insulin specifically labelled by semisynthesis at glycine-A1, *Biochem. J.* **231**:389-392.

Davies, J. G., Muir, A. V., and Offord, R. E., 1986, Identification of some cleavage sites of insulin by insulin proteinase, *Biochem. J.* **240**:609-612.

Davies, J. G., Rose, K., Bradshaw, C. G., and Offord, R. E., 1987, Enzymatic semisynthesis of insulin specifically labelled with tritium at position B-30, *Protein Engineering* **1**:407-411.

Davies, J. G., Muir, A. V., Rose, K., and Offord, R. E., 1988, Identification of radioactive insulin fragments liberated by insulin proteinase during the degradation of semisynthetic [[^{3}H]Gly^{A1}]insulin and [[^{3}H]Phe^{B1}]insulin, *Biochem. J.* **249**:209-214.

Dixon, H. B. F., and Fields, R., 1972, [33] Specific modification of NH_2-terminal residues by transamnination, *Methods Enzymol.* **25**:409-419.

Duckworth, W. C., and Kitabchi, A. E., 1981, Insulin metabolism and degradation, *Endocrinol. Rev.* **2**:210-233.

Gavish, M., Zakut, R., Wilchek, M., and Givol, D., 1978, Preparation of a semisynthetic antibody, *Biochemistry* **17**:1345-1351.

Glass, J., and Pelzig, M., 1974, Enzymes as reagents in peptide synthesis. Enzyme-labile protection for carboxyl groups, *Proc. Natl. Acad. Sci. USA* **74**:2739-2741.

Hagenmaier, H., Ohms, J. P., Jahns, J., and Anfinsen, C. B., 1978, Studies on the semisynthesis of myoglobin from natural fragments, in: *Semisynthetic Peptides and Proteins* (R. E. Offord and C. Di Bello, eds.), pp. 23-35, Academic Press, London.

Halban, P. A., and Offord, R. E., 1975, The preparation of a semisynthetic tritiated insulin with a specific radioactivity of up to 20 curies per millimole, *Biochem. J.* **151**:219-225.

Haneda, M., Kobayashi, M., Maegawa, H., Watanabe, B., Takata, Y., Ishibashi, O., Shigeta, A., and Inouye, K., 1985, Decreased biologic activity and degradation of human [Ser^{B24}]-insulin, a second mutant insulin, *Diabetes* **34**:568-573.

Harris, D. E., and Offord, R. E., 1977, A functioning complex between tryptic fragments of cytochrome *c*, *Biochem. J.* **161**:21-25.

Hefta, S. A., Lyle, S. B., Busch, M. R., Harris, D. E., Matthew, J. B., and Gurd, F. R., 1988, Site-specific semisynthetic variant of human hemoglobin, *Proc. Natl. Acad. Sci. USA* **85**:709-713.

Hunter, M. L., and Ludwig, M. J., 1962, The reaction of imidoesters with proteins and related small molecules, *J. Am. Chem. Soc.* **84**:3491-3504.

Inouye, K., Watanabe, K., Morihara, K., Tochino, T., Kanaya, T., Emura, J., and Sakakibara, S., 1979, Enzyme-assisted semisynthesis of human insulin, *J. Am. Chem. Soc.* **101**:751-752.

Jones, R. M. L., and Offord, R. E., 1982, The proteinase-catalysed synthesis of peptide hydrazides, *Biochem. J.* **203**:125-129.

Jones, R. M. L., Rose, K., and Offord, R. E., 1987, Semisynthetic human [[^{3}H$_2$]Phe^1]proinsulin, *Biochem. J.* **247**:785-788.

Kaiser, E. T., and Lawrence, D. S., 1984, Chemical mutation of enzyme active sites, *Science* **226**:505-511.

Kihara, H., Ishimaru, K., and Ohno, M., 1981, Cleavage of *Trimeresurus flavoridis* phospholipase A_2 with cyanogen bromide: Sequence of the short peptide fragment and formation of a non-covalently bonded complex from the fragments, *J. Biochem.* **90**:363-370.

Knoblauch, H., Rüterjans, H., Bloemhoff, W., and Kerling, K. E. T., 1988, ^{15}N- and ^{1}H-NMR investigations of the active-site amino acids in semisynthetic RNase S' and RNase A, *Eur. J. Biochem.* **172:**485–497.

Kobayashi, M., Ohgaku, S., Iwasaki, M., Maegawa, H., Shigeta, Y., and Inouye, K., 1982, Supernormal insulin: [D-Phe^{B24}]-insulin with increased affinity for insulin receptors, *Biochem. Biophys. Res. Commun.* **107:**329–336.

Kobayashi, M., Haneda, M., Maegawa, H., Watanabe, N., Takada, Y., Shigeta, Y., and Inouye, K., 1984, Receptor binding and biological activity of [Ser^{B24}]-insulin, an abnormal mutant Insulin, *Biochem. Biophys. Res. Commun.* **119:**49–57.

Kobayashi, M., Takata, Y., Ishibashi, O., Sasaoka, T., Iwasaki, M., Shigeta, Y., and Inouye, K., 1986, Receptor binding and negative cooperativity of a mutant insulin [Leu^{A3}]-insulin, *Biochem. Biophys. Res. Commun.* **137:**250–257.

Kubiak, T., Whitney, D. B., and Merrifield, R. B., 1987, Synthetic peptides V$_H$(27–68) and V$_H$(16–68) of the myeloma immunoglobulin M603 heavy chain and their association with the natural light chain to form an antigen binding site, *Biochemistry* **26:**7849–7855.

Kullmann, W., 1987, *Enzymatic Peptide Synthesis,* 140 pp. CRC Press, Boca Raton, FL.

Laskowski, M., Jr., 1978, The use of proteolytic enzymes for the synthesis of specific peptide bonds in globular proteins, in: *Semisynthetic Peptides and Proteins* (R. E. Offord and C. Di Bello, eds.), pp. 255–262, Academic Press, London.

Leary, T. R., and Laskowski, M., Jr., 1973, Enzymatic replacement of Arg^{63} by Trp^{63} in the reactive site of soybean trypsin inhibitor (Kunitz)—An intentional change from tryptic to chymotryptic specificity, *Fed. Proc.* **32:**465.

Li, C. H., Blaker, J., and Hayashida, T., 1978, Human somatotropin: Semisynthesis of the hormone by noncovalent interaction of the NH$_2$-terminal fragment with synthetic analogs of the COOH-terminal fragment, *Biochem. Biophys. Res. Commun.* **82:**217–222.

Ludwig, M. L., and Byrne, R., 1962, Reversible blocking of protein amino groups by the acetimidyl group, *J. Am. Chem. Soc.* **84:**4160–4162.

Mahrenholz, A. M., Flanders, K. C., Hoosein, N. M., Gurd, F. R-N., and Gurd, R. S., 1987, Semisynthetic D-His^1,N^ϵ-acetimidoglucagon: Structure–function relationships, *Arch. Biochem. Biophys.* **257:**379–386.

Markussen, J., 1980, Process for preparing insulin esters, Danish Patent Application 574/80.

Means, G. E., and Feeney, R. E., 1971, *Chemical Modification of Proteins,* 254 pp., Holden-Day, San Francisco.

Muir, A. V., Offord, R. E., and Davies, J. G., 1986, The identification of a degradation of insulin by insulin proteinase, *Biochem. J.* **237:**631–637.

Nakagawa, S. H., and Tager, H. S., 1986, Role of the phenylalanine B25 side chain in directing insulin interaction with its receptor, *J. Biol. Chem.* **261:**7332–7341.

Neet, K. E., and Koshland, D. E., 1966, The conversion of serine at the active site of subtilisin to cysteine: A "chemical mutation," *Proc. Natl. Acad. Sci. USA* **56:**1606–1611.

Offord, R. E., 1966, Electrophoretic mobilities of peptides on paper and their use in the determination of amide groups, *Nature* **211:**591–593.

Offord, R. E., 1980, *Semisynthetic Proteins,* 235 pp., Wiley, Chichester, New York.

Offord, R. E., 1987, Review—Protein engineering by chemical means? *Protein Engineering* **1:**151–157.

Offord, R. E., and Rose, K., 1986a, Press-stud protein conjugates, in: *Protides of the Biological Fluids XXXIV* (H. Peeters, ed.), pp. 35–38, Pergamon Press, Oxford.

Offord, R. E., and Rose, K., 1986b, Polypeptide and protein derivatives and process for their preparation, G.B. Patent Application No. 8721875.

Offord, R. E., Pochon, S., and Rose, K., 1986, Press-stud protein conjugates, in: *Peptides 1986* (D. Theodoropoulos, ed.), pp. 279–281, de Gruyter, Berlin.

Offord, R. E., and Rose, K., 1987, Polypeptide and protein derivatives and process for their preparation, U.S. Patent Application Ser. No. 043530.

Polgar, L., and Bender, M. L., 1966, Thiol-substitution, *J. Am. Chem. Soc.* **88**:3153–3154.

Proudfoot, A. E. I., and Wallace, C. J. A., 1987, Semisynthesis of cytochrome *c* and analogues, *Biochem. J.* **248**:965–967.

Proudfoot, A. E. I., Wallace, C. J. A., Harris, D. E., and Offord, R. E., 1986, A new non-covalent complex of semisynthetically modified tryptic fragments of cytochrome *c*, *Biochem. J.* **239**:333–337.

Richards, F. M., 1958, Ribonuclease S, *Proc. Natl. Acad. Sci. USA* **44**:162–166.

Rose, K., Pochon, S., and Offord, R. E., 1984, ^{18}O-labelled human insulin: Semisynthesis and mass-spectrometric analysis, in: *Peptides 1984* (U. Ragnarson, ed.) Almquist and Wiksell International, Stockholm.

Rose, K., Herrero, C., Proudfoot, A. E. I., Wallace, C. J. A., and Offord, R. E., 1987, Enzyme-assisted semisynthesis of polypeptide active esters for subsequent spontaneous coupling, pp. 219–220, W. de Gruyter, Berlin. in: *Peptides 1986* (D. Theodoropoulos, ed.), pp. 219–222.

Rose, K., Herrero, C., Proudfoot, A. E. I., Offord, R. E., and Wallace, C. J. A., 1988a, Enzyme-assisted semisynthesis of polypeptide active esters and their use, *Biochem. J.* **249**:83–88.

Rose, K., Jones, R. M. K., Sundaram, G., and Offord, R. E., 1988b, Attachment of linker groups to carboxyl termini using enzyme-assisted reverse proteolysis, pp. 272–276. in: *Proc. 20th Eur. Pep. Symp.* (G. Jung, ed.).

Rose, K., Savoy, L. A., Muir, A. V., Davies, J. G., Offord, R. E., and Turcatti, G., 1988c, Insulin proteinase liberates from glucagon a fragment known to zhave enhanced activity against Ca^{2+} + Mg^{2+}-dependent ATPase, *Biochem. J.* **256**:847–851.

Rosenberg, S., and Terry, W. D., 1977, Passive immunotherapy of cancer in animals and man, *Adv. Cancer Res.* **25**:323–388.

Savoy, L-A., Jones, R. M. L., Pochon, S., Davies, J. G., Muir, A. V., Offord, R. E., and Rose, K., 1988, Identification by fast atom bombardment mass spectrometry of insulin fragments produced by insulin proteinase, *Biochem. J.* **249**:215–222.

Schwartz, G. P., Thompson Burke, G., and Katsoyannis, P. G., 1987, A superactive insulin: [B10-aspartic acid]insulin(human), *Proc. Natl. Acad. Sci. USA* **84**:6408–6411.

Sealock, R. W., and Laskowski, M. Jr., 1969, Enzymatic replacement of the arginyl by a lysyl residue in the reactive site of soybean trypsin inhibitor, *Biochemistry* **8**:3703–3710.

Seetharam, R., and Seetharama Acharya, A. 1986, Synthetic potential of *Staphylococcus aureus* V8-protease: An approach toward semisynthesis of covalent analogs of α-chain of hemoglobin S, *J. Cell. Biochem.* **30**:87–99.

Sheppard, R. C., 1980, Partial synthesis of peptides and proteins, in: *The peptides* (E. Gross and S. Maienhofer, eds.), Vol. 2, pp. 441–484. Academic Press, London.

Shoelson, S. E., Polonsky, K. S., Zeidler, A., Rubenstein, A. H., and Tager, H. S., 1983, Identification of a mutant human insulin predicted to contain a serine-for-phenylalanine substitution, *Proc. Natl. Acad. Sci. USA* **80**:7390–7394.

Shoelson, S. E., Polonsky, K. S., Zeidler, A., Rubinstein, A. H., and Tager, H. S., 1984, Human insulin B24 (Phe → Ser). Secretion and metabolic clearance of the abnormal insulin in man and in a dog model, *J. Clin. Invest.* **73**:1351–1358.

Slaby, I., and Holmgren, A., 1975, Reconstitution of *Escherichia coli* thioredoxin from complementing peptide fragments obtained by cleavage at methionine-37 or arginine-73, *J. Biol. Chem.* **25**:1340–1347.

Taniuchi, H., Afinsen, C. B., and Sodja, A., 1967, Nuclease-T: An active derivative of staphylococcal nuclease composed of two noncovalently bonded peptide fragments, *Proc. Natl. Acad. Sci. USA* **58**:1235–1242.

Tschesche, H., and Kupfer, S., 1976, Hydrolysis–resynthesis equilibrium of the lysine-15-alanine-16 peptide bond in bovine trypsin inhibitor (Kunitz), *Hoppe-Seyler's Z. Physiol. Chem.* **357**:769–776.

Tschesche, H., Beckmann, J., Mehlich, A., Schnabel, E., and Wenzel, H. R., 1987, Semisynthetic engineering of proteinase inhibitor homologues, *Biochem. Biophys. Acta.* **913**:97–101.

van Binsbergen, J., Slotboom, A. J., and de Haas, G. H., 1988, Modification of phospholipase A_2, in: *Proc. 20th Eur. Peptide Symposium* (G. Jung, ed.) (in press).

Varley, J. M., Davies, J. G., Shire, D., Offord, R. E., and Timmis, K. N., 1988, Engineered rat insulin I analogue having a B16 Tyr/Asp replacement exhibits unchanged susceptibility to cleavage by insulin proteinase, *Eur. J. Biochem.* **171**:351–354.

Wallace, C. J. A., 1987, Functional consequences of the excision of an ω-loop. Residues 40–55, from mitochondrial cytochrome *c*, *J. Biol. Chem.* **262**:16767–16770.

Wallace, C. J. A., and Corthésy, B. E., 1986, Protein engineering of cytochrome *c* by semisynthesis: substitutions at glutamic acid 66, *Protein Engineering* **1**:23–27.

Wallace, C. J. A., and Harris, D. E., 1984, The preparation of fully *N*-ε-acetimylidated cytochrome *c*, *Biochem. J.* **217**:589–594.

Wallace, C. J. A., and Proudfoot, A. E. I., 1987a, Semisynthesis of a deletion mutant, of cytochrome *c* by condensation of enzymically activated fragments, in: *Proc. 10th Amer. Peptide Symposium* (C. M. Deber, V. J. Hruby, and K. D. Kopple, eds.), pp. 372–375.

Wallace, C. J. A., and Proudfoot, A. E. I., 1987b, On the relationship between oxidation-reduction potential and biological activity in cytochrome *c* analogues, *Biochem. J.* **245**:773–779.

Wallace, C. J. A., Corradin, G., Borin, G., and Marchiori, F., 1986, Cytochrome *c* chimerae from natural and synthetic fragments: Significance of the biological properties, *Biopolymers* **25**:2121–2132.

Widmer, F., and Johansen, J. F., 1979, Enzymatic peptide synthesis. Carboxypeptidase Y catalysed formation of peptide bonds, *Carlsberg Res. Commun.* **44**:37–46.

Widmer, F., Breddam, K., and Johansen, J. T., 1981, Influence of the structure of the amine component on carboxypeptidase Y catalysed peptide bond formation, *Carlsberg Res. Commun.* **46**:97–106.

Wieczorek, M., Park, S. J., and Laskowski, M. Jr., 1987, Covalent hybrids of ovomucoid third domains made from one synthetic and one natural peptide chain, *Biochem. Biophys. Res. Commun.* **144**:499–504.

Wilson, M. E., and Whitesides, G. M., 1978, Conversion of a protein to a homogeneous asymmetric hydrogenation catalyst by site-specific modification with a diphosphinerhodium (I) moiety, *J. Am. Chem. Soc.* **100**:306–307.

Yagisawa, S., 1981, Studies on protein semisynthesis. I. Formation of esters, hydrazides, and substituted hydrazides of peptides by the reverse reaction of trypsin, *J. Biochem. (Tokyo)* **89**:491–501.

Zahn, H., Naithani, V. K., Gattner, H. G., Büllesbach, E. E., and Thamm, Pl.M., 1981, Protein-Semisynthese mit Hilfe gemischter Anhydride und Enzyme, *Naturwissenschaften* **68**:56–62.

New Engineered Proteins for Use in Therapy and Vaccine Design

J. P. LECOCQ, M. COURTNEY, E. DEGRYSE, S. JALLAT, M. P. KIENY, P. MEULIEN, and A. PAVIRANI

1. Introduction

Protein engineering has opened a new door in biotechnology. With localized mutagenesis of the corresponding DNA, new proteins with altered activity and better stability can be obtained. This new technology has stimulated the imagination of scientists, and in the past few years spectacular results have been obtained.

We will summarize some results recently obtained in our laboratory with molecules involved in the coagulation cascade. Protein engineering is also useful in improving the antigenicity of naturally poor antigen and we will describe the modifications of the envelope protein of HIV.

2. New Generation of Factor VIII Molecules for the Treatment of Hemophilia A

Factor VIII (FVIII), a large plasma glycoprotein, is an essential cofactor in the blood clotting cascade in the step in which activated factor IX (FIXa) converts factor X to activated FX (FXa). The chromosome X-linked bleeding disorder hemophilia A is due to the absence or malfunction of FVIII. In plasma, FVIII is noncovalently linked to the von Willebrand factor (vWF), which stabilizes this

J. P. LECOCQ, M. COURTNEY, E. DEGRYSE, S. JALLAT, M. P. KIENY, P. MEULIEN, and A. PAVIRANI • Transgene S.A., 67082 Strasbourg Cedex, France.   *Present address for M. C.:* Delta Biotechnology, Nottingham NG7 AFD, United Kingdom.

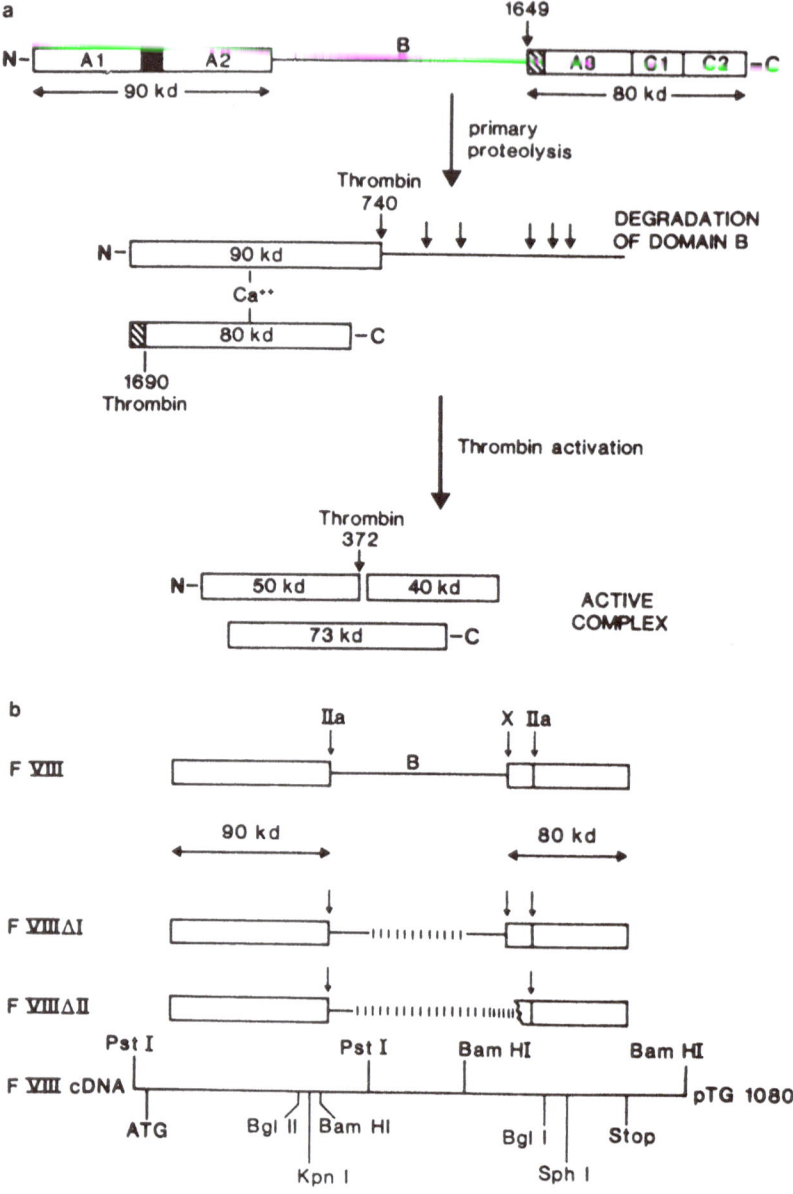

Figure 1. (a) Proteolytic processing of FVIII, after Toole *et al.* (1984) and Eaton *et al.* (1986a). The primary cleavage by an unknown protease occurs between amino acid 1648 and 1649. Degradation of the B domain is then thought to occur prior to activation by thrombin (IIa). The association between heavy and light chains is mediated by a metal cation, probably Ca^{2+}. (b) Representation of FVIII and two deletion derivatives, FVIIIΔI, in which DNA-encoding amino acids 868–1582 are missing, and FVIIIΔII, which lacks the DNA-encoding amino acids 771–1666. Deletions are denoted by

labile molecule. In the absence of vWF, for example in patients with severe von Willebrand disease, levels of FVIII activity are severely affected.

The cloning of the human FVIII cDNA by three independent groups (Wood *et al.*, 1984; Toole *et al.*, 1984; Truett *et al.*, 1985) has shed light on some structural and functional relationships of this complex protein. FVIII is now known to contain 2332 amino acids (aa) and has internal homologies predicting a domain structure of A1-A2-B-A3-C1-C2 (Toole *et al.*, 1984; Vehar *et al.*, 1984). The large unique B domain that separates the heavy and light chains is encoded by a single 3.1-kb exon (Gitschier *et al.*, 1984).

The proteolytic processing of the FVIII protein, schematized in Figure 1a, is associated with the appearance of different molecular forms on SDS gels. The primary cleavage, which may well occur inside the cell, is caused by an unknown protease (X). Further cleavages by thrombin (IIa) activate the molecule. It is now clear from several studies correlating different molecular species with FVIII activity that the central portion (B domain) is not required for procoagulant function, as it does not partake in the active complex (Toole *et al.*, 1984; Vehar *et al.*, 1984; Fulcher *et al.*, 1983; Eaton *et al.*, 1986a).

This hypothesis was further confirmed using molecular biology techniques: several groups have demonstrated the expression from recombinant mammalian cell culture systems of active FVIII variant molecules that partially lack the B domain. This was achieved either by coexpressing DNA constructs encoding the heavy and light chain separately (Burke *et al.*, 1986; Pavirani *et al.*, 1987a), or by expressing molecules deleted for part of the B domain (Toole *et al.*, 1986; Eaton *et al.*, 1986b).

In our laboratory, we have used the vaccinia virus expression system as a means of studying the expression of FVIII in different cell types, investigating ways of stabilizing this fragile molecule and testing various new FVIII variant molecules (Pavirani *et al.*, 1987b; Meulien *et al.*, 1988). Two such molecules are depicted in Figure 1b. Whereas FVIIIΔI contains all the major cleavage sites associated with FVIII processing, FVIIIΔII lacks the site at aa1648/1649 cleaved by the unidentified protease. In the vaccinia virus system FVIIIΔII is expressed at a level five times that of complete FVIII and over twice that of FVIIIΔI.

In order to more fully characterize FVIIIΔII, recombinant cell lines expressing either complete FVIII or FVIIIΔII were established by transfecting CHO cells with an expression vector capable of expressing heterologous genes in mammalian cells (Balland *et al.*, 1988), into which the appropriate cDNA was integrated. The

---

(| | | | |). Cleavage sites are shown for thrombin (IIa) and the unknown protease (X) that cleaves between amino acids 1648–1649 (the latter deleted in FVIIIΔII). At the bottom the FVIII cDNA plasmid pTG1080 (Pavirani *et al.*, 1987b) is represented showing restriction enzyme sites used in the various constructions.

expressed protein was characterized with respect to both its electrophoretic migration on SDS-gel and its thrombin activation profile. Whereas full-length FVIII migrates as several bands on SDS-gel, ranging from 200 to 80 kDa, FVIIIΔII migrates as a single polypeptide species of ~165 kDa. This was to be expected, since the deletion abolished the primary cleavage site at aa1648/1649. One possible drawback to manipulating the cDNA close to the thrombin cleavage sites is that a molecule could be created that would no longer be sensitive to thrombin activation. As shown in Figure 2, FVIIIΔII is activated to a two- to fourfold greater extent than complete FVIII (recombinant or plasma derived). This thrombin activation profile may therefore reflect either a more stable molecule or a molecule with a higher potential specific activity. The fact that FVIIIΔII migrates as a single-chain species on SDS-gels could indicate that FVIIIΔII is a more stable molecule and would therefore be more resistant to proteolytic inactivation than the complete FVIII molecule, which is known to be partially activated when isolated from plasma.

Because of these unique features, as well as the observation that this new molecule still retains the ability to bind to vWF, FVIIIΔII may well constitute an interesting candidate for a second-generation hemophilia A therapeutic.

3. Hirudin: Variants with Improved Antithrombic Activity

Hirudins are polypeptides of 64–66 amino acids synthesized by the medicinal leech *Hirudo medicinalis*. They share 85–90% homology and contain a sulfated tyrosine at position 63–64. Hirudins are capable of a highly specific binding to the central blood coagulation protease α-thrombin, with its concomitant complete inhibition, *in vitro*. Moreover, low doses prevent thrombus formation *in vivo* in a number of animal models, and evidence is accumulating that high doses are innocuous to humans. These observations stimulated the efforts of several groups to use genetic engineering techniques in order to produce such molecules for their use in pharmacological applications (Harvey *et al.*, 1986; Dodt *et al.*, 1986; Fortkamp *et al.*, 1986).

The cDNA coding for hirudin was cloned (Harvey *et al.*, 1986) and expressed in yeast (Loison *et al.*, 1988). The recombinant protein (variant HV2) was characterized (Riehl-Bellon *et al.*, 1989). In order to improve the thrombin inhibitory properties of recombinant hirudin, localized mutagenesis of the cDNA was performed and the corresponding proteins were analyzed.

We observed that a change of lys35 to thr35 had no effect on activity; thus lys35 is probably not part of the "active site" of hirudin (Degryse *et al.*, 1989), even though lys35 is present in an external finger domain (residues 31–36) (Clore *et al.*, 1987).

On the contrary, when asn47, an amino acid believed to be part of the

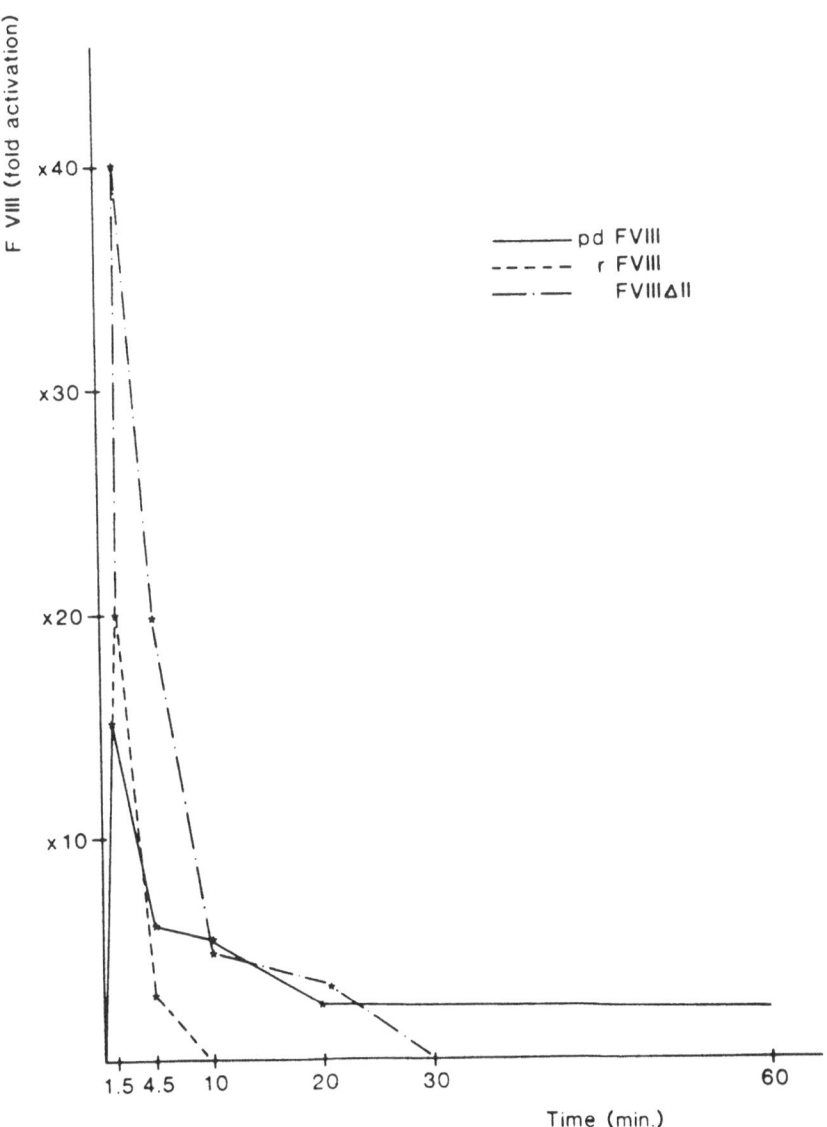

Figure 2. Thrombin activation profile (average of 10 determinations) of pdFVIII-, rFVIII, and FVIIIΔII, showing increased activation in the case of FVIIIΔII compared to plasma-derived or recombinant molecules.

"active site" of hirudin, was replaced with lys47 or arg47, the obtained variants of HV2 displayed an increased antithrombin activity both *in vitro* and *in vivo* (Degryse *et al.*, 1989). Figure 3 shows a plot of the thrombin inhibition curves obtained with HV2 and HV2 arg47, respectively. HV2 lys47 and HV2 arg47 are capable of a higher thrombin inhibition than HV2, at the same concentration. Surprisingly, the same correlation with the *in vitro* results can be demonstrated *in vivo* (Degryse *et al.*, 1989).

Our kinetic analysis demonstrated the presence of two kinetic components: Kis and Kii. The first component, Kis, is substrate dependent and points to the presence of a competitive element between the inhibitor and substrate for binding on the active site. The second component, Kii, is substrate independent. It points to an element in the thrombin–hirudin inhibition which is independent of the substrate binding and might therefore indicate a second binding site for hirudin on thrombin. As thrombin has preference for cleavage after lysine and arginine residues, it becomes tempting to speculate that the asn47 → lys47 or arg47 exchange occurred at the "active site" of hirudin. However, the contribution to the overall binding energy is only on the order of 1–1.5 kcal/mole (Degryse *et al.*, 1989), which is comparable to that of the sulfate group of the penultimate tyrosine (Dodt *et al.*, 1987; Stone and Hofsteenge, 1986).

The relatively small contribution to the binding strength of lys47, present in

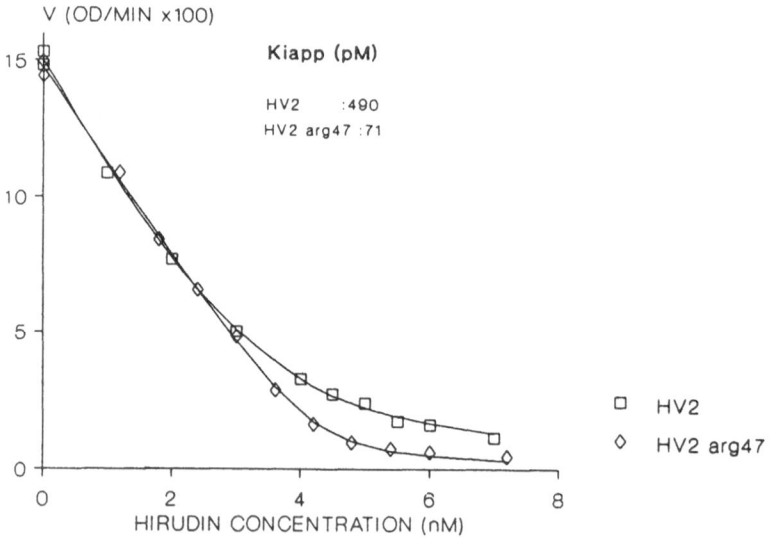

Figure 3. Thrombin inhibition curve by recombinant hirudins measured as described (Degryse *et al.*, 1989). The substrate, Chromozyn PL, was used at 500 μM for HV2 and 460 μM for HV2 arg47.

a domain that shows a marked homology with thrombin substrates, contrasts with the major loss of thrombin inhibitory capacity on the deletion of nine C-terminal amino acids (resulting in a reduction on the order of 5–6 kcal/mole). We have therefore proposed that hirudin binds thrombin at its active site and that additional binding energy is obtained at another thrombin binding site, through the hirudin C-terminus. Hirudin would thus act as a clamp obstructing the thrombin active site. The fact that sufficient binding energy can be obtained other than through the lys47 or arg47 interaction with the basic pocket on thrombin explains why HV2 still acts as an excellent thrombin inhibitor. This kind of information could be used to design third-generation products with modified activity profiles.

4. α_1-Antitrypsin

α_1-Antitrypsin (α_1-AT) is a potent neutrophil elastase inhibitor that functions in protecting the human lung from proteolytic damage; a hereditary autosomal deficiency leads to development of severe emphysema in homozygotes. Oxidation of a Met residue (Met-358) at the active site prevents formation of the α_1AT–elastase inhibition complex, thus inactivating the molecule. Oxidative inactivation occurs *in vivo* and is thought to play an important pathological role in smokers' emphysema, adult and neonatal respiratory distress syndrome, and inflammatory joint diseases. α_1AT replenishment thus represents a rational therapy for these conditions. The possibility of producing an oxidation-resistant α_1AT, in addition to affording a real advantage in the preparation and storage of the product, should permit enhanced efficacy and thus a lower effective dose.

Using site-directed mutagenesis, the oxidation-sensitive Met-358 has been replaced with "inert" residues that fit the elastase active site (i.e., residues that are recognized as a cleavage site by elastase). α_1AT (Leu-358), (Val-358), (Ala-358), and (Ile-358) were produced in *Escherichia coli;* each inhibits elastase and is resistant to oxidation (Courtney *et al.*, 1985; Jallat *et al.*, 1986). α_1AT (Leu-358) is of particular interest, since it displays the most efficient elastase inhibition kinetics and it alone remains an effective inhibitor of cathepsin G, the second elastolytic protease from neutrophils.

The inhibitory activity of α_1AT has been further modified by substituting the active site Met-358 residue with Arg (Courtney *et al.*, 1985). This mutation produced a switch in the specificity of inhibition from elastase to Arg-specific proteases, including thrombin and the contact-phase protease, plasma kallikrein (Schapira *et al.*, 1986). α_1AT blocks thrombin formation in animal models and, furthermore, because of its anticontact phase function, acts as a powerful antihypotensive in a rat model. The specificity can be tuned even more finely. The kallikrein inhibition kinetics (Table I) can be enhanced at the expense of thrombin by introducing a second mutation, giving α_1AT (Ala-357, Arg-358), thus

Table I. Kinetic Constants for Inactivation of Kallikrein β-Factor XIIa and Thrombin by Recombinant α_1-Antitrypsin Mutants and Natural Cl Inhibitor[a]

	Thrombin	Plasma kallikrein	β-FXIIa
αAT (Val^{358})	3,000	100	100
αAT (Arg^{358})	360,000	69,000	35,000
αAT (Ala^{357}, Arg^{358})	73,000	360,000	9,200
Cl INH	ND	17,000	3,100

[a]Values are in M^{-1} sec^{-1}.

augmenting homology with the active site of Cl inhibitor, the natural kallikrein inhibitor (Schapira *et al.*, 1987). This variant no longer affects coagulation parameters, *in vivo*, whereas strong antihypotensive activity is retained. α_1AT (Ala-357, Arg-358) is now being assessed for the treatment of septic shock and angioedema attacks associated with Cl inhibitor deficiency.

5. HIV Envelope Protein: Improved Antigenicity by Cleavage Site Removal

The major antigen present at the surface of HIV particles and on the HIV-infected cell is the envelope (*env*) glycoprotein, and efforts to develop a vaccine against HIV have concentrated on this protein. The HIV-1 *env* gene encodes an 861 amino acid polypeptide; a signal sequence at the N-terminus directs entry of the precursor into a secretory pathway where the signal peptide is removed and the protein is extensively N-glycosylated. The 160-kDa glycoprotein gp160 is subsequently cleaved internally to generate the N-terminal gp120 extracellular protein and the C-terminal gp41 transmembrane protein. These two proteins remain loosely associated at the cell surface by a mechanism that does not appear to involve inter-protein disulfide links (McDougal *et al.*, 1986).

In generating anti-HIV immunity, we and others have used recombinant techniques to express *env* in different hosts (Kieny *et al.*, 1986; Hu *et al.*, 1986, 1987; Chakrabarti *et al.*, 1986; Cabradilla *et al.*, 1986; Lasky *et al.*, 1986; Barr *et al.*, 1987; Adams *et al.*, 1987; Rusche *et al.*, 1987). However, vaccines based on recombinant *env* protein or on *env* expressed from recombinant vaccinia virus (VV) have elicited low titers of HIV-reactive antibodies in experimental animals. In view of our results obtained with the soluble rabies glycoprotein, the shedding of gp120 from the cell surface may be partly responsible for the low immunogenicity of vaccinia recombinants (Kieny *et al.*, 1986). We thus sought to improve their immunogenicity by constructing recombinants expressing *env* variants with modified transmembrane anchor regions and altered cleavage sites (Fig. 4). The production of a VV recombinant expressing the intact *env* glycoprotein from the HIV LAV-Bru isolate was previously described (Kieny *et al.*, 1986). We have subsequently developed variants in which the cleavage site separating the gp120 and gp41 regions has been eliminated.

Cleavage of the gp160 precursor to yield gp120 and gp41 occurs at nucleotide 7303–7314 in Wain-Hobson *et al.* (1985). We sought to prevent the shedding of gp120 from the cell surface by replacing conserved basic residues in the cleavage sequence with other amino acids. Site-directed mutagenesis allowed the construction of a recombinant in which the sequence REKR is replaced by NEHQ. However, cells infected by the recombinant showed only a slight decrease in the level of gp160 cleavage, in contrast to a recent report (McCune *et al.*, 1988) in which modification appears to block cleavage. Examination of the amino acid sequence identified a second conserved cluster of basic amino acids to the N-terminal side of the cleavage region (nt. 7254–7281); site-directed

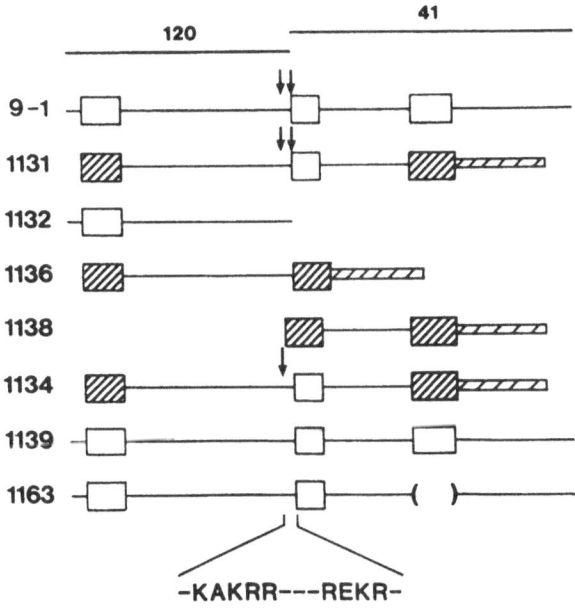

Figure 4. Schematic representation of *env* glycoprotein variants separately encoded by vaccinia recombinant viruses. The cleavage sites separating the gp120 and gp41 moieties (overlined) are indicated by arrows. Boxes, Hydrophobic domains; lines, hydrophilic domains. Zones derived from the rabies virus glycoprotein are cross-hatched. The bracketed area in recombinant 1163 denotes a deletion. The amino acid sequence at the two putative cleavage site(s) is indicated below.

Table II. Properties of Vaccinia-env Recombinant Viruses

	Cell surface fluorescence	Shedding of gp120 into culture supernatant	Capacity to promote		Western recognition		
			Aggregation	Cell fusion	gp160	gp120	gp41
9–1	+	++	++	++	$1/50$	$1/50$	$1/25$
1132	−	++	−	NA	$<1/25$	$<1/25$	NA
1136	+	++	++	−	$<1/25$	$1/50$	NA
1138	+	NA	−	NA	$1/100$	NA	$<1/25$
1134	++	+	++	−	$1/200$	$1/200$	$<1/25$
1139	+++	−	++	−	$1/500$	$1/500$	$1/200$
1163	(+)	NA	−	NA	$<1/25$	$<1/25$	$<1/25$

mutagenesis (KAKRR to KAQNH) was used to eliminate this second region. Immunoprecipitation analysis of extracts of cells infected with the second variant revealed almost complete inhibition of gp160 cleavage (Kieny et al., 1986) and the intensity of anti-env cell-surface fluorescence was markedly increased, demonstrating that altering the cleavage site(s) can block release of gp120 from the cell surface.

To determine whether altered properties of env variants might affect the immunogenicity of VV recombinants, mice inoculated with the live recombinants were examined for serum immunoglobulin capable of reacting with HIV proteins. Whereas parental recombinant yielded low titers of antibody reacting with gp41, gp120, or gp160, as reported previously (Kieny et al., 1986), inoculation with recombinants in which the proteolytic cleavage site(s) within env had been modified gave a significant increase in antibody titer (Table II). An analogous construction lacking the transmembrane anchor region, and thus resulting in the production of a soluble env, was negative in this assay, as expected.

Therefore, suppression of cleavage sites of gp160 leads to an improved immunogenicity of the protein.

ACKNOWLEDGMENTS. The authors thank all their collaborators who participated in the projects described in this chapter. The program on FVIII is supported by CNTS, the program on HIV by Pasteur Vaccins. The secretarial assistance of N. Monfrini was greatly appreciated.

References

Adams, S. E., Dawson, K. M., Gull, K., Kingsman, S. M., and Kingsman, A. J., 1987, The expression of hybrid HIV:Ty virus-like particles in yeast, Nature 329:68.
Balland, A., Faure, T., Carvallo, D., Cordier, P., Ulrich, P., Fournet, B., De La Salle, H., and

Lecocq, J-P., 1988, Characterization of two differently processed forms of human recombinant factor IX synthesized in CHO cells transformed with a polycistronic vector, *Eur. J. Biochem.* **172:**565–572.

Barr, P. J., Steimer, K. S., Sabin, E. A., Parkes, D., George-Nascimento, C., Stephans, J. C., Powers, M. A., Gyenes, A., Van Nest, G. A., Miller, E. T., Higgins, K. W., and Luciw, P. A., 1987, Antigenicity and immunogenicity of domains of the human immunodeficiency virus (HIV) envelope polypeptide expressed in the yeast *Saccharomyces cerevisiae*, *Vaccine* **5:**90.

Burke, R. L., Pachl, C., Quiroga, M., Rosenberg, S., Haigwood, N., Nordfang, O., and Ezban, M., 1986, The functional domains of coagulation factor VIII:C, *J. Biol. Chem.* **261:**12574–12578.

Cabradilla, C. D., Groopman, J. E., Lanigan, J., Renz, M., Lasky, L. A., and Capon, D. J., 1986, Serodiagnosis of antibodies to the human AIDS retrovirus with a bacterially synthesized *env* polypeptide, *Bio/Technol.* **4:**128.

Chakrabarti, S., Robert-Guroff, M., Wong-Staal, F., Gallo, R. C., and Moss, B., 1986, Expression of the HTLV-III envelope gene by a recombinant vaccinia virus, *Nature* **320:**535.

Clore, G. M., Sukumaran, D. K., Nilges, M., Zarbock, J., and Gronenborn, A. M., 1987, The conformations of hirudin in solution: A study using nuclear magnetic resonance, distance geometry and restrained molecular dynamics, *EMBO J.* **6:**529–537.

Courtney, M., Jallat, S., Tessier, L. H., Benavente, A., Crystal, R. G., and Lecocq, J-P., 1985, Synthesis in *E. coli* of alpha₁-antitrypsin variants of therapeutic potential for emphysema and thrombosis, *Nature* **313:**149–151.

Degryse, E., Acker, M., Defreyn, G., Bernat, A., Maffrand, J. P., Roitsch, C., and Courtney, M., 1989, Point mutations modifying the thrombin inhibition kinetics and antithrombotic activity *in vivo* of recombinant hirudin, *Protein Engineering* **2:**459–465.

Dodt, J., Schmitz, T., Schäfer, T., and Bergmann, C., 1986, Expression, secretion and processing of hirudin in E. coli using the alkaline phosphatase signal sequence, *FEBS Lett.* **202:**373–377.

Dodt, J., Seemüller, U., and Fritz, H., 1987, Influence of chain shortening on the inhibitor properties of hirudin and eglin c, *Biol. Chem. Hoppe-Seyler* **368:**1447–1453.

Eaton, D., Rodriguez, H., and Vehar, G. A., 1986a, Proteolytic processing of human factor VIII. Correlation of specific cleavages by thrombin, factor Xa, and activated protein C with activation and inactivation of factor VIII coagulant activity, *Biochemistry* **25:**505–512.

Eaton, D. L., Wood, W. I., Eaton, D., Hass, P. E., Hollingshead, P., Wion, K., Mather, J., Lawn, R. M., Vehar, G. A., and Gorman, C., 1986b, Construction and characterization of an active factor VIII variant lacking the central one-third of the molecule, Biochemistry **25:**8343–8347.

Fortkamp, E., Rieger, M., Heisterberg-Moutses, G., Schweitzer, S., and Sommer, R., 1986, Cloning and expression in *Escherichia coli* of a synthetic DNA for hirudin, the blood coagulation inhibitor in the leech, *DNA* **5:**511–517.

Fulcher, C. A., Roberts, J. R., and Zimmerman, T. S., 1983, Thrombin proteolysis of purified factor VIII procoagulant protein: Correlation with generation of a specific polypeptide, *Blood* **61:**807–811.

Gitschier, J., Wood, W. I., Goralka, T. M., Wion, K. L., Chen, E. Y., Eaton, D. H., Vehar, G. A., Capon, D. J., and Lawn, R. M., 1984, Characterization of the human factor VIII gene, Nature **312:**326–330.

Harvey, R. P., Degryse, E., Stefani, L., Schamber, F., Cazenave, J-P., Courtney, M., Tolstoshev, P., and Lecocq, J-P., 1986, Cloning and expression of a cDNA coding for the anticoagulant hirudin from the bloodsucking leech, *Hirudo medicinalis, Proc. Natl. Acad. Sci. USA* **83:**1084–1088.

Hu, S. L., Kosowski, S. G., and Dalrymple, J. M., 1986, Expression of AIDS virus envelope gene in recombinant vaccinia viruses, *Nature* **320:**537.

Hu, S. L., Kosowski, S. G., and Schaaf, K. F., 1987, Expression of envelope glycoproteins of human immunodeficiency virus by an insect virus vector, *J. Virol.* **61:**3617–3620.

Jallat, S., Caravallo, D., Tessier, L. H., Roecklin, D., Roitsch, C., Ogushi, F., Crystal, R. G., and

Courtney, M., 1986, Altered specificities of genetically engineered α_1-antitrypsin variants, *Protein Engineering* 1:29–35.

Kieny, M. P., Rautmann, G., Schmitt, D., Dott, K., Wain-Hobson, S., Alizon, M., Girard, M., Chamaret, S., Laurent, A., Montagnier, L., and Lecocq, J. P., 1986, AIDS virus *env* protein expressed from a recombinant vaccinia virus, *Bio/Technol.* 4:790.

Lasky, L. A., Groopman, J. E., Fennie, C. W., Benz, P. M., Capon, D. J., Dowbenko, D. J., Nakamura, G. R., Nunes, W. M., Renz, M. E., and Berman, P. W., 1986, Neutralization of the AIDS retrovirus by antibodies to a recombinant envelope glycoprotein, *Science* 233:209–212.

Loison, G., Findeli, A., Bernard, S., Nguyen-Juilleret, M., Marquet, M., Riehl-Bellon, N., Carvallo, D., Guerra-Santos, L., Brown, S. W., Courtney, M., Roitsch, C., and Lemoine, Y., 1988, Expression and secretion in *S. cerevisiae* of biologically active leech hirudin, *Bio/Technol.* 6:72–77.

McCune, J. M., Rabin, L. B., Feinberg, M. B., Lieberman, M., Kosek, J. C., Reyes, G. R., and Weissman, I., 1988, Endoporteolytic cleavage of gp160 is required for the activation of human immunodeficiency virus, *Cell* 53:55.

McDougal, J. S., Kennedy, M. S., Sligh, J. M., Cort, R. C., Mawie, A., and Nicholson, J. K. A., 1986, Binding of HTLV-III/LAV to T4 + T cells by a complex of the 110 k viral protein and the T4 molecule, *Science* 231:382.

Meulien, P., Faure, T., Mischler, F., Harrer, H., Ulrich, P., Bouderbala, B., Dott, K., Sainte-Marie, M., Mazurier, C., Wiesel, M.-L., Van de Pol, H., Cazenaze, J.-P., Courtney, M., and Pavirani, A., 1988, A new recombinant procoagulant protein derived from the cDNA encoding human factor VIII, *Prot. Engineer.* 2:301–306.

Pavirani, A., Meulien, P., Harrer, H., Dott, K., Mischler, F., Wiesel, M-L., Mazurier, C., Cazenave, J-P., and Lecocq, J-P., 1987a, Two independent domains of factor VIII co-expressed using recombinant vaccinia viruses have procoagulant activity, *Biochem. Biophys. Res. Commun.* 145:234–240.

Pavirani, A., Meulien, P., Harrer, H., Schamber, F., Dott, K., Villeval, D., Cordier, Y., Wiesel, M-L., Mazurier, C., Van der Pol, H., Piquet, Y., Cazenave, J-P., and Lecocq, J-P., 1987b, Choosing a host cell for active recombinant factor VIII production using vaccinia virus, *Bio/Technol.* 5:389–392.

Riehl-Bellon, N., Carvallo, D., Acker, M., Van Dorsselaer, A., Marquet, M., Loison, G., Lemoine, Y., Brown, SW., Courtney, M., and Roitsch, C., 1989, Purification and biochemical characterization of recombinant hirudin produced by *S. cerevisiae*, *Biochemistry* 28:2941–2949.

Rusche, J. R., Lynn, D. L., Robert-Guroff, M., Langlois, A. J., Lyerly, H. K., Carson, H., Krohn, K., Ranki, A., Gallo, R. C., Bolognesi, D. P., Putney, S. D., and Matthews, T. J., 1987, Humoral immune response to the entire human immunodeficiency virus envelope glycoprotein made in insect cells, *Proc. Natl. Acad. Sci. USA* 84:6924–6928.

Schapira, M., Ramus, M. A., Jallat, S., Carvallo, D., and Courtney, M., 1986, Recombinant α_1-antitrypsin Pittsburgh (Met[358] → Arg) is a potent inhibitor of plasma kallikrein and activated factor XII fragment, *J. Clin. Invest.* 77:635–637.

Schapira, M., Ramus, M. A., Waeber, B., Brunner, H. R., Jallat, S., Carvallo, D., Roitsch, C., and Courtney, M., 1987, Protection by recombinant α_1-antitrypsin Ala[357] Arg[358] against arterial hypotension induced by factor XII fragment, *J. Clin. Invest.* 80:582–585.

Stone, S. R., and Hofsteenge, J., 1986, Kinetics of the inhibition of thrombin by hirudin, *Biochemistry* 25:4622–4628.

Toole, J. J., Knopf, J. L., Wozney, J. M., Sultzman, L. A., Buecker, J. L., Pittman, D. D., Kaufman, R. J., Brown, E., Shoemacker, C., Orr, E. C., Amphlett, G. W., Foster, W. B., Coe, M. L., Knutson, G. J., Fass, D. N., and Hewick, R. M., 1984, Molecular cloning of a cDNA encoding human antihaemophilic factor. *Nature* 312:342–347.

Toole, J. J., Pittman, D. D., Orr, E. C., Murtha, P., Wasley, L. C., and Kaufman, R. J., 1986, A large region (~95 kDa) of human factor VIII is dispensable for *in vitro* procoagulant activity, *Proc. Natl. Acad. Sci. USA* **83**:5939–5942.

Truett, M. A., Blacher, R., Burke, R. L., Caput, D., Chu, C., Dina, D., Hartog, K., Kuo, C. H., Masiarz, F. R., Merryweather, J. P., Najarian, R., Pachl, C., Potter, S. J., Puma, J., Quiroga, M., Rall, L. B., Randolph, A., Urdea, M. S., Valenzuela, P., Dahl, H. H., Favalaro, J., Hansen, J., Nordfang, O., and Ezban, M., 1985, Characterization of the polypeptide composition of human factor VIII:C and the nucleotide sequence and expression of the human kidney cDNA, *DNA* **4**:333–349.

Vehar, G. A., Keyt, B., Eaton, D., Rodriguez, H., O'Brien, D. P., Rotblat, F., Oppermann, H., Keck, R., Wood, W. I., Harkins, R. N., Tuddenham, E. G. D., Lawn, R. M., and Capon, D. J., 1984, Structure of human factor VIII, *Nature* **312**:337–342.

Wain-Hobson, S., Sonigo, P., Danos, O., Cole, S., and Alizon, M., 1985, Nucleotide sequence of the AIDS virus, LAV, *Cell* **40**:9–17.

Wood, W. I., Capon, D. J., Simonsen, C. C., Eaton, D. L., Gitschier, J., Keyt, B., Seeburg, P. H., Smith, D. H., Hollingshead, P., Wion, K. L., Delwart, E., Tuddenham, E. G. D., Vehar, G. A., and Lawn, R. M., 1984, Expression of active human factor VIII from recombinant DNA clones, *Nature* **312**:330–337.

Receptor-Based Design of Dihydrofolate Reductase Inhibitors

LEE F. KUYPER

1. Introduction

The enzyme dihydrofolate reductase (DHFR) has become a well-established target for drug action since it was identified about 30 years ago. Clinically useful drugs whose activity stems from DHFR inhibition include the antibacterial agent trimethoprim (TMP, 1) (see Finland *et al.*, 1982), and methotrexate (MTX, 2) (see Roth and Cheng, 1982), a compound used in the treatment of certain forms of cancer.

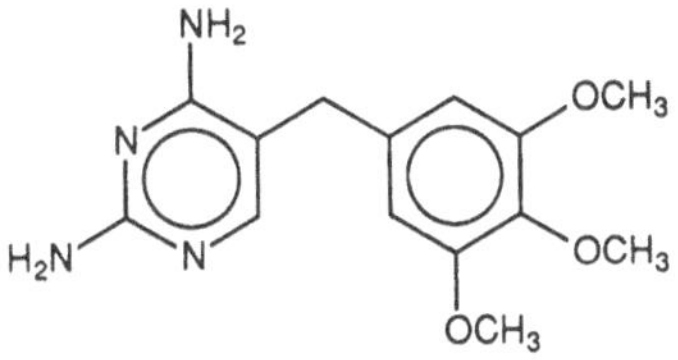

1 TMP

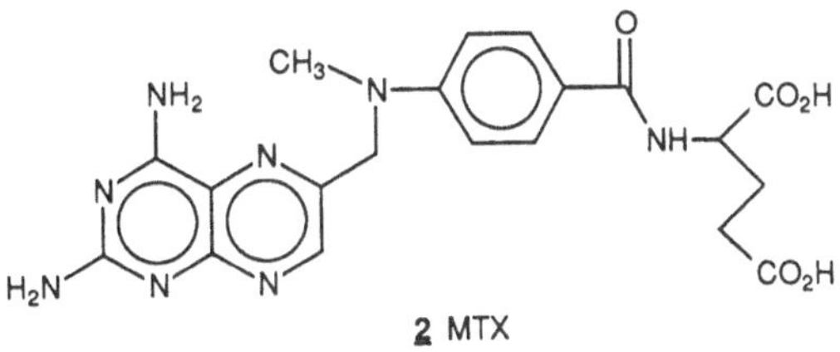

2 MTX

LEE F. KUYPER • Wellcome Research Laboratories, Research Triangle Park, North Carolina 27709.

The literature covering DHFR-related research is extensive and has been recently reviewed (see Kraut and Matthews, 1987; Champness *et al.*, 1986a; Blakley, 1984; Freisheim and Matthews, 1984; Beddell, 1984). Among recent developments, the determination of the three-dimensional molecular structure of the enzyme (see Oefner *et al.*, 1988, and references therein) is clearly the most interesting to the medicinal chemist. This structural information will, hopefully, provide an understanding of protein–ligand interactions and the insight necessary for the rational design of novel ligands. This chapter chronicles some of our attempts to design analogs of trimethoprim based on X-ray crystal structures of DHFR.

2. The Enzyme

2.1. Role in Cellular Metabolism

DHFR catalyzes the NADPH-dependent reduction of the 5,6-double bond of dihydrofolate (FH_2) as illustrated in Fig. 1. The product of this reduction,

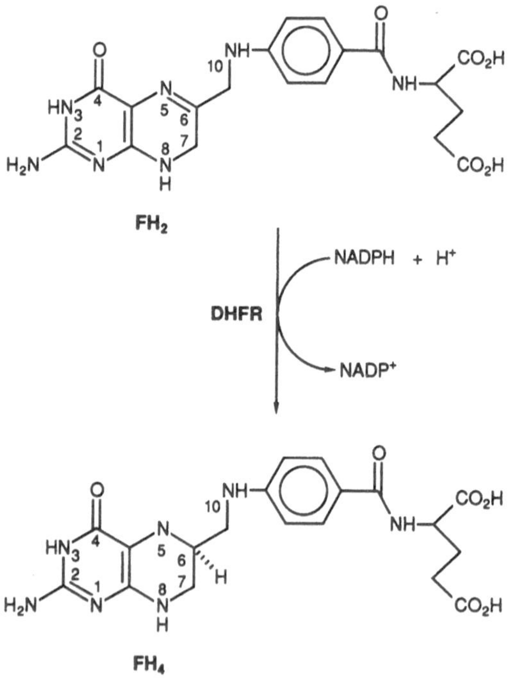

Figure 1. Reduction of dihydrofolate to tetrahydrofolate by NADPH as catalyzed by dihydrofolate reductase.

tetrahydrofolate (FH$_4$), is used in a number of important biosynthetic pathways, in which it serves as a carrier of various one-carbon functionalities (Hitchings, 1983). For example, in the biosynthesis of methionine, homocysteine is methylated using the methyl group of 10-methyl FH$_4$. Similarly, two-ring carbon atoms of adenosine and guanosine are derived from the 10-formyl and 5,10-methenyl derivatives of FH$_4$. In each instance, FH$_4$ functions simply as a shuttle for the one-carbon unit and exits the process unscathed and ready to be used again.

The fate of FH$_4$ is different in the biosynthesis of thymidylate, as shown in Fig. 2. In that case the 5,10-methylene derivative of FH$_4$ acts as a cofactor for thymidylate synthase in its methylation of deoxyuridylate. The methyl group of thymidylate is formed from the 5,10-methylene moiety of the cofactor, along with the hydrogen at carbon 6, and FH$_2$ is produced in the process. This reaction is the only known use of an FH$_4$ derivative in which the cofactor is oxidized to the dihydro form and is also the only recognized *de novo* source of thymidylate, an essential building block of DNA. Therefore DNA synthesis consumes FH$_4$, and without DHFR the cell's supply of FH$_4$ and its derivatives would be depleted. Obviously, DHFR is crucial to cell function. Its inhibition by compounds

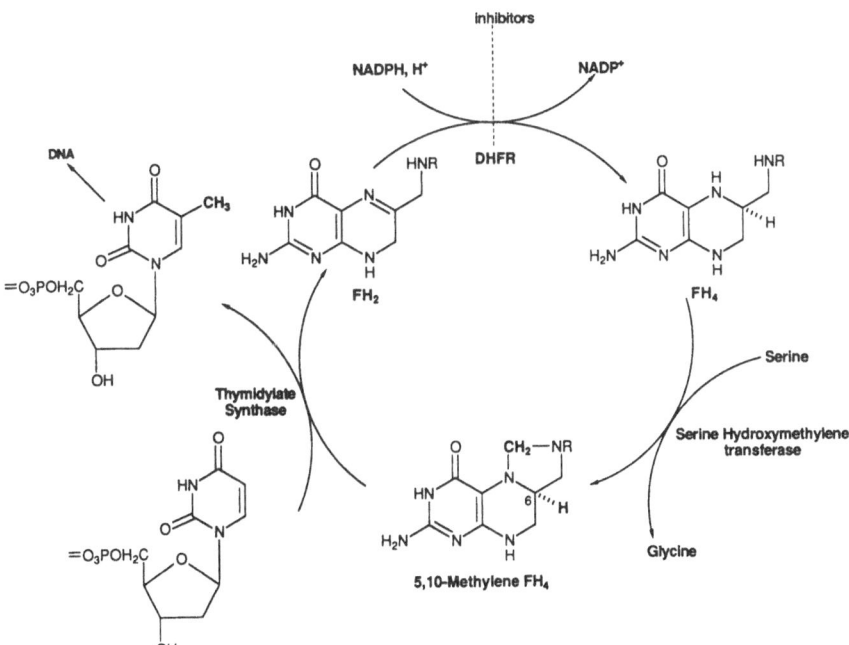

Figure 2. Role of dihydrofolate reductase in the biosynthesis of thymidylate.

Table I. Amino Acid Sequences of DHFR from Selected Organisms[a,b]

Block I

Ec	—	—	1M	2I	3S	4L	5I	6A	7A	8L	9A	10V	11D	12R	13V	14I	15G	16M	17E	18N	19A	20M	21P	22W	23N	24L	25P	—
Lc	—	—	—	1T	2A	3F	4L	5W	6A	7Q	8N	9R	10D	11G	12L	13I	14G	15K	16D	17G	18H	19L	20P	21W	22H	23L	24P	—
Cl	1V	2R	3S	4L	5N	6S	7A	8V	9A	10V	11C	12Q	13N	14M	15G	16I	17G	18K	19D	20G	21N	22L	23P	24W	25P	26P	27L	28R
Hu	1V	2G	3S	4L	5N	6C	7A	8V	9A	10V	11S	12Q	13N	14M	15G	16I	17G	18K	19N	20G	21D	22L	23P	24W	25P	26P	27L	28R
Mo	1V	2R	3P	4L	5N	6C	7A	8V	9A	10V	11S	12Q	13N	14M	15G	16I	17G	18K	19N	20G	21D	22L	23P	24W	25P	26P	27L	28R

Block II

Ec	—	—	26A	27D	28L	29A	30W	31F	32K	33R	34N	35T	36L	—	—	—	—	37N	38K	39P	40V	41I	42M	43G	44R	45H	46T	—
Lc	—	—	25D	26D	27L	28H	29Y	30F	31R	32A	33Q	34T	35V	—	—	—	—	36G	37K	38I	39M	40V	41V	42G	43R	44R	45T	—
Cl	29N	30E	31Y	32K	33Y	34F	35Q	36R	37M	38T	39S	40T	41S	42H	43V	44E	45G	46K	47Q	48N	49A	50V	51I	52M	53G	54K	55K	56T
Hu	29N	30E	31Y	32R	33Y	34F	35Q	36R	37M	38T	39T	40T	41S	42S	43V	44E	45G	46K	47Q	48N	49L	50L	51I	52M	53G	54K	55K	56T
Mo	29N	30E	31F	32K	33Y	34F	35Q	36R	37M	38T	39T	40T	41S	42S	43V	44E	45G	46K	47Q	48N	49L	50L	51I	52M	53G	54R	55R	56T

Block III

Ec	—	—	47W	48E	49S	50I	51G	—	—	52P	53P	54L	55P	56G	57R	58K	59N	60I	61I	62L	63S	64S	65Q	66P	67G	68T	69D	70D
Lc	—	—	46Y	47E	48S	49F	50P	—	51K	52R	53P	54L	55P	56E	57R	58T	59N	60V	61V	62L	63T	64H	65Q	66E	68Y	69Q	70A	71Q
Cl	57W	58F	59S	60I	61P	62E	63K	64N	65R	66P	67L	68K	69D	70R	71I	72N	73L	74V	75L	76S	77R	78E	79L	80K	81E	82A	83P	84K
Hu	—	57W	58F	59S	60I	61P	62E	63K	64N	65R	66P	67L	68K	69G	71I	72N	73L	74V	75L	76S	77R	78E	79L	80K	81E	82P	83P	84Q
Mo	57W	58F	59S	60I	61P	62E	63K	64N	65R	66P	67L	68K	69D	70R	71I	72N	73L	74V	75L	76S	77R	78E	79L	80K	81E	82P	83P	84R

Block IV

Ec	71R	—	72V	73T	74W	75V	76K	77S	78V	79D	80E	81A	82I	83A	84A	85C	86G	87D	88V	89P	—	—	—	—	—	—	—	—
Lc	72G	—	73A	74V	75V	76V	77H	78D	79V	80A	81A	82V	83F	84A	85Y	86A	87K	88Q	89H	90L	91D	92Q	—	—	—	—	—	—
Cl	85G	86A	87H	88Y	89L	90S	91K	92S	93L	94D	95D	96A	97L	98A	99L	100S	101D	102S	103P	104E	105L	106K	107S	108K	109V	110D	111M	112V
Hu	85G	86A	87H	88F	89L	90S	91D	92Q	93F	94D	95D	96A	97L	98L	99L	100T	101E	102Q	103P	104E	105L	106A	107A	108K	109V	110D	111M	112V
Mo	85G	86A	87H	88F	89L	90A	91D	92Q	93F	94D	95D	96A	97L	98R	99L	100L	101E	102Q	103P	104E	105L	106L	107S	108K	109V	110D	111M	112V

Block V

Ec	90E	91I
Lc	93E	94L
Cl	—	—
Hu	—	—
Mo	—	—

Ec 92M 93V 94I 95G 96G 97G 98R 99V 100Y 101E 102Q 103F 104L — — 105P 106K 107A 108Q 109K 110L 111Y 112L 113T 114H 115I 116D 117A

Lc 95V 96I 97A 98G 99G 100A 101Q 102I 103F 104T 105A 106F 107K — — 108D 109D 110V 111D 112T 113L 114L 115V 116I 117R 118L 119A 120G

Cl 113W 114I 115V 116G 117G 118T 119A 120V 121Y 122K 123A 124A 125M 126E 127K 128P 129I 130N 131H 132R 133L 134F 135V 136T 137R 138I 139L 140H

Hu 113W 114I 115V 116G 117G 118S 119S 120V 121Y 122K 123E 124A 125M 126N 127H 128P 129G 130H 131L 132K 133L 134F 135V 136T 137R 138I 139M 140Q

Mo 113W 114I 115V 116G 117G 118S 119S 120V 121Y 122E 123Q 124A 125M 126N 127E 128P 129G 130H 131L 132R 133L 134F 135V 136T 137R 138I 139M 140Q

Ec 118Q 119V 120E 121G 122D 123T 124H 125F 126P 127D 128Y 129E 130P 131D 132D 133W 134E 135S 136V 137F 138S — — — — 139E

Lc 121S 122F 123E 124G 125D 126T 127K 128M 129I 130P 131L 132N 133W 134D 135D 136F 137T 138K 139V 140S 141S — — — — 142R

Cl 141E 142F 143E 144S 145D 146T 147F 148F 149P 150E 151I 152D 153Y 154K 155S 156T 157K 158L 159L 160T 161E 162Y 163P 164G 165V 166P 167A 168D

Hu 141D 142F 143E 144S 145D 146T 147F 148F 149P 150E 151I 152D 153L 154E 155L 156K 157K 158L 159L 160P 161E 162Y 163P 164G 165V 166L 167S 168D

Mo 141E 142F 143E 144S 145D 146T 147F 148F 149P 150E 151I 152D 153L 154G 155L 156K 157K 158L 159L 160P 161E 162Y 163P 164G 165V 166L 167S 168E

Ec 140F 141H 142D 143A 144D 145A 146Q 147N 148S 149H 150S 151Y 152C 153F 154E 155I 156L 157E 158R 159R —

Lc 143T 144V 145E 146D 147T 148N 149P 150A 151L 152T 153H 154T 155Y 156E 157V 158W 159Q 160K 161K 162A —

Cl 169I 170Q 171E 172E 173D 174G 175I 176Q 177Y 178K 179F 180F 181V 182Y 183Q 184S 185S 186V 187L 188A 189Q

Hu 169V 170Q 171E 172E 173K 174G 175I 176Y 177Y 178K 179F 180F 181V 182Y 183E 184K 185N 186D —

Mo 169V 170Q 171E 172E 173D 174G 175I 176K 177Y 178K 179F 180F 181V 182Y 183E 184K 185K 186D —

[a]Ec = E. coli (Baccanari et al., 1981); Lc = L. casei (Bitar et al., 1977); Cl = chicken liver (Kumar et al., 1980); Hu = human (Masters and Attardi, 1983); Mo = mouse (Stone et al., 1979). A—Alanine, R—Arginine, N—Asparagine, D—Aspartic Acid, C—Cysteine, Q—Glutamine, E—Glutamic Acid, G—Glycine, H—Histidine, I—Isoleucine, L—Leucine, K—Lysine, M—Methionine, F—Phenylalanine, P—Proline, S—Serine, T—Threonine, W—Tryptophan, Y—Tyrosine, V—Valine. Sequence alignment for the Ec, Lc, and Cl enzymes follows that of Volz et al. (1982), which was based on the corresponding three-dimensional structures. Alignment of the
[b]Hu and Mo sequences were based on maximal homology.

such as TMP causes a general disruption in the cell's biosynthetic machinery and can lead to cell death.

2.2. Primary Structure

DHFR is a ubiquitous protein and has been isolated from sources ranging from bacteria and protozoa to plants and animals. The amino acid sequences of many different species of DHFR are known (Champness *et al.*, 1986a); a representative set is shown in Table I.

Sequence similarity among the isozymes of animal origin is generally very high, in the range of 75–90%. On the other hand, enzymes from bacterial sources have less similarity to one another, with sequence conservation of only 25–40%. The similarity between animal and bacterial DHFR is 20–30%, and enzyme from animals is generally about 25 residues longer than the bacterial proteins. As seen in Table I, those extra residues of the animal protein are accommodated as insertions at various locations throughout the sequence.

In spite of the sequence dissimilarity, these various species of DHFR exhibit three-dimensional architectures that are much the same, as determined by X-ray crystallography (Freisheim and Matthews, 1984). The insertion/deletion differences between these enzymes appear to have little effect on the overall geometry of their structures. The substrate binding cleft is highly conserved: linear sequence alignment of DHFR from 11 different sources shows that 12 residues are strictly conserved, and each of them is located in or near the active site (Kraut and Matthews, 1987).

Comparison of DHFR structures is important to inhibitor design, because the *selectivity* of an inhibitor for the enzyme of a pathogen is generally critical to the effectiveness of the compound as a drug. The level of difficulty in designing an inhibitor of appropriate selectivity depends on the similarity of the two enzymes of interest. Fortunately, there is sufficient difference between the bacterial and human forms of the enzyme for selective binding of the antibacterial drug TMP (Hitchings *et al.*, 1989). These differences also fuel hope for the possibility of designing other novel and perhaps more selective agents. That subject will be further addressed later in this chapter.

2.3. Three-Dimensional Structure

The X-ray crystal structures of DHFR from *Escherichia coli, Lactobacillus casei*, chicken, mouse, and, more recently, human have been solved as complexes with a variety of inhibitors, including TMP and MTX (see Oefner *et al.*, 1988, and references therein). Many of these structures have been refined at high resolution, and they provide a wonderfully detailed picture of this enzyme and its interactions with drug molecules. This section describes the general nature of the enzyme structure; specific interactions with inhibitors will be discussed in the sections on TMP and MTX.

The dominant feature of the DHFR structure is a central eight-stranded β-sheet flanked on either side by alpha helices, as illustrated in Fig. 3. This same general folding pattern is observed in each of the five different forms of DHFR for which crystal structures are known. [An X-ray crystal structure of a plasmid-

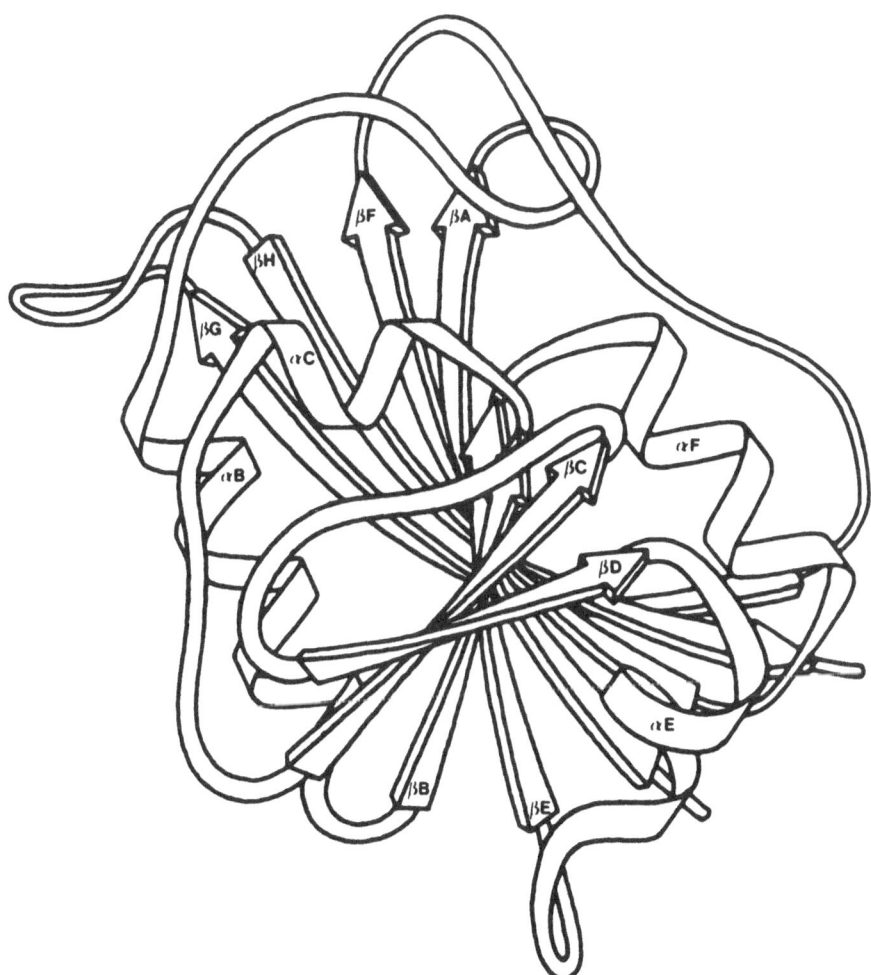

Figure 3. Ribbon representation of *L. casei* DHFR from Richardson (1981) showing the central eight-stranded β-sheet and the flanking α-helices.

encoded DHFR has also been solved and is structurally unrelated to chromosomal DHFR (Matthews, *et al.*, 1986).]

The enzyme surface is marked by a large cavity between the B and C helices, and it is there that inhibitors and substrates bind. The binding of MTX and NADPH in that cleft is illustrated in Fig. 4. The cofactor occupies the lower half of the active site, with its nicotinamide ring presumably positioned for hydride transfer to FH_2 in the catalyzed reduction (Filman *et al.*, 1982). In its extended conformation, the coenzyme is bound with the diphosphate bridge between the N-termini of helices C and F near the lower edge of the β-sheet and with the adenosine moiety on the opposite side of the sheet.

MTX is bound in the upper part of the cleft in much the same manner observed recently for folic acid (Oefner *et al.*, 1988). The *p*-aminobenzoylglutamate portions of the two ligands interact similarly with DHFR, and the pteridine ring system of each compound binds in the same region of protein. Interestingly, however, the pteridine rings are flipped approximately 180° about the C-2 to 2-amino bond with respect to one another (see also Charlton *et al.*, 1985, and Bolin *et al.*, 1982). This difference in binding mode is apparently due to differences in hydrogen-bonding capabilities between the two ligands (4-oxo vs. 4-amino) and the observation that, unlike folate or FH_2, MTX is protonated in its complex with DHFR and can interact ionically with an active site carboxyl group (a side chain from an aspartate or a glutamate). Additional details of MTX binding are discussed below.

Based on comparison of inhibitor data, it has been assumed that there is strong similarity among vertebrate DHFR, and DHFR from sources such as rat, chicken, mouse, and pigeon are commonly used as convenient surrogates for the more difficult-to-obtain human enzyme. This assumption appears to be reasona-

Figure 4. Alpha-carbon representation of the *L. casei* DHFR ternary complex with MTX and NADPH. Carbon atoms are shown as small open circles, oxygens by larger open circles, nitrogens by blackened circles, and phosphorus by larger blackened circles.

bly justified based on a comparison of linear amino acid sequences (see Table I) and the three-dimensional structures of human, chicken, and mouse DHFR. However, some differences in structural detail are suggested by the crystal structure data (Oefner *et al.*, 1988). With recombinant DNA techniques now providing large quantities of human DHFR (Prendergast *et al.*, 1988), it is the enzyme of choice for inhibitor studies. In the discussions that follow, however, data from chicken, mouse, and rat DHFR are assumed to be qualitatively representative of the human enzyme unless otherwise noted.

3. Enzyme–Inhibitor Interactions

3.1. Overview

TMP and MTX were first synthesized over three decades ago, and literally thousands of analogs have been reported during the ensuing years (Blaney *et al.*, 1984). In spite of this immense synthetic effort, these two drugs have yet to be supplanted—a testimony to the difficulties of drug design.

Although drug design must encompass molecular properties other than receptor interactions, such as pharmacokinetics, adverse toxicity, and toxicity to the targeted cells or organisms, the properties that influence binding to the receptor are of obvious, fundamental importance. An understanding of how TMP and MTX bind to DHFR should improve our ability to design more effective inhibitors and should presumably increase the odds of developing a useful therapeutic agent. The following two sections will focus on the salient structural features of the TMP and MTX complexes with DHFR as seen primarily through X-ray crystallographic analyses. The discussion will serve as a backdrop to the final sections on inhibitor design.

3.2. DHFR–Methotrexate Complexes

MTX complexes with DHFR from *E. coli*, *L. casei*, mouse (L1210 cells), and human have been studied by X-ray crystallography. The structures reported for DHFR from *E. coli* and human are binary complexes of inhibitor and enzyme, and those from *L. casei* and mouse are ternary forms that include NADPH. In general, MTX binds to each of these proteins in much the same way. The only exception is the complex with mouse DHFR, in which the glutamate moiety is modeled in an unexpected orientation with little enzyme interaction (Stammers *et al.*, 1987). There is currently no satisfactory explanation for this oddity, and it is beyond the scope of this chapter to explore that structure in any detail. For our purposes, we will focus on the *L. casei* DHFR structure that was briefly referred to in Section 2.3.

The *L. casei* DHFR-NADPH-MTX crystal structure was solved and refined at 0.17 nm resolution (Filman *et al.*, 1982), and it displays striking complementarity between inhibitor and enzyme. The inhibitor binding site of the protein can be partitioned into three regions: (1) a set of hydrogen bond acceptors, including a carboxylate, deep inside the cleft; (2) a central hydrophobic core; and (3) the guanidinium group of an arginine residue buried in a shallow hydrophobic pocket. MTX takes advantage of each region, as illustrated in Fig. 5.

The pteridine ring of MTX is protonated at N-1 (Cocco *et al.*, 1983) and forms an ionic bond with the carboxyl group of Asp-26. That interaction is mediated through hydrogen bonds donated by the N-1 hydrogen and a hydrogen of the 2-amino group. The second hydrogen of the 2-amino group associates with a fixed water molecule that has been observed in every refined DHFR structure studied to date, which appears to be an important structural component for substrate binding. In addition, the structure is consistent with two hydrogen bonds between the 4-amino group and two backbone carbonyl oxygen atoms. Considering this extraordinary complemental array of hydrogen bonds, the fact that essentially all potent inhibitors of DHFR contain a diaminopyrimidine ring, or a closely related heterocycle, is not surprising.

The central hydrophobic part of MTX binds in the middle of a ring of

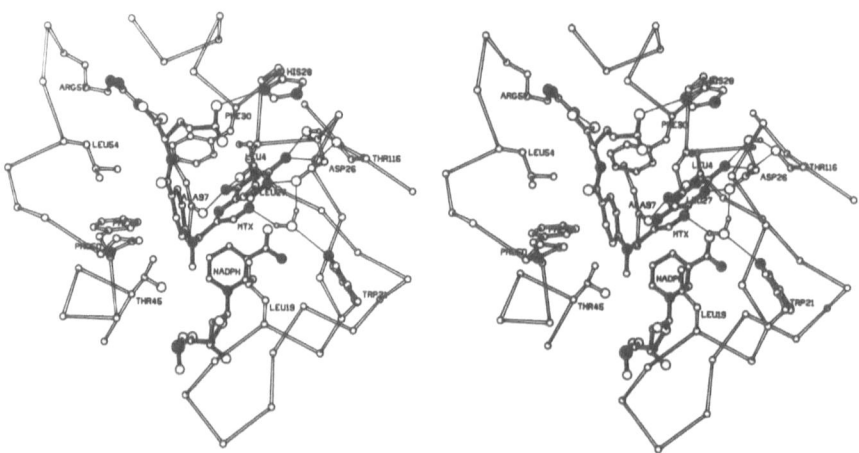

Figure 5. Binding site of MTX in *L. casei* DHFR. Ligands (MTX and NADPH) are depicted with filled bonds and protein with open bonds. The protein is represented by an alpha-carbon backbone with a selected set of amino acid side chains. Two backbone carbonyl groups that hydrogen-bond to the 4-amino group of MTX are also included. Hydrogen atoms, shown only for the two water molecules, are represented by very small open circles, carbon atoms by small open circles, oxygens by larger open circles, nitrogens by blackened circles, and phosphorus by larger blackened circles. Hydrogen bonds are indicated by single lines between the appropriate atoms.

several lipophilic amino acid residues, such as Leu-19, Leu-27, Phe-30, Phe-49, and Leu-54, which line the outer lip of the binding cleft. At the upper left corner of the hydrophobic cavity, a sequestered arginine residue interacts ionically with the alpha-carboxyl group of the inhibitor through a pair of hydrogen bonds. In the *L. casei* enzyme the drug's gamma-carboxyl group is within hydrogen-bonding distance of His-28 (Bolin *et al.*, 1982), and nuclear magnetic resonance (NMR) data are consistent with that interaction (Antonjuk *et al.*, 1984); in the other DHFR complexes with MTX the gamma-carboxyl appears to contribute little to binding. Inhibitory potencies of the alpha- and gamma-monoamide derivatives of MTX also suggest that the alpha-carboxyl group is much more important than the gamma-acid in the binding of MTX to DHFR (Piper *et al.*, 1982).

The dissociation constant of MTX from its ternary complex with DHFR is on the order of 10^{10} (Hood and Roberts, 1978), which corresponds to about 14 kcal/mole of binding energy (see also Subramanian and Kaufman, 1978). Based on structural information, that overall energy of binding must arise from a composite of many different kinds of intermolecular interaction. Even from the static view of an X-ray structure, ignoring dynamics, entropy, solvation, and other factors, the MTX–DHFR association is complex. Some insight into this complexity is being gained by interesting site-directed mutagenesis studies that alter active-site residues that interact with MTX and analyze the MTX binding properties of the mutant enzyme (Villafranca *et al.*, 1983; Appleman *et al.*, 1988; Howell *et al.*, 1986). New theoretical methods may also begin to play an important role in understanding protein–ligand interactions (Singh, 1988; Pettitt and Karplus, 1986; Wong and McCammon, 1986; Bash *et al.*, 1987).

3.3. DHFR–Trimethoprim Complexes

3.3.1. Binding Properties

The usefulness of TMP as an antibacterial agent results in part from its differential affinity for bacterial vs. human DHFR (Roth, 1983). The measured ratio of binding constants (K_i values) of TMP for DHFR from *E. coli* and human is 12,000 : 1 (Appleman *et al.*, 1988), which corresponds to a difference in binding energy of about 6 kcal/mole. An understanding of the molecular interactions responsible for this selectivity should help to provide a basis for the design of novel and perhaps even more selective inhibitors.

Many hundreds of analogs of TMP have been synthesized and tested as DHFR inhibitors. Substituents on the benzyl group can have a marked effect on binding to DHFR and also influence selectivity (Roth, 1983). As an illustration of these effects, Table II shows the DHFR activities of several close analogs of TMP. For this series of compounds, the activity of inhibitor against *E. coli* DHFR is clearly related to the number of methoxy groups on the phenyl ring.

Table II. DHFR Inhibition Data for TMP and Its
Desmethoxy Analogs[a]

NH$_2$

[structure of 2,4-diaminopyrimidine linked via CH$_2$ to a benzene ring bearing R]

	DHFR I$_{50}$ (10^{-8} M)		
R	*E. coli*	Mouse (SRI)	Selectivity
H	410.0	18000.0	44.0
3-OCH$_3$	45.0	19000.0	420.0
4-OCH$_3$	48.0	16000.0	330.0
3,4-diOCH$_3$	5.1	7300.0	1400.0
3,5-diOCH$_3$	5.3	22000.0	4200.0
3,4,5-triOCH$_3$ (TMP)	0.8	28000.0	35000.0

[a]Data from Baccanari *et al.*, 1982.

The unsubstituted parent compound is the least active, and inhibitory power is increased about 10-fold with the addition of each methoxy group. In contrast, methoxy substitution has little effect on binding to the mouse enzyme, and the net result is a dramatic increase in selectivity with the increase in number of methoxy groups.

3.3.2. Mode of DHFR Binding

X-ray crystal structures of TMP bound to DHFR from *E. coli*, chicken, mouse, and human have been reported. Complexes with DHFR from *L. casei*, *E. coli*, and mouse have also been studied in solution by NMR spectroscopy (Cheung *et al.*, 1986; Birdsall *et al.*, 1983; Searle *et al.*, 1988). Although DHFR–TMP complexes have been subjected to many other methods of investigation, the X-ray and NMR data have given the most detailed picture of these structures.

3.3.2a. Binding to Bacterial DHFR. The X-ray structure of the binary complex of *E. coli* DHFR and TMP has been solved and refined at 0.17 nm resolution. The analogous ternary structure with bound NADPH is also known, but was solved at 0.30 nm resolution and has not been refined (Champness *et al.*, 1986b). For this discussion, the latter structure will illustrate TMP binding. The presence of cofactor appears to have little effect on the geometry of binding for

TMP, but it is known that NADPH shows a cooperative effect on TMP affinity for *E. coli* DHFR, and the ternary complex is apparently the biologically important form of the inhibited enzyme (Baccanari *et al.*, 1982). Figure 6 shows the *E. coli* DHFR–NADPH–TMP complex.

The diaminopyrimidine ring of TMP is associated with the *E. coli* enzyme in a manner analogous to the binding of that ring in MTX to *L. casei* DHFR. The pyrimidine ring is protonated at N-1 when bound to DHFR: this has been shown unambiguously by ^{15}N NMR experiments in which the signal from the ^{15}N-labeled 1-nitrogen appears as a proton-coupled doublet (Bevan *et al.*, 1985). The X-ray structure is consistent with five hydrogen bonds between diaminopyrimidine and protein; the 4-amino group is hydrogen-bonded to the backbone carbonyl oxygens of Ile-5 and Ile-94, the 2-amino donates hydrogen bonds to the conserved, buried water molecule and to the carboxyl group of Asp-27, and the hydrogen at N-1 also interacts with Asp-27. The ionic interaction with the active-site acidic residue is clearly a key feature of binding for the diaminopyrimidine-type inhibitor.

Like the phenyl ring of MTX, the trimethoxyphenyl group of TMP is positioned in the hydrophobic core of the binding cavity. Nearby residues include Met-20, Leu-28, Phe-30, Ile-50, and Leu-54. To orient the benzyl group in this fashion, TMP adopts a conformation similar to that observed in a crystal structure of TMP as a hydrobromide salt (Phillips and Bryan, 1969). This suggests

Figure 6. Binding site of TMP in *E. coli* DHFR ternary complex. Molecule and atom representations are as described in the legend of Fig. 5, with the addition that sulfur is shown as the largest open circle.

that the DHFR-bound form of TMP is a relatively low-energy conformation. Essentially the same conformation has been assigned for TMP in its complexes with DHFR from *L. casei* and *E. coli* in solution, using elegant NMR techniques (Birdsall *et al.*, 1983).

3.3.2b. Binding to Vertebrate DHFR. High-resolution, refined X-ray structures of TMP bound in ternary complex with mouse (Stammers *et al.*, 1987) and chicken DHFR (Matthews *et al.*, 1985a,b) have been reported. More recently, the structure of human DHFR–TMP has been solved at relatively low resolution and was not refined (Oefner *et al.*, 1988). The mode of TMP binding is similar among these three complexes, but interestingly is different from that in the bacterial enzyme. Here the chicken DHFR structure, shown in Fig. 7, will be used to illustrate the manner in which TMP binds to the vertebrate protein.

Unlike its interaction with *E. coli* DHFR, the diaminopyrimidine ring of TMP is positioned deeper in the active site of vertebrate DHFR so as to allow only four hydrogen bonds. The ionic link to the active-site acidic residue, glutamate in the case of vertebrate DHFR (see Table I), is retained, as is the hydrogen bond to the buried water molecule. However, the 4-amino group hydrogen-bonds only to the carbonyl group of Ile-7. The X-ray structure is not consistent with the expected hydrogen bond between the 4-amino group and the backbone carbonyl of Val-115. The lack of this hydrogen bond is one relatively clear-cut structural detail that appears to contribute to the differential DHFR affinity of TMP.

The reason for the altered mode of binding is not entirely clear. A comparison of the *E. coli* and mouse DHFR structures (see Figs. 5 and 6) shows that

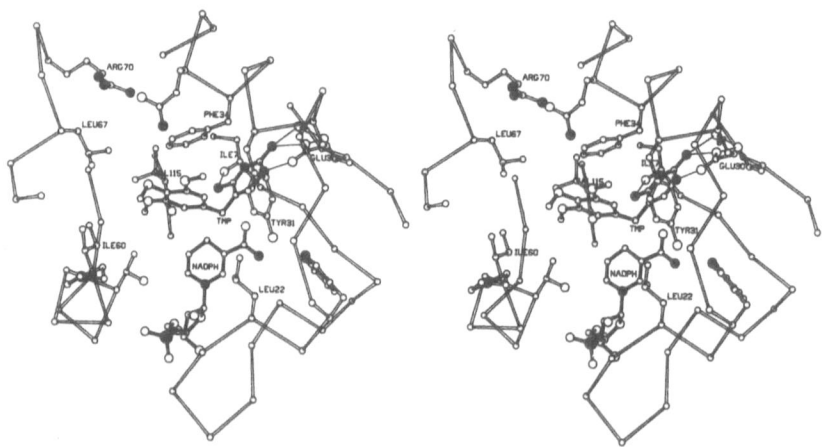

Figure 7. Binding site of TMP in chicken DHFR ternary complex. Molecule and atom representations are as described in the legend of Fig. 5.

TMP adopts two very different conformations in these two complexes, and that the trimethoxyphenyl ring is oriented in different regions of the binding cleft. Based on a thorough comparison of the *E. coli* and chicken DHFR complexes with TMP, Matthews *et al.* (1985a,b) suggested that the differences in binding geometry and affinity may arise from differences in the width of the binding cleft. In the two structures compared, the active-site cavity of the vertebrate enzyme is 0.15–0.20 nm wider than the cleft of the bacterial enzyme (Matthews *et al.*, 1985a,b). The authors argued that the geometry of TMP is such that optimal binding of the pyrimidine ring does not allow for simultaneously favorable binding of the trimethoxyphenyl ring in the larger hydrophobic cavity of chicken DHFR. The hydrogen bond between Val-115 and the 4-amino group of TMP must be sacrificed to allow favorable positioning of the trimethoxy phenyl group in an upper niche in the cleft. The validity of this argument was questioned recently in a report based on the study of the high-resolution refined structure of human DHFR in complex with folic acid (Oefner *et al.*, 1988). The cleft width of the human structure is similar to that of the *E. coli* enzyme. A complete understanding of TMP binding and selectivity remains a challenge.

4. Examples of Inhibitor Design

The three-dimensional structures of DHFR–inhibitor complexes offer the medicinal chemist an exciting opportunity to learn about receptor–drug interactions and to approach inhibitor design more rationally. The following discussion concerns a few examples of our work on the design and synthesis of TMP analogs, in which we tried to take advantage of the available DHFR structural data.

4.1. Trimethoprim Analogs with Acidic Substituents

The first DHFR structure reported was the MTX complex with *E. coli* DHFR (Matthews *et al.*, 1977). The unrefined version of that structure was the basis for our first attempts to use three-dimensional knowledge of DHFR in the design of TMP analogs. With the binding mode of TMP still open to question at that time, our initial goal was to design analogs with a dual purpose: to enhance binding to the bacterial enzyme and to provide insight into TMP binding geometry.

Our first series of analogs was designed to interact ionically with Arg-57, a strictly conserved residue and the binding site of the alpha-carboxyl group of MTX (Kuyper *et al.*, 1985) (a similar study has been reported by Kompis and Then, 1984). The corresponding residue of human DHFR interacts with the alpha-carboxyl of folic acid (Oefner *et al.*, 1988). The question was how to substitute TMP with a carboxyl group to gain a similar interaction. Simple modeling suggested that a 3′-carboxyalkyl substituent might be appropriate. A model of TMP appeared to be reasonably accommodated in DHFR with its diaminopyrimidine superimposed over that ring of MTX and the conformation of TMP adjusted to match that observed in a crystal structure of TMP·HBr (Phillips and Bryan, 1969). This particular conformation had previously been implicated by other data (Roth et al., 1980). In that model for TMP binding, the 3′-position projected toward Arg-57 and a carboxyalkyl group of appropriate length could be modeled to bind to that site. Modeling suggested that a carboxypropoxy substituent would have the minimum length necessary to interact optimally with the guanidinium group of Arg-57. The series of compounds shown in Table III was prepared to test these modeling predictions (Kuyper et al., 1985).

The binding data in Table III supported the ideas derived from modeling. The most active analogs (K_i = 24–60 pM) all had substituent chain lengths long enough to reach Arg-57. The compounds with shorter chains were significantly less active. Although the molecular modeling experiments were simple, the results were exciting. The first analogs of TMP to be designed using the three-dimensional structure of DHFR were by far the most potent DHFR inhibitors among the many hundreds of TMP analogs known at the time, and the data were helpful in understanding how TMP might bind to DHFR.

The binding interactions of TMP and the carboxyl-containing analogs were subsequently determined experimentally by X-ray crystallography (Kuyper et

Table III. Binding Constants of *E. coli* DHFR
for TMP and Several Carboxylic Acid Analogs[a]

n	10^9 Ki, M
— (TMP)	1.3
1	2.6
2	0.37
3	0.035
4	0.066
5	0.024
6	0.050

[a]Data from Kuyper et al., 1985.

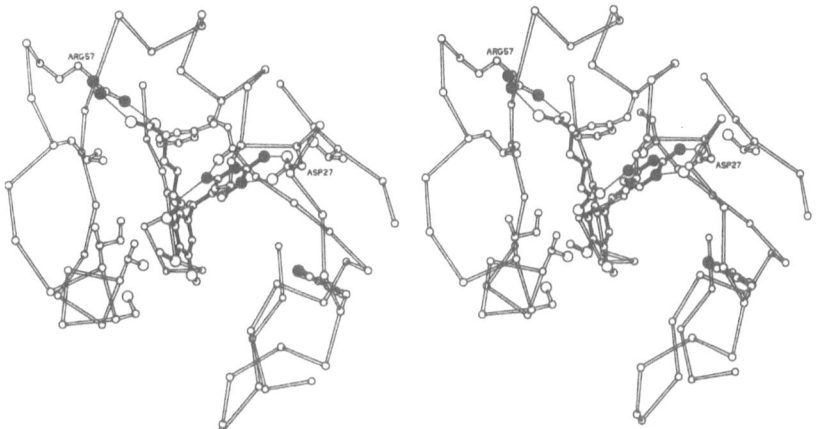

Figure 8. E. coli DHFR binding site of meta-carboxypentoxy analog of TMP showing the interaction with Arg-57. Molecule and atom representations are as described in the legend of Fig. 5.

al., 1985). The DHFR–TMP complex was discussed in Section 3.3.2a (see Fig. 6), and Figs. 8 and 9 illustrate the structures of *E. coli* DHFR complexed with two of the analogs. As shown in Fig. 8, the carboxypentoxy analog, the most potent inhibitor in the series, is closely associated with Arg-57 through two hydrogen bonds. The carboxyethoxy analog, which is only marginally more effective than TMP as an inhibitor of *E. coli* DHFR, is able to interact with Arg-57 with just one hydrogen bond.

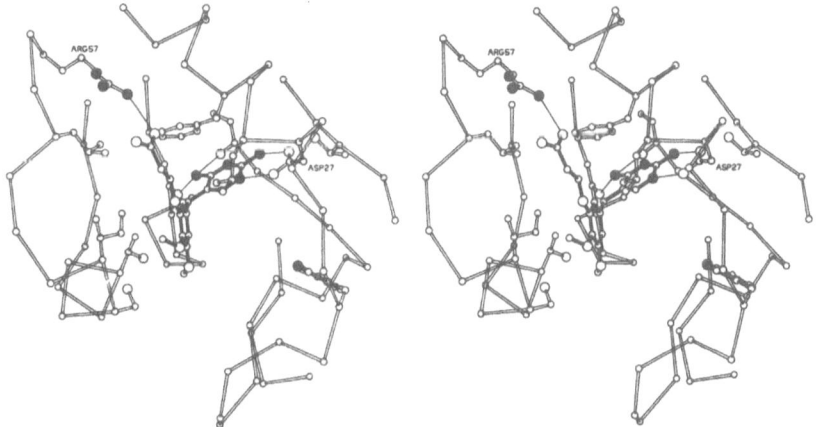

Figure 9. E. coli DHFR binding site of meta-carboxypropoxy analog of TMP showing the interaction with Arg-57. Molecule and atom representations are as described in the legend of Fig. 5.

With information in hand on the binding conformation of TMP, our attention turned to ideas that might take advantage of that knowledge. Some of that
work is described below.

4.2. Conformationally Restricted Analogs of Trimethoprim

As mentioned earlier, an important property of TMP is its selective affinity
for bacterial DHFR. Although TMP is relatively inactive as an inhibitor of
human DHFR, that activity is not totally inconsequential, and it does impose
certain restrictions on the clinical use of the drug. A compound of even greater
selectivity would be expected to have an advantage. Therefore, a goal of our
research has been to design analogs of TMP with improved DHFR selectivity,
and one approach to that objective was conformational restriction.

Trimethoprim is a relatively flexible molecule with a number of different
shapes available to it. A working hypothesis for analog design was that TMP
might bind to bacterial and human DHFR in different conformations. At the time
no experimental evidence existed to support this idea, but it was a convenient
postulate because we did have information on TMP's conformation of binding to
E. coli DHFR. The design strategy was to limit the conformational flexibility of
TMP so it would greatly favor the conformation observed in the *E. coli* DHFR–
TMP complex. Planning for the necessary molecular modifications would take
into account the geometry of the enzyme active site. A rigid analog with the
desired conformation and compatibility with the bacterial enzyme might exhibit
high selectivity.

The first class of rigid analogs we pursued involved a one- or two-carbon
bridge between the 4-amino and methylene groups of TMP. This type of compound was chosen based on modeling to unrefined structures of *E. coli* DHFR in
complex with MTX and TMP. In those models, the 4-amino group of inhibitor
was involved in only one hydrogen bond (Matthews *et al.*, 1977), that to the
backbone carbonyl of Ile-5, and the amino hydrogen oriented toward the methylene group of TMP was modeled in a hydrophobic pocket with no close contacts. If the models were correct, then substitution of that hydrogen with carbon
would be expected to have a favorable effect on binding. Although unrefined
protein crystal structures can be misleading in such detail, Compounds 3 and 4
were designed and synthesized based on those models.

As shown in Fig. 10, the calculated low-energy conformations of Compounds 3 and 4 were similar to the binding conformation of TMP. Nonetheless, these analogs were significantly less active than TMP as inhibitors of bacterial DHFR: Compounds 3 and 4 were 35- and 1000-fold less inhibitory than TMP, respectively. A logical explanation for at least part of the decreased activity came from refined X-ray structures of *E. coli* DHFR–inhibitor complexes.

The unrefined structures used to design Compounds 3 and 4 were indeed misleading. According to the refined versions of those structures, the 4-amino group of the inhibitor hydrogen bonds not only to the backbone carbonyl group of Ile-5, but also to the carbonyl of Ile-94, as described in previous sections of this chapter. Thus replacement with carbon of the 4-amino hydrogen that binds to Ile-94 would be expected to have detrimental effects on enzyme affinity. The observed mode of binding, as seen in X-ray crystal structures of 3 and 4 bound to *E. coli* DHFR, was, as expected, similar to that of TMP, but quantitative interpretation of those structures in terms of binding energy differences was difficult. Aspects other than the missing hydrogen bond must also contribute, both

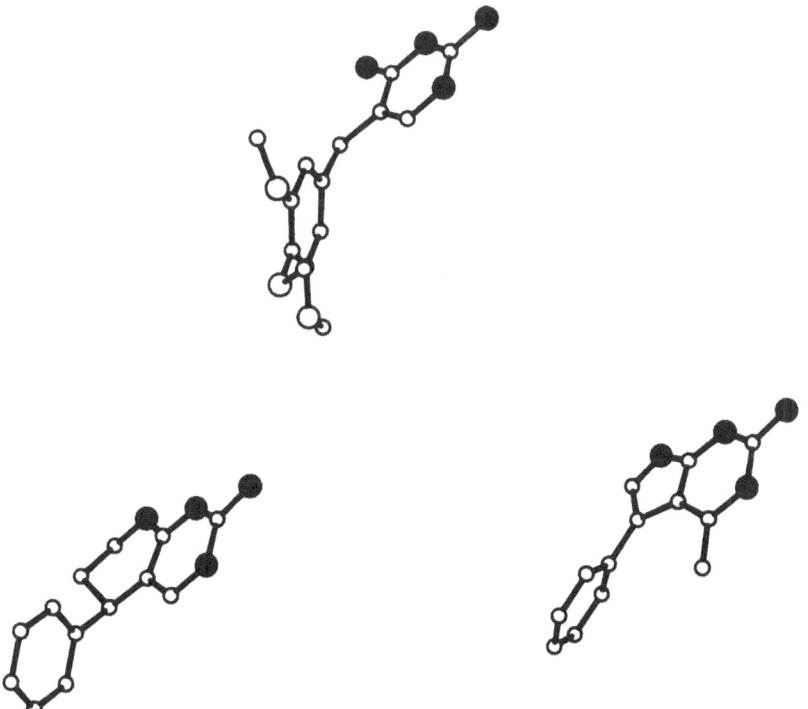

Figure 10. Comparison of the *E. coli* DHFR-bound conformation of TMP (top center) and the calculated low-energy conformations of Compounds 3 (lower left) and 4 (lower right).

favorably and unfavorably, to the overall DHFR affinity of these inhibitors. For example, the pKa of Compound 3 (6.3) is one unit less than that of TMP (7.2). Because these inhibitors bind to DHFR in their protonated form, the lower basicity of Compound 3 should be detrimental to its apparent affinity. Differences in geometry, entropy, and other factors must also play a role in the observed differences in binding constants. Whatever the rationale for poor activity, the results led us to try other means of restricting the conformation of TMP. (This work illustrates the need for refined, high-resolution X-ray structures for the detailed design of enzyme inhibitors. The nontrivial investment in synthetic effort might have been saved if our design had been based on refined structures.)

Another approach to the conformational restriction of TMP is to link the 6-position of the pyrimidine ring to the ortho position of the phenyl ring. Molecular modeling of various possible linking units suggested that a chain of three methylene groups might provide the desired properties. Thus molecular mechanics modeling showed two types of low-energy conformations for Structure 5 (see Ollis *et al.*, 1974), shown in Fig. 11 as forms A and B, one of which (A) reasonably matched the *E. coli* DHFR-bound form of TMP. (Mirror-image partners of A and B must also be considered, but for the sake of simplicity those forms are not shown.) At that time, the significance of conformation B was not known. Without information on the human DHFR-bound form of TMP, all we could do was hope that the shape of B was not appropriate for the human enzyme.

5 : R = H

6 : R = OCH₃

Simple modeling of Structure 5 in the active site of the refined structure of *E. coli* DHFR was encouraging, but not so convincing that we immediately embarked on a synthetic effort. Instead we turned to a modeling procedure reported by Blaney *et al.* (1982), which offered the potential of semiquantitative estimates of binding energy. In their method, differences in free energy of binding for a series of closely related inhibitors were approximated by differences in the internal energy of the enzyme–inhibitor complex [$\Delta\Delta E$(calc.)], as calculated by molecular mechanics, minus differences in the solvation free energies of the inhibitors [$\Delta\Delta G$(inhibitor solv.)], as indicated in Eq. (1). Using that procedure they were able to correctly rank the binding free energies of a series of thyroxine

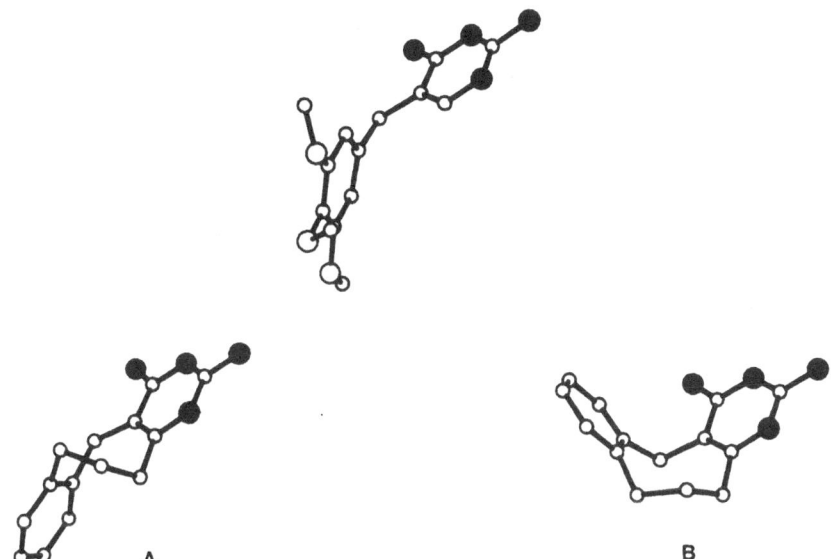

Figure 11. Comparison of the *E. coli* DHFR-bound conformation of TMP (top center) and the calculated low-energy conformations of Structure 5.

analogs for the protein prealbumen. Much of the error in such a simple approximation is apparently minimized by looking at differences between similar compounds.

$$\Delta\Delta G \simeq \Delta\Delta E(\text{calc.}) - \Delta\Delta G(\text{inhibitor solv.}) \tag{1}$$

In a test of method, we first attempted to estimate differences in free energy of binding to *E. coli* DHFR for a series of known TMP analogs (all of the compounds in Table II except the 3,5-dimethoxy analog) (Kuyper, 1985). For each inhibitor the internal energy of the corresponding complex with *E. coli* DHFR-NADPH was calculated in the following way. The diaminopyrimidine ring of the inhibitor was assumed to bind to bacterial DHFR through five hydrogen bonds in a manner identical to that observed for TMP. Because the benzyl group appeared to have room within the enzyme active site for some conformational flexibility, especially for the unsubstituted analog, a number of sterically reasonable conformations were generated as starting points for molecular mechanics energy minimization. We employed the AMBER program (Weiner and Kollman, 1981) for those calculations. Each of the 15–20 starting structures for each complex was energy minimized, and the resulting structure of lowest energy was assumed to be the "global" minimum for the complex. The calculated minimum energy of each inhibitor, in the absence of enzyme, was subtracted from the energy of the corresponding enzyme–inhibitor complex in order to compare energies of the various complexes.

Differences in free energy of solvation for this series of compounds were estimated from measured values for model compounds. Solvation energies were experimentally determined for mono-, di- and trimethoxy-substituted benzenes, and the energy differences among those compounds were assumed to be representative of differences in solvation energy among the corresponding TMP analogs. The data indicated that an aromatic methoxy group contributes approximately -1.5 kcal/mole to solvation free energy.

A comparison of estimated [using Eq. (1)] relative binding energies and observed binding constants for this test series of inhibitors is illustrated in Fig. 12. The good correlation prompted us to attempt to use these test data as a calibration for predicting the relative activity of hypothetical inhibitors, particularly Structure 5. (Structure 5, rather than any of its methoxy-substituted analogs, was our initial target because of practical considerations, both synthetic and computational.) The energy of the DHFR complex with 5 was calculated, and the relative solvation energy of the inhibitor was estimated from the group contribution tables of Hine and Mookerjee (1975). The resulting prediction of the relative binding energy of Structure 5, shown in Fig. 12, was encouraging. The predicted binding energy of 5 was somewhat better than that of the unsubstituted benzylpyrimidine, suggesting a reasonable complementarity with the enzyme active site. Based on this prediction, Structure 5 was synthesized.

The calculated properties of Compound 5 were in good agreement with those determined experimentally. Its conformational behavior, as observed by ^{1}H NMR, was consistent with the molecular modeling results, and its measured binding to *E. coli* DHFR was reasonably close to that predicted, as shown in Table IV.

However, Compound 5 displayed essentially no DHFR selectivity. The ratio of *E. coli* and rat liver DHFR I_{50} values was considerably lower than that of the unsubstituted benzylpyrimidine. A possible explanation of this observation was found in the subsequently reported crystal structure of TMP complexed with chicken DHFR (Matthews *et al.*, 1985a). As discussed in Section 3.3.2b, TMP binds to the chicken enzyme in a conformation that differs significantly from the *E. coli* DHFR-bound form of the drug. As it turns out, the form of TMP that binds to chicken DHFR is similar to conformation B of Compound 5. That conformation is compared to the chicken DHFR-bound form of TMP in Fig. 13. Unfortunately, the flexibility restrictions imposed by the trimethylene bridge of Compound 5 were inappropriate for achieving the desired selectivity.

As in the benzylpyrimidine series of inhibitors, methoxy substitution of Structure 5 had a dramatic effect on DHFR selectivity. The dimethoxy analog 6 was 300-fold more selective than parent 5. The difference in affinity for *E. coli* DHFR observed for Compounds 5 and 6 was predicted by the computational

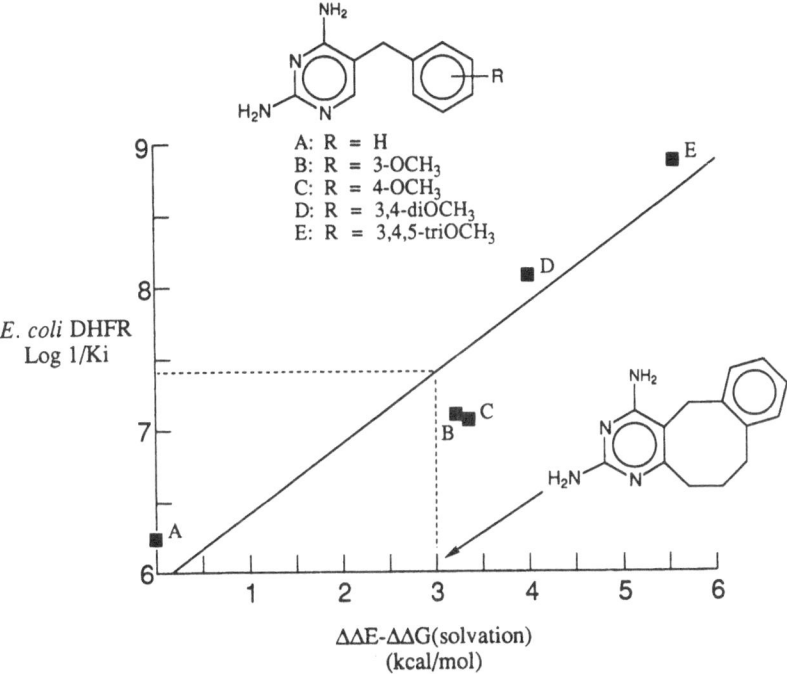

Figure 12. Prediction of *E. coli* DHFR binding constant for Structure 5. The calculated [Eq. (1)] relative binding energies for the indicated series of methoxy-substituted benzylpyrimidines are plotted vs. the observed binding constants. The correlation was used to predict the binding constant of hypothetical Structure 5, using its calculated relative binding energy.

Table IV. DHFR Inhibition Data for TMP and Several Conformationally Restricted Analogs

Compound	DHFR I_{50} (10^{-8} M)		
	E. coli	Rat liver	Selectivity
5	17.0	67.0	3.9
	(5.0 calc)		
6	1.3	1660.0	1300.0
	(1.0 calc)		
8	8.2	1800.0	220.0
	(2.0 calc)		
1 (TMP)	0.8	35000.0	44000.0

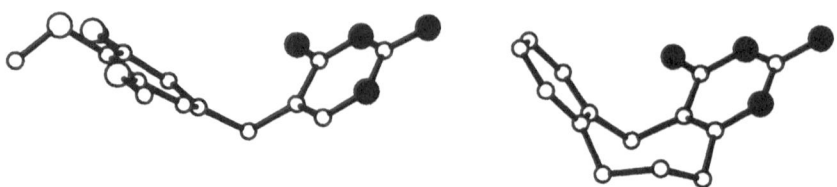

Figure 13. Comparison of chicken DHFR-bound conformation of TMP (left) with form B of Compound 5.

method described earlier. As shown in Tables II and IV, compound 6 is equivalent to the dimethoxy-substituted benzylpyrimidine in terms of selectivity and is more active as an inhibitor of the *E. coli* enzyme, suggesting that perhaps other analogs such as trimethoxy derivatives might show selectivity ratios in the range observed for TMP. However, our goal was to prepare compounds with significantly higher selectivity than that of TMP, and we chose to pursue other parent structures with altered conformational properties, as discussed below.

Knowledge of both binding conformations of TMP put us in·a better position to design conformationally restricted analogs. With that information in hand, the goal was to design a rigid compound that could mimic the shape of TMP in its complex with bacterial enzyme, but that could not adopt the form that is apparently appropriate for vertebrate DHFR. One obvious idea was to look for modifications of Structure 5 that would disfavor conformation B, and molecular modeling suggested a rather simple possibility. The introduction of a double bond in the three-carbon bridge, as in Structure 7, imposes additional rigidity, and that structure has only one type of conformation available to it (a pair of mirror-image forms), according to molecular modeling calculations. Structure 7 displayed a similarity to the *E. coli* DHFR-bound form of TMP and also appeared to be compatible with the bacterial enzyme binding site. Synthesis of Structure 7 was initiated, but encountered a number of difficulties, and alternative target structures were considered. One of those alternatives was Structure 8, in which a phenyl group provides the double bond in the three-carbon bridge. Each of the two calculated mirror-image conformations of 8 displayed an overall resemblance to the bacterial DHFR-bound form of TMP, but showed little similarity to the alternative conformation of the drug, as shown in Fig. 14. Although the conformational properties of Structure 8 appeared to be acceptable, enzyme compatibility was questionable because of the additional bulk due to the extra phenyl ring. Molecular mechanics modeling was again used to evaluate that concern.

Because the success of the computational method for predicting differences in binding energy was expected to depend on the similarity of the hypothetical

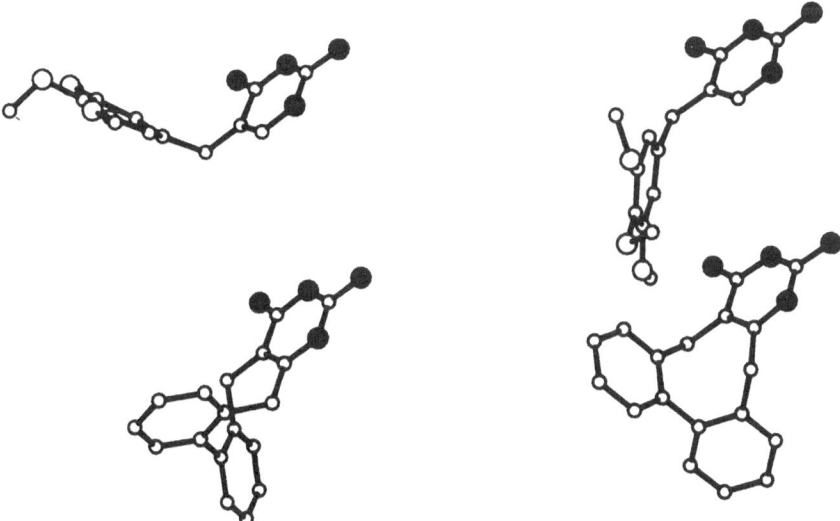

inhibitor in question and the series of known inhibitors used for calibration, a new set of benchmark compounds, shown in Fig. 15, were chosen based on resemblance to Structure 8. The prediction of binding energy for Structure 8 is illustrated in Fig. 15, and the hypothetical model of Structure 8 bound to *E. coli* DHFR is shown in Fig. 16. The extra phenyl group appears to be accommodated in a groove between Leu-28 and Phe-31, and the predicted binding energy was sufficiently encouraging to justify a synthetic effort.

Compound 8 was successfully prepared, and again the calculated properties were in good agreement with experimental results. X-ray crystal analysis of 8 confirmed the predicted conformation, and ^{1}H NMR studies were consistent with the expected conformational equilibrium between the mirror-image forms of 8. The predicted relative affinity for *E. coli* DHFR was in accord with the measured value shown in Table IV. More important, Compound 8 showed rather low

Figure 14. Comparison of *E. coli* DHFR-bound conformation (upper right) and chicken DHFR-bound conformation (upper left) of TMP with the mirror-image pair of conformations of Structure 8.

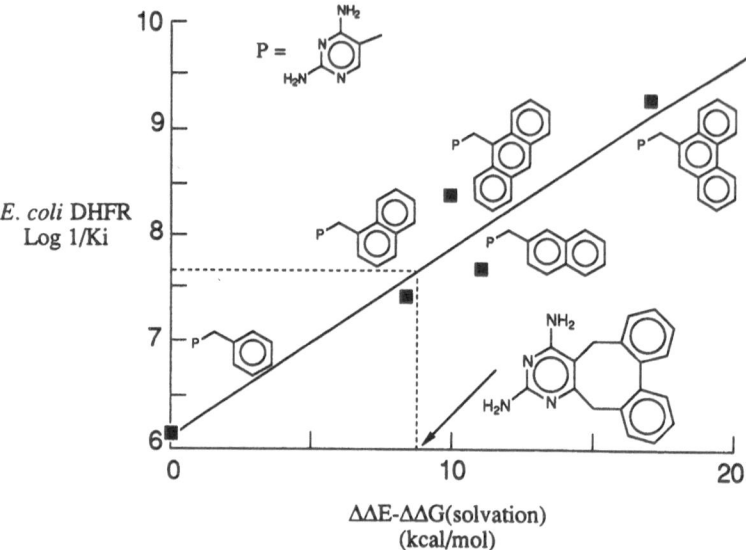

Figure 15. Prediction of *E. coli* DHFR binding constant for Structure 8. The calculated [Eq. (1)] relative binding energies for the indicated series of substituted benzylpyrimidines are plotted vs. the observed binding constants. The correlation was used to predict the binding constant of hypothetical Structure 8, using its calculated relative binding energy.

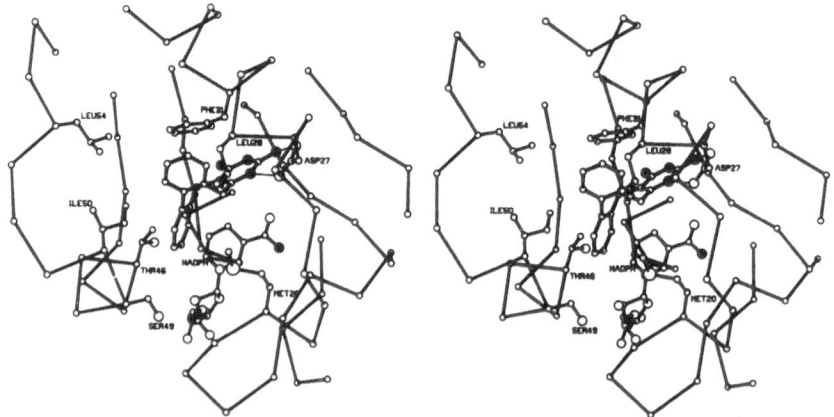

Figure 16. Hypothetical model of Structure 8 bound to *E. coli* DHFR. Molecule and atom representations are as described in the legend of Fig. 5.

affinity for vertebrate DHFR and was 50-fold more selective than Compound 5, in agreement with the conformational considerations discussed earlier.

Although no analogs of Compound 8 have yet been made, the potential for 8 to serve as the parent structure for a series of highly selective inhibitors seems good. For example, if dimethoxy substitution would increase DHFR selectivity as it did for Compound 6, then the corresponding dimethoxy analog of 8 would exhibit selectivity exceeding that of TMP. Perhaps an additional methoxy group would further increase selectivity, as it does in the TMP series of inhibitors. The rigid nature of Compound 8 and the number of available sites for substitution also provide reasons to believe that appropriately substituted analogs of Compound 8 might meet our goals of DHFR selectivity.

5. Conclusion

Our understanding of the interactions that control the complexation of DHFR and drug molecules has increased dramatically in recent years. In particular, X-ray crystallography has provided a wealth of detailed, three-dimensional information about DHFR–drug interactions that can clearly serve as a basis for the rational design of novel inhibitors. Hopefully, the few examples of inhibitor design described here illustrate the potential usefulness of enzyme structure for the design of inhibitors.

Although the DHFR structural data are clearly applicable to inhibitor design, it is equally clear that inhibitor design based on this information alone is still fraught with difficulties and many unknowns. Even if one assumes that the knowledge of enzyme will lead to novel ideas for potential inhibitors, it is generally difficult to accurately evaluate such ideas prior to synthesis. The ability to predict DHFR affinity of hypothetical inhibitors is an obvious need for an efficient program of design and synthesis, but the complexities associated with solvation, entropy, conformational flexibility, and other factors make such predictions highly questionable. Computational modeling offers some hope and was useful in the work described here. However, present methods, including the exciting technique of free-energy perturbation (for example, see Bash *et al.*, 1987), are limited to the comparison of closely related systems. Hopefully, more powerful methods of modeling will be developed soon. In the meantime, it is important to understand the currently available techniques and to use them appropriately.

ACKNOWLEDGMENTS. DHFR crystal structure data were provided by the group of D. Matthews and J. Kraut, and also C. Beddell, J. Champness, and D. Stammers. Much of the synthetic work was carried out by S. Davis, J. Garvey, and H. LeBlanc. D. Baccanari and R. Ferone were responsible for the enzymology. The AMBER package of software and advice on how to use it were furnished by P. Kollman. Computer programs TEXSAN GRAPHICS from Molecular Structure Corp. and MacroModel from the group of C. Still were used to compose and plot many of the figures in this chapter. We thank A. Jones for editorial comments, J. Bentley for general assistance in computer-related activities, and M. Cory for his continued support.

References

Antonjuk, D. J., Birdsall, H. T., Cheung, A., Clore, G. M., Feeney, J., Gronenborn, A., Roberts, G. C. K., and Tran, T. Q., 1984, A ¹H NMR study of the role of the glutamate moiety in the binding of methotrexate to *Lactobacillus casei* dihydrofolate, *Br. J. Pharm.* **81**:309–315.

Appleman, J. R., Prendergast, N., Delcamp, T. J., Freisheim, J. H., and Blakley, R. L., 1988, Kinetics of the formation and isomerization of methotrexate complexes of recombinant human dihydrofolate reductase, *J. Biol. Chem.* **263**:10304–10313.

Baccanari, D. P., Stone, D., and Kuyper, L., 1981, Effect of a single amino acid substitution on *Escherichia coli* dihydrofolate reductase catalysis and ligand binding, *J. Biol. Chem.* **256**:1738–1747.

Baccanari, D. P., Daluge, S., and King, R. W., 1982, Inhibition of dihydrofolate reductase: Effect of reduced nicotinamide adenine dinucleotide phosphate on the selectivity and affinity of di-aminobenzylpyrimidines, *Biochemistry* **21**:5068–5075.

Bash, P. A., Singh, U. C., Brown, F. K., Langridge, R., and Kollman, P. A., 1987, Calculation of the relative change in binding free energy of a protein–inhibition complex, *Science* **235**:574–576.

Beddell, C. R., 1984, Dihydrofolate reductase: Its structure, function, and binding properties, in: *X-Ray Crystallography and Drug Action,* (A. S. Horn and C. J. DeRanter, eds.), pp. 169–193, Oxford University Press, New York.

Bevan, A. W. Roberts, G. C. K., Feeney, J., and Kuyper, L. F., 1985, ¹H and ¹⁵N NMR studies of protonation and hydrogen-bonding in the binding of trimethoprim of dihydrofolate reductase, *Eur. Biophys. J.* **11**:211–218.

Birdsall, B., Roberts, G. C. K., Feeney, J., Dann, J. G., and Burgen, A. S. V., 1983, Trimethoprim binding to bacterial and mammalian dihydrofolate reductase: A comparison by proton and carbon-13 nuclear magnetic resonance, *Biochemistry* **22**:5597–5604.

Bitar, K. G., Blankenship, D. T., Walsh, K. A., Dunlap, R. B., Reddy, A. V., and Freisheim, J. H., 1977, Amino acid sequence of dihydrofolate reductase from an amethopterin-resistant strain of *Lactobacillus casei, FEBS Lett.* **80**:119–122.

Blakley, R. L., 1984, Dihydrofolate Reductase, in: *Folates and Pterins,* Vol. 1 (R. L. Blakley and S. J. Benkovic, eds.), pp. 191–253, Wiley, New York.

Blakley, R. L., and Benkovic, S. J. (eds.), 1984, *Folates and Pterins,* Vol. 1, Wiley, New York.

Blaney, J. M., Weiner, P. K., Dearing, A., Kollman, P. A., Jorgensen, E. C., Oatley, S. J., Burridge, J. M., and Blake, C. C. F., 1982, Molecular mechanics simulation of protein-ligand interactions: Binding of thyroid hormone analogues to prealbumin, *J. Am. Chem. Soc.* **104**:6424–6434.

Blaney, J. M., Hansch, C., Silipo, C., and Vittoria, A., 1984, Structure–activity relationships of dihydrofolate reductase inhibitors, *Chem. Rev.* **84**:333–407.

Bolin, J. T., Filman, D. J., Matthews, D. A., Hamlin, R. C., and Kraut, J., 1982, Crystal structures of *Escherichia coli* and *Lactobacillus casei* dihydrofolate reductase refined at 1.7Å resolution. I. General features and binding of methotrexate, *J. Biol. Chem.* **257**:13650–13662.

Champness, J. N., Kuyper, L. F., and Beddell, C. R., 1986a, Interaction between dihydrofolate reductase and certain inhibitors, in : *Topics in Molecular Pharmacology*, Vol. 3 (A. S. V. Burgen, G. C. K. Roberts, and M. S. Tute, eds.), pp. 335–362, Elsevier, New York.

Champness, J. N., Stammers, D. K., and Beddell, C. R., 1986b, Crystallographic investigation of the cooperative interaction between trimethoprim, reduced cofactor and dihydrofolate reductase, *FEBS Lett.* **199**:61–67.

Charlton, P. A., Young, D. W., Birdsall, B., Feeney, J., and Roberts, G. C. K., 1985, Stereochemistry of reduction of the vitamin folic acid by dihydrofolate reductase, *J. Chem. Soc. Perkin Trans. I*, 1349, 1353.

Cheung, H. T. A., Searle, M. S., Feeney, J., Birdsall, B., Roberts, G. C. K., Kompis, I., and Hammond, S. J., 1986, Trimethoprim binding to *Lactobacillus casei* dihydrofolate reductase: A ^{13}C NMR study using selectively ^{13}C-enriched trimethoprim, *Biochemistry* **25**:1925–1931.

Cocco, L., Roth, B., Temple, C., Jr., Montgomery, J. A., London, R. E., and Blakley, R. L., 1983, Protonated state of methotrexate, trimethoprim, and pyrimethamine bound to dihydrofolate reductase, *Arch. Biochem. Biophys.* **226**:567–577.

Filman, D. J., Bolin, J. T., Matthews, D. A., and Kraut, J., 1982, Crystal structures of *Escherichia coli* and *Lactobacillus casei* dihydrofolate reductase refined at 1.7Å resolution. II. Environment of bound NADPH and implications for catalysis, *J. Biol. Chem.* **257**:13663–13672.

Finland, M., Kass, E. H., and R. Platt, eds., 1982, Trimethoprim–sulfamethoxazole revisited, *Rev. Infect. Dis.* **4**:185–618.

Freisheim, J. H., and Matthews, D. A., 1984, The comparative biochemistry of dihydrofolate reductase, in: *Folate Antagonists as Therapeutic Agents*, Vol. 1 (F. M. Sirotnak, J. J. Burchall, W. B. Ensminger, and J. A. Montgomery, eds.), pp. 69–131, Academic Press, New York.

Hine, J., and Mookerjee, P. K., 1975, The intrinsic hydrophilic character of organic compounds. Correlations in terms of structural contribution, *J. Org. Chem.* **40**:292–298.

Hitchings, G. H., 1983, Functions of tetrahydrofolate and the role of dihydrofolate reductase in cellular metabolism, in: *Handbook of Experimental Pharmacology*, Vol. 64 (G. H. Hitchings, ed.), pp. 11–23, Springer-Verlag, Berlin.

Hitchings, G. H., Kuyper, L. F., and Baccanari, D. P., 1989, Selective inhibitors of dihydrofolate reductase, in: *Design of Enzyme Inhibitors as Drugs* (M. Sandler and H. J. Smith, eds.), Oxford University Press, New York, pp. 343–362.

Hood, K., and Roberts, G. C. K., 1978, Ultraviolet difference-spectroscopic studies of substrate and inhibitor binding to *Lactobacillus casei* dihydrofolate reductase, *Biochem. J.* **171**:357–366.

Howell, E. E., Villafranca, J. E., Warren, M. S., Oatley, S. J., and Kraut, J , 1986, Functional role of aspartic acid-27 in dihydrofolate reductase revealed by mutagenesis, *Science* **231**:1123–1128.

Kompis, I., and Then, R. L., 1984, Rationally designed brodimoprim analogues: Synthesis and biological activities, *Eur. J. Med. Chem.—Chim. Ther.* **19**:529–534.

Kraut, J., and Matthews, D. A., 1987, Dihydrofolate reductase, in: *Biological Macromolecules and Assemblies*, Vol. III (F. Jurnak and A. McPherson, eds.), pp. 1–71, Wiley, New York.

Kumar, A. A., Blankenship, D. T., Kaufman, B. T., and Freisheim, J. H., 1980, Primary structure of chicken liver dihydrofolate reductase, *Biochemistry* **19**:667–678.

Kuyper, L. F., 1985, Molecular mechanics modeling of dihydrofolate reductase-inhibitor complexes: Correlation between calculated energy and observed affinity, Abstracts of Papers, 189th ACS National Meeting, Miami Beach, FL, April 28-May 3, Washington, DC, Abstr. MEDI 88.

Kuyper, L. F., Roth, B., Baccanari, D. P., Ferone, R., Beddell, C. R., Champness, J. N., Stammers, D. K., Dann, J. G., Norrington, F. E., Baker, D. J., and Goodford, P. J., 1985, Receptor-based design of dihydrofolate reductase inhibitors: Comparison of crystallographically determined enzyme binding with enzyme affinity in a series of carboxy-substituted trimethoprim analogues, *J. Med. Chem.* **28**:303–311.

Masters, J. N., and Attardi, G., 1983, The nucleotide sequence of the cDNA coding for the human dihydrofolic acid reductase, *Gene* **21**:59–63.

Matthews, D. A., Alden, R. A., Bolin, J. T., Freer, S. T., Hamlin, R., Xuong, N., Kraut, J., Poe, M., Williams, M., and Hoogsteen, K., 1977, Dihydrofolate reductase: X-ray structure of the binary complex with methotrexate, *Science* **197**:452–455.

Matthews, D. A., Bolin, J. T., Burridge, J. M., Filman, D. J., Volz, K. W., Kaufman, B. T., Beddell, C. R., Champness, J. N., Stammers, D. K., and Kraut, J., 1985a, Refined crystal structures of *Escherichia coli* and chicken liver dihydrofolate reductase containing bound trimethoprim, *J. Biol. Chem.* **260**:381–391.

Matthews, D. A., Bolin, J. T., Burridge, J. M., Filman, D. J., Volz, K. W., and Kraut, J., 1985b, Dihydrofolate reductase. The stereochemistry of inhibitor selectivity, *J. Biol. Chem.* **260**:392–399.

Matthews, D. A., Smith, S. L., Baccanari, D. P., Burchall, J. J., Oatley, S. J., and Kraut, J., 1986, Crystal structure of a novel trimethoprim-resistant dihydrofolate reductase specified in *Escherichia coli* by R-plasmid R67, *Biochemistry* **25**:4194–4204.

Oefner, C., D'arcy, A., and Winkler, F. K., 1988, Crystal structure of human dihydrofolate reductase complexed with folate, *Eur. J. Biochem.* **174**:377–385.

Ollis, W. D., Stoddart, J. F., and Sutherland, I. O., 1974, The conformational behaviour of some medium-sized ring systems, *Tetrahedron* **30**:1903–1921.

Pettitt, M., and Karplus, M., 1986, Interaction energies: their role in drug design, in: *Topics in Molecular Pharmacology*, Vol. 3 (A. S. V. Burgen, G. C. K. Roberts, and M. S. Tute, eds.), pp. 75–113, Elsevier, New York.

Phillips, T., and Bryan, R. F., 1969, X-ray crystal structures of the antimalarial agents daraprim and trimethoprim, *Acta Crystallogr. Sect. A* **A25**:S200.

Piper, J. R., Montgomery, J. A., Sirotnak, F. M., and Chello, P. L., 1982, Syntheses of α- and γ-substituted amides, peptides, and esters of methotrexate and their evaluation as inhibitors of folate metabolism, *J. Med. Chem.* **25**:182–187.

Prendergast, N. J., Delcamp, T. J., Smith, P. L., and Freisheim, J. H., 1988, Expression and site-directed mutagenesis of human dihydrofolate reductase, *Biochemistry* **27**:3664–3671.

Richardson, J. S., 1981, The anatomy and taxonomy of protein structure, *Adv. Prot. Chem.* **34**:167–339.

Roth, B., 1983, Selective inhibitors of bacterial dihydrofolate reductase: Structure-activity relationships, in: *Handbook of Experimental Pharmacology*, Vol. 64 (G. H. Hitchings, ed.), pp. 107–127, Springer-Verlag, Berlin.

Roth, B., and Cheng, C. C., 1982, Recent progress in the medicinal chemistry of 2,4-diaminopyrimidines, in: *Progress in Medicinal Chemistry*, Vol. 19 (C. P. Ellis and G. B. West, eds.), pp. 269–331, Elsevier Biomedical Press, Amsterdam.

Roth, B., Aig, E., Lane, K., and Rauckman, B. S., 1980, 2,4-Diamino-5-benzylpyrimidines as antibacterial agents. 4. 6-Substituted trimethoprim derivatives from phenolic Mannich intermediates. Application to the synthesis of trimethoprim and 3,5-dialkylbenzyl analogues, *J. Med. Chem.* **23**:535–541.

Searle, M. S., Forster, M. J., Birdsall, B., Roberts, G. C. K., Feeney, J., Cheung, H. T. A., Kompis, I., and Geddes, A. J., 1988, Dynamics of trimethoprim bound to dihydrofolate reductase, *Proc. Natl. Acad. Sci. USA* **85**:3787–3791.

Singh, U. C., 1988, Probing the salt bridge in the dihydrofolate reductase–methotrexate complex by using the coordinate-coupled free-energy perturbation method, *Proc. Natl. Acad. Sci. USA* **88**:4280–4284.

Stammers, D. K., Champness, J. N., Beddell, C. R., Dann, J. G., Eliopoulos, E., Geddes, A. J.,

Ogg, D., and North, A. C., 1987, The structure of mouse L1210 dihydrofolate reductase–drug complexes and the construction of a model of human enzyme, *FEBS Lett.* **218**:178–184.

Stone, D., Paterson, S. J., Raper, J. H., and Phillips, A. W., 1979, The amino acid sequence of dihydrofolate reductase from the mouse lymphoma L1210, *J. Biol. Chem.* **254**:480–488.

Subramanian, S., and Kaufman, B. T., 1978, Interaction of methotrexate, folates and pyridine nucleotides with dihydrofolate reductase: Calorimetric and spectroscopic binding studies, *Proc. Natl. Acad. Sci. USA* **75**:3201–3205.

Villafranca, J. E., Howell, E. E., Voet, D. H., Strobel, M. S., Ogden, R. C., Abelson, J. N., and Kraut, J., 1983, Directed mutagenesis of dihydrofolate reductase, *Science* **222**:782–788.

Volz, K. W., Matthews, D. A., Alden, R. A., Freer, S. T., Hansch, C., Kaufman, B. T., and Kraut, J., 1982, Crystal structure of avian dihydrofolate reductase containing phenyltriazine and NADPH, *J. Biol. Chem.* **257**:2528–2536.

Weiner, P. K., and Kollman, P. A., 1981, AMBER: Assisted model building with energy refinement. A general program for modeling molecules and their interactions, *J. Comp. Chem.* **2**:287–303.

Wong, C. F., and McCammon, J. A., 1986, Dynamics and design of enzymes and inhibitors, *J. Am. Chem. Soc.* **108**:3830–3832.

VII

MACROMOLECULES AND TARGETED DRUG DELIVERY

16

Control of the Biological Dispersion of Therapeutic Proteins

E. TOMLINSON

1. Introduction

Proteins have immense structural variety and variability, *ergo* their consideration and use of therapeutic agents able to elicit a pharmacological response. As for conventional drugs, much of today's activity with therapeutic proteins in classical research and development environments ignores the issues of site(s) of action and routes, rates, and frequency of administration when selecting a drug to be developed. Selective drug delivery and targeting seeks to achieve the optimal arrival of a drug at its site of action in a manner that is appropriate for the disease and the drug, and which leads to a significant reduction in the possibility of drug side effects. This chapter examines how the control of the biological dispersion of proteins may be achieved using protein remodeling, protein hybridization, and synthetic adduction, as well as by modulating their administration.

2. Therapeutic Proteins

2.1. Type

Hormones, serum proteins, and enzymes have long been part of the clinician's armamentarium. In the past they had been variously produced via isolation from plants, animals, microorganisms, and human fluids and tissues, and augmented by synthetically produced material. In recent years the control of gene

---

E. TOMLINSON • Advanced Drug Delivery Research, Ciba-Geigy Pharmaceuticals, Horsham, West Sussex RH12 4AB, England;   *present address:* Somatix Corporation, Cambridge, Massachusetts 02139.

expression has enabled the production of genetically engineered proteins intended for therapeutic use. These may be homologous or heterologous. The most important indications for these are in cardiovascular disorders, cancer, autoimmune diseases, and infections. Almost 500 human proteins have had their primary structure elucidated by gene cloning; most are now available in abundant quantities using expression in host cells. Four broad classes of structure are of interest: hormones, pharmacological and operational receptors, enzymes, and components of the immune system. Of those of therapeutic interest, at least 15 proteins involved in hemostasis, ranging from tPA to Factor VIII and streptokinase, have been isolated and cloned, as have many immunomodulators (cytokines) and growth and differentiation factors. (For a comprehensive review of the endogeneous therapeutic proteins that have been cloned and expressed see Blohm *et al.*, 1988.)

In mid-1988 the U.S. Office of Technological Assessment estimated that in addition to the nine drugs and vaccines approved by the Food and Drug Administration (FDA), an additional 400 human biotechnology therapeutic products (defined as those produced either by cells that express cloned genes or by hybridomas) were at some stage of clinical development. Although comprising about 2% of all Investigational New Drugs (INDs) currently on file at FDA, it must be appreciated that in 1987 nearly 50% of all INDs were for such biotechnology products!

Proteins proposed for therapy usually have regulatory or homeostatic functions, and must also be considered for both pathophysiological events and the healing process. Apart from homologous polypeptides and proteins, they also include derivatives of those produced by *inter alia* site-directed mutagenesis, proteolysis, gene fusion, protein aggregation, and/or conjugation with other biologically active effector functions. They include also poly- and monoclonal antibodies, engineered enzymes, and genetically defined site-specific delivery systems such as fusion hybrid proteins.

2.2. Physical and Chemical Properties

Many proteins intended for therapeutic use are glycoproteins. The biological dispersion of therapeutic glycoproteins is due to three prime determinants: their chemical and metabolic stability, size and shape, and surface features. This chapter examines how these features can be adjusted so as to control this dispersion. The influence of symmetrical and asymmetrical input rate and frequency and type of administration on dispersion are also briefly examined.

2.2.1. Stability

The biological half-life of most proteins is short. For paracrine-like and autocrine-like mediators this is largely due to their degradation by peptidases and proteinases in the vascular endothelium, liver, and so forth. In addition, like the signal sequences of amino acids that confer to a protein its ability to enter distinct cellular compartments, the amino acids of a terminal chain of proteins can control their intracellular metabolic stability, and hence their intracellular residence time (Bachmair *et al.*, 1986). Much current attention is being given to adjusting the reactivity of proteins intended for intracellular deposition so as to improve their stability and alter their potency. For example, protein-engineering tools have been used to replace labile amino acids, such as the oxidation-resistant amino acids alanine, serine, or threonine, or to produce proteins having differing foldings—potentially leading to proteins that are protected from inactivation. Synthetic approaches may also be considered for introducing amino acids that are not genetically encoded.

2.2.2. Size and Ability to Extravasate

Therapeutic (glyco) proteins can be flexible and/or globular, and have a size of up to 300 kDa in the nonaggregated state. This affects their ability to diffuse across endo- and epithelial membranes, although, for example, when placed into the vasculature they are able to interact with the surfaces of blood and endothelial cells, as well as with exposed parenchymal cells of organs with discontinuous endothelia (e.g., liver and spleen). Many of the (patho)physiological and anatomical opportunities and constraints for the movement of macromolecules within the body have been analyzed (Tomlinson, 1987a), including their ability to extravasate and to pass continuous and noncontinuous endothelia in normal and diseased areas. Size selectivities due to region and pathology are evident. Extravasation, which can be due to passive and/or cell-mediated (e.g., neutrophil) passage through (disrupted) endothelium, is a key criterion in the selection of therapeutic proteins to be developed.

2.2.3. Polymeric State

Protein engineering can be used to alter the polymeric state of a therapeutic protein. This may affect its biological dispersion, as shown by recent work on monomeric insulins (Brange *et al.*, 1988). Most interestingly, monomeric insulin is absorbed into the body two to three times faster than its polymeric forms after subcutaneous administration.

3. Clinical Use

The clinical development of therapeutic proteins depends on successfully demonstrating their identity, purity, safety, and efficacy. While recombinant DNA and hybridoma technologies have focused on producing proteins in abundant and pure amounts, the study of the pharmacology of therapeutic proteins has largely followed the traditional route taken in pharmaceutical research and development, i.e., parenteral administration, for dose-ranging studies. Little regard has been given to how this relates to the site of action of these drugs, the chronicity of the disease, and, as importantly, to chronobiological influences on disposition and target location and responsiveness.

Although for a few proteins there is little relation between dose applied and effect, with most it is highly critical, particularly as nonlinear dose–effect relationships are often described. Bell-shaped, and even double bell-shaped, dose-response relationships have been found clinically (Talmadge, 1986), e.g., for parathyroid hormone, substance P, and δ-sleep inducing peptide.

3.1. Site and Mode of Action

Initial approaches to the development of protein drugs have given little rational consideration to the site of their action, i.e., whether the putative therapeutic protein is to act systemically, or whether it is the mimic of an endogeneous molecule that is normally produced to act locally. In addition, the development of therapeutic proteins must in many cases be related to the way in which the target biological processes develop and can be regulated. For example, as pointed out by Blohm *et al.* (1988), in the control of hemostasis, the therapeutic effect of a protein depends on whether it is acting as an agonist, an antagonist, or a combined agonist–antagonist (e.g., plasminogen activator–inhibitor) in the complex interactions regulating blood clotting and fibrinolysis. Consider also the cells of the immune system as targets for clinical intervention, and the differing roles of growth factors and immunomodulators. Hemopoiesis leads to both growth and differentiation of numerous varied cells that are acted upon by growth factors, and to resting phase cells that may be activated by immunomodulators. Many of the molecules that act on both of these two types of cells produce numerous similar biological effects (e.g., tumor necrosis factor and interleukin-1, and α- and β-interferons), with the resultant effect differing only in magnitude. Since dysregulation of biological control would lead to a plethora of harmful events, we argue here that it is vital that both the (physico)chemical structure and the administration of therapeutic proteins are tailored to take into account their required spatial localizations.

3.2. Endocrine-Like and Para-/Autocrine-Like Mediators

There is relatively little problem with the parenteral administration of endogeneous proteins that would normally circulate in the blood, such as hormones, since these are produced naturally to act over long distances from their site of manufacture; they are also stable in blood, and, if relevant, their size and surface character enable their (specific) extravasation. However, many endogeneous proteins are produced and released to act locally; they have very short chemical half-lives, which ensures that they do not give rise to untoward effects on nontarget neighboring cells. These are often produced at sites of inflammation, tumors, and injuries (e.g., the angiogenesis factors—fibroblast growth factor and angiogenenin). Hence, if administered parenterally as drugs, not only may such proteins be readily destroyed in the blood, but because of their size and surface features they are unlikely to be able to extravasate through normal endothelia either passively or actively, in order to reach an intended extravascular intra- or extracellular site of action. Since the specificity of action of such molecules is due to their local release and action, and (as described above) since these types of molecules can produce different effects on the same cells at different stages of their life cycle, their inappropriate route and type of administration can explain why they have an extremely complex pharmacology (Talmadge, 1986). Clearly, the production and use of the latter class of molecules is contraindicated unless means are found for adjusting their pharmacodisposition (and stability).

Although the intrinsic interaction between proteins and their targets could be responsible for observed nonlinear dose–effect relationships, such behavior could also be due to nonlinearity in dispersion, caused by, *inter alia*, poor ability to enter into (saturable) receptor-mediated transport processes; chemical instability; incorrect sequence of administration with other mediators; and/or wrong time of administration due to inappropriate temporal location and responsiveness of the target (cells). Certainly the sequential use of mediators can be used to modulate the intracellular dispersion of therapeutic proteins. This has been shown *in vitro* with γ-interferon, which appears to influence the cytotoxic action of both tumor necrosis factors TNF-α and TNF-β, by up-regulating TNF receptors on the surface of TNF-resistant cells (Aggarawal *et al.*, 1985). Examine also the influence of timing and chronobiology on response, as given by the work of Castro and Gras (1984), who showed that the kinetics of IgM and IgG responses are strongly influenced by the rhythm of antigen administration. Thus, for the clinical use of many therapeutic proteins, the additional issues of preactivation of target cells and their temporal localization and responsiveness need consideration.

4. Protectants

Numerous approaches may be taken to alter the dispersion of proteins to enhance their activity and reduce their potential for side effects. For therapeutic proteins needing to remain in the blood compartment, two prime methods have been adopted: increasing the apparent size of the protein, and/or reducing its (untoward) interactions with blood and tissue components.

Immunosurveillance is mediated through physiochemical interaction between a (therapeutic) protein and components of the immune system. Frequently, opsonization by fibrinogen, fibronectin, and other blood components is a prelude to recognition and then removal by cells of the formed complex; antigen–antibody interaction and Fc-mediated removal also occur. Opsonized materials are taken into cells by engulfment after adherence to, and vesiculation of, phagocytosing cell membranes. Both opsonization and adherence can be diminished if the attractive forces between the interacting therapeutic protein, blood macromolecule, and, for example, a cell-surface macromolecule are diminished. Adsorption and adhesion are complex phenomena that are controlled by many factors including hydration and electrostatic, dispersion, and steric forces, and by other short-range interactions Norde (1984). Interfacial adsorption is dependent on a balance between these forces. Colloidal particles will attract each other through van der Waals interactions (short-range), and repel each other through longer-range, repulsive (e.g., Coulombic) forces. As proteins approach one another there is a net attraction, with a potential energy barrier to interaction at closer proximities, and with strong interaction at very short ranges. Interaction can be avoided by creating a high potential energy barrier. Although *in vitro* this may be achieved by charge/charge effects, this is likely to be diminished *in vivo*. Napper and Netschey (1971) have argued that for (particulate) colloids, a high potential energy barrier can be formed by creating a sterically stabilized surface upon introducing a hydrated (i.e., hydrophilic) polymer at the surface of the colloid. The hydration effect is enthalpic in origin, with the stabilization effect being manifested by both osmotic effects and chain entanglements—both of which are entropic in origin (Ottewill, 1977). The size of any repulsive barrier should be determined by both the thickness of the polymer layer and its density, as well as by polymer–polymer interactions caused by specific interactions along the polymer chain. It is probable that steric stabilization is akin to the mechanism whereby blood cells and various bacteria and parasites escape detection by the mononuclear phagocyte system (MPS). Surface modifications to proteins can be made to improve their tolerance within the vasculature—due largely to the formation of a surface that makes it energetically unfavorable for other macromolecules to approach.

Table I gives examples of hydrophilic (bio)polymeric protectants that have been described for conjugation to therapeutic proteins. Synthetic and biological materials have been used or suggested, and include polyethylene glycols, poloxamers, poloxamines, albumin, immunoglobulin G, carboxymethycellulose, natural xanthans and sorbitans, and so forth. Conjugations of proteins with hydrophilic polymers have often been reported as being very successful in altering their potency as well as reducing their immunogenicity and increasing their duration of action. Abuchowski *et al.* (1977) were the first to adopt this approach for stabilizing therapeutic proteins by forming protein conjugates with hydrophilic polyethylene glycol chains. Others have increasingly used this approach,

Table I. Protectants for Therapeutic Proteins[a]

Protein	Proposed use for conjugate
Polyethylene glycols	
Islet-activating protein	Isulinogenic activity
Superoxide dismutase	Kidney transplantation, burns, re-perfusion damage (Phase II)[b]
L-Asparaginase	Malignant hematological disorders (Phase III)[b]
Adenosine deaminase	AD deficiency
Urokinase	Coagulant, fibrinolytic
Proteins	Radioprotection
Interleukin 2	Cancer
Uricase	Altered antigenicity
Catalase	
Interferons	
Insulin	
Growth hormones	
Immunoglobulins	
Trypsin inhibitors	
Proteases and peptidases	
Octylphenoxy polyethoxy ethanols	
Interleukin 2	
Poloxyethylene sorbitans	
Interleukin 2	
Dextran	
Urokinase	Fibrinolytic
Purine nucleosides	Inhibitors of adenosine deaminase
α-1,4-Glucosidase *neo*glycoproteins	Enzyme replacement therapy
Carboxypeptidase G_2	Enzyme replacement therapy
β-Galactosidase	Enzyme replacement therapy
L-Asparaginase	Cancer (lower antigen reactivity and increased circulatory persistence)
Superoxide dismutase	
Albumin	
Asparaginase	Cancer
Poly-D-alanyl peptides	
Asparaginase	Cancer
N-(2-Hydroxypropyl)methacrylamide (HPMA)	
Antibodies	Cancer-seeking agents

[a]Described in the (patent) literature.
[b]Clinical trial status as of July 1988.

and modifications of it, for lengthening the blood half-lives of a number of peptidergic mediators, such as interleukin-2 (Katre *et al.*, 1987), and enzymes, including catalase, asparaginase, and urokinase, while still maintaining their reactive functionalities (Table I). Significant changes to biological dispersion can be made, as demonstrated by the recent studies of Ho *et al.* (1988), which showed that pegoylated asparaginase in rabbits has an increase in its circulating half-life over the native enzyme increasing from about 20 hr to between approximately 125 and 160 hr, with a greatly reduced clearance from the blood of from about 100 ml.kg^{-1}.day^{-1} to about 3 ml.kg^{-1}.day^{-1}, and with an almost 40-fold increase in the plasma area under the curve. It has been reported that a pegoylated–asparaginase conjugate is well tolerated in patients who had already been treated with native asparaginase and had neutralizing antiasparaginase antibodies. Steric stabilization (or polymer-excluded volume) approaches to avoiding opsonization have recently received attention for modifying antibodies (Ríhová *et al.*, 1986); the modification of antibodies with hydrophilic polymers can be additionally utilized to enable the chelation of small inorganics for diagnostic purposes (Torchilin *et al.*, 1986).

A decrease in immunogenicity of therapeutic proteins may result from a reduction in their aggregation, or simply from a masking of any antigenic determinants. It has been found that the primary and secondary IgE antibody responses to protein may be suppressed by chemically conjugating protein with a derivative of polyaspartic acid, i.e., α,β-poly[(2-hydroxyethyl)-DL-aspartamide] (chosen since it had been used as a plasma expander without apparent toxicity, see Okada *et al.*, 1985). Although the mechanism of this effect is not fully defined, it has been shown that the suppressive effect of protein modified with polyethylene glycol (Lee *et al.*, 1981) or fatty acid (Segawa, 1981) is due to the induction of suppressor T cells, and that conjugation with a copolymer of D-glutamate and D-lysine leads to suppression via tolerance of B cells (Katz *et al.*, 1972).

Conjugation of proteins with hydrophilic polymers can also increase their chemical and physicochemical stability and is known to enhance resistance to proteolysis and heat denaturation. Wileman *et al.* (1986) have found that oxidized dextrans conjugated to asparaginase give the enzyme increased circulatory half-times and a lowered antigen reactivity. Their studies showed that dextrans bound to asparaginase protect the enzyme from inactivation by proteases and enzyme-specific antibody, as well as reducing the antigen reactivity of the enzyme *in vivo*. This increased stability could be due to a number of factors, including modification of the lysine residues present in the antigenic sites of the enzyme or at trypsin binding sites, and/or the fact that soluble dextran could sterically hinder the interaction with these destructive elements. This important study found that *in vivo*, antigen reactivity of the conjugate generally fell with an increase in the size of the dextran; however, the *in vivo* disposition of protein–

dextran conjugates can show a complex pattern. For example, with dextrans of molecular weight greater than 50 kDa there is little glomerular excretion of the dextran—this moiety being susceptible to uptake by both cells of the MPS and the parenchymal cells of the liver; however, conjugation with proteins leads to a reduction in the degradation (of an enzyme) by liver cells, although the dextran moiety is degraded more rapidly than its unconjugated form (Melton *et al.*, 1987).

A marked variability exists in the relative abilities of the various protectants to affect either the action and/or the immunogenicity and antigenicity of therapeutic proteins. Recent work has described the adduction of *Serratia* Mn superoxide dismutase (SOD) with polyethylene glycol (5 kDa) and with dextran (80 kDa) (Miyata *et al.*, 1988). SOD was modified at 34% of its free amino acids by dextran; although the Dx–SOD conjugate retained 67% of the native activity of SOD, it was found to have a higher immunogenicity than SOD *in vivo*. With PEG, 24% of the free amino groups were modified. PEG–SOD conjugates retained 52% of the original activity of SOD, and had much lower antigenicity and immunogenicity. (Dx–SOD and PEG–SOD antigenicities were 2% and 10% of SOD, respectively, but Dx–SOD—in contrast to PEG–SOD—was found to be very immunogenic.) The blood half-life of PEG–SOD in the rat *in vivo* was about 10 times greater than that found for SOD. Miyata *et al.* claim that pegoylation of SOD enhances its pharmacological activities as an anti-inflammatory and radioprotective agent, and list the advantages of pegoylation as an increase in (chemical) stability and antioxidant activity, a reduction in antigenicity and immunogenicity, and an increase in circulation half-life, together with an increase in anti-inflammatory activity, an increase in radioprotection, and a reduction in the incidence of ventricular arrhythmia induced by ischemia!

The avoidance of uptake by the MPS should enable therapeutic proteins to remain for considerably longer periods within the circulation. This can be considered useful for increasing the statistical probability that a competing process will occur (e.g., extravasation or interaction with an intravascular target cell), and/or for developing a long-term circulating depot of active protein.

5. Protein (Re)glycosylation

As described earlier, the dispersions of proteins are largely controlled by their size, surface character, and chemical reactivity. Numerous groups (e.g., Baenziger, 1985) have demonstrated that numerous endogeneous glycoproteins (i.e., serum glycoproteins, lysosomal enzymes, and perhaps also sulfated pituitary glycoproteins such as chorionic gonadotropin) interact through their specific carbohydrate residues complexing with (oligosaccharide-specific) recognition systems on the plasma surfaces of target cells. Hence, glycosylation patterns are signals used by the body to regulate the dispersion of its own glycoproteins, as implicated for both enzyme and hormone disposition as well as immune surveillance, coagulation, and so forth. For example, the intracellular translocation of lysosomal enzymes to lysosomes is due to the phospho-D-mannopyranosyl moiety of lysosomal enzymes, such that phosphorylation of the D-mannose residues is essential for their uptake and intracellular transport to lysosomes (Madiyalakan *et al.*, 1986). Therefore, changing the oligosaccharide content of proteins has potential for controlling the fate of therapeutic glycoproteins.

5.1. Oligosaccharide Variability and Recognition

In contrast to possible changes in the amino acid composition of a protein, the numerous variations possible in linking simple sugars together afford glycoproteins an almost limitless variability and diversity in structure. Baenziger (1985) has pointed out that a relatively typical decasaccharide consisting of mannose, galactose, *N*-acetylglucosamine, and sialic acid could give rise to 10^{24} structures. Additional modifications, such as removal or addition of peripheral sugars and/or other functional groups such as acetyl, methyl, sulfate, and phosphate, are also possible. Oligosaccharides may be *N*-glycosidically linked (to the peptide at Asn), or *O*-glycosidically linked (attached to Ser and Thr). The oligosaccharides of the plasma glycoproteins are linked to protein primarily through *L*-asparagine-*N*-acetyl-D-glucosamine. Other prime linkages are L-serine-*N*-acetyl-D-galactosamine and L-threonine-*N*-acetyl-D-galactosamine (as in mucus glycoproteins and immunoglobulins). The following monosaccharides are the major components of the side chains of these oligosaccharides: L-fucose, sialic acids, D-galactose, D-mannose, *N*-acetyl-D-glucosamine, and *N*-acetyl-D-galactosamine. In mammals, the β-D-galactose residues occur as the penultimate units of the glycoprotein glycans; if exposed, this results in the glycoprotein being efficiently taken up by galactose-specific recognition systems. Carbohydrate recognition is an extremely selective event, with single changes resulting in nonrecognition. For example, with the following two structures the oligosaccharide specificity of rat hepatocyte Gal/Gal/NAc-specific receptor on isolated hepatocytes structure is demonstrated; with Structure 1 being well taken up by these cells, whereas Structure 2 is not. This indicates that the hepatocyte Gal/GalNAc-specific receptor recognizes oligosaccharides only when they have a very specific spatial relationship to each other (Baenziger, 1985).

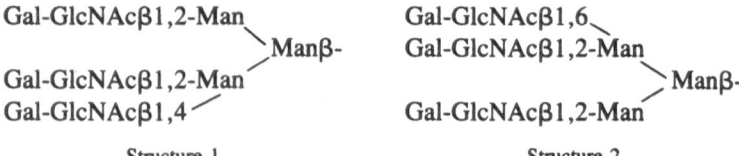

Structure 1   Structure 2

Numerous other structure/binding relationships are emerging. Asialoglyco-proteins with oligosaccharides biantennary and triantennary are preferentially adsorbed by leukocyte lectins, whereas those having oligosaccharide triantennary and tetraantennary are more rapidly bound to the plasma membrane of liver hepatocytes. Other studies also indicate that it is the spatial corelationship between residues, and not their number, that is important in cell recognition.

Under normal physiological conditions the glycoproteins subject to oligosaccharide-specific recognition are mannose-6-phosphate lysosomal enzymes; few glycoproteins in human serum have terminal galactose. Interestingly, Baenziger (1985) and colleagues have shown that the hinge region of monomeric IgA$_1$ contains five O-glycosidic saccharides. Although these result in IgA$_1$ being well taken up by *isolated* hepatocytes, under normal physiological conditions it appears that such binding moieties are protected, leading to high levels of IgA$_1$ remaining in the serum. (It is speculated that when antigen binding occurs, the structure of IgA$_1$ is perturbed to reveal the hinge region saccharides—which then causes the antigen–immunoglobulin complex to be removed from the general circulation.) Galactose-specific recognition systems have been identified on hepatocytes (for D-galactose and N-acetyl-D-galactosamine), on lymphocytes, and more recently on macrophages (Kelm and Schauer, 1986). Similarly, mannose-containing oligosaccharides are taken up by both the endothelial cells and the Kupffer cells of the liver; which leads to the suggestion that reductive *mannosamination* may provide a means of directing therapeutic proteins to such cells (Wilson *et al.*, 1979). Carbohydrates often play no direct role in the biological activity of the glycoprotein; instead, as structural elements, they may influence the *stability* and *conformation* of glycoproteins. Also, they frequently impart good aqueous solubility to glycoproteins because of their hydrophilic character.

Three biological properties of glycoproteins may be adjusted by altering the distribution of carbohydrates on their surface: circulating blood half-life (and potentially duration of action), immunogenicity, and ability to access (cellular) sites of action. Clearly, oligosaccharide structure has enormous potential in controlling the dispersion of therapeutic proteins. Numerous studies with both high- and low-molecular-weight drugs have shown the potential for this. For example, covalent linkage of low density lipoprotein to 200–250 residues of lactose per particle results in their entry into hepatocytes via the galactose receptor and not the normal low-density-lipoprotein receptor. The incorporation of oligosaccharides into macromolecular drug carriers for the targeting of small molecules has been studied extensively, as typified by the conjugation of adenine-9-β-D-arabinofuranoside (ara-C) to asialofetuin and/or lactosaminated serum albumin (both galactosyl-terminating glycoproteins), resulting both in successfully directing the ara-C to its site of action and in increasing its potency (Fiume *et al.*, 1982). For synthetic polymer drug carriers, the incorporation of galactosamine has been successfully used to enhance their extraction from the blood by the liver (Cartlidge *et al.*, 1987).

Glycosylation of therapeutic proteins may occur either during expression-cell processing, or postexpression via synthetic conjugation. Recent work has shown that by conjugating fragment A of diphtheria toxin to a galactose-containing oligosaccharide such as asialofetuin or asialoorosomucoid, the resultant conjugate targets to hepatocytes and is up to 3 orders of magnitude more toxic than the native fragment A. If galactosyl residues are to be used clinically to target proteins to hepatocytes, it is important to appreciate that the *in vivo* uptake of galactosylated neoglycoproteins has been shown to be highly dose-dependent (Vera *et al.*, 1984).

5.2. Biotechnological Processes

Glycosylation does not usually occur in prokaryotic cells such as *Escherichia coli*. Some glycosylations, such as in yeast and mammalian cells, do occur in eukaryotes, however, the resultant glycosylation patterns can often be different, depending on the cells used. Also, although recombinant bacteria are able to produce large amounts of protein, these are often changed by bacterial proteases; within expression cells it is possible that expressed proteins of low solubility denature and form aggregates. Mammalian cells are particularly suitable both for expressing proteins with complex modifications such as γ-carboxylation of glutamyl residues, and for obtaining homologous (human) glycosylation patterns. Thus, it is apparent that biotechnology processes can have a marked effect on the biological disposition and efficacy of therapeutic proteins.

5.3. Protein Remodeling

Much recent progress has been made in the remodeling of expressed glycoproteins intended for therapeutic use in order to affect their selectivity for target cells. Because of the complexity of these reactions, conventional carbohydrate chemistry is of little use, and interest is centered on the modifications of proteins with oligosaccharides via enzymatic synthesis. Various chemical approaches to remodeling proteins postexpression are being developed (e.g., Akiyama *et al.*, 1987; Sandoz, 1988). An important new approach has recently been described, which employs enzymes to elongate and terminate peripheral glycan chains of glycoproteins (Berger *et al.*, 1986, 1987). Mammalian glycoproteins expressed in yeasts are likely to be substituted by mannans, and this group has been able to incorporate sialic acid into endo-β-*N*-acetylglucosaminidase H-treated oligomannose glycoproteins (Berger *et al.*, 1986). The technology is of potential use in reducing any mannose-receptor uptake of these glycoproteins by cells of the MPS. Although the approach needs optimizing in terms of enzyme activity, it demonstrates the successful incorporation of sialic acid into glycoproteins of the oligomannose type. Berger *et al.* (1987) have also suggested the use of purified galactosyltransferase for the galactosylation of glycoproteins prior to their sialylation or their chemical linkage with oligosaccharides, and argue that this approach appears to be promising for the *in vitro* remodeling of glycan chains in heterologous glycoproteins such as tissue plasminogen activator and α_1-antitrypsin.

5.4. Enzyme-Storage Diseases

Postexpression protein remodeling has been used to resurface some of the enzymes indicated in lysosomal storage diseases. Patient molecular biology for the 30 known types of such human diseases show that there is considerable genetic heterogeneity in the biosynthesis of the necessary enzymes. For example, in mucoliposes Type II the mannose-6-phosphate moiety of the active enzyme, which is the signal for hepatic extraction and translocation (via endosomes) to the lysosomes where substrates await their breakdown, is erroneously expressed. In Gaucher's disease, which is most prevalent among eastern European (Ashkenazi) Jews and appears to affects 20,000–40,000 patients worldwide, a genetic disorder leads to a lack of the enzyme glucocerebrosidase. This normally breaks down lipids in erythrocytes, and its deficiency leads to a buildup of lipid in liver, spleen, and bone marrow, resulting in numerous debilitating conditions, including broken bones, and can lead to early death. Homologous glucocerebrosidase, administered at the doses required to effect a limited response, suffers from the fact that the enzyme has a fairly short plasma residence time and elicits the production of antienzyme antibodies. Attempts at achieving the correct dispersion of this and other lysosomal enzymes include the use of synthetic macromolecular carriers (Tomlinson, 1987b) and glycoprotein remodeling. For the latter case, modification of the carbohydrate portion of the protein, changes in its disposition to target cells, and recognition by elements of the immune system result. Immunogenicity is of particular importance for repeated and/or chronic dosing with proteins, e.g., human growth hormone or the antianemia agent erythropoietin (EPO). Human trials in the United States with Ceredase (modified glucocerebrosidase produced by Genzyme) are now underway. Other studies are planned for the use of similarly remodeled ceramidedetrihexosidase (CTH) in the treatment of Fabry's disease, which is caused by a toxic buildup of CTH's substrate, trihexosyl ceramide, and which leads to symptoms including pain, eye and skin lesions, and cardiac and renal disease. In this case, glycoprotein remodeling is intended to modify carbohydrate structures in order to produce a therapeutic enzyme that has a significantly increased circulating blood half-life.

If clinical trials show promise, it is possible that postexpression glycoprotein reglycosylation could be adopted for other therapeutic proteins. Candidate proteins reported amenable for this kind of remodeling include tissue-plasminogen activator, factor VIII, EPO, colony-stimulating factors, β- and γ-interferons, interleukins I–III, and even antibodies, and so on (Table II).

Table II. Reglycosylated Therapeutic Proteins in or Suggested for Clinical Trial

In clinical trial
Glucocerebrosidase (Gaucher's)
Ceramidedetrihexosidase (Fabry's)
Suggested
Human growth hormone
Erythropoietin
Factor VIII
Colony-stimulating factors
α- and β-interferons
Interleukins I–III
Antibodies
Insulin
Growth hormones releasing factor

5.5. Deglycosylation

Conversely, protein deglycosylation should also affect dispersion (and stability and solubility) of a therapeutic glycoprotein. Indeed, deglycosylation (via chemical means using a mixture of sodium metaperiodate and sodium cyanoborohydride at low pH and temperature) has been shown to virtually eliminate the uptake of the toxin ricin A chain by the liver (Blakey and Thorpe, 1986). Ricin A chain is an oligomannosidic glycoprotein that has a very short half-life due to rapid removal by the mononuclear phagocyte system, and deglycosylation of the A chain may reduce removal by Kupffer cells. The chemical method described results in "destruction" of approximately 50% of the mannose and most of the fucose residues, while leaving the N-acetylglucosamine and most of the xylose residues undisturbed. Although chemical deglycosylation only marginally delays the clearance of deglycosylated ricin A from the blood compartment (in mice), this is probably due to the production of a smaller macromolecule that can be effectively filtered and excreted through the kidneys. Blakey and Thorpe assume that linking the deglycosylated ricin A chain to an immunoglobulin would diminish this latter effect. Ricin A chain lacking carbohydrate side chains has also been produced using recombinant technology (Vitetta *et al.*, 1987).

6. Protein Hybrids

The large amount of information currently emerging on the role operational receptors in the body play in controlling the extracellular and intracellular dispersion of proteins is leading to the design of heterologous hybrid proteins with the combined or reordered features of one or more proteins, and which have both effector functions as well as protection and recognition moieties.

6.1. Cell Processing

It has been argued that in designing site-specific drug delivery approaches for both low-molecular-weight and macromolecular drugs, a knowledge of the minimum amount of structural information required to cause a particular routing (or tropism) is often required (Tomlinson, 1986, 1987a). Mechanisms of normal cell processes that are being elucidated currently include simple secretory and receptor-mediated events, both of which require signal recognition and often feedback control. These events are often mediated through leader sequences, which can act in a number of ways, either through their basicity/acidity properties or via some function of their primary and secondary structure. Intracellular and transcellular processing of both endogeneous and heterologous proteins is being increasingly understood (e.g., Von Heijne, 1985; Zimmermann and Meyer, 1986), which can lead to new strategies for altering the dispersion of proteins at the intracellular level due to a combination of recognition and specific release dependent on a unique intra- or extracellular feature (Keogh-Bennett *et al.*, 1986). Intracellular recognition signal structures have been used for designing synthetic peptides able to mimic (secretory) events (e.g., Goldfarb *et al.*, 1986). Oligonucleotide-directed mutagenesis, etc., to produce hybrid proteins is becoming common in elucidating cell function (e.g., the intracellular dispersion of secretory proteins, Moore and Kelly, 1986). (Parenthetically, it is noted that these and similar studies are amplifying concerns that under conditions in which the functionality of operational receptors is affected by low-molecular-weight drugs, genetic disorder, and so forth, the use of these routings to control the dispersions of therapeutic proteins could be contraindicated.)

Site-specific hybrid proteins may be produced either by synthetically linking protein fragments (e.g., Offord and Rose, 1987), or by using ligated gene fusion processes. Table III gives examples of hybrid proteins that have been produced or suggested in the (patent) literature for therapeutic use.

6.2. Gene-Fusion Hybrids

Gene-fusion techniques may be used to produce distinct therapeutic proteins combining the varied properties of parent proteins. This is exemplified by a recently proposed novel strategy for targeting toxins to specific cells within the blood compartment in the treatment of graft vs. host rejection disease (which can occur after organ transplantation). The cytotoxic T cells of the immune system that come into play at times of rejection do so with their surface receptor for interleukin-2 (IL-2) exposed. Using a fused gene, that portion of IL-2 which

Table III. Hybrid Protein Delivery Systems

Fragments of:	
Recognition portion	Effector portion
Gene fusion products	
Interleukin 2	Diphtheria toxin
Growth factor	Toxin (e.g., ricin A)
Cell-specific polypeptide (α/β MSH; substance P)	Restructured diptheria toxin
Antitumor Fab immunoglobulin	Fragment A of diptheria toxin
	γ-Interferon and β-tumor necrosis factor [mechanism of recognition/effector function currently unknown (Feng *et al.*, 1988)]
Chemical linkage of fragments	
Human placental lactogen hormone	Diphtheria toxin A chain
β Chain of *h* chorionic gonadotropin hormone	Ricin A chain
Insulin	Diphtheria A
Epidermal growth factor	Ricin A
Antibody fragments	(Deglycosylated) ricin A
IgG(2a) fragments	Gelonin
	Diphtheria toxin
	Pseudomonas toxin
Anticollagen antibody	Toxin
HIV-specific Ab	Ricin A
Antifibrin Ab	Tissue plasminogen activator
Anti-T-cell antibody	Ricin A chain
Antibody fragments	Gelonin toxin
Antiendothelia IgG	Glucose oxidase
Antiepithelia Ig/fragments	Ricin and other toxins
Anti-sIgM-IgM	Saponin-6

interacts with its receptor on the T cell has been linked to a toxic material (e.g., a portion of diphtheria toxin) to create a cytotoxic site-specific hybrid protein, which recognizes, enters, and then destroys activated T cells. A similar strategy has been adopted for targeting toxins to specific cells as hybrid fusion proteins created by ligating toxin and growth factor genes (Chaudhary *et al.*, 1987). The latter approach relies on the deletion of the toxin gene sequence encoding the cell-binding site, allowing the hybrid-fusion protein to display the cell specificity of the growth factor. The resultant molecule is smaller, which could enable it to penetrate tumors better, and to possibly be less immunogenic.

Hybrid protein approaches may involve not just adduction of a protein (fragment) to a recognition moiety, but also a reordering of the structure of the effector portions of therapeutic proteins in order to enhance their pharmacological action. For example, Murphy (1987) has described a hybrid protein of, in sequence (and has joined by peptide bonds), the enzymatically active fragment A of diphtheria toxin, a fragment including the cleavage domain l_1 adjacent to fragment A, a fragment that includes (at least) a portion of the hydrophobic domain of fragment B (although it does not include the generalized eukaryotic binding site of fragment B), and finally a fragment that includes a portion of a cell-specific polypeptide able to bind the conjugate delivery system to a specific cell feature (see Table III).

6.3. Synthetically Linked Hybrid Conjugates

Remarkable changes in the biological dispersion of proteins have been reported upon their chemical linkage to other protein (fragments). For example, when the toxin gelonin, which has a circulation half-life in mice estimated at 3.5 min, is conjugated to immunoglobulin (fragments), it has a terminal phase blood half-life on the order of days, with only a slight variation in this time as the conjugated immunoglobulin (fragment) is changed (Scott *et al.,* 1987). The immunotoxin field provides many relationships between protein structure and deposition. Greenfield *et al.* (1987) have produced point mutations in the B polypeptide chain of diphtheria toxin that block nonspecific binding to nontarget cells. Upon linking this entity to an anti-T-cell monoclonal antibody, they demonstrated that because of a change in the nontarget tissue distribution of the toxin, it becomes orders of magnitude less toxic than the native toxin to nontarget cells (*in vitro*).

6.4. Antibodies

The numerous concerns that arise with the use of antibodies for targeting drugs (including proteins) to sites of action relate both to their ability to extravasate (if required) in adequate amounts, and to the immunogenicity of non-human-immunoglobulin delivery systems. For immunotoxins, consider also the devastating effect that Fc-mediated uptake by cells of the MPS would have on the immune system. These features have led both to the use of variable-region fragments of antibodies, and to the development of humanized antibodies. However, concern has recently been expressed that the idiotypic portion of the reshaped human IgG_1 antibody, etc., may be immunogenic. Similar doubts relate also to both heavy- and light-chain Gm and Km allotypes, which, although they themselves may be safe, by their presence may enhance anti-idiotype immune responses (Mage, 1988). This has led to the suggestion that after Gm and Km matching recipients with such humanized antibodies, it may be necessary to introduce point mutations to produce antibodies with the alternative allotypic forms of, say, human IgG_1 and κ-light chains. The hope is that a reduction in anti-idiotypic response would increase the blood circulation half-life of these reagents.

7. Administration

7.1. Route, Rate, and Frequency of Input

The nature of the input of a therapeutic protein into the body is a strong determinant of its resultant pharmacokinetic and pharmacodynamic behavior. The influence of controlled input on pharmacology and toxicology of small-molecular-weight drugs is well indicated (see Benet, 1978; Urquhart and Nicholls, 1985). Urquhart and Nicholls (1985) have described three prime input factors that have a marked influence on drug action: duration of drug-free interval, attainment of critical thresholds in plasma concentrations, and the rate of increase in drug plasma concentration. There are similar findings for polypeptide and protein drugs. These are well reflected in the erudite studies of Robinson and his colleagues, who examined the growth responses to intravenous growth hormone administered to hypophysectomized animals as various pulsitile infusions, and showed that a pulsatile growth hormone produces a greater response than does (unphysiological) continuous infusions (Clark *et al.*, 1985). Second, since growth hormone release is profoundly affected by its inhibitory factor, somatostatin, this group also examined the influence of rates and frequency of administration of somatostatin on growth in normal animals. Their findings show that the effects of somatostatin on growth depend on the pattern in which it is administered: when it is administered in an intermittent manner that promotes a more pulsitile growth hormone secretory pattern, this powerful inhibitory peptide paradoxically stimulates weight gain (Clark and Robinson, 1988).

The route of administration can affect the pharmacodynamics of the response to therapeutic proteins as well as the site of action. In animals, for example, orally administered insulin acts mainly in the liver, but when given parenterally it generally acts on peripheral muscle cells.

7.2. Parenteral/Interstitial Administration

Apart from the parenteral administration of endocrine-like therapeutic proteins both as solutions or in long-acting dosage forms, opportunities exist for the interstitial administration of both classes of therapeutic protein. Numerous groups are developing controlled-release formulations that would enable varied forms of protein release from implants. Biodegradable polymers have been studied to obtain a constant and prolonged input of therapeutic proteins (see Baker *et al.*, 1984; de Nijs *et al.*, 1988). One concern with this latter route of administration is the effect the presence of tissue proteases may have on the stability of proteins; for example, Berger *et al.* (1979) have found that proteins such as insulin can degrade substantially at their site of administration.

Although it is highly likely that the next few years will see dosing mainly via parenteral and/or interstitial administration, it is relevant to consider what opportunities there are for their administration via routes that are more convenient to patient and physician alike. It has been considered unlikely that this would be possible with administration to nonmucosal barriers, although interestingly, recent reports claim that insulin can be administered transdermally using electrical current as an inducer (Meyer *et al.*, 1988).

7.3. Gastrointestinal Tract

The epithelial membrane of the gastrointestinal tract comprises an anatomically continuous barrier of cells that permits the passage of low-molecular-weight material by simple diffusion and various (nutrient) carrier processes. Macromolecules may be absorbed from the lumen by cellular vesicular processes via either fluid-phase pinocytosis or specialized (receptor-mediated) endocytic processes. Also, specialized cells (M cells) exist within the Peyer's patches of the gastrointestinal tract epithelium that can take up soluble and particulate macromolecules and transport these into the Peyer's patches of the gut-associated lymphoid tissue. Although the capacity of this system for transport appears to be extremely low, since mucosal immune responses are initiated within this lymphoidal tissue, it may be that the M cell could be used to bring an appropriate antigen, for example an immunomodulatory protein, to the attention of the mucosal immune system.

A priori, the physical and chemical properties of therapeutic proteins make them unlikely candidates for oral administration. That is, they are large, often chemically unstable—and hence degrade in the environment of the lumen of the gastrointestinal tract—and they have a poor ability to pass biological membranes via either a passive transcellular or a paracellular route. Numerous attempts are being made to overcome each of these difficulties, with much attention currently being given to which are rate-controlling (see Matuszewska *et al.*, 1988).

7.3.1. Intralumenal Stability

Apparent chemical stability can be enhanced either by derivatizing proteins to become protease-resistant, or by simultaneously administering antiproteolytic enzyme inhibitors, such as Trasylol. Various attempts have also been made to use particulate carriers (e.g., microcapsules and liposomes), or to develop protease-resistant dosage forms that can release protein in a specific region of the gastrointestinal tract (e.g., using an azoaromatic polymer film; Saffran *et al.*, 1986).

7.3.2 *Transepithelial Transport*

Transepithelial transport of proteins occurs to a significant degree in neonates. The evidence suggests that this occurs to a very small extent in adults (Warshaw *et al.*, 1974; O'Hagan *et al.*, 1987). Some proteins and enzymes have been reported to be absorbed through the intestinal tract membrane of numerous species, including chymotrypsin, insulin, horseradish peroxidase, pancreatic lipase, and *Serratia* protease (Miyata *et al.*, 1980); immunoglobulins, intrinsic factor, and various hormones are known to be absorbed via a receptor-mediated endocytosis process allowing specific and rapid uptake of a significant fraction of applied dose. Some groups are attempting to utilize natural transport processes for the effective transepithelial passage of polypeptides and proteins (e.g., the cobalamin/intrinsic factor complex; Russell-Jones and de Aizpurua, 1988). Although small, the amounts of protein that can traverse the tract membrane are often described as being significant in terms of pharmacological response. For example, the oral administration of urokinase in adult humans induces a plasma fibrinolytic staging, suggesting the transport of urokinase across the intestinal tract membrane in amounts sufficient to at least stimulate the synthesis and/or release of endogeneous urokinase-type proteins (Toki *et al.*, 1985). The process of epithelial transport is often saturable (as described for numerous proteins including *Serratia* protease, serum albumin, lipase, and iodinated elastase).

7.3.3. *Intracellular Stability*

Stability of proteins within the epithelial cells also requires consideration in determining whether oral administration is feasible. This is well demonstrated by the studies of Heyman *et al.* (1987) on the fate of a glycoprotein extracted from *Klebsiella pneumoniae*, i.e., heterogeneous compound composed of two subunits with molecular weights of 350 kDa and 95 kDa. This is active when given orally at a dose 10^3 higher than when administered either peritoneally or parenterally. It is found that the compound is absorbed across both the absorptive cells of the duodenum in rabbit and Peyer's patches of isolated epithelium. Although its flux across the Peyer's patches of approximately 350 $ng \cdot hr^{-1} \cdot cm^{-2}$ is about five times lower than the flux across duodenal epithelium, this is largely of high-molecular-weight glycoprotein, whereas duodenal epithelial transport results in extensive lysosomal degradation of the macromolecular compound. Compared with the fluid-phase marker horseradish peroxidase (HRP), these data interestingly show that although HRP transports across duodenal epithelium at a rate 3.5 times higher than the heterogenous glycoprotein, values for transport of both across the Peyer's patch are similar.

7.3.4. Enhancement of Macromolecular Epithelial Absorption

The epithelial absorption of macromolecules may be enhanced using various adjuvants. These include water–oil–water multiple emulsions, ionic and nonionic surfactants, mixed bile salt micelles, and surfactant-lipid mixed micelles, all of which may cause a transient reduction in the resistivity of the surface of the apical membrane of the epithelial cells. For example, sodium glycocholates or polyethylene-9-dodecyl ether enhance the transmucosal permeation of insulin (Desai *et al.*, 1986). Similarly, a marked increase in plasma levels of human epidermal growth factor (EGF) is observed when EGF is administered rectally as a microenema containing hydrophilic polymers such as sodium caprolate and sodium lauroylphenylalanine: claims of 60% and 90% bioavailability in rats, for rectal and nasal administration, respectively, have been made (Murakami *et al.*, 1987). Other enhancers include salicylates (for growth hormone) (Moore *et al.*, 1986). In addition, studies on the enteric absorption of α and β human interferons in the rat large intestine have shown that the addition of apricot kernel oil and polyethylene glycol 6 to a citrate-buffered solution favors the blood bioavailability of α- but not of β-interferon. This could be due to an increase in both the stability of α-interferon and/or its penetration (Bocci *et al.*, 1986); this is similar to previous work by Yoshikawa *et al.* (1984), which showed that lipid surfactant mixed micelles promote the (lymphatic) absorption of β-interferon from the large intestine. Few of these types of studies have yet examined the possible enhancement in the coabsorption of enterotoxins.

7.4. Nasal and Buccal Mucosae

Recent attention has been given to the nasal administration of macromolecular polypeptide and protein drugs. Molecules as large as insulin, interferon, and calcitonin have been found to be pharmacologically active after being applied to the nasal mucosa. Present evidence suggests that nasal administration for polypeptides results in bioavailabilities of between 1 and 20% of applied dose, depending on their molecular weight and physicochemical properties; numerous additional factors may also play a role in controlling the entry and penetration of nasal epithelia, including local enzymatic degradation in the nasal cavity, various pathologies and their symptoms (e.g., rhinitis, colds), the formulation and administration systems used, and the deposition and clearance of same. It is clear, however, that tolerance, toxic events, and even unique immune events can be demonstrated after the nasal administration of polypeptides and proteins (e.g., Levy *et al.*, 1988).

The buccal mucosa has often been considered for the entry and penetration of small drugs, although it has only recently received attention as a possible route of administration of polypeptides and proteins (Siddiqui and Chien, 1987).

8. *Concluding Remarks*

This chapter has pointed to some of the features of protein (physico)-chemical structure, as well as to the nature of their administration, that can affect the pharmacodynamics, pharmacokinetics, and toxicity of therapeutic proteins (Table IV). From the above, it is apparent that the first of the homologous proteins that can be successfully developed as therapeutic modalities will be those that will mainly act intravascularly at extra- and intracellular sites, except in those instances when the protein is able to readily access tissue parenchyma, perhaps through discontinuous endothelia. Given the rate at which new molecules are being produced and characterized, and since many of these have several biological functions, it is apparent that enormous financial and human resources will be required to bring these to the market. However, it is that as we know more about how the body regulates itself, and how and why endogeneous and exogenous macromolecules interact, then a new class of therapeutic proteins will emerge. These will have the ability to recognize discrete sites, be protected from the body (and vice versa), and be active only at desired sites of pharmacological

Table IV. Methods for Adjusting the Biological Dispersion of Therapeutic Proteins

Approach	Examples
Site-directed mutagenesis	α_1-Antitrypsin, subtilisin (substitution of methionine amino acids to improve stability)
(Re)glycosylation	Glucocerebrosidase (improved circulatory half-life)
Hybrid proteins	
Covalent	Immunoglobulin fragment toxin
Fused gene products	Growth factor toxin
Protectants	
Polyethylene glycols	Interleukin, superoxide dismutase
Dextrans	L-Asparaginase
Administration	
Frequency and rate	Growth hormone
Route	Insulin and site of action
Staging	Interferons and tumor necrosis factor
Timing	Growth factors
Lymph/plasma ratio	Auto-/paracrine and endocrine-like mediators

action. A major priority in determining the commercial success of such site-specific therapeutic proteins will be understanding and evaluating their potencies and (potential) novel toxicities. Unquestionably much of this will relate to two features: the development of sensitive and specific analytical procedures, and the ability to examine for the efficacy and safety of species-specific products. Further, it is apparent for endocrine- and auto-/paracrine-like mediators that as we know more about their action in relation to their spatial and kinetic dispersion in the body, then the way in which we can control this dispersion, perhaps through one or more of the approaches described in this chapter, will be vital. It is also clear that our current practices in this area are often very naïve. Thus, in rethinking our approaches to drug research and clinical development and usage as they apply to therapeutic proteins (as well as to conventional drugs), we should endeavor to give deeper consideration to the need for symmetrical and/or asymmetrical administration—by giving the substances via the correct route, and at an appropriate amount, rate, frequency, and duration, as well with a proper staging of application with (other) proteinaceous mediators.

References

Abuchowski, A., Van Es, T., Palczuk, N. C., and Davis, F. F., 1977, Alteration of immunological properties of bovine serum albumin by covalent attachment of polyethylene glycol, *J. Biol. Chem.* **252:**3578–3581.

Aggarawal, B. B., Eessalu, T. E., and Hass, P. E., 1985, Characterisation of receptors for human tumour necrosis factor and their regulation by γ-interferon, *Nature* **318:**665–667.

Akiyama, A., Bednarski, M., Kim, M-J., Simon, E. S., Waldmann, H., and Whitesides, G. M., 1987, Enzymes in organic synthesis, *Chem. Brit.* **23**(7):645–654.

Bachmair, A., Finley, D., and Varshavsky, A., 1986, *In vivo* half-life of a protein is a function of its amino-terminal residue, *Science* **234:**179–186.

Baenziger, J. U., 1985, The role of glycosylation in protein recognition, *Am. J. Physiol.* **121:**382–391.

Baker, R. W., Tuttle, M. E., and Helwing, R., 1984, Novel erodible polymers for the delivery of macromolecules, *Pharm. Technol.* **8:**26–30.

Benet, L. Z., 1978, Effect of route of administration and distribution on drug action, *J. Pharmacokin. Biopharm.* **6:**559–585.

Berger, M., Halban, P. A., Girardier, L., Seydoux, J., Offord, R. E., and Renold, A. E., 1979, Absorption kinetics of subcutaneously injected insulin. Evidence for degradation at the injection site, *Diabetologia* **17:**97–99.

Berger, E. G., Greber, U. F., and Mosbach, K., 1986, Galactosyltransferase-dependent sialylation of complex and endo-*N*-acetylglucosaminidase H-treated core *N*-glycans *in vitro*, *FEBS Lett.* **203:**64–68.

Berger, E. G., Müller, U., Aegerter, E., and Strous, G. J., 1987, Biology of galactosyltransferase: Recent developments, *Biol. Chem. Trans.* **15:**610–613.

Blakey, D. C., and Thorpe, P. E., 1986, Effect of chemical deglycosylation on the *in vivo* fate of ricin A-chain, *Cancer Drug Delivery* **3:**189–196.

Blohm, D., Bollschweiler, C., and Hillen, H., 1988, Pharmaceutical proteins, *Angew. Chem.* **100:**213–231.

Bocci, V., Corradeschi, F., Naldini, A., and Lencioni, E., 1986, Enteric absorption of human interferons α and β in the rat, *Int. J. Pharm.* **34**:111–114.

Brange, J., Ribel, U., Hansen, J. F., Dodson, G., Hansen, M. T., Havelund, S., Melberg, S. G., Norris, F., Norris, K., Snel, L., Sørensen, A. R., and Voigt, H. O., 1988, Monomeric insulins obtained by protein engineering and their medical implications, *Nature* **333**:679–682.

Cartlidge, S. A., Duncan, R., Lloyd, J. B., Kopecková-Rejmanová, P., and Kopecek, J., 1987, Soluble crosslinked *N*-(2-hydroxypropyl)methacrylamide copolymers as potential drug carriers 3. Targeting by incorporation of galactosamine residues. Effect of route of administration, *J. Controlled Res.* **4**:265–278.

Castro, M. R., and Gras, J., 1984, The importance of rhythym of antigen administration in the kinetics of the IgM and IgG responses, *Immunologia* **3**:157–160.

Chaudhary, V. K., Fitzgerald, D. J., Adhya, S., and Pastan, I., 1987, Activity of a recombinant fusion protein between transforming growth factor type α and *Pseudomonas* toxin, *Proc. Natl. Acad. Sci. USA* **84**:4538–4542.

Clark, R. G., and Robinson, I. C. A. F., 1988, Paradoxical growth-promoting effects induced by patterned infusions of somatostatin in female rats, *Endocrinology* **122**:2675–2682.

Clark, R. G., Jansson, J-O., Isaksson, O., and Robinson, I. C. A. F., 1985, Intravenous growth hormone: Growth responses to patterned infusions in hypophysectomized rats, *J. Endocr.* **104**:53–61.

de Nijs, H., Bouwman, T. R. M., and Eenink, M. J. D., 1988, Controlled peptide delivery using biodegradable microcapsule formulations, *Pharm. Weekblad Sci. Ed.* **10**:49.

Desai, D. S., Tojo, K., Huang, Y. C., and Chien, Y. W., 1986, Transmucosal permeation of macromolecular drug: insulin, *Pharm. Res.* Suppl. **3**:55S.

Feng, G-S., Gray, P. W., Shepard, H. M., and Taylor, M. W., 1988, Anti-proliferative activity of a hybrid protein between interferon-gamma and tumor necrosis factor-β, *Science* **241**:1501–1503.

Fiume, L., Busi, C., and Mattioli, A., 1982, Lactosaminated human serum albumin as hepatotropic drug carrier. Rate and uptake by mouse liver, *FEBS Lett.* **146**:42–46.

Goldfarb, D. S., Gariépy, J., Schoolnik, G., and Kornberg, R. D., 1986, Synthetic peptides as nuclear localization signals, *Nature* **322**:641–644.

Greenfield, L., Johnson, V. G., and Youle, R. J., 1987, Mutations in diphtheria toxin separate binding from entry and amplify immunotoxin selectivity, *Science* **238**:536–539.

Heyman, M., Bonfils, A., Fortier, M., Crain-Denoyelle, A. M., Smets, P., and Desjeux, J. F., 1987, Intestinal absorption of RU 41740, an immunomodulating compound extracted from *Klebsiella pneumoniae*, across duodenal epithelium and Peyer's patches of the rabbit, *Int. J. Pharm.* **37**:33–39.

Ho, D. H. W., Wang, C-Y., Lin, J-R., Brown, N., Newman, R. A., and Krakoff, I. H., 1988, Polyethylene glycol-L-asparaginase and L-asparaginase studies in rabbits, *Drug Metab. Dispos.* **16**:27–29.

Katre, N. V., Knauf, M. J., and Laird, W. J., 1987, Chemical modification of recombinant interleukin 2 by polyethylene glycol increases its potency in the murine Meth A sarcoma model, *Proc. Natl. Acad. Sci. USA* **84**:1487–1491.

Katz, D. H., Hamaoka, T., and Benacerraf, B., 1972, Immunological tolerance in bone marrow-derived lymphocytes. I. Evidence for an intracellular mechanism of inactivation of hapten-specific precursors of antibody-forming cells, *J. Exp. Med.* **136**:1410–1429.

Kelm, S., and Schauer, R., 1986, The galactose-recognizing system of rat peritoneal macrophages. Receptor-mediated binding and uptake of glycoproteins, *Biol. Chem.* **367**:989–998.

Keogh-Bennett, J. M., Matthews, I. T., Beesley, J. S., Paul, F., Wilson, G., and Taylor, P. W., 1986, Intracellular delivery and release of fluorescein from an acid-labile transferrin conjugate, *Biochem. Soc. Trans.* **15**:P444–445.

Lee, W. Y., Sehon, A. H., and AÅkerblom, E., 1981, Suppression of reaginic antibodies with modified allergens. IV. Induction of suppressor T cells by conjugates with polyethylene glycol (PEG) and monomethoxy PEG with ovalbumin, *Int. Arch. Allergy Appl. Immunol.* **64**:100–114.

Levy, F., Muff, R., Dotti-Sigrist, M. A., and Fischer, J. A., 1989, Formation of neutralizing antibodies during intranasal synthetic salmon calcitonin treatment of Paget's disease, *J. Clin. Endocrinol. Metab.* **67**:541–545.

Madiyalakan, R., Chowdhary, M. S., Rana, S. S., and Matta, K. L., 1986, Lysosomal-enzyme targeting: The phosphorylation of synthetic D-mannosyl saccharides by UDP-*N*-acetylglucosamine: Iysosomal-enzyme *N*-acetylglucosaminephosphotransferase from rat-liver microsomes and fibroblasts, *Carbohydrate Res.* **152**:183–194.

Mage, R. G., 1988, Designing antibodies for human therapies, *Nature* **333**:807–808.

Matuszewska, B., Liversidge, G. G., Ryan, F., Dent, J., and Smith, P. L., 1988, In vitro study of intestinal absorption and metabolism of 8-L-arginine vasopressin and its analogues, *Int. J. Pharm.* **46**:111–120.

Melton, R. G., Wiblin, C. N., Baskerville, A., Foster, R. L., and Sherwood, R. F., 1987, Covalent linkage of carboxypeptidase G$_2$ to soluble dextrans. II. *In vivo* distribution and fate of conjugates, *Biochem. Pharmacol.* **36**:113–121.

Meyer, B. R., Katzeff, H., Eschbach, J. C., Trimmer, J., Zacharias, S., and Rosen, S., 1988, Successful transdermal delivery of human insulin to rabbits with alloxan-induced diabetes mellitus, *Clin. Res.* **36**:367A.

Miyata, K., Hirai, S., Yashiki, T., and Tomoda, K., 1980, Intestinal absorption of *Serratia* protease. *J. Appl. Biochem.* **2**:111–116.

Miyata, K., Nakagawa, Y., Nakamura, M., Ito, T., Sugo, K., Fujita, T., and Tomoda, K., 1988, Altered properties of Serratia superoxide dismutase by chemical modification, *Agric. Biol. Chem.* **52**:1575–1581.

Moore, H. H., and Kelly, R. B., 1986, Re-routing of a secretory protein by fusion with human growth hormone sequences, *Nature* **321**:443–446.

Moore, J. A., Pletcher, S. A., and Ross, M. J., 1986, Absorption enhancement of growth hormone from the gastrointestinal tract of rats, *Int. J. Pharm.* **34**:35–43.

Murakami, T., Kishimoto, M., Kawakita, H., Higashi, Y., Yata, N., Amagase, H., Nojima, N., and Fuwa, T., 1987, Enhanced rectal and nasal absorption of human epidermal growth factor by the presence of absorption promoters in rats, *J. Pharm. Sci.* **76**:S85.

Murphy, J. R., 1987, Hybrid protein, U.S. Patent, 4,675,382.

Napper, D. H., and Netschey, A., 1971, Studies of the steric stabilisation of colloidal particles, *J. Colloid Interface Sci.* **37**:528–535.

Norde, W., 1984, Physicochemical aspects of the behaviour of biological components at solid/liquid interfaces, in: *Microspheres and Drug Delivery: Pharmaceutical, Immunological and Medical Aspects* (S. S. Davis, L. Illum, J. G. McVie, and E. Tomlinson, eds.), pp. 25–37, Elsevier, Amsterdam.

Offord, R. E., and Rose, K., 1987, New protein and polypeptide derived conjugates especially containing a reporter group or cytotoxic agent linked through specific *N*-containing groups, European Patent Applic., 87106113.1.

O'Hagan, D. T., Palin, K. J., and Davis, S. S., 1987, Intestinal absorption of proteins and macromolecules and the immunological response, *CRC Crit. Rev. Therap. Drug Carrier Systs.* **4**:197–220.

Okada, M., Matsushima, A., Katsuhata, A., Aoyama, T., Ando, T., and Inada, Y., 1985, Suppression of IgE antibody response against ovalbumin by the chemical conjugate of ovalbumin with a polyaspartic acid derivative, *Int. Arch. Allergy Apply. Immunol.* **76**:79–81.

Ottewill, R. H., 1977, Stability and instability in disperse systems, *J. Colloid Interface Sci.* **58**:357–373.

Říhová, B., Kopecek, J., Kopecková-Rejmanová, P. Strohalm, J., Plocová, D., and Šemorádová, H., 1986, Bioaffinity therapy with antibodies and drugs bound to soluble synthetic polymers, *J. Chromatogr.* **376**:221–233.

Russell-Jones, G. J., and de Aizpurua, H. J., 1988, Vitamin B12: A novel carrier for orally presented antigens, *Proc. Int. Symp. Control. Rel. Bioact. Mater.* **15**:142–143.

Saffran, M., Kumar, G. S., Savariar, C., Burnham, J. C., Williams, F., and Neckers, D. C., 1986, A new approach to the oral administration of insulin and other peptide drugs, *Science* **233**:1081–1084.

Sandoz, A. G., 1988, Peptide derivatives, Int. Patent Applic., WO 88/02756.

Scott, C. F., Lambert, J. M., Goldmacher, V. S., Blattler, W. A., Sobel, R., Schlossman, S. F., and Benacerraf, B., 1987, The pharmacokinetics and toxicity of murine monoclonal antibodies and of gelonin conjugates of these antibodies, *Int. J. Immunopharm.* **9**:211–225.

Segawa, A., Borges, M. S., Yokota, Y., Matsushima, A., Inada, Y., and Tada, T., 1981, Suppression of IgE antibody response by the fatty acid-modified antigen, *Int. Arch. Allergy Appl. Immunol.* **66**:189–199.

Siddiqui, O., and Chien, Y. W., 1987, Nonparenteral administration of peptide and protein drugs, *CRC Crit. Rev. Therap. Drug Carrier Syst.* **3**:195–208.

Talmadge, J. E., 1986, Biological response modifiers: realising their potential in cancer therapeutics, *Trends Pharmacol. Sci.* **7**:277–281.

Toki, N., Sumi, H., Sasaki, K., Boreisha, I., and Robbins, K. C., 1985, Transport of urokinase across the intestinal tract of normal human subjects with stimulation of synthesis and/or release of urokinase-type proteins, *J. Clin. Invest.* **75**:1212–1222.

Tomlinson, E., 1986, (Patho)physiology and the temporal and spatial aspects of drug delivery, in: *Site-Specific Drug Delivery*, (E. Tomlinson and S. S. Davis, eds.), pp. 1–27, Wiley, Chichester.

Tomlinson, E., 1987a, Theory and practice of site-specific drug delivery, *Adv. Drug Delivery Rev.* **1**:87–198.

Tomlinson, E., 1987b, Biological opportunities for site-specific drug delivery using particulate carriers, in: *Drug Delivery Systems: Fundamentals and Techniques* (P. Johnson and G. Lloyd-Jones, eds.), London, pp. 32–65, Ellis Horwood.

Torchilin, V. P., Khaw, B. A., Klibanov, A. L., Slinkin, M. A., Haber, E., and Smirnov, V. N., 1986, Modification of monoclonal antibodies by polymers possessing chelating properties, *Bull. Exp. Biol. Med.* **102**:946–948.

United States Office of Technological Assessment, 1988, U.S. investment in biotechnology, in: *New Developments in Biotechnology*, July 12, 1988, Washington, D.C.

Urquhart, J., and Nicholls, K., 1985, Delivery systems and pharmacodynamics in new drug research and development, *World Biotech. Rep.* **2**:321–331.

Vera, D. R., Krohn, K. A., Stadalnik, R. C., and Scheibe, P. O., 1984, Tc-99m-galactosyl-neoglycoalbumin: *in vivo* characterisation of receptor-mediated binding to hepatocytes, *Radiology* **151**:191–199.

Vitetta, E. S., Fulton, R. J., May, R. D., and Uhr, J. W., 1987, Redesigning nature's poisons to create anti-tumor reagents, *Science* **238**:1098–1104.

Von Heijne, G., 1985, Structural and thermodynamic aspects of the transfer of proteins into and across membranes, *Curr. Topics Transport* **24**:151–179.

Warshaw, A. L., Walker, W. A., and Isselbacher, K. J., 1974, Protein uptake by the intestine, *Gastroenterology* **66**:987–992.

Wileman, T. E., Foster, R. L., and Elliott, P. N. C., 1986, Soluble asparaginase-dextran conjugates show increased circulatory persistence and lowered antigen reactivity, *J. Pharm. Pharmacol.* **38**:264–271.

Wilson, G., Eidelberg, M., and Michalak, V., 1979, Selective hepatic uptake of synthetic glycoproteins. Mannosaminated ribonuclease A dimer and serum albumin, *J. Gen. Physiol.* **74**:495–509.

Yoshikawa, H., Takada, K., Muranishi, S., Satoh, Y-I , and Naruse, N., 1984, A method to potentiate enteral absorption of interferon and selective delivery into lymphatics. *J. Pharmacobio-Dyn.* **7**:59–62.

Zimmermann, R., and Meyer, D. I., 1986, 1986: A year of new insights into how proteins cross membranes, *Trends Biochem. Sci.* **11**:512–515.

17

Antibody-Mediated Drug Delivery

JOHN N. WEINSTEIN and KENJI FUJIMORI

1. Introduction

Antibodies are quintessentially "information macromolecules." Since the information resides in their antigen-binding sites, it is not surprising that most of the contributions on antibodies in this volume have focused on binding. That is, they have focused on structures and processes that influence the last few nanometers before molecular docking. Here we will consider the prior process, the pharmacokinetic path by which an antibody molecule reaches the proximity of its target antigen. That this perspective is important to the rational design of next-generation ligand molecules should require no argument. Although the title of this chapter emphasizes antibody–drug conjugates, we will focus on generic issues of pharmacology that apply to any of the forms in which antibodies (or, for that matter, other biological ligands) are used *in vivo*.

For the last several years, we have been trying to understand the pharmacology of monoclonal antibodies (MAb) and other ligands by splicing together information from several hierarchical levels:

1. Global (i.e., whole-body)	The "classical" pharmacokinetic description.
2. Regional	The way in which restricted anatomical or physiological compartments condition the problem of selective delivery.
3. Microvascular	The issues related to blood flow and passage of a molecule across the capillary wall.

JOHN N. WEINSTEIN and KENJI FUJIMORI • Theoretical Immunology Section, Laboratory of Mathematical Biology, DCBD, National Cancer Institute, National Institutes of Health, Bethesda, Maryland 20892.

4. Tissue level   The "percolation" problem.
5. Cellular binding   The way in which targets and nontargets bind ligand.
6. Cellular handling   Uptake and metabolism of the ligand; metabolism and reexpression of the antigen.

2. Global and Regional Pharmacokinetics

Our work on the global pharmacology of MAb in humans (Eger *et al.*, 1987) and in other animals (Covell *et al.*, 1986; Holton *et al.*, 1987) has been described elsewhere. Similarly, our work on regional delivery via the lymphatics for diagnosis or treatment of cancer in the lymph nodes of humans (Lotze *et al.*, 1986; Keenan *et al.*, 1987a,b) and other animals (Weinstein *et al.*, 1982, 1983; Steller *et al.*, 1986; Covell *et al.*, 1986; Mulshine *et al.*, 1987; Parker *et al.*, 1987; Black *et al.*, 1988) has been the subject of previous review (Weinstein *et al.*, 1985, 1986, 1987a,b). Here we will emphasize the microscopic issues (points 3–6 above). From that perspective, the global pharmacology can be viewed as an input to the microscopic problem. That is, plasma profiles such as

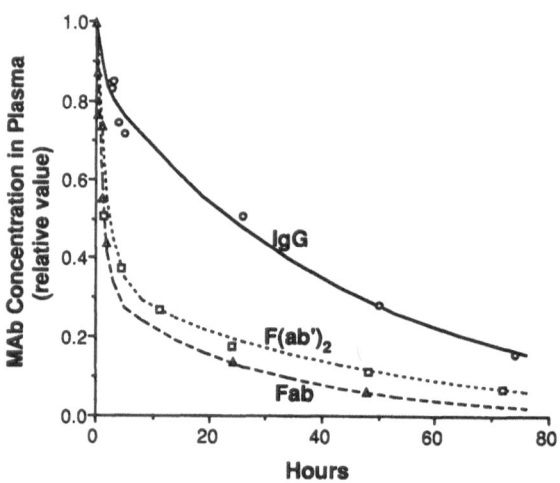

Figure 1. Plasma concentration profiles of IgG (——), F(ab')$_2$ (---), and Fab (— —) after bolus i.v. injection in humans. These curves define the levels of MAb available for passage across tumor capillary walls to reach tumor antigen. The curves represent fits of a linear two-compartment global pharmacokinetic model to data from clinical trials on murine monoclonal antibodies and their fragments directed against human melanoma. From Fujimori *et al.* (1989).

those shown in Fig. 1 determine what concentration of antibody is available at each point in time to cross the wall of a capillary and percolate through the extravascular space to reach antigens there. Because of our own parochial interests and because most of the available data come from applications of MAb in oncology, we will focus here on the context of tumors. Early stages of this work have been presented elsewhere (Weinstein *et al.*, 1986, 1987a,b; Fujimori *et al.*, 1989).

Over the last few years, a number of clinical and preclinical studies have shown microscopic and macroscopic heterogeneity of MAb distribution in tumors after intravenous administration. That heterogeneity results in part from inhomogeneous distribution of antigen among tumor cells and among regions of a tumor (Schlom *et al.*, 1983; Del Vecchio *et al.*, 1988). However, non-uniform access from the bloodstream also appears to contribute, as we will explore in the following sections.

3. Microvascular Transport

If passage across the endothelial layer or basement membrane of a capillary is the rate-limiting step in penetration of a tumor, then one must take into account the barriers and possible mechanisms of transcapillary transport depicted in Fig. 2. There are four essentially different types of blood capillaries, based on the architecture of their endothelium and underlying basement membrane: continuous, fenestrated, sinusoidal, and tight (reviewed in Weiss and Greep, 1977; Poste and Kirsch, 1983; Poste, 1985; Weinstein, 1987). Continuous capillaries are found in muscle, connective tissue, skin, and most other tissues of the body. In continuous endothelia, adjacent cells are apposed to form a more or less continuous lining, and there is a fully formed basement membrane. Fenestrated capillaries are found in many glands, in the gastrointestinal tract, and in the renal glomerulus. The endothelium contains fenestrae ranging from about 30 to 80 nm in diameter. Fenestrae in the glomerulus appear open, but in other locations they are filled by a diaphragm 4–6 nm thick. The basement membrane of a fenestrated capillary is continuous. Sinusoids (typically with 100-nm apertures) are found in the liver, spleen, and bone marrow. Tight endothelia, characteristic of the blood–brain barrier, contain fewer vesicles in their cytoplasm, and their intercellular junctions are closed off by belts of tight junction.

Several possible mechanisms for transcapillary transport have been identified, as shown schematically in Fig. 2. Despite generations of concerted study, there is still not general agreement about the mechanisms or pathways by which macromolecules such as the immunoglobulins cross continuous capillary walls (Poznansky and Juliano, 1984). In the case of sinusoidal capillaries, the openings are large enough to admit macromolecules with ease, and at the other end of the

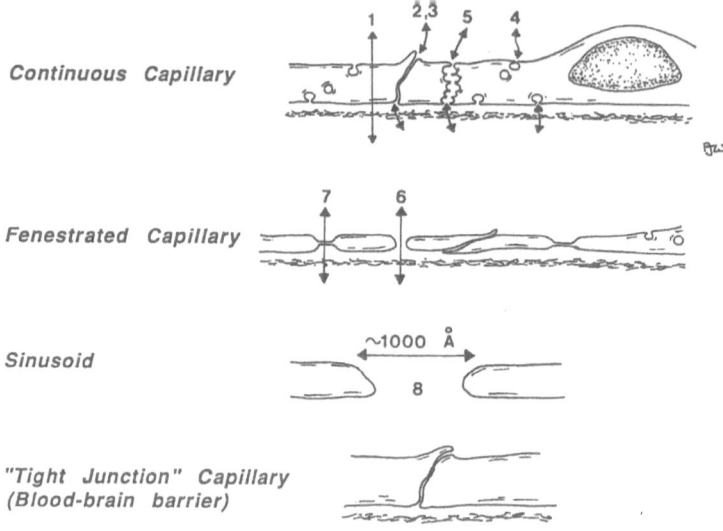

Figure 2. Schematic representation of four types of blood capillaries, indicating possible pathways for vascular transport of macromolecules such as the immunoglobulins. Panels showing continuous and fenestrated capillaries were modified from Renkin (1977). (1) Transcellular; (2) junctional: "small pore"; (3) junctional: "large pore"; (4) vesicular; (5) chains of vesicles; (6) open fenestrae; (7) closed fenestrae; (8) sinusoidal openings.

spectrum, macromolecules cannot pass through the blood–brain barrier (unless specifically transported). Functionally, the properties have been interpreted in terms of a "small pore" pathway for molecules less than about 9 nm in diameter and a "large pore" pathway for materials up to about 70 nm (reviewed in Bundgaard, 1980; Poste, 1985).

The problem in tumors can be even more complex. Tumors are characterized by unregulated or poorly regulated growth, and the resulting heterogeneity (Poste and Fidler, 1980; Poste and Kirsh, 1983) is expressed in their vasculature. Structural defects have been seen in the endothelium and basement membranes of a variety of tumors (Peterson, 1979). During angiogenesis, there appears to be a window of time during which macromolecules can extravasate more easily, and within some tumors there are vessels with half-formed endothelia or even blood channels lined by tumor cells without any defined endothelium. The various types of defective vessels occur sporadically, both within and between tumors. Despite this kaleidoscopic diversity of properties, however, it is clear at least that molecules the size of serum proteins do not have free access to the interstitial space in tumors (Peterson, 1979).

4. Tissue Level: The Percolation Problem

Our next point of focus in this discussion is the process of antibody distribution after transcapillary passage—what we have called the "percolation" problem. Consider the stages of tumor growth shown schematically in Fig. 3: (1) The tumor nodule (or, analogously, a tumor spheroid *in vitro*) grows to about 1 mm in diameter. ("Carcinoma *in situ*" is a nonvascularized aggregate such as this, entirely sequestered within an epithelial layer.) (2) With further growth, cells at the center of the nodule become progressively less healthy. (3) From about 1–3

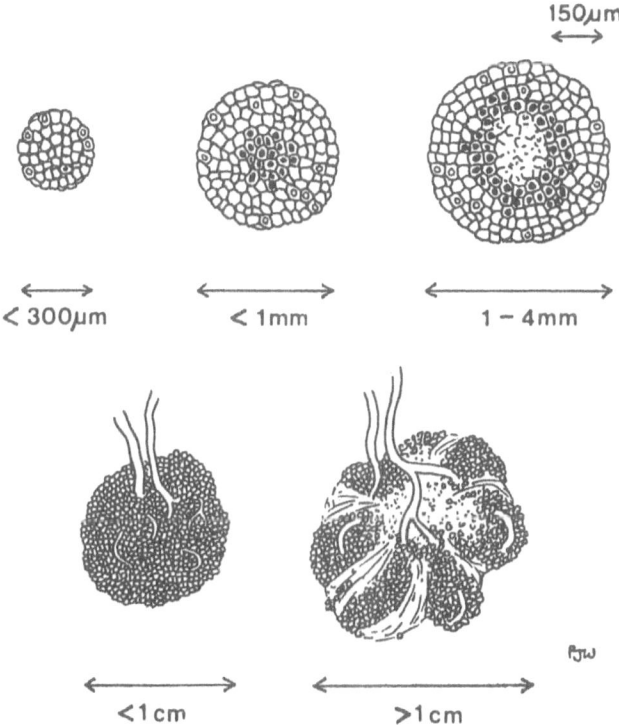

Figure 3. Schematic representation of tumor growth through prevascular (top row) and vascular (bottom row) phases. See text for discussion of each stage.

mm, growth is accompanied by death and necrosis in the center, due to poor oxygenation and/or nutrition. (4) For growth beyond about 3 mm, vascularization is required. New vascular buds recruited from the surrounding tissues ("angiogenesis") permit fresh, explosive growth of the tumor. (5) Beyond about 1 cm, the center tends to become necrotic and/or fibrotic, in large part because of internal hydrostatic pressure gradients generated by the lack of local lymphatics. Although the tumor recruits a blood vascular supply, it does not similarly provide itself with effective lymphatic drainage. Lymph capillaries are responsible for scavenging macromolecular waste from the tissues and returning it (after filtration through lymph nodes) to the blood circulation. In the absence of lymphatics in tumors, these osmotically active molecules accumulate and generate regions of high pressure. That is one of the major reasons (in addition to fibrosis) that solid tumors are hard. When a solid tumor (for example, in the liver) is cut through at autopsy, it will often bulge outward from the surface of the surrounding tissue and even exude fluid. High osmotic and hydrostatic pressure tend to shut down nutrition in any of several ways: by decreasing flow to the local blood vessels, by shutting them off, and by preventing filtration of molecules from the vessels. The last of those effects has been nicely analyzed by Jain (1988) and Jain and Baxter (1988). These mechanisms tend to reduce delivery of an administered antibody to the central region of a solid tumor.

5. Percolation Calculations and the Binding Site Barrier

Having now summarized some of the important factors in IgG pharmacology, we will illustrate how information from the various hierarchical levels can be combined. For simplicity, we consider an intermediate stage of tumor growth, focusing on a simple, uniform cylinder of cells through which antibody must percolate radially from a central vessel (see, for example, Fletcher, 1978). This model corresponds to the cord of viable tumor cells around a newly formed tumor capillary. The question is: How uniform will the concentration of antibody be as a function of time and as a function of various parameters of the system? Elsewhere, we will consider more complex cases in which the tumor is intrinsically nonuniform, in which convection dominates the transport properties (cf. Swabb et al., 1974), and in which the geometry is not cylindrical or spherical.

Regardless of the specific geometry, a central consideration is what we have termed the "binding site barrier"—the prediction that bindable immunoglobulin (or any other ligand) will be retarded in its diffusion or convection through the tumor by the very fact of its successful binding. Nonuniformity of distribution will tend to be increased by a low diffusion coefficient, a high density of antigenic sites, and high affinity of binding. To illustrate the pattern of this analysis, we will focus here on the effect of one parameter, the binding affinity, on IgG distribution. We will then show the effect of removing the capillary wall as a barrier to penetration.

Calculations were performed on a CRAY X-MP computer using a program package that we call "PERC." PERC uses a collocation algorithm for solving the ordinary and partial differential equations of diffusion/convection/reaction in the substance of a tumor. Global pharmacology and capillary wall characteristics enter through the boundary conditions. We have cross-checked solutions obtained from PERC by using a second program based on a modified Crank–Nicolson algorithm. These two very different methods give identical results. Baseline parameter values were obtained from clinical studies of antibody administration at the National Institutes of Health, from clinical and preclinical studies elsewhere, from the literature on tumor physiology, and from calculations based on *in vitro* antibody binding experiments (Dower *et al.*, 1984; Weinstein *et al.*, 1987b). The computer programs, parameter sets, and calculations are described in detail elsewhere (Fujimori *et al.*, 1989).

The following are some of the parameters arising from various hierarchical levels of information listed earlier and used by PERC as input to the modeling of antibody distribution:

1. Global: Plasma concentration of IgG (as a function of time), as determined by fitting clinical data using a two-compartment kinetic model.
2. Regional: No explicit values, but the lymphatic system plays an implicit role.
3. Microvascular: Effective capillary permeation constant (related to diffusive and/or convective flow across the membrane). When required, a nonequilibrium thermodynamic analysis can be introduced to split the permeation into explicit convective and diffusive components coupled through the reflection coefficient (cf. Patlak *et al.*, 1963; Rippe and Haraldsson, 1987).
4. Tissue level: Geometric specifications, effective diffusion coefficient, volume flow profile, solute-drag, partition in and out of proteoglycanrich regions, and so on.
5. Cellular binding: Stoichiometry of binding, forward rate constant, reverse rate constant, affinity.
6. Cellular handling: Rates of internalization, metabolism, and regeneration of surface antigen.

Figure 4 illustrates the effect of increasing antibody affinity (K_a) from 10^8 to 10^9 M^{-1} while holding all other parameters constant at their baseline values. The average concentration at all time points increases somewhat, but the nonuniformity of penetration is greatly increased. This small calculation illustrates a larger point: that binding sites can effectively prevent uniform distribution of antibody molecules. If the aim of therapy with an antibody-conjugated drug,

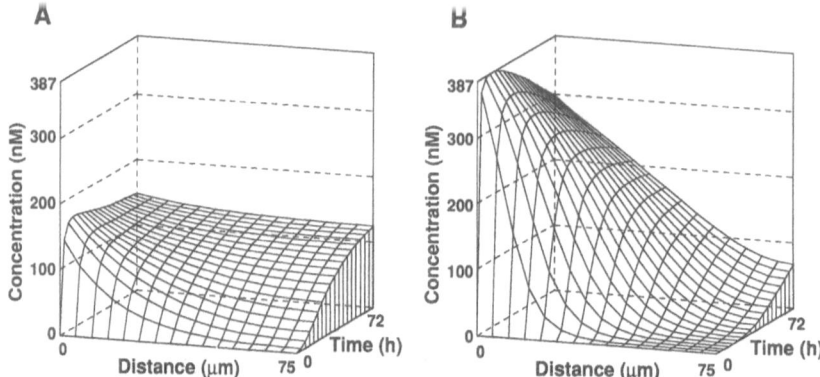

Figure 4. PERC calculation of the effect of binding affinity on the distribution profiles in a cylindrical cord of tumor cells after bolus injection of intact monoclonal antitumor IgG. (A) Affinity of binding (bivalent) = 10^8 M^{-1}. (B) Affinity of binding (bivalent) = 10^9 M^{-1}. Under these conditions, the higher affinity leads to increased concentration near the capillary wall (left side of each panel), but less uniformity of penetration and lower absolute concentrations 75 μm from the capillary wall (right side of each panel). This calculation illustrates the "binding site barrier" to percolation. The input parameters used for this sample PERC calculation are explained in Fujimori *et al.* (1989).

toxin, or alpha emitter is to damage the tumor blood vessels and immediately surrounding cells, poor percolation may be an advantage; if, on the other hand, the aim is to reach all tumor cells, including those distant from the vascular supply, poor percolation may be a problem. The dose of antibody can always be raised high enough to swamp this "binding site barrier," but the trade off is a loss of specificity.

Although the calculation here relates to a specific geometry and set of parameter values, the general principles remain quite robust as those factors are altered. For example, the results are similar for PERC calculations on spherical and on Cartesian geometries. Elsewhere (Fujimori *et al.*, 1989), we have defined a range of antibody characteristics that may provide a useful compromise with respect to affinity, dose, and the other parameters of the system. We were frankly surprised to find such a favorable regime in the parameter space.

Figure 5 shows the effect of removing the capillary as a barrier; in other words, allowing free flow of IgG into the tumor substance. Percolation is greatly accelerated.

6. Concluding Remarks

The spirit of this analysis should be clearly understood. Many of the factors that determine antibody distribution are still not known. Empirical data were

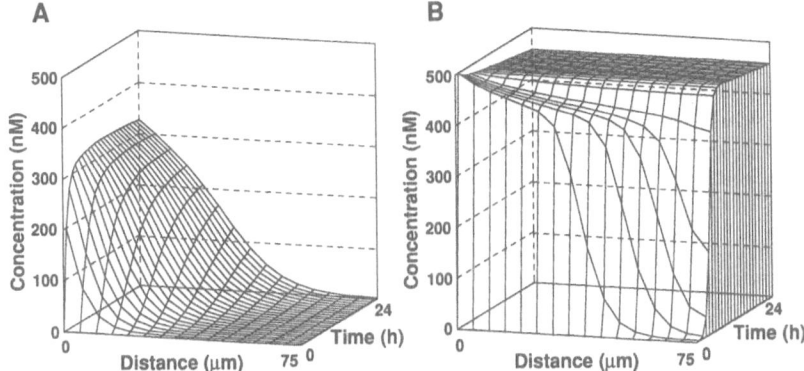

Figure 5. PERC calculation illustrating the effect of the capillary wall on IgG distribution in the tumor cord after bolus injection of antitumor IgG. (A) represents the same standard calculation as that in Figure 4B, except for changes in the concentration and time scales. (B) shows the effect of assuming the vascular wall to be infinitely permeable to IgG. Without the vascular barrier, MAb concentration becomes essentially uniform throughout the tumor within 6 hr. The input parameters used for this sample PERC calculation are explained in Fujimori *et al.* (1989).

used to establish the parameter values used for calculation, but the results are validated by only fragmentary experimental information. Over the last several years, a number of experimental studies have demonstrated at least qualitatively the nonuniformity of percolation (Flessner *et al.*, 1985; Meeker *et al.*, 1985; Abrams and Oldham, 1985; Jones *et al.*, 1986; Sutherland *et al.*, 1987; Blasberg *et al.*, 1987; McFadden and Kwok, 1988). In collaboration with V. Langmuir, R. Sutherland, and their collaborators, we have used the PERC programs to analyze experimental influx and efflux data from spheroids of LS174 tumor cells. That analysis (manuscript in preparation) shows good correspondence with calculated values for penetration. Nonetheless, the results from PERC should be considered principally as aids to concept development and as a set of null hypotheses with which to guide experiment. Experiments and simulations will continue in tandem.

We have considered here the problem of a "binding site barrier" and the role of the capillary wall. Even these sample calculations clearly reinforce the idea that no single "figure-of-merit" and no particular type of immunoglobulin-based ligand species will be optimal for all applications. Rather, a complex interplay of factors should guide the design of next-generation molecules and methods. It should be noted as well that these models and concepts are quite general in form: The global pharmacokinetics, transcapillary transport, percolation, binding kinetics, and cellular handling can be specified (and modeled using PERC) for macromolecular ligands other than immunoglobulins—and, indeed, for low-molecular-weight species as well. Further analysis of these issues may improve our ability to design therapeutic ligand molecules and also to understand the endogenous ones.

ACKNOWLEDGMENTS. We are grateful to Dr. David G. Covell, Dr. John E. Fletcher, and Mr. David Boyd for their help on the computer programs used for these calculations.

References

Abrams, P. G., and Oldham, R. K., 1985, Monoclonal antibody therapy of solid tumors, in: *Monoclonal Antibody Therapy of Human Cancer*, (K. A. Foon and A. C. Morgan, Jr., eds.), pp. 103–120, Martinus Nijhoff, Boston.

Black, C. D. V., Atcher, R. W., Barbet, J., Brechbiel, M. W., Holton, O. D., III, Hines, J. J., Gansow, O. A., and Weinstein, J. N., 1988, Selective ablation of B lymphocytes in vivo by an alpha emitter, ^{212}bismuth, chelated to a monoclonal antibody, *Antibody Immunoconjugates, Radiopharm.* 1:43–53.

Blasberg, R. G., Nakagawa, H., Bourdon, M. A., Groothuis, D. R., Patlak, C. S., and Binger, D. D., 1987, Regional localization of a glioma-associated antigen defined by monoclonal antibody 81C6 *in vivo:* Kinetics and implications for diagnosis and therapy, *Cancer Res.* 47:4432–4443.

Bundgaard, M., 1980, Transport pathways in capillaries—In search of pores, *Annu. Rev. Physiol.* 42:325–336.

Covell, D. G., Barbet, J., Holton, O. D., III, Black, C. D. V., and Weinstein, J. N., 1986, The pharmacokinetics of monoclonal IgG1, F(ab')$_2$, and Fab' in mice, *Cancer Res.* 46:3969–3978.

Del Vecchio, S., Reynolds, J. C., Blasberg, R. G., Neumann, R. D., Carrasquillo, J. A., Hellstrom, I., and Larson, S. M., 1988, Measurement of local Mr 97,000 and 250,000 protein antigen concentration in sections of human melanoma tumor using *in vitro* quantitative autoradiography, *Cancer Res.* 48:5475–5481.

Dower, S. K., Ozato, K., and Segal, D. M., 1984, The interaction of monoclonal antibodies with MHC class I antigens on mouse spleen cells I. Analysis of the mechanism of binding, *J. Immunol.* 132:751–758.

Eger, R. R., Covell, D. G., Carrasquillo, J. A., Abrams, P. G., Foon, K. A., Reynolds, J. C., Schroff, R. W., Morgan, A. C., Larson, S. M., and Weinstein, J. N., 1987, Kinetic model for the biodistribution of an ^{111}In-labeled monoclonal antibody in humans, *Cancer Res.* 47:3328–3336.

Flessner, M. F., Dedrick, R. L., and Schultz, J. S., 1985, Exchange of macromolecules between peritoneal cavity and plasma, *Am. J. Physiol.* 248:H15–H25.

Fletcher, J. E., 1978, Mathematical modeling of the microcirculation, *Math. Biosci.* 38:159–202.

Fujimori, K., Covell, D. G., Fletcher, J. E., and Weinstein, J. N., 1989, Modeling analysis of IgG, F(ab')$_2$, and Fab distribution in tumors, *Cancer Res.* 49:5656–5663.

Holton, O. D., III, Black, C. D. V., Parker, R. J., Covell, D. G., Barbet, J., Sieber, S. M., Talley, M. J., and Weinstein, J. N., 1987, Biodistribution of monoclonal IgG1, F(ab')$_2$, and Fab' in mice after intravenous injection: A comparison between anti-B cell (anti-LyB8.2) and irrelevant (MOPC-21) antibodies, *J. Immunol.* 139:3041–3049.

Jain, R. K., 1988, Determinants of tumor blood flow: a review, *Cancer Res.* 48:2641–2658.

Jain, R. K., and Baxter, L. T., 1988, Mechanisms of heterogeneous distribution of monoclonal antibodies and other macromolecules in tumors: Significance of elevated interstitial pressure, *Cancer Res.* 48:7022–7032.

Jones, P. L., Gallagher, B. M., and Sands, H., 1986, Autoradiographic analysis of monoclonal antibody distribution in human colon and breast tumor xenografts, *Cancer Immunol. Immunother.* **22**:139–143.

Keenan, A. M., Weinstein, J. N., Carrasquillo, J. A., Bunn, P. A., Jr., Reynolds, J. C., Foon, K. A., Smarte, N. C., Ghosh, B., Fejka, R. M., Larson, S. M., and Mulshine, J. L., 1987a, Immunolymphoscintigraphy and the dose-dependence of indium-111-labeled T101 monoclonal antibody in patients with cutaneous T-cell lymphoma, *Cancer Res.* **47**:6093–6099.

Keenan, A. M., Weinstein, J. N., Mulshine, J. L., Carrasquillo, J. A., Bunn, P. A., Jr., Reynolds, J. C., Foon, K. A., Perentesis, P., Ghosh, B., and Larson, S. M., 1987b, Evaluation of lymphoma by immunolymphoscintigraphy: Subcutaneous injection of indium-111-labeled T101 monoclonal antibody, *J. Nucl. Med.* **28**:42–46.

Lotze, M. T., Carrasquillo, J. A., Weinstein, J. N., Bryant, G. J., Perentesis, P., Reynolds, J. C., Matis, L. A., Eger, R. R., Keenan, A. M., Hellstrom, I., Hellstrom, K-E., and Larson, S. M., 1986, Monoclonal antibody imaging of human melanoma: Radioimmunodetection by subcutaneous or systemic injection, *Ann. Surg.* **204**:223–235.

McFadden, R., and Kwok, C., 1988, Mathematical model of simultaneous diffusion and binding of antitumor antibodies in multicellular human tumor spheroids, *Cancer Res.* **48**:4032–4037.

Meeker, T. C., Lowder, J., Maloney, D. G., Miller, R. A., Thielemans, K., Warnke, R., and Levy, R., 1985, A clinical trial of anti-idiotype therapy for B cell malignancy, *Blood* **65**:1349–1363.

Mulshine, J. L., Keenan, A. M., Carrasquillo, J. A., Walsh, T., Linnoila, R. I., Holton, O. D., Harwell, J. Larson, S. M., Bunn, P. A., and Weinstein, J. N., 1987, Immunolymphscintigraphy of pulmonary and mediastinal lymph nodes: A new approach to lung cancer imaging, *Cancer Res.* **47**:3572–3576.

Parker, R. J., Weinstein, J. N., Keenan, A. M., Dower, S. K., Steller, M. A., Holton, O. D., III, and Sieber, S. M., 1987, Targeting of radiolabelled monoclonal antibodies in the lymphatics, *Cancer Res.* **47**:2073–2076.

Patlak, C. S., Goldstein, D. A., and Hoffman, J. F., 1963, The flow of solute and solvent across a two-membrane system, *J. Theoret. Biol.* **5**:426–442.

Peterson, H. I., 1979, *Tumor Blood Circulation*, CRC Press, Boca Raton, FL.

Poste, G., 1983, Liposome targeting *in vivo:* Problems and opportunities, *Biol. Cell* **47**:19–38.

Poste, G., 1985, Drug targeting in cancer therapy, in: *Receptor-Mediated Targeting of Drugs* (G. Gregoriadis, G. Poste, J. Senior, and A. Trouet, eds.), pp. 427–474, Plenum Press, New York.

Poste, G., and Fidler, I. J., 1980, The pathogenesis of cancer metastasis, *Nature* **283**:139–146.

Poste, G., and Kirsh, R., 1983, Site-specific (targeted) drug delivery in cancer therapy, *Biotechnology* **1**:869–878.

Poznansky, M. J., and Juliano, R. L., 1984, Biological approaches to the controlled delivery of drugs: A critical review, *Pharmacol. Rev.* **36**:277–335.

Renkin, E. M., 1977, Multiple pathways of capillary permeability, *Circ. Res.* **41**:735–743.

Rippe, B., and Haraldsson, B., 1987, Fluid and protein fluxes across small and large pores in the microvasculature. Application of two-pore equations, *Acta Physiol. Scand.* **131**:411–428.

Schlom, J., Colcher, D., Hand, P. H. Wunderlich, D., Nuti, M., and Teramoto, Y. A., 1983, Antigenic heterogeneity, modulation and evolution in breast cancer lesions as defined by monoclonal antibodies, in: *Understanding Breast Cancer: Clinical and Laboratory Concepts*, (M. Rich, J. C. Hager, and P. Furmanski, eds.), pp. 315–358, Marcel Dekker, New York.

Steller, M. A., Parker, R. J., Covell, D. G., Holton, O. D., III, Keenan, A. M., Sieber, S. M., and Weinstein, J. N., 1986, Optimization of monoclonal antibody delivery via the lymphatics: The dose-dependence, *Cancer Res.* **46**:1830–1834.

Sutherland, R., Buchegger, F., Schreyer, M., Vacca, A., and Mach, J., 1987, Penetration and binding of radiolabeled anti-carcinoembrionic antigen monoclonal antibodies and their antigen binding fragments in human colon multicellular tumor spheroids, *Cancer Res.* **47**:1627–1633.

Swabb, E. A., Wei, J., and Gullino, P. M., 1974, Diffusion and convection in normal and neoplastic tissues, *Cancer Res.* **34**:2814–2822.

Weinstein, J. N., 1987, Liposomes in the diagnosis and treatment of cancer, in: *Liposomes* (M. Ostro, ed.), pp. 277–338, Marcel Dekker, New York.

Weinstein, J. N., Parker, R. J., Keenan, A. M., Dower, S. K., Morse, H. C., 3rd, and Sieber, S. M., 1982, Monoclonal antibodies in the lymphatics: Toward the diagnosis and therapy of tumor metastases, *Science* **218**:1334–1337.

Weinstein, J. N., Steller, M. A., Keenan, A. M., Covell, D. G., Key, M. E., Sieber, S. M., Oldham, R. K., Hwang, K. M., and Parker, R. J., 1983, Monoclonal antibodies in the lymphatics: Selective delivery to lymph node metastases of a solid tumor, *Science* **222**:423–427.

Weinstein, J. N., Parker, R. J., Holton, O. D., III, Keenan, A. M., Covell, D. G., Black, C. D. V., and Sieber, S. M., 1985, Lymphatic delivery of monoclonal antibodies: Potential for detection and treatment of lymph node metastases, *Cancer Invest.* **3**:85–95.

Weinstein, J. N., Black, C. D. V., Barbet, J., Eger, R. R., Parker, R. J., Holton, O. D., III, Mulshine, J. L., Keenan. A. M., Larson, S. M., Carrasquillo, J. A., Sieber, S. M., and Covell, D. G., 1986, Selected issues in the pharmacology of monoclonal antibodies, in: *Site-Specific Drug Delivery*, (E. Tomlinson and S. S. Davis, eds.), pp. 81–91, Wiley, New York.

Weinstein, J. N., Covell, D. G., Barbet, J., Eger, R. R., Holton, O. D., III, Talley, M. J., Parker, R. J., and Black, C. D. V., 1987a, Local and cellular factors in the pharmacology of monoclonal antibodies, in: *Membrane Mediated Toxicity* (B. Bonavida and R. J. Collier, eds.), pp. 279–289, Alan R. Liss, New York.

Weinstein, J. N., Eger, R. R., Covell, D. G., Black, C. D. V., Mulshine, J., Carrasquillo, J. A., Larson, S. M., and Keenan, A. M., 1987b, The pharmacology of monoclonal antibodies, *Ann. NY Acad. Sci.* **507**:199–210.

Weiss, L., and Greep, R. O., 1977, *Histology*, McGraw-Hill, San Francisco.

Semisynthetic Catalytic Antibodies

SCOTT J. POLLACK and PETER G. SCHULTZ

Our ability to carry out selective chemical transformations on biologically important molecules—including proteins, nucleic acids, and sugars—has been limited until recently to the specific reactions catalyzed by existing enzymes and the relatively indiscriminate reactions carried out by chemical reagents. The rational design of catalysts for the selective modification of structurally complex molecules would greatly impact on biochemistry, molecular biology, and medicine. Moreover, the synthesis and characterization of such catalysts should provide additional insight into the molecular basis for ligand–receptor recognition and catalysis. Crucial to the design of selective catalysts is the generation of highly selective binding sites. By taking advantage of the natural immune system, monoclonal antibody technology (Kohler and Milstein, 1975; Seiler *et al.*, 1985) has made it possible to generate homogeneous ligand binding sites with enzymelike affinities and specificities. Antibodies have been selectively generated against biopolymers, such as nucleic acids, proteins, and polysaccharides, and smaller multifunctional molecules, such as steroids and prostaglandins. Antibodies bind ligands ranging in size from about 6 to 34 Å with association constants in the range of 10^4–10^{14} M^{-1} (Pressman and Grossberg, 1968; Goodman, 1975; Nisonoff *et al.*, 1975). Antibody–ligand specificity is well illustrated by the following examples: antibodies generated against *cis* N-phenylmaleic acid monoamide bound the *trans* isomer with 10^3 lower affinity (Landsteiner and van der Scheer, 1934); antibodies against 3,17-androstenedione bound 3a, 17-dihydroxyandrostene with 10^3 lower affinity (Milewich *et al.*, 1975).

A number of strategies have recently been developed whereby the high binding affinity and specificity of antibodies can be exploited in the design of novel catalysts with tailored specificities. The binding energy of antibodies to

SCOTT J. POLLACK and PETER G. SCHULTZ • Department of Chemistry, University of California, Berkeley, California 94720.

specific recognition elements has been used to selectively stabilize transition state configurations on reaction pathways (Pollack *et al.*, 1986; Pollack and Schultz, 1987; Jacobs *et al.*, 1987; Tramontano *et al.*, 1986; Durfor *et al.*, 1988; Janda *et al.*, 1988) or to overcome entropic barriers involved in orienting reaction partners (Napper *et al.*, 1987; Jackson *et al.*, 1988; Hilvert *et al.*, 1988). Antibody–hapten complementarity has been exploited to introduce catalytically active side chains into antibody combining sites (Cochran *et al.*, 1988) and antibodies have been generated with cofactor binding sites (Shokat *et al.*, 1988; Raso and Stollar, 1975). Using these approaches, antibodies have been shown to selectively catalyze several classes of reactions, including transacylation, lactonization, elimination, pericyclic, redox, and photochemical reactions. Finally, site-directed mutagenesis and genetic selection are currently being applied to the development of catalytic antibodies.

Alternatively, it should be possible to introduce catalytic groups directly into antibody combining sites, by either chemical or genetic modification. This approach combines the high binding affinity and exquisite specificity of antibodies and the extremely efficient, diverse catalysts available from synthetic chemistry. Kaiser and co-workers demonstrated that enzymes can be selectively modified with a catalytic cofactor: the active site thiol of papain was derivatized with a redox-active flavin (Kaiser and Lawrence, 1984). We describe here a chemical strategy whereby antibody combining sites can be selectively modified with additional chemical functionality to produce a new class of semisynthetic antibodies. Cleavable affinity labels are used to site-specifically incorporate a derivatizable thiol into the combining sites of antibodies of unknown structure (Pollack *et al.*, 1988). Importantly, this thiol acts as a handle whereby additional functionalities can be introduced into the antibody combining site. For example, selective derivatization of the thiol with an imidazole provided a specific catalyst for ester hydrolysis. Derivatization of antibodies with other natural or synthetic catalytic groups, such as transition metal complexes, bases, nucleophiles, or natural cofactors, should enable us to generate antibodies that selectively catalyze a wide array of reactions. Moreover, the introduction of other groups should lead to antibodies with a variety of novel functions. Introducing fluorophores or redox-active molecules should afford homogeneous sensors, and the introduction of oxidizing or hydrolytic agents or cytotoxic molecules should lead to selective immunoglobulin-derived therapeutics. It should also be possible to apply this same strategy to the selective modification of other biopolymers, including those for which the three-dimensional structure is unavailable.

1. Design Considerations

The key to the generation of semisynthetic catalytic antibodies is the development of mild methods for introducing derivatizable groups of unique reactivity into or near the combining site. These groups can then be modified in a second step to incorporate the chemical functionality (e.g., cofactor, metal–ligand complex, fluorophore) of interest. One such derivatizable group is a free thiol, which, because of its high nucleophilicity and ease of oxidation, can be selectively modified via disulfide exchange or electrophilic reactions. Moreover, most antibodies contain no free surface-accessible thiols to interfere with these desired reactions. Site-directed mutagenesis has been used to selectively introduce a cysteine thiol into an enzyme in order to subsequently introduce an oligonucleotide binding site (Corey and Schultz, 1987; Zuckerman *et al.*, 1988). We are currently investigating the genetic modification of antibodies with derivatizable cysteine residues. However, site-directed mutagenesis requires knowledge of the three-dimensional structure of the protein, which is not available for most proteins of interest.

In the absence of detailed structural information, affinity-labeling reagents can be used to selectively modify proteins at their ligand-binding sites. If the affinity label is designed with a cleavable cross-link such as a thiol ester or a disulfide, cleavage of the label after site-specific derivatization incorporates a free thiol at the binding site. A thiol ester cross-link can be cleaved after derivatization with hydroxylamine, and a disulfide can be reduced with dithiothreitol. We chose to incorporate disulfides into our affinity labeling reagents. Under the mild conditions used, the label disulfides are reduced without reducing the buried protein disulfides. Thiol esters have the disadvantage of acylating various amino acid side chains such as lysines. Moreover, thiol esters are not stable to some protein modification, purification, and characterization conditions.

In addition to thiols, there are other possible groups of unique reactivity that can be incorporated into antibody binding sites using the cleavable affinity label strategy. Another candidate for the cleavable linkage is an acetal, which would lead to a protein-bound aldehyde that could be derivatized by reductive amination. However, the acidic conditions required for cleavage could lead to inactivation of acid-sensitive proteins. An azobenzene cross-link could be reductively cleaved to give an aromatic amine, which could then be selectively derivatized at low pH. Since polyfunctional disulfides are more readily obtainable synthetically, and since they are less sterically obtrusive than the phenyl ring of azobenzene, we decided to start with these cross-links. Moreover, the types of compounds that can be used to derivatize thiols (e.g., activated disulfides) are more likely to be compatible with functionality on the prosthetic group than are the activated acyl compounds needed to react with aminobenzene.

A number of factors went into the choice of the reactive functionality in the affinity-labeling reagents. The reactive group should react slowly enough so that it will react primarily in the binding site of the protein, after binding equilibrium is established. The reaction of a label with more than one type of amino acid residue can be advantageous for maximizing the possibilities for successful derivatization, yet at the same time it could complicate the characterization of the labeled protein. Ideally, a unique residue should be modified to give a homogeneous adduct. Another important consideration is the synthesis of the labeling reagents. It should be possible to easily vary the length of the reagent in order to optimize incorporation of label into the protein. The affinity label should also be easily synthesized in radiolabeled form, for ease in quantitating label incorporation. Finally, the label should be stable to subsequent protein manipulation conditions, including the degradative methods used in protein sequence determination.

Two possible reactive labeling groups are diazoketones and aromatic azides. These groups can be transformed into reactive species in the protein binding site by photochemical irradiation and therefore have the advantage that they can be "turned on" in the binding site of a protein. In the past, however, there have been difficulties in identifying the labeled residue (Givol and Wilchek, 1977), presumably because of the high and nonspecific reactivity of the active species.

Several other reactive chemical functionalities can be considered. Diazonium groups react specifically with tyrosine side chains. However, the limited reactivity of diazonium compounds and difficulty in preparing these reagents limit their

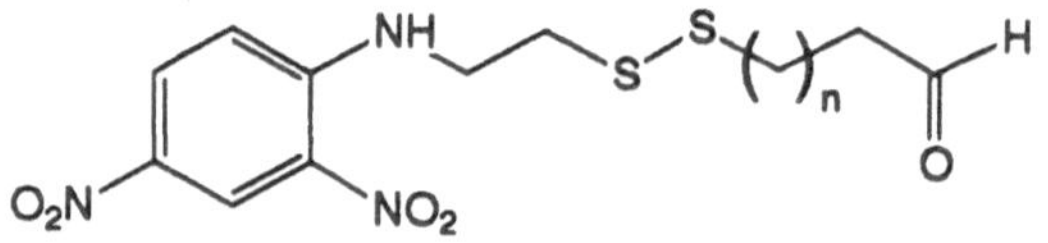

1: n = 1
2: n = 2

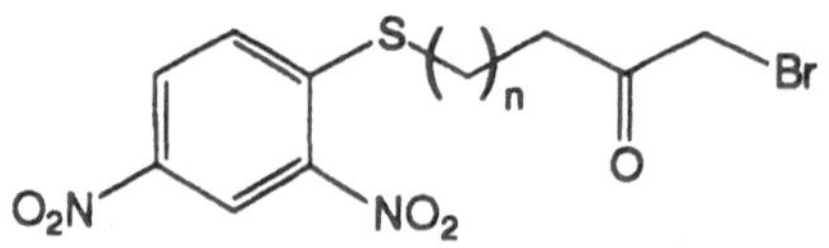

3: n = 1
4: n = 2
5: n = 3

Figure 1. Structures of affinity labels.

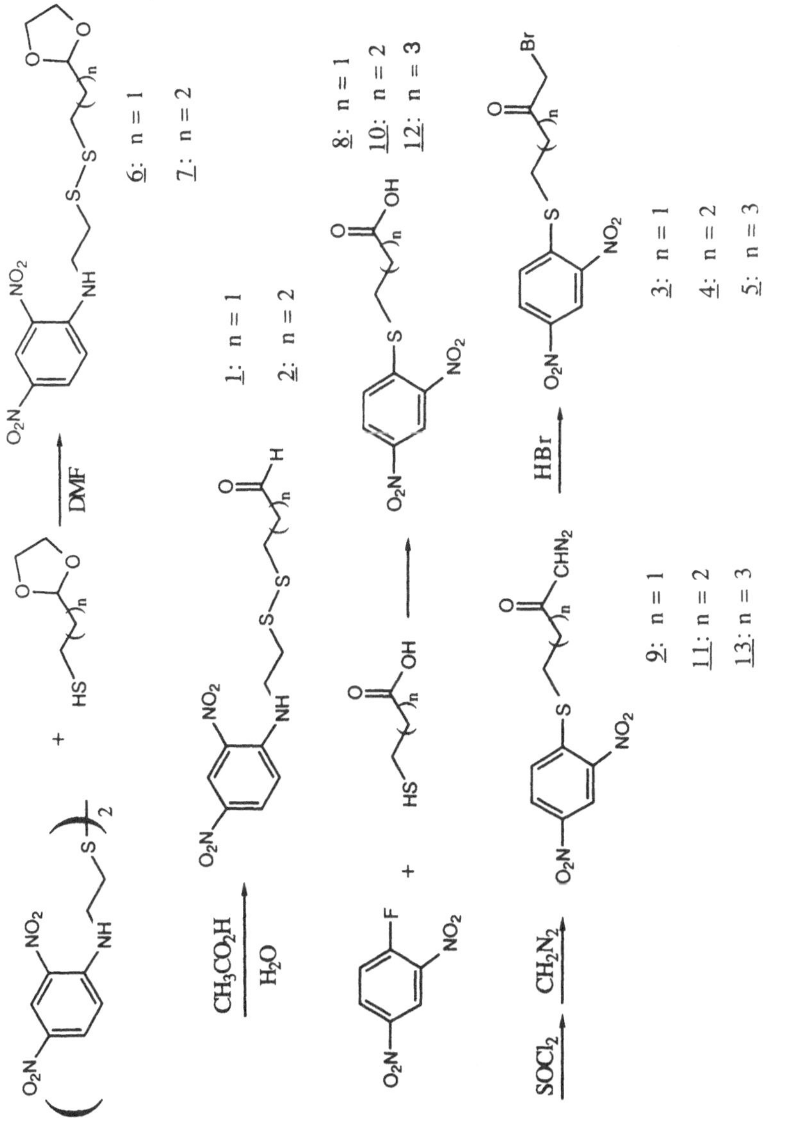

Figure 2. Synthesis of affinity labels.

applicability. In addition, the tyrosine adducts are not stable to Edman degradation in protein sequencing (Givol and Wilchek, 1977). Epoxide groups react with various nucleophilic side chains including lysines and carboxylates. Aldehydes react with lysine residues to form Schiff base intermediates, which can be reduced with sodium cyanoborohydride. The use of tritiated $NaCNB^3H_3$ leads to tritium incorporation, facilitating the characterization of the labeled protein. Finally, α-bromoketones have been shown to be versatile affinity-labeling reagents (Givol and Wilchek, 1977). Their reactivity is sufficiently slow that they can label proteins nearly stoichiometrically in their binding sites. Moreover, they react with a number of nucleophilic side chains, including carboxylates, lysines, histidines, and tyrosines.

We chose to carry out our initial affinity-labeling studies with aldehydes and α-bromoketones. In addition to the advantages discussed, both types of reagents are easily synthesized and linker lengths can easily be varied. Our initial protein modification studies were carried out on the IgA MOPC 315, which binds substituted 2,4-dinitrophenyl (DNP) ligands with association constants ranging from 5×10^4 to 10^6 M^{-1} (Haselkorn *et al.*, 1974). Although a three-dimensional structure is not available, the antibody combining site has been characterized by spectroscopic methods [ultraviolet, fluorometry, nuclear magnetic resonance (NMR)], chemical modification, and amino acid sequencing of the variable region. Moreover, earlier affinity-labeling studies with reagents of varying structures (Givol and Wilchek, 1977; Givol *et al.*, 1971; Strausbauch *et al.*, 1971) defined a number of reactive amino acid side chains in the vicinity of the combining site.

We have incorporated cleavable tethers into affinity labels specific for the MOPC 315 combining site (Fig. 1). These affinity-labeling reagents contain the DNP group linked to electrophilic aldehyde or α-bromoketone groups via cleavable disulfide or thiophenyl linkages. Covalent attachment of the label to the antibody, followed by cleavage of the cross-link and removal of the free ligand, results in site-specific incorporation of a free thiol into the antibody combining site. The geometry of affinity-labeling reagents *1–5* varies with regard to the distance between the DNP group and electrophilic moiety, since the position of a nucleophilic lysine, histidine, or tyrosine side chain is not precisely known. The syntheses of affinity labels *1–5* are outlined in Fig. 2.

2. Synthesis of Affinity Labels

(N-2,4-Dinitrophenyl)-2-aminoethyl 2-(1,3-dioxolanyl)-ethyl disulfide *6*
To a suspension of *N,N'*-bis-(2,4-dinitrophenyl)-cystamine (Chan *et al.*, 1979) (4.84 g, 10 mmoles) in dimethylformamide (200 ml) with triethylamine (50 μl) was added 2-(2-mercaptoethyl)-1,3-dioxolane (2.45 g, 18 mmoles), and

the mixture was stirred at 60°C for 24 hr. Hydrogen peroxide (2 ml of a 30% solution) was added, and the dimethylformamide was removed *in vacuo*. The residual oil was dissolved in methylene chloride (100 ml). The methylene chloride layer was washed with water (3 × 50 ml), dried over MgSO$_4$, and concentrated *in vacuo* to an oil. Silica gel column chromatography with 3:2 hexanes: ethyl acetate (R_f = 0.31) afforded *6* (1.37 g, 3.65 mmoles, 20%) as an orange oil. IR (thin film): 3353, 2923, 1623, 1525, 1342, 1138 cm^{-1}; ^{1}H-NMR, CDCl$_3$: δ 2.08 (m, 2H), 2.82 (t, 2H, J = 7.1 Hz), 3.00 (t, 2H, J = 6.6 Hz), 3.8– 4.0 (m, 6H), 4.95 (t, 1H, J = 4.3 Hz), 7.00 (d, 1H, J = 9.5 Hz), 8.30 (d, 1H, J = 9.6 Hz), 8.80 (s, 1H), 9.16 (d, 1H, J = 2.6 Hz); mass spectrum (EI) 375 (M$^+$). Anal. Calcd. for C$_{13}$H$_{17}$N$_3$O$_6$S$_2$: C, 41.60; H, 4.53; N, 11.20; S, 17.07. Found: C, 41.57; H, 4.67; N, 11.10; S, 16.99.

(*N*-2,4-Dinitrophenyl)-2-aminoethyl 3-oxopropyl disulfide *1*

A solution of dioxolane *6* (0.50 g, 1.33 mmoles) in water (2 ml), acetonitrile (2 ml), and acetic acid (8 ml) was heated at reflux for 20 hr. At this time the mixture was cooled to room temperature and added slowly to 150 ml of cold saturated aqueous NaHCO$_3$. The aqueous layer was extracted with methylene chloride (3 × 50 ml). The combined extracts were dried over MgSO$_4$ and concentrated *in vacuo* to an orange oil. Silica gel chromatography using a gradient of 30–40% ethyl acetate in hexanes afforded *1* (0.30 g, 0.91 mmole, 68%) as an orange semisolid: R_f = 0.22 in 3:2 hexanes: ethyl acetate; IR (KBr pellet) 3353, 1722, 1623, 1525, 1504, 1426, 1342, 1131 cm^{-1}; ^{1}H-NMR (CDCl$_3$) δ 2.90–3.04 (m, 6H), 3.79 (t, 2H, J = 6.5 Hz), 7.01 (d, 1H, J = 9.5 Hz), 8.31 (d, 1H, J = 9.5 Hz), 8.80 (m, 1H), 9.15 (d, 1H, J = 2.6 Hz), 9.83 (s, 1H); mass spectrum (EI) 331 (M$^+$). Anal. Calcd. for C$_{11}$H$_{13}$N$_3$O$_5$S$_2$: C, 39.88; H, 3.93; N, 12.69; S, 19.34. Found: C, 40.01; H, 3.91; N, 12.66; S, 19.34.

(*N*-2,4-Dinitrophenyl)-2-aminoethyl 3-(1,3-dioxolanyl)-propyl disulfide *7*

This compound was prepared as described above for *6*, starting with *N*,*N*'-bis-(2,4-dinitrophenyl)-cystamine (2.42 g, 5 mmoles) and 3-(3-mercaptopropyl)-1,3-dioxolane (1.48 g, 10 mmoles), to give *7* (0.60 g, 1.54 mmoles, 15%) as an orange oil: R_f = 0.43 in 3:2 hexanes: ethyl acetate; IR (thin film) 3360, 2923, 1623, 1592, 1528, 1426, 1335, 1131 cm^{-1}; ^{1}H-NMR (CDCl$_3$): δ 1.80– 1.95 (m, 6H), 2.78 (t, 2H, J = 7.0 Hz), 2.99 (t, 2H, J = 6.7 Hz), 3.8–4.0 (m, 6H), 4.87 (t, 1H, J = 4.2 Hz), 7.01 (d, 1H, J = 9.5 Hz), 8.32 (d, 1H, J = 9.5 Hz), 8.79 (s, 1H), 9.16 (d, 1H, J = 2.6 Hz); mass spectrum (EI) 389 (M), 368, 359, 294, 279, 256, 225, 210, 196. Anal. Calcd. for C$_{14}$H$_{19}$N$_3$O$_6$S$_2$: C, 43.19; H, 4.88; N, 10.80; S, 16.45. Found: C, 43.27; H, 4.65; N, 10.79; S, 16.40.

(*N*-2,4-Dinitrophenyl)-2-aminoethyl 4-oxobutyl disulfide *2*

This compound was prepared as described for *1*, starting with dioxolane *6* (389 mg, 1.0 mmole), to give *2* (340 mg, 0.99 mmole, 99%) as an orange solid: mp 64–65°C; R_f = 0.25 in 3:2 hexanes: ethyl acetate; IR (KBr pellet) 3331, 3092, 2923, 2846, 1722, 1630, 1589, 1525, 1415, 1342, 1247, 1146 cm^{-1}; ^{1}H-

NMR (CDCl$_3$) δ 2.06 (t, 2H, J = 7.3 Hz), 2.63 (t, 2H, J = 7.0 Hz), 2.75 (t, 2H, J = 7.1 Hz), 3.00 (t, 2H, J = 6.6 Hz), 3.81 (t, 2H, J = 6.1 Hz), 7.01 (d, 1H, J = 9.5 Hz), 8.31 (d, 1H, J = 9.5 Hz), 8.79 (s, 1H), 9.16 (d, 1H, J = 2.6 Hz), 9.80 (s, 1H); mass spectrum (EI) 345 (M). Anal. Calcd. for C$_{12}$H$_{15}$N$_3$O$_5$S$_2$: C, 41.74; H, 4.35; N, 12.17; S, 18.55. Found: C, 41.71; H, 4.38; N, 12.11; S, 18.49.

3-(S-DNP)-mercaptopropanoic acid 8

To a solution of 3-mercaptopropanoic acid (1.59 g, 15 mmoles) in 70 ml of 2.0 M aqueous sodium acetate (pH 5.2) was added a solution of 2,4-dinitro-fluorobenzene (3.32 g, 18.0 mmoles) in 50 ml of absolute ethanol over 15 min with stirring at room temperature under nitrogen. After 24 hr, the pale-yellow solid precipitate was collected and washed with water to give 8 (3.40 g, 12.5 mmoles, 83%): mp 154.5–157.5°C; IR (KBr) 3630, 3476 (br), 3113, 3082, 1718, 1589, 1511, 1344, 1055, 922 cm^{-1}; ^{1}H-NMR (acetone-d$_6$) δ 2.71 (t, 2H, J = 7.5 Hz), 3.40 (t, 2H, J = 7.5 Hz), 3.40 (t, 2H, J = 7.4 Hz), 3.40 (s, 1H), 7.87 (d, 1H, J = 9.1 Hz), 8.32 (dd, 1H, J = 2.5, 9.0 Hz), 8.76 (d, 1H, J = 2.5 Hz); mass spectrum (FAB$^-$) 271 (M-H)$^-$, 249, 199, 183, 141, 113, 109. Anal. Calcd. for C$_9$H$_8$N$_2$O$_6$S: C, 39.71; H, 2.96; N, 10.29; S, 11.78. Found: C, 39.92; H, 2.96; N, 10.15; S, 11.64

1-Diazo-2-oxo-4-(S-DNP)-mercaptobutane 9

To a solution of 8 (0.70 g, 2.5 mmoles) in dry dioxane (10 ml) was added thionyl chloride (5 ml). The mixture was stirred under nitrogen for 12 hr. The volatiles were removed *in vacuo* and the yellow residue was dissolved in dry dioxane (25 ml). A solution of diazomethane (0.6 M) in diethyl ether (25 ml) was added. The mixture was stirred at 0°C for 4 hr, after which the volatiles were removed *in vacuo*. The residue was purified by silica gel chromatography using 9:1 methylene chloride: ethyl acetate as the eluent to give 9 (0.29 g, 0.92 mmoles, 36%) as a yellow solid: mp 117–119°C; R_f = 0.52 in above eluent; IR (KBr pellet) 3458 (br), 3090, 2111, 1637, 1583, 1508, 1340, 1096 cm^{-1}; ^{1}H-NMR (CDCl$_3$) δ 2.75 (t, 2H, J = 7.1 Hz), 3.35 (t, 2H, J = 7.3 Hz), 5.32 (s, 1H), 7.61 (d, 1H, J = 9.0 Hz), 8.48 (dd, 1H, J = 2.5, 9.0 Hz), 9.07 (d, 1H, J = 2.5 Hz); mass spectrum (FAB$^-$) 296 (M$^-$), 199, 183, 141, 109. Anal. Calcd. for C$_{10}$H$_8$N$_4$O$_5$S: C, 40.54; H, 2.72; N, 18.91; S, 10.82. Found: C, 40.81; H, 2.89; N, 18.59; S, 10.79.

1-Bromo-2-oxo-4-(S-DNP)-mercaptobutane 3

To a solution of 9 (95 mg, 0.32 mmoles) in dry dioxane (5 ml) at 25°C under nitrogen was added a saturated solution of HBr in dioxane (2 ml). The mixture was stirred for 10 min, and the volatiles were then removed completely *in vacuo* to give pure 3 (130 mg, 0.373 mmole, 86%) as a yellow solid: mp 99–101°C; R_f = 0.78 in 9:1 methylene chloride: ethyl acetate; IR (KBr) 3106, 3097, 3012, 2958, 1724, 1592, 1510, 1339, 1068, 1053 cm^{-1}; ^{1}H-NMR (acetone-d$_6$) δ 3.24 (t, 2H, J = 7.0 Hz), 3.46 (t, 2H, J = 7.0 Hz), 4.28 (s, 2H), 7.98 (d, 1H, J = 9.0 Hz), 8.48 (dd, 1H, J = 2.5, 9.0 Hz); mass spectrum (FAB$^-$) 268 (M-HBr)$^-$,

233, 199, 183. 109. Anal. Calcd. for $C_{10}H_9BrN_2O_5S$: C, 34.40; H, 2.60; Br, 22.89; N, 8.02; S, 9.18. Found: C, 34.50; H, 2.60; Br, 22.70; N, 7.88; S, 9.04.

4-(S-DNP)-mercaptobutanoic acid *10*

This compound was prepared as described for *8*, starting with 4-mercaptobutanoic acid (1.80 g, 15 mmoles), giving *10* (3.21 g, 11.2 mmoles, 75%) as a pale-yellow solid: mp 126.5–130°C; IR (KBr) 3528 (br), 3117, 3090, 2948, 1709, 1587, 1518, 1341 cm^{-1}; ^{1}H-NMR (DMSO-d$_6$) δ 1.85 (m, 2H), 2.34 (t, 2H, J = 7.0 Hz), 3.18 (t, 2H, J = 7.5 Hz), 3.78 (s, 1H), 7.95 (d, 1H, J = 9.1 Hz), 8.41 (dd, 1H, J = 2.4, 9.0 Hz), 8.84 (d, 1H, J = 2.4 Hz); mass spectrum (FAB$^-$) 285 (M-H)$^-$, 249, 199, 139. Anal. Calcd. for $C_{10}H_{10}N_2O_6S$: C, 41.96; H, 3.52; N, 9.79; S, 11.20. Found: C, 41.94; H, 3.50; N, 9.60; S, 10.97.

1-Diazo-2-oxo-5-(S-DNP)-mercaptopentane *11*

This compound was prepared as described for *9*, starting with *10* (1.00 g, 3.49 mmoles), to give *11* (0.68 g, 2.20 mmoles, 63%) as a yellow solid: mp 84–86°C; R_f = 0.52 in 9:1 methylene chloride: ethyl acetate; IR (KBr) 3106, 2939, 2866, 2104, 1626, 1589, 1513, 1343 cm^{-1}; ^{1}H-NMR (CDCl$_3$) δ 2.10 (m, 2H), 2.55 (t, 2H, J = 5.7 Hz), 3.10 (t, 2H, J = 7.5 Hz), 5.29 (s, 1H), 7.74 (d, 1H, J = 9.0 Hz), 8.38 (dd, 1H, J = 2.5, 9.0 Hz), 9.06 (d, 1H, J = 2.5 Hz); mass spectrum (FAB$^-$) 310 (M$^-$), 249, 199, 183, 139, 107. Anal. Calcd. for $C_{11}H_{10}N_4O_5S$: C, 42.58; H, 3.25; N, 18.06; S, 10.33. Found: C, 42.79; H, 3.39; N, 17.93; S, 10.14.

1-Bromo-2-oxo-5-(S-DNP)-mercaptopentane *4*

This compound was prepared as described for *3*, starting with *11* (101 mg, 0.326 mmoles) to give *4* (116 mg, 0.32 mmoles, 98%) as a yellow solid: mp 87–90°C; R_f = 0.78 in 9:1 methylene chloride: ethyl acetate; IR (KBr) 3113, 3009, 2953, 1723, 1592, 1512, 1343 cm^{-1}; ^{1}H-NMR (DMSO-d$_6$) δ 1.89 (m, 2H), 2.79 (t, 2H, J = 7.1 Hz), 3.17 (t, 2H, J = 7.4 Hz), 4.36 (s, 2H), 7.88 (d, 1H, J = 9.1 Hz), 8.43 (dd, 1H, J = 2.5, 9.0 Hz), 8.85 (d, 1H, J = 2.6 Hz). Anal. Calcd. for $C_{11}H_{11}BrN_2O_5S$: C, 36.38; H, 3.05; Br, 2.00; N, 7.71; S, 8.83. Found: C, 36.57; H, 3.17; Br, 21.83; N, 7.49; S, 9.01.

5-(S-DNP)-mercaptopentanoic acid *12*

This compound was prepared as described for *8*, starting with 4-mercaptopentanoic acid (2.00 g, 15 mmoles) giving *12* (4.20 g, 14.0 mmoles, 93%) as a pale-yellow solid: mp 159–161°C; IR (KBr) 3117, 2948 (br), 1707, 1587, 1517, 1341 cm^{-1}; ^{1}H-NMR (acetone-d$_6$) δ 1.83 (m, 4H), 2.39 (t, 2H, J = 6.8 Hz), 2.83 (s, 1H), 3.26 (t, 2H, J = 7.0 Hz), 7.95 (d, 1H, J = 9.0 Hz), 8.45 (dd, 1H, J = 2.5, 9.0 Hz), 8.96 (d, 1H, J = 2.5 Hz); mass spectrum (FAB$^-$) 299 (M-H)$^-$, 199, 183, 141, 113, 109. Anal. Calcd. for $C_{11}H_{12}N_2O_6S$: C, 44.00; H, 4.03; N, 9.33; S, 10.68. Found: C, 43.83; H, 3.92; N, 9.42; S, 10.59.

1-Diazo-2-oxo-6-(S-DNP)-mercaptohexane *13*

This compound was prepared as described for *9*, starting with *12* (1.00 g, 3.33 mmoles), to give *13* (0.70 g, 2.15 mmoles, 65%) as a yellow solid: mp

120–122°C, R_f = 0.30 in 9:1 methylene chloride: ethyl acetate, IR (KBr) 3093, 2937, 2110, 1637, 1584, 1508, 1336, 1109 cm^{-1}; ^{1}H-NMR (CDCl$_3$) δ 1.83 (m, 4H), 2.39 (t, 2H, J = 5.8 Hz), 3.04 (t, 2H, J = 7.4 Hz), 5.25 (s, 1H), 7.53 (d, 1H, J = 9.0 Hz), 8.37 (dd, 1H, J = 2.5, 9.0 Hz), 9.07 (d, 1H, J = 2.5 Hz); mass spectrum (FAB$^-$) 324, 307, 267, 199, 183, 141, 109. Anal. Calcd. for C$_{12}$H$_{12}$N$_4$O$_5$S: C, 44.44; H, 3.73; N, 17.28; S, 9.89. Found: C, 44.60; H, 3.74; N, 16.99; S, 9.65.

1-Bromo-2-oxo-6-(S-DNP)-mercaptohexane 5

This compound was prepared as described for 3, starting with 13 (119 mg, 0.366 mmoles) to give 5 (0.127 g, 0.336 mmoles, 92%) as a yellow solid: mp 80–83°C; R_f = 0.78 in 9:1 methylene chloride: ethyl acetate; IR (KBr) 3121, 2937, 2872, 1719, 1586, 1513, 1339, 1095, 1053; ^{1}H-NMR (acetone-d$_6$) δ 1.81 (m, 4H), 2.79 (t, 2H, J = 6.6 Hz), 3.24 (t, 2H, J = 6.8 Hz), 4.18 (s, 2H), 7.93 (d, 1H, J = 9.0 Hz), 8.45 (dd, 1H, J = 2.5, 9.0 Hz), 8.95 (d, 1H, J = 2.5 Hz); mass spectrum (FAB$^-$) 378 (M$^-$), 307, 263, 199, 183, 141, 109. Anal. Calcd. for C$_{12}$H$_{13}$BrN$_2$O$_5$S: C, 38.21; H, 3.47; Br, 21.18; N, 7.43; S, 8.50. Found: C, 38.40; H, 3.45; Br, 21.34; N, 7.34; S, 8.46.

3. Antibody Modification

3.1. Affinity Labeling

Our strategy for introducing a thiol into the binding site of MOPC 315 is outlined in Fig. 3. These steps were carried out with the Fab fragment, generated by treating the reduced and alkylated IgA with papain followed by affinity chromatography on DNP-coupled Sepharose 4B (Goetzl and Metzger, 1970). The Fab fragment (at 10 μM) was treated with 1.5 equivalents of aldehydes 1 or 2 for 1 hr followed by 7.5 equivalents of NaCNBH$_3$, in 0.2 M sodium phosphate, pH 7.0, at 37°C for 20 hr, or with 1.3 equivalents of α-bromoketones 3, 4, or 5 in 0.1 M sodium bicarbonate, pH 9.0, at 37°C for 16 hr. In both cases, the labeled Fab was purified by chromatography on Sephadex G-50 in 0.1 M sodium phosphate, pH 7.3, and, after lyophilization, by affinity chromatography on DNP-coupled Sepharose 4B (Goetzl and Metzger, 1970) to remove unlabeled or nonspecifically labeled Fab. The extent of derivatization was quantitated spectrophotometrically (for labels 1 and 2, λ_{max} = 362 nm, ε = 16,000 M^{-1} cm^{-1}; for labels 3–5, λ_{max} = 360 nm, ε = 12,800 M^{-1} cm^{-1}) and additionally, for labels 1 and 2, by incorporating tritium from NaCNB^{3}H$_3$ (using 12 mCi / millimole NaCNB^{3}H$_3$). With labels 2 and 4, over 90% of the label was incorporated (85% yield after affinity chromatography) (Table I). In both cases, less than 10% nonspecific labeling occurred in the presence of 10 mM of the competitive inhibitor DNP-glycine.

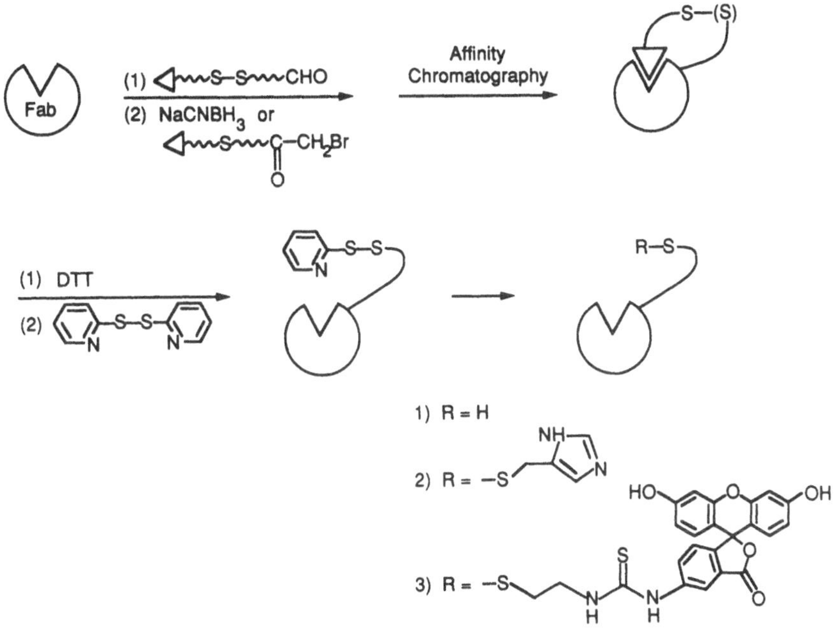

Figure 3. Antibody modification scheme.

3.2. Cleavage of Labels and Isolation of Stable S-Thiopyridyl Adducts

Cleavage of the Fab affinity-labeled with *2* or *4* (2.0 mg) with 50 mM dithiothreitol in 2 mM EDTA, 0.1 M sodium phosphate, pH 8.0 (2 ml) for 12 hr at 37°C afforded the free thiol. The thiol-containing Fab was collected by fast-desalting chromatography using a Pharmacia FPLC column in 0.1 M sodium phosphate, pH 7.3, directly into 1.0 ml of a 4.5 mM solution of 2,2'-dithiodi-pyridine (0.1 M sodium phosphate, pH 5.5, containing 15% acetonitrile). This mixture (final volume of 10 ml) was allowed to react at 20°C for 12 hr, after

Table I. Incorporation of Affinity Labels into Fab

Compound	% Fab labeled (without DNP-gly)	% Fab labeled (with DNP-gly)
1	22	5
2	95	10
3	0	0
4	85	0
5	15	0

which excess 2,2′-dithiodipyridine was removed by exhaustive dialysis. The thiolated antibody was derivatized in greater than 90% yield (65% recovery in the case of 2), based on the absorbance of thiopyridone at 343 nm ($\epsilon = 7060$ M^{-1} cm^{-1}) (Grassetti and Murray, 1967) after cleavage with 10 mM dithiothreitol, and the protein absorbance at 280 nm ($E_{0.1\%} = 1.37$, mol. wt. = 50,000 for Fab).

4. Determination of Modified Site

Fab fragments labeled with 2 or 4 were subjected to tryptic digestion and peptide mapping in order to determine the selectivity of thiol incorporation (Fig. 4). Fab was affinity-labeled in the presence of label 2 and NaCNB^{3}H$_3$ (12 mCi / mmole), or with ^{3}H-labeled 4 (4 mCi / mmole) as described earlier. ^{3}H-labeled 4 was prepared as described for 4, starting with 3,5-^{3}H-2,4-dinitrofluorobenzene (Amersham). The labeled Fab (1.0 mg / ml) was then denatured in 8 M urea, 0.1 M Tris-HCl, pH 8.0, reduced with 20 mM dithiothreitol (1 hr at 37°C), and alkylated with 60 mM iodoacetamide (1 hr at 37°C). After dialysis against 7 M urea, 20 mM Tris-HCl, pH 8.0, the heavy and light chains were separated by anion exchange chromatography on a Pharmacia Mono Q FPLC column. Separation was achieved at 1.0 ml/min in 7 M urea, 20 mM Tris-HCl, pH 8.0, with a

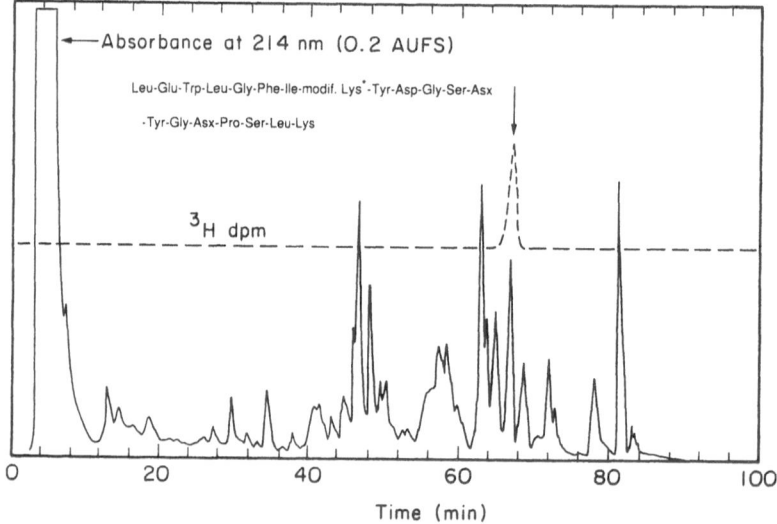

Figure 4. HPLC profile of the tryptic digest of the heavy chain of Fab labeled with 2, monitored by UV absorbance (solid line) and radioactivity (dashed line).

linear gradient of 40–200 mM sodium chloride over 30 min. In the case of Fab labeled with *2*, the heavy chain was found to contain over 95% of the incorporated tritium; with Fab labeled with *4*, over 95% of the tritium label was on the light chain. In each case, the radiolabeled chain was dialyzed against 2 M urea/100 mM ammonium bicarbonate, pH 8.2, and then treated with trypsin (1:25 w/w trypsin: protein) in the presence of 0.1 mM $CaCl_2$ at 37°C in the dark. After 8 hr, the reactions were quenched with 10% vol/vol acetic acid. The radiolabeled peptides were purified by reverse phase HPLC with a 70-min linear gradient from 0–50% acetonitrile in water at 1.0 ml/min [0.1% and 0.06% (vol/vol) trifluoroacetic acid was added to the water and acetonitrile, respectively]. Peptides were detected by their absorbance at 214 nm, fractions were collected at 1.0-ml intervals, and aliquots were counted for radioactivity (Fig. 4).

Fractions containing radioactivity were rechromatographed using a 70-min linear gradient from 0–50% 2-propanol in water at 0.75 ml/min. The water and the 2-propnaol contained 0.1% and 0.06% trifluoroacetic acid, respectively.

The amino acid sequences of the pure peptides were determined on an Applied Biosystems 477A Protein Sequencer. With heavy chain obtained from Fab labeled with *2*, all of the detectable derivatization was on lysine 52 H. With the light chain from Fab labeled with *4*, all of the detectable derivatization was on tyrosine 34 L. These residues are identical to those labeled by Givol and co-workers (Haimovich *et al.*, 1972) with the reagents bromoacetyl-N^ϵ-DNP-L-lysine and *N*-bromoacetyl-N'-DNP-ethylenediamine.

5. Ester Cleavage Using Thiol-Derivatized Antibody

In addition to acting as a handle for selectively derivatizing the antibody with catalytic groups, the thiol itself can act as a nucleophile in the thiolysis of appropriate substrates. DNP-containing esters *14* and *15* were chosen as substrates for the reaction with thiolated Fab labeled with *2* or *4*, respectively (Fig. 5). The position of the cleavable linkage in affinity label *2* approximates that of the ester in the corresponding substrate *14*, ensuring that the thiol is positioned appropriately in the combining site to attack the ester. Likewise, the thiol in Fab labeled with *4* should be positioned to attack ester *15*. Moreover, these substrates contain fluorescent coumarin leaving groups, which can readily be detected at nanomolar concentrations.

5.1. Synthesis of Substrates

(*N*-DNP-3-Aminopropanoic acid 7-hydroxycoumarin ester *14*

A mixture of *N*-DNP-3-aminopropanoic acid (1.28 g, 5 mmoles) and 7-hydroxycoumarin (0.81 g, 5 mmoles) in phosphorous oxychloride (8 ml) was

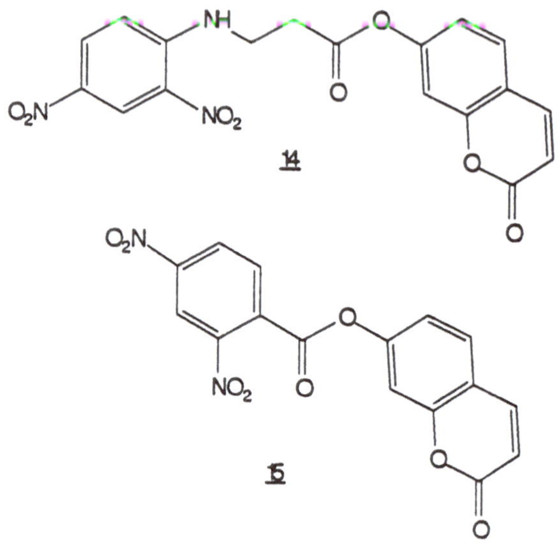

Figure 5. Ester substrates.

heated at reflux under nitrogen for 2 hr. The mixture was cooled to room temperature and added to 60 ml of cold water. The brown solid was filtered, washed with cold water, and dried *in vacuo*. Trituration with acetone gave *14* (0.79 g, 2.0 mmoles, 40%) as a light brown solid: mp 155–156°C, IR (KBr pellet) 3374, 3107, 2368, 1750, 1722, 1624, 1532, 1426, 1349, 1159, 1127 cm^{-1}; ^{1}H-NMR (DMSO-d$_6$) δ 3.04 (t, 2H, J = 6.7 Hz), 3.90 (t, 2H, J = 6.5 Hz), 6.48 (d, 1H, J = 9.6 Hz), 7.17 (d, 1H, J = 8.4 Hz), 7.30 (s, 1H), 7.37 (d, 1H, J = 9.7 Hz), 7.78 (d, 1H, J = 8.5 Hz), 8.07 (d, 1H, J = 9.3 Hz), 8.29 (d, 1H, J = 9.6 Hz), 8.87 (s, 1H), 8.99 (m, 1H); mass spectrum (FAB$^+$) 400 (MH$^+$). Anal. Calcd. for C$_{18}$H$_{13}$N$_3$O$_8$: C, 54.14; H, 3.26; N, 10.53. Found: C, 54.05; H, 3.26; N, 10.23.

2,4-Dinitrobenzoic acid, 7-hydroxycoumarin ester *15*

A mixture of 2,4-dinitrobenzoic acid (1.06 g, 5.0 mmoles) and 7-hydroxy-coumarin (0.81 g, 5.0 mmoles) in phosphorous oxychloride (4 ml) was heated at reflux with a calcium sulfate drying tube for 40 min. The mixture was cooled to room temperature and added to water (40 ml). The brown solid was filtered, washed with cold water and dried *in vacuo* to give *15* (1.32 g, 3.7 mmoles, 74%) as a red-brown solid: mp 200–205°C; IR (KBr) 3107, 3057, 2364, 1736, 1619, 1544, 1348, 1246, 1122 cm^{-1}; ^{1}H-NMR (DMSO-d$_6$) δ 6.53 (d, 1H, J = 9.6 Hz), 7.34 (dd, 1H, J = 2.2, 8.5 Hz), 7.47 (d, 1H, J = 2.1 Hz), 7.89 (d, 1H, J = 8.5 Hz), 8.12 (d, 1H, J = 9.6 Hz), 8.44 (d, 1H, J = 8.4 Hz), 8.78 (dd, 1H, J = 2.2, 8.4 Hz), 8.93 (d, 1H, J = 2.2 Hz); mass spectrum (EI) 356 (M$^+$), 195, 162, 134, 122.

Anal. Calcd. for $C_{16}H_8N_2O_8$: C, 53.94; H, 2.26; N, 7.86. Found: C, 52.18; H, 2.29; N, 7.51.

5.2. Ester Cleavage Assays

Cleavage of DNP coumarin esters *14* and *15* by the thiol-containing anti-bodies were assayed in the presence and absence of 0.1 μM Fab in 50 mM NaCl, 50 mM sodium phosphate, pH 7.0, with 24 μM dithiothreitol at 10°C. The release of free coumarin was quantitated fluorometrically, exciting at 355 nm and measuring emission at 455 nm. The *S*-thiopyridyl Fab was first dialyzed against assay buffer; 30 min prior to the assays, an aliquot of the thiopyridyl Fab (0.15 mg in 0.19 ml) was reduced with 3.8 mM dithiothreitol at 20°C. For each assay, reduced, thiolated Fab (15 μg in 18.8 μl) was diluted with assay buffer (2.95 ml) so that the net concentrations of thiolated Fab and of dithiothreitol were 0.1 μM and 24 μM, respectively. After equilibrating, the substrate was added (30 μl of a stock solution in acetonitrile) and the solution was mixed for 10 sec before fluorescence change was monitored. Antibody rates were corrected by subtract-ing the rate of cleavage in the absence of Fab.

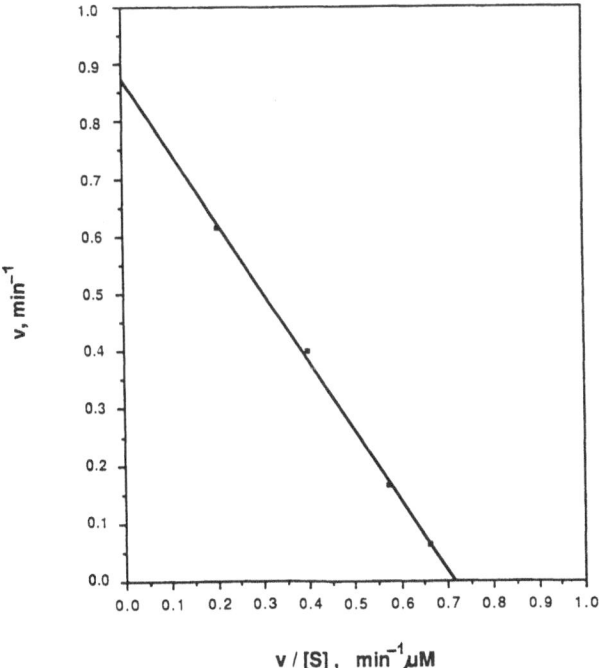

Figure 6. Eadie–Hofstee plot of cleavage of ester *14* by thiolated Fab labeled with *2*.

The antibody affinity-labeled with *2* was found to accelerate the cleavage of ester *14* by a factor of 6×10^4 over the cleavage reaction in equimolar di-thiothreitol. The reaction kinetics are consistent with the formation of a Michaelis complex (Fig. 6). The kinetic constants k_{cat} and K_m for the reaction are 0.87 min^{-1} and 1.2 µM, respectively. The thiolysis reaction was competitively inhibited

$$\text{Ig} + 14 \underset{k_{-1}}{\overset{k_1}{\rightleftharpoons}} \text{Ig} \cdot 14 \overset{k_{cat}}{\rightarrow} \text{Ig} - \text{P} + \text{coumarin}$$

by DNP-glycine with a K_i of 8 µM. Neither the uncleaved affinity-labeled anti-body nor the iodoacetamide-alkylated antibody accelerated the rate of the thio-lysis reaction above the background rate. The stoichiometry of product release corresponded to 1.0 coumarin to 1.0 Fab-SH. Upon reaction with *14* and after fast desalting chromatography, the esterified antibody did not have a titratable thiol. However, subsequent treatment with hydroxide (pH 13, 30 min) led to the release of one free thiol. The introduction of both a thiol and a general base into the label, resulting in structures such as *16*, may lead to a catalytic system in which transacylation by the thiol is followed by hydrolysis of the thiol ester intermediate by the base (Bruice, 1959; Street *et al.*, 1985). Interestingly, thio-lated Fab labeled with bromoketone *4* did not cleave esters *14*, *15*, or DNP-acetate. Fluorescence quenching experiments revealed that the derivatized Fab did not bind the substrate with appreciable affinity, presumably as a result of steric congestion of the antibody combining site.

16

5.3. Substrate Specificity and Enantioselectivity

The substrate specificity of the thiolysis reaction with Fab labeled with *2* was determined by preparing homologous esters of longer and shorter lengths (Table II).

N-2,4-Dinitrophenyl (DNP) glycine, 7-hydroxycoumarin ester *17*

A mixture of *N*-DNP-glycine (603 mg, 2.5 mmoles) (Sigma) and thionyl chloride (1.0 ml) in dry THF was stirred at room temperature under nitrogen for

Table II. Specificity of Thiolysis Reaction

Compound		$k_{cat}(min^{-1})$	$K_m(\mu M)$

		$k_{cat}(min^{-1})$	$K_m(\mu M)$
	15	0.14	1.9
n = 1	**17**	1.4	2.5
n = 2	**14**	0.88	1.2
n = 4	**18**	0.53	2.1
n = 5	**19**	0.12	0.9

12 hr. The volatiles were removed *in vacuo.* Twice, dry THF (10·ml) was added and removed *in vacuo* to ensure complete removal of excess thionyl chloride. The residue was dissolved in dry THF (10 ml) and treated with 7-hydroxy-coumarin (405 mg, 2.5 mmoles) followed by triethylamine (0.70 ml, 5.0 mmoles). The mixture was stirred at room temperature for 6 hr, when the solvent was removed *in vacuo.* The residue was washed with cold water and dried *in vacuo* to give *17* (390 mg, 1.01 mmole, 41%) as a yellow solid: mp 220–222°C; IR (KBr pellet) 3349, 3100, 1729, 1624, 1525, 1398, 1349, 1237, 1117 cm^{-1}; ^{1}H-NMR (DMSO-d$_6$) δ 4.75 (d, 2H, J = 4.1 Hz), 6.50 (d, 1H, J = 9.6 Hz), 7.23 (d, 1H, J = 8.4 Hz), 7.30 (s, 1H), 7.37 (d, 1H, J = 9.7 Hz), 7.78 (d, 1H, J = 8.5 Hz), 8.07 (d, 1H, J = 9.3 Hz), 8.29 (d, 1H, J = 9.6 Hz), 8.87 (s, 1H), 9.10 (m, 1H); mass spectrum (FAB$^+$) 386 (MH$^+$). Anal. Calcd. for C$_{17}$H$_{11}$N$_3$O$_8$: C, 52.99; H, 2.86; N, 3.64. Found: C, 52.81; H, 2.92; N, 3.50.

N-DNP-aminovaleric acid, 7-hydroxycoumarin ester *18*

A mixture of *N*-DNP-aminovaleric acid (566 mg, 2.0 mmoles) (Sigma) and 7-hydroxycoumarin (324 mg, 2.0 mmoles) in methylene chloride (20 ml) was cooled to 0°C. Dicyclohexylcarbodiimide (454 mg, 2.2 mmoles) was added. The mixture was stirred under nitrogen at 0°C for 2 hr and at room temperature for 24 hr. The dicyclohexyl urea was filtered off and the filtrate was concentrated *in vacuo.* Trituration with 1:1 hexanes: ethyl acetate afforded *18* (384 mg, 0.90 mmole, 45%) as a yellow solid: mp 124–125°C; IR (KBr pellet) 3340, 3107, 2831, 1729, 1624, 1525, 1504, 1419, 1349, 1258, 1159, 1124 cm^{-1}; ^{1}H-NMR (CDCl$_3$) δ 1.60 (m, 2H), 1.95 (m, 2H), 2.73 (t, 2H, J = 6.7 Hz), 3.50 (t, 2H, J = 6.5 Hz), 6.41 (d, 1H, J = 9.5 Hz), 6.94 (d, 1H, J = 9.5 Hz), 7.04 (d, 1H, J

= 8.4 IIɪ), 7.26 (s, 1H), 7.50 (d, 1H, J = 8.4 IIɪ), 7.70 (d, 1H, J = 9.5 IIɪ), 8.29 (d, 1H, J = 9.5 Hz), 8.59 (m, 1H), 9.15 (s, 1H); mass spectrum (FAB+) 428 (MH+). Anal. Calcd. for $C_{20}H_{17}N_3O_8$: C, 56.21; H, 3.98; N, 9.84. Found: C, 55.88; H, 3.88; N, 9.50.

N-DNP-aminocaproic acid, 7-hydroxycoumarin ester 19

This compound was prepared as described for 18, starting with N-DNP-aminocaproic acid (Sigma), to give 19 (36%) as a yellow solid: mp 110–112°C; IR (KBr pellet) 3339, 3107, 2931, 1735, 1624, 1525, 1426, 1335, 1278, 1117 cm^{-1}; ^{1}H-NMR (CDCl$_3$) δ 1.60 (m, 2H), 1.85 (m, 4H), 2.67 (t, 2H, J = 7.2 Hz), 3.48 (t, 2H, J = 6.4 Hz), 6.41 (d, 1H, J = 9.5 Hz), 6.94 (d, 1H, J = 9.5 Hz), 7.04 (d, 1H, J = 8.4 Hz), 7.26 (s, 1H), 7.50 (d, 1H, J = 8.4 Hz), 7.70 (d, 1H, J = 9.5 Hz), 8.29 (d, 1H, J = 9.5 Hz), 8.59 (m, 1H), 9.15 (s, 1H); mass spectrum (FAB+) 442 (MH+). Anal. Calcd. for $C_{21}H_{19}N_3O_8$: C, 57.14; H, 4.31; N, 9.52. Found: C, 56.99; H, 4.40; N, 9.36.

In general, the rate of the antibody-accelerated ester cleavage by thiolated Fab labeled with 2 was lower with shorter or longer ester substrates (Table II), presumably due to steric constraints in the binding site with the shorter substrates and higher entropic barriers with the longer esters. One exception to this trend is the relatively high k_{cat} for the ester of N-DNP-glycine, 17 (1.4 min^{-1}). However, this ester also has a high rate of hydrolysis in the absence of antibody, and the factor of acceleration by the antibody (vs. equimolar dithiothreitol) is 1.0×10^4, sixfold lower than that for ester 14.

We also investigated the enantioselectivity of the ester cleavage reaction, using coumarin esters of D and L N-DNP-valine and of N-DNP-alanine. Since the thiolated MOPC 315 Fab contains a chiral binding cavity, it is conceivable that it could selectively bind and cleave one enantiomer of the ester more efficiently than the other. These ester substrates were synthesized as described for esters 18 and 19, starting with the N-DNP-amino acids. With N-DNP-valine 7-hydroxy-coumarin ester, the antibody-accelerated cleavage reaction was much slower than that observed for ester 14, with no measurable difference in rates between the D and L enantiomers. The cleavage of N-DNP-alanine 7-hydroxycoumarin ester by the thiolated antibody was also not significantly enantioselective. The kinetic constants k_{cat} and K_m for the reaction were 1.54 min^{-1} and 2.05 μM, respectively, for the D enantiomer, and 1.44 min^{-1} and 1.60 μM for the L enantiomer. The combining site of MOPC 315 is probably sufficiently open to accommodate both enantiomers in a nonselective fashion. However, the generation of a semi-synthetic antibody from a monoclonal antibody generated specifically to a chiral substrate should afford an enantioselective esterase.

6. Derivatization with Imidazole

The question arises as to whether the combining site thiol, in addition to acting as a nucleophile for ester thiolysis, can be derivatized with various catalytic groups to generate a new class of semisynthetic catalytic antibodies. In order to demonstrate the feasibility of this method, an imidazole was introduced into the binding site of MOPC 315 Fab (Fig. 3) (Pollack and Schultz, 1988). An imidazole can act as a general base or as a nucleophile in ester hydrolysis. In order to derivatize the thiolated antibody with an imidazole, the thiopyridyl disulfide adduct of the antibody (Pollack *et al.*, 1988) (2.5 mg, 50 nmoles) was treated with 4-mercaptomethylimidazole (Street *et al.*, 1985) (0.5 μmole) in 10 mM sodium phosphate, 150 mM NaCl, pH 7.4 (PBS) at 20°C. Incorporation of imidazole was assayed by monitoring thiopyridone release spectrophotometrically at 343 nm ($\epsilon = 7060 \text{ M}^{-1}\text{cm}^{-1}$) (Grassetti and Murray, 1967). After 30 min, the reaction was complete, with quantitative release of thiopyridone. The mixture was dialyzed exhaustively against 0.01 M sodium phosphate, 0.15 M NaCl, pH 7.4 (PBS) and once against assay buffer. To verify imidazole incorporation, a sample of the imidazole–antibody adduct (1.0 mg) was reduced with 20 mM dithiothreitol and subjected to C18 analytical reverse-phase high-performance liquid chromatography. A linear gradient was used from 0–20% acetonitrile in aqueous 0.1 M triethylammonium acetate, pH 7.5 over 20 min, monitoring absorbance at 230 nm. A peak in the elution profile was observed with identical retention time (4.6 min) to that for an authentic sample of 4-mercaptomethylimidazole.

Ester Hydrolysis Assays with Imidazole-Containing Antibody

In order to determine the catalytic activity of the semisynthetic antibody in ester hydrolysis, the same DNP-containing coumarin esters described earlier were used as substrates. The hydrolysis of coumarin esters *14* and *17–19* (Fig. 7) by the semisynthetic antibody was assayed in the presence and absence of 1 mM derivatized Fab at varying pH, at 30°C. Morpholineethanesulfonic acid (100

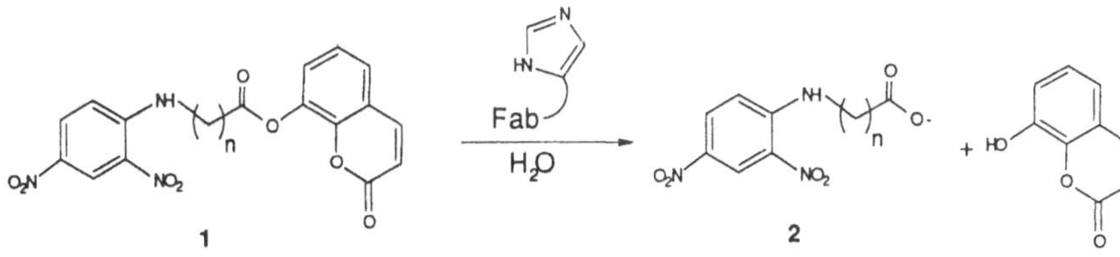

	n	k_{cat} (min^{-1})
1a:	1	0.008
1b:	2	0.051
1c:	4	0.011
1d:	5	0.000

Figure 7. Hydrolysis reaction by the imidazole–antibody adduct.

mM) was used as a buffer in the range of pH 5–7. Sodium phosphate (100 mM) was used in the range of pH 6–8 and tris-HCl (100 mM) was used in the range of pH 7–9. The release of free coumarin was quantitated fluorometrically, exciting at 355 nm and measuring emission at 455 nm. Antibody rates were corrected by subtracting the rate of cleavage in the absence of antibody. From an Eadie–Hofstee plot of initial rate data, the kinetic constants, k_{cat} and K_m, for the hydrolysis of ester *14* ($n=2$) were determined to be 0.051 ± 0.005 min^{-1} and 1.9 ± 0.2 µM, respectively (pH 7.0). Multiple turnovers were observed. The hydrolysis reaction is competitively inhibited by *N*-DNP-glycine with a K_i of 4 ± 1 µM (pH 7.0). This value is almost identical to the K_D (5 µM) for the binding of *N*-DNP-glycine to underivatized MOPC 315 (determined by fluorescence quenching in PBS) (Haselkorn *et al.*, 1974). The catalytic activity of the imidazole-derivatized antibody shows a pH dependence consistent with a titratable residue with pKa 7.7 ± 0.2 (Fig. 8), which is similar to the pKa of 4-methylimidazole, 7.5 (Bruice and Schmir, 1958). At pH > 8, the unmodified Fab slightly accelerates the cleavage of ester *14* in a stoichiometric reaction, presumably involving modification of an amino acid side chain near the binding site. However, this reaction is not observed in the catalysis by the imidazole-derivatized antibody, where the reactive antibody side chain is probably inaccessible. The catalytic activity of the semisynthetic antibody is destroyed by treatment with diethylpyrocarbonate, an imidazole-specific reagent (Holbrook and Ingram, 1973). These data are consistent with the presence of a catalytic imidazole acting either as a general base or directly as a nucleophile in the hydrolysis of ester *14*. We are currently carrying out further studies to determine

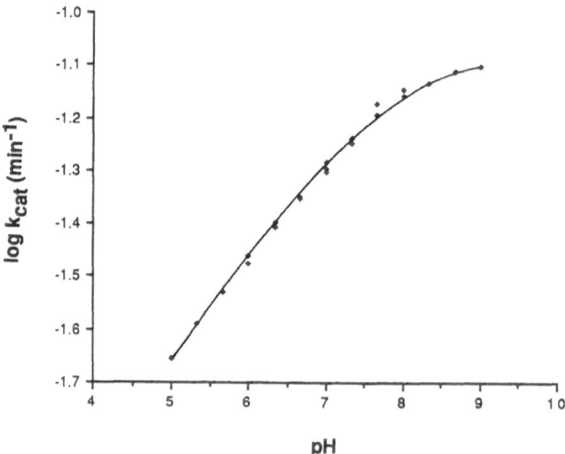

Figure 8. Dependence of log (k_{cat}) for the antibody-catalyzed reaction on pH. These experiments were carried out at 30°C in the presence of 20 µM ester *14*.

the mechanism of the imidazole-assisted catalysis. The rate of the antibody-catalyzed reaction decreased when shorter or longer esters were used as substrates (Fig. 7), again probably due to steric constraints in the binding site with the shorter substrates and higher entropic barriers with the longer esters. The second-order rate constant for the antibody-catalyzed reaction, k_{cat}/K_m, was compared to the rate constant for the reaction catalyzed by 4-methylimidazole at pH 7. The ratio $[k_{cat}/K_m]/k_{4\text{-methylimidazole}}$ gives an acceleration factor of $1.3 \pm 0.1 \times 10^3$ for hydrolysis by the antibody.

7. Derivatization with a Spectroscopic Probe

In addition to catalytic groups, spectroscopic probes are potentially useful ligands to place in the binding sites of antibodies or other proteins. The introduction of fluorophores, redox-active metals, or other reporter molecules into antibody combining sites should provide antibodies with built-in sensors for bound ligand. Such antibodies may find applications as chemical sensors and diagnostics. In order to demonstrate that the thiolated Fab can be derivatized with a reporter molecule, *N*-fluoresceinthioureido-2-mercaptoethylamine (Zuckermann *et al.*, 1987) (Fig. 3) was selectively introduced via a disulfide exchange reaction. The *S*-thiopyridyl Fab labeled with 2 (9.0 nmole in 0.50 ml of 0.1 M sodium phosphate, pH 7.3) was treated with *N*-fluoresceinthioureido-2-mercaptoethylamine (Zuckermann *et al.*, 1987) (52 nmoles in 0.50 ml of 0.1 M triethylammonium acetate, pH 7.5, containing 10% acetonitrile) at 23°C. The reaction was monitored by release of thiopyridone at 343 nm (Grassetti and Murray, 1967). After 3 hr, no additional thiopyridone was released and the mixture was dialyzed exhaustively against assay buffer (50 mM NaCl, 50 mM sodium phosphate, pH 7.0). The resulting adduct was isolated in 84% overall yield, with greater than 90% incorporation of the fluorophore.

Fluorescence Quenching Binding Assay

Addition of the ligand *N*-DNP-glycine to the fluorescein–Fab adduct resulted in a decrease in fluorescence, providing a direct assay of ligand binding (Fig. 9). Fluorescence quenching experiments were carried out at 10°C using 492 nm for excitation and measuring emission at 521 nm. The fluorescein–Fab adduct was diluted with assay buffer (50 mM NaCl, 50 mM sodium phosphate, pH 7.0) to 0.10 μM. Aliquots of *N*-DNP-glycine were added and, after mixing, the fluorescence was observed. The association constant of *N*-DNP-glycine to the fluorescein-Fab adduct was determined to be 3.0×10^5 M^{-1}, almost identical to the K_A of *N*-DNP-glycine to underivatized MOPC 315, 2.0×10^5 M^{-1} (Haselkorn *et al.*, 1974). The addition of *N*-DNP-glycine to a solution of one

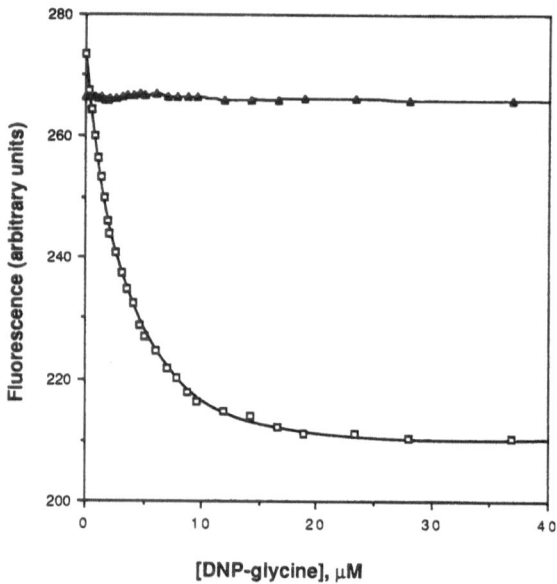

[DNP-glycine], µM

Figure 9. Fluorescence quenching binding assay for *N*-DNP-glycine with the fluorescein–Fab adduct (squares) vs. control with fluorescein + underivatized Fab (triangles).

equivalent free *N*-fluoresceinthioureido-2-mercaptoethylamine and underivatized Fab (each at 0.10 µM) resulted in no detectable fluorescence change (Fig. 9).

8. Prospects

We have shown that a nucleophilic thiol can be site-specifically introduced into the combining site of the antibody MOPC 315, and that the resulting semisynthetic antibody accelerated ester thiolysis by a factor of 6×10^4. Further-

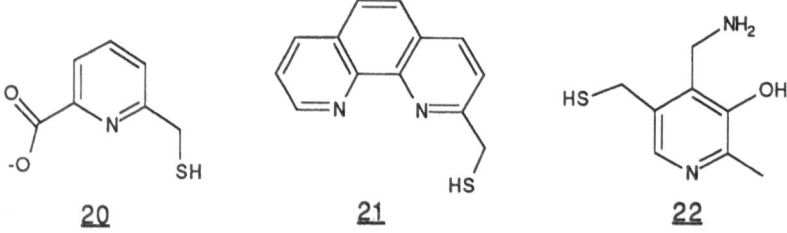

Figure 10. Catalytic groups currently being studied for semisynthetic antibody design.

more, the thiol was selectively modified with an imidazole which acted as a catalytic group in the hydrolysis of esters. The thiol was also derivatized with a fluorescent group, providing a direct spectroscopic assay for antibody–ligand complexation. Using the same chemical strategy, it should be possible to introduce additional functionalities, such as enzymatic cofactors, metal–ligand complexes, redox-active compounds, or drugs, into antibody combining sites. Some of the functionalities we are currently investigating are shown in Fig. 10. The application of this same strategy to antibodies specific for peptides, oligosaccharides, or nucleic acids could lead to the generation of new catalysts for the modification of biological macromolecules.

References

Bruice, T. C., 1959, Imidazole catalysis. The intramolecular nucleophilic catalysis of the hydrolysis of an acyl thiol, *J. Am. Chem. Soc.* **81**:5444–5449.

Bruice, T. C., and Schmir, G. L., 1958, Imidazole catalysis. The reaction of substituted imidazoles with phenyl acetates in aqueous solution, *J. Am. Chem. Soc.* **80**:148–156.

Chan, E. Y., Crampton, R. G., and Clapp, L. B., 1979, Benzothiazepinones, related compounds, and the Smiles rearrangement, *Phosphorous Sulfur* **7**:41–45.

Cochran, A., Sugasawara, R., and Schultz, P. G., 1988, Photosensitized cleavage of a thymine dimer by an antibody, *J. Am. Chem. Soc.* **110**:7888–7890.

Corey, D. R., and Schultz, P. G., 1987, Generation of a hybrid sequence-specific single-stranded deoxyribonuclease, *Science* **238**:1401–1403.

Durfor, C. N., Bolin, R. J., Sugasawara, R., Massey, R. J., Jacobs, J. W., and Schultz, P. G., 1988, Antibody catalysis in reverse micelles, *J. Am. Chem. Soc.* **110**:8713–8714.

Givol, D., and Wilchek, M., 1977, Affinity labeling of antibody combining sites as illustrated by anti-dinitrophenyl antibodies, *Methods Enzymol.* **46**:479–492.

Givol, D., Strausbauch, P. H., Hurwitz, E., Wilchek, M., Haimovich, J., and Eisen, H. N., 1971, Affinity labeling and cross-linking of the heavy and light chains of a myeloma protein with anti-2,4-dinitrophenyl activity, *Biochemistry* **10**:3461–3466.

Goetzl, E. J., and Metzger, H., 1970, Affinity labeling of a mouse myeloma protein which binds nitrophenyl ligands, *Biochemistry* **9**: 1267–1278.

Goodman, J. W., 1975, in: *The Antigens* (M. Sela, ed.), Academic Press, New York, pp. 127–187.

Grassetti, D. R., and Murray, J. S., 1967, Determination of sulfhydryl groups with 2,2'- or 4,4'-dithiodipyridine, *Arch. Biochem. Biophys.* **119**:41–49.

Haimovich, J., Eisen, H. N., Hurwitz, E., and Givol, D., 1972, Localization of affinity-labeled residues on the heavy and light chain of two myeloma proteins with anti-hapten activity, *Biochemistry* **11**:2389–2398.

Haselkorn, D., Friedman, S., Givol, D., and Pecht, I., 1974, Kinetic mapping of the antibody combining site by chemical relaxation spectrometry, *Biochemistry* **13**:2210–2222.

Hilvert, D., Carpenter, S. H., Nared, K. D., and Auditor, M. M., 1988, Catalysis of concerted reactions by antibodies: The Claisen rearrangement, *Proc. Natl. Acad. Sci. U.S.A.* **85**:4953–4955.

Holbrook, J. J., and Ingram, V. A., 1973, Ionic properties of an essential histidine residue in pig heart lactate dehydrogenase, *Biochem. J.* **131**:729–738.

Jackson, D. Y., Jacobs, J. W., Sugasawara, R., Reich, S. H., Bartlett, P. A., and Schultz, P. G., 1988, An antibody-catalyzed Claisen rearrangement, *J. Am. Chem. Soc.* **110**:4841–4842.

Jacobs, J. W., Schultz, P. G., Sugasawara, R., and Powell, M., 1987, Catalytic antibodies, *J. Am. Chem. Soc.* **109**:2174–2176.

Janda, K. D., Lerner, R. A., and Tramontano, A., 1988, Antibody catalysis of bimolecular amide formation, *J. Am. Chem. Soc.* **110**:4835–4837.

Kaiser, E. T., and Lawrence, D. S., 1984, Chemical mutation of enzyme active sites, *Science* **226**:505–511.

Kohler, G., and Milstein, C., 1975, Continuous cultures of fused cells secreting antibody of pre-defined specificity, *Nature* **256**:495–497.

Landsteiner, K., and van der Scheer, J., 1934, Serological studies on azoproteins, *J. Exp. Med.* **59**:751–768.

Milewich, L., Sanchez, C. G., MacDonald, P. C., and Siiteri, P. K., 1975, Radioimmunoassay of androstenedione: The steroid molecule as a probe for antibody specificity, *J. Steroid Biochem.* **6**:1381–1387.

Napper, A. D., Benkovic, S. J., Tramontano, A., and Lerner, R. A., 1987, A stereospecific cyclization catalyzed by an antibody, *Science* **237**:1041–1043.

Nisonoff, A., Hopper, J., and Spring, S., 1975, *The Antibody Molecule*, Academic Press, New York.

Pollack, S. J., and Schultz, P. G., 1987, Antibody catalysis by transition state stabilization, *Cold Spring Harbor Symp. Quant. Biol.* **52**:97–104.

Pollack, S. J., and Schultz, P. G., 1989, A semisynthetic catalytic antibody, *J. Am. Chem. Soc.* **111**:1929–1931.

Pollack, S. J., Jacobs, J. W., and Schultz, P. G., 1986, Selective chemical catalysis by an antibody, *Science* **234**:1570–1573.

Pollack, S. J., Nakayama, G. R., and Schultz, P. G., 1988, Introduction of nucleophiles and spectroscopic probes into antibody combining sites, *Science* **242**:1038–1041.

Pressman, D., and Grossberg, A., 1968, *The Structural Basis of Antibody Specificity*, Benjamin, New York.

Raso, V., and Stollar, B. D., 1975, The antibody–enzyme analogy. Characterization of antibodies to phosphopyridoxyltyrosine derivatives, *Biochemistry* **14**:584–591.

Seiler, F. R., Gronski, P., Kurrle, R., Luben, G., Harthus, H., Ax, W., Bosslet, K., and Schwick, H., 1985, Monoclonal antibodies: their chemistry, functions, and possible uses, *Angew. Chem. Int. Ed. Engl.* **24**:139–226.

Shokat, K. M., Leumann, C. H., Sugasawara, R., and Schultz, P. G., 1988, An antibody-mediated redox reaction, *Angew. Chem. Int. Ed. Eng.* **27**:1172–1174.

Strausbauch, P. H., Weinstein, Y., Wilchek, M., Shaltiel, S., and Givol, D., 1971, A homologous series of affinity labeling reagents and their use in the study of antibody binding sites, *Biochemistry* **10**:4342–4348.

Street, J. P., Skorey, K. I., Brown, R. S., and Ball, R. G., 1985, Biomimetic models for cysteine proteases, *J. Am. Chem. Soc.* **107**:7669–7679.

Tramontano, A., Janda, K. D., and Lerner, R. A., 1986, Catalytic antibodies, *Science* **234**:1566–1570.

Zuckermann, R. N., Corey, D. R., and Schultz, P. G., 1987, Efficient methods for attachment of thiol specific probes to the 3′-ends of synthetic oligodeoxyribonucleotides, *Nucleic Acids Res.* **15**:5305–5321.

Zuckermann, R. N., Corey, D. R., and Schultz, P. G., 1988, Site-selective cleavage of RNA by a hybrid enzyme, *J. Am. Chem. Soc.* **110**:1614–1615.

VIII

*STRATEGIES FOR DIRECTED
MUTAGENESIS OF PROTEINS*

Engineering Novel, Prolonged-Acting Insulins

JAN MARKUSSEN

1. Introduction

1.1. Background and Aims

Insulin-dependent diabetic patients need a constant supply of basal insulin to control blood glucose between meals, in addition to meal-related bolus insulin to counter the peak in glucose following a meal. Intensive basal/bolus insulin regimens have been introduced in order to improve blood glucose control, particularly in Scandinavia and England. The long-acting insulins used to provide the basal supply come in the form of neutral, crystalline suspensions, obtained by crystallization using either Zn^{2+} (Lente-type preparations) or protamine (NPH-type preparations). These suspensions require thorough shaking prior to injection in order to ensure homogeneity of the suspension and correct dosage. Furthermore, the inter- and intra-patient absorption of insulin suspensions is extremely variable: the $T_{50\%}$ (i.e., the time taken for 50% of the insulin to disappear from the injection site) ranges from 8 to 72 hr (Lauritzen et al., 1979). Whereas the homogeneity problem of suspensions could be overcome if a soluble, prolonged-acting insulin was made available, the variability in absorption caused by the physiological conditions, like blood flow at the site of injection, could hardly be totally eliminated by applying new principles of prolongation.

A further problem with insulin preparations is the chemical stability. Depending on the pharmaceutical formulation, reactions such as deamidation, dimerization, and, in the case of protamine-containing preparations, coupling of insulin to protamine take place (Brange et al., 1987).

JAN MARKUSSEN • Novo Research Institute, Novo Allé, DK-2880 Bagsvaerd, Denmark.

In summary, it would be highly desirable to have a soluble, prolonged-acting insulin with a more reproducible absorption following first-order absorption kinetics, a half-life of 24–48 hr independent of dose, and improved chemical stability.

1.2. Rationale of Design and Mode of Action of Prolonged-Acting, Soluble Insulins

Prolonged-acting, soluble insulin preparations such as globin insulin, protamine-zinc insulin (PZI), and surfen insulin have been used in the past. (Surfen is 1,3-bis(4-amino-2-methyl-6-quinolyl.) The principle was that after injecting an acid, unbuffered solution of insulin, and the basic compound, a complex forms that is sparingly soluble at neutral reaction. One reason for the abandonment of these preparations was that their duration of action was too short (see Schlichtkrull *et al.*, 1975).

Another principle would be to alter the acidic insulin, isoelectric pH (pI) ~ 5.4, to a neutral molecule by adding charges. It can be inferred from Fig. 1 that adding one, two, and three charges will increase pI to 6.2, 6.8, and 7.7, respectively. Charges can be added by substituting acidic amino acids with neutral residues, by blocking α-COOH groups, by substituting neutral residues with lysine or arginine, and by adding basic groups to the termini of the chains. There

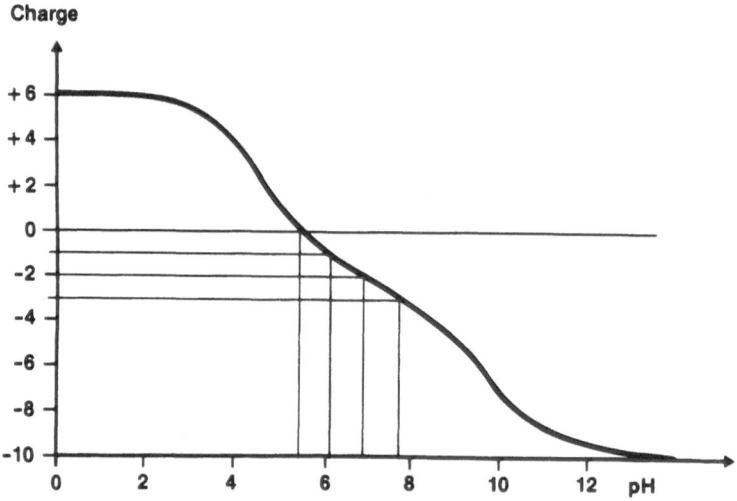

Figure 1. The net charge of insulin (human, porcine, and bovine) as a function of pH. The isoelectric pH shifts from 5.4 to 6.2, 6.8, and 7.7 by addition of one, two, and three charges, respectively. (After Markussen *et al.*, 1987a, with permission.)

are few limitations to the type of substitutions that can be made by combining genetic engineering and semisynthesis (Markussen *et al.*, 1987a,b, 1988a,b). However, it was found that the mere displacement of pI to ~7 and isoelectric precipitation will not necessarily result in significantly prolonged absorption; only insulins that crystallize instantly as pH approaches neutrality showed sufficiently prolonged absorption to be potentially useful in the clinic.

The unit cell of insulin crystals is a hexamer, in both the 2 and 4 Zn^{2+}/hexamer structures (Hodgkin *et al.*, 1983). An enhanced crystallizability could result either from substitutions that stabilize the hexamer or from substitutions that contribute to hexamer–hexamer interactions. Modeling on the basis of the 2 Zn^{2+}/hexamer enabled rational design of a stabilized hexamer. Substitutions located on the surface of the hexamer that enhanced the prolonged action (i.e., promoted the rate of crystallization) were found by trial and error; backtrack rationalization by modeling failed to yield obvious explanations of the findings. Different crystal forms were found among insulins with marginal differences in primary structures, which obviously reflect small differences in energy between various stackings of insulin hexamers. Hence, prediction of the crystallization behavior of insulin analogs based on the known structures is not possible at present.

2. Synthesis of Insulin Analogs

2.1. The Tryptic Transpeptidation Reaction

The tryptic transpeptidation reaction in organic-aqueous media was first developed to convert porcine insulin to an ester of human insulin (Markussen, 1982), exploiting the fact that the penultimate residue of the B-chain, B29, is a lysine (i):

$$\text{(i)} \quad \text{insulin(Ala}^{B30}) + \text{Thr-OR} \xrightarrow{\text{trypsin}} \text{insulin(Thr}^{B30}\text{-OR)} + \text{Ala}$$
$$\qquad\qquad \text{porcine insulin} \qquad\qquad\qquad \text{human insulin ester}$$

The formation of by-products due to a reaction at Arg^{B22} can be suppressed by carrying out the reaction at low concentrations of water and in the presence of acid (Markussen, 1987). The reaction renders itself to synthesis of B30-substituted insulins (Markussen *et al.*, 1987a, 1988b), but it also catalyzes the formation of an intramolecular peptide bond from Lys^{B29} to Gly^{A1} (see Fig. 2 Reaction ii), leading to single-chain des-(B30) insulin (SCI) (Markussen *et al.*, 1985). SCI can in turn be transpeptidized in the presence of an amino component, X, leading to B30-substituted insulins (see Fig. 2, Reaction iii). If X is a threonine ester, a human insulin ester is obtained (Markussen, 1987). Since the

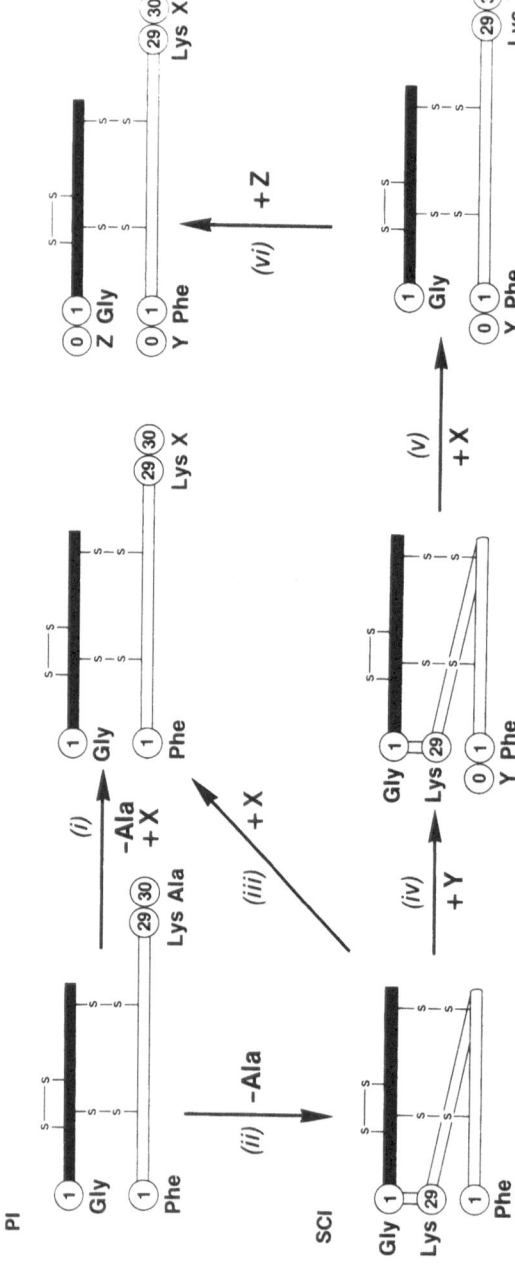

Figure 2. Trypsin-catalyzed reactions. Transpeptidation of porcine insulin (PI) in the presence of an amino component (X) leads to B30-substituted insulins [Reaction (i)]. Transpeptidation without addition of an amino component leads to the single-chain des-(B30) insulin (SCI) by forming a peptide bond from Lys^{B29} to Gly^{A2} [Reaction (ii)]. SCI can be obtained by fermentation of plasmid-transformed yeast and found in the broth. Transpeptidation of SCI in the presence of amino components (X) leads to B30-substituted insulins [Reaction (iii)]. The amino group of B1 may be acylated by an activated carboxyl component (Y) [Reaction (iv)], followed by transpeptidation in the presence of an amino component (X), leading to B0- and B30-substituted insulins [Reaction (v)]. Finally, the amino group of Gly^{A1} may be selectively acylated by an activated carboxyl component (Z), leading to A0, B0, and B30-subsituted insulins [Reaction (vi)]. Solid bars represent the A-chain, residues (2–21); open bars represent the B-chain, residues (2–28).

amino group of Gly^{A1} is blocked in SCI, that of the Phe^{B1} can be selectively acylated yielding B-chain elongation derivatives (see Fig. 2, Reaction iv). Trans-peptidation in the presence of an amino component (X) eventually leads to B0 and B30-substituted insulins (see Fig. 2, Reaction v) (Markussen *et al.*, 1988b). If Y represents an *N*-protected amino acid, the amino group of Gly^{A1} may be acylated yielding A0-, B0-, and B30-substituted insulins (see Reaction vi).

2.2. Single-Chain Des-(B30) Insulins

The SCI molecule is biologically inactive despite the fact that its conforma-tion, as determined by X-ray crystallography, is identical to that of porcine insulin with the exception of the two connected residues, Lys^{B29} and Gly^{A1}. However, it has two properties that render it a useful precursor of insulin in biosynthesis in yeast. First, it is devoid of pairs of adjacent, basic amino acids as found in proinsulin, nature's own single-chain precursor. Attempts to express human proinsulin in yeast resulted mostly in the production of a cleavage prod-uct, C-peptide, as yeast possesses a protease with specificity for pairs of adjacent basic amino acids (Thim *et al.*, 1986). Low levels of insulin indicated that cleavage had taken place before disulfide bond formation. Second, it was demon-strated that SCI could be recovered in better yields than proinsulin after having been subjected to a reduction of the disulfide bonds followed by oxidation of the thus generated cysteine residues (Markussen, 1985), suggesting that the SCI correctly folds up more efficiently than porcine proinsulin. As correct and expe-dient folding must be an advantage in the biosynthesis of disulfide-containing proteins, SCI appeared to be a useful candidate for biosynthetic insulin pre-cursors.

2.3. Biosynthesis in Yeast

Biosynthesis in yeast uses single-chain insulin precursors without adjacent pairs of basic amino acids. The link between the chains may be either a direct Lys^{B29}–Gly^{A1} peptide bond or a small peptide spacer inserted between these two residues, Ala–Ala–Lys or Ser–Lys (Markussen *et al.*, 1987c). A leader se-quence is used to promote export to the medium, and the membrane-bound protease with specificity for pairs of basic amino acids is exploited to cleave the leader from the precursor by introducing a Lys–Arg sequence at the joint. The processes conducted by yeast are summarized in Fig. 3. Insulin analogs sub-stituted in the B-chain sequence 1–29 and in the A-chain sequence 1–21 are made by site-specific mutagenesis in the single-chain insulin precursor gene (Markussen *et al.*, 1987b, 1988a).

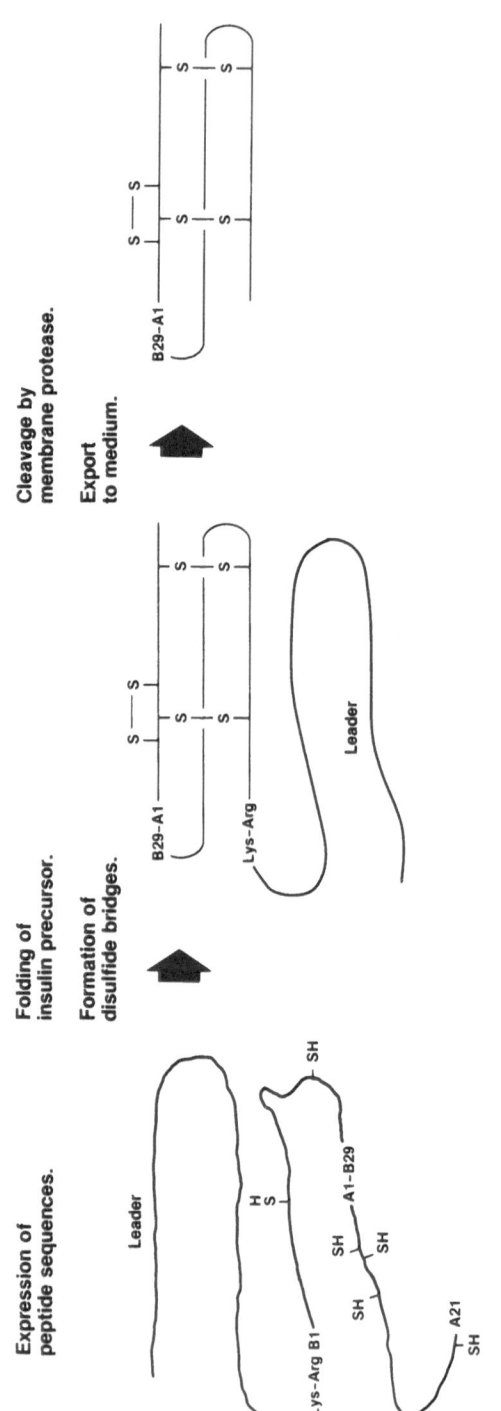

Expression of peptide sequences.

Folding of insulin precursor.

Formation of disulfide bridges.

Cleavage by membrane protease.

Export to medium.

Figure 3. The processes performed by yeast. The plasmid encodes for a sequence consisting of a leader followed by Lys-Arg-B(1-29)-A(1-21), where A and E denote the chains of insulin. The expressed sequence folds up correctly in the endoplasmic reticulum and the disulfide bridges of insulin (A6-A11, A7-B7, A20-B19) form. Subsequently, cleavage by a membrane bound protease in the Golgi apparatus at the Lys-Arg sequence releases the single-chain des-(B30) insulin from the leader. Finally, export to the medium occurs, probably in Golgi-derived vesicles.

2.4. Substitutions that Increase pI

An increase in charge can be achieved by replacing an acidic amino acid residue with a neutral one. Insulin features four glutamic acid residues in positions A4, A17, B13, and B21. Substitution with glutamine has currently been carried out in the latter three (Markussen *et al.*, 1987b, 1988a). Furthermore, the C-terminal carboxyl group of the B-chain can be permanently blocked if the amino component of the transpeptidation reaction (X in Fig. 2) is an amino acid amide. Presently, the only way to selectively block the C-terminal carboxyl group of the A-chain would be to add a glycine in position A22 by genetic engineering and then digest the biosynthetic single-chain precursor with the C-terminal-glycine dehydrogenase (Bradbury *et al.*, 1982).

Addition of charge can also be achieved by substituting a neutral amino acid residue by a basic one. However, there are limits to how much this strategy can be exploited using the present route of biosynthesis and semisynthesis. Trypsin would be expected to cleave the single-chain insulin precursor in the transpeptidation reaction if lysine or arginine residues are inserted in accessible positions. The presence of arginine in position B22 causes by-product formation in the tryptic transpeptidation of porcine insulin and single-chain precursors. However, in this position by-product formation can be suppressed by selecting proper conditions of reaction. *A priori*, the only safe position for basic amino acid substitution is position B27, because B28 is a proline residue (Markussen *et al.*, 1987b). Positioning a basic residue C-terminally will cause, not cleavage, but coupling of the amino component in the reaction mixture to the basic residue. This was seen both in the A-chain, where Thr-NH_2 coupled to Arg^{A21} (Markussen *et al.*, 1988a), and in the B-chain, where Arg-NH_2 polymerized to Lys^{B29} (Markussen *et al.*, 1988b). Extending the B-chain N-terminally by basic residues is possible via reactions (iv) plus (v) of Fig. 2. Thus $Lys^{B(-1)}$-$Lys^{B(0)}$ human insulin was synthesized using Boc-Lys(Boc)-Lys(Boc)-OSu as acylating reagent in reaction (iv) of Fig. 2 (Markussen *et al.*, 1987a). Finally, the A-chain of insulin can be selectively extended by acylation of N-protected B-chain [reaction (vi) of Fig. 2]. Thus Z may represent Boc-Arg-OH and Y may be a Boc protecting group. This route of synthesis has not yet been exploited in the search for prolonged-acting insulins, because A1-substituted insulins feature a reduced biological potency (Gliemann and Gammeltoft, 1974).

2.5. Substitutions that Stabilize Insulin

The chemical reactions known to occur in pharmaceutical preparations are deamidations of asparagine and glutamine residues and dimerizations between an N-terminal amino group of one molecule and a carboxamide function of another (Brange *et al.*, 1987). At neutral conditions these reactions may occur via succinimide (Geiger and Clarke, 1987) and glutarimide intermediates, respectively.

It was found that an internal asparagine, Asn^{B3}, was most prone to deamidate (Brange *et al.*, 1987). In acid solution imide formation is suppressed, but de-amidation of Asn^{A21} and formation of A21 β-aspartyl bonded dimers dominate (Sundby, 1962; Helbig, 1976). As the mode of action of the desired prolonged-acting insulins involves an acid solution from which the insulin crystallizes upon neutralization in the tissue, one residue to substitute would be Asn^{A21}. This advance seemed *a priori* to be prohibited by the fact that Asn^{A21} has been conserved throughout evolution, at least in all the sequenced insulins. Neverthe-less, nucleotides encoding for Gly, Ser, Thr, Asp, His, and Arg were inserted in position A21 of single-chain precursor genes (Markussen *et al.*, 1988a).

3. Biological Evaluation

3.1. Free Fat Cell Bioassay

The free fat cell (FFC) assay, in which the incorporation of tritium from tritiated glucose into lipids is measured as the insulin-sensitive parameter (Moody *et al.*, 1974), has been the most popular way of assessing the potency of insulins (Gliemann and Gammeltoft, 1974). The potency in this assay is defined as the concentration required for an analog to produce half the maximal effect, as compared to the concentration for human insulin required to produce the same effect. It was shown for a series of analogs that the FFC potency correlates closely with receptor binding affinity (Gliemann and Gammeltoft, 1974). Poten-cies over 100% in the FFC assay may be found: the classic example is chick-en/turkey insulin, in which a His^{A8} residue causes a potency of 300%. The popularity of the FFC assay is due to its simplicity and the small variance of the estimated potencies.

3.2. Mouse Blood Glucose Assay

The blood glucose (BG) lowering effect of insulins was earlier assessed by the mouse convulsion assay, but the direct measurement of BG yields higher precision with fewer animals (*European Pharmacopoeia*, 1984). Since the po-tency calculation is based on a single blood sample taken when BG is at nadir for the standard of human insulin (usually at 15 min), insulins absorbed both faster and slower will appear weaker than the standard, even if their potencies in the FFC assay are 100%. The huge range of potencies in the FFC assay is narrowed down considerably in the mouse BG (MBG) test, and values above 100% have not been published using this *in vivo* method. This potency leveling *in vivo* is best explained by the correlation between insulin receptor–mediated degradation and receptor binding. Hence, a lower receptor binding results in slower degradation and, consequently, in longer plasma half-life (Kobayashi *et al.*, 1985). Thus, once an insulin molecule is bound its effect in number of glucose molecules removed from the circulation appears to be independent of the binding constant. All insulins are then in a sense equipotent; that is, they feature full intrinsic activity.

3.3. Index of Prolongation

The prolonged absorption of insulin preparations has traditionally been assessed by the prolonged glucose lowering effect in rabbits (*British Pharmacopoeia*, 1980). Glucose profile curves for three typical insulin formulations currently in clinical use are shown in Fig. 4. Using discriminant analysis on linear expressions of the glucose values, an index of prolongation (IP) was defined and used in the screening (Markussen *et al.*, 1987a,b). The IP was scaled to a value of 0 for the short-acting Actrapid and to 100 for the long-acting Ultralente (bovine insulin). The influence of the various substitutions on the IP of solutions of insulins, viz. ΔIP, was calculated by a multiple regression analysis, in which the Zn^{2+} concentration of the solutions was included as a factor. Additivity of ΔIPs from multiple substitutions was assumed, although the calculated fit in the multiple regression analysis was far from perfect.

3.4. Absorption Kinetics in Pigs Using External γ-Counting

The disappearance from the site of injection was followed by external γ-counting. The insulin preparation contained a [mono-^{125}I] insulin tracer of the analog in question. Injections were given subcutaneously in both sides of the neck in fasted pigs (Markussen *et al.*, 1987b). Besides the influence of substitutions on the rate of absorption, galenic factors such as insulin concentration (240/600 μM), Zn^{2+} concentration, dose, and inclusion of buffers in the preparation were investigated by this technique. Clinical studies of Lente and NPH-type preparations using similar techniques had demonstrated large day-to-day variations within patients, an absorption deviating from first-order kinetics and an initially decreased absorption rate, i.e., a time lag (Binder, 1969; Lauritzen *et al.*, 1979; Lauritzen, 1985).

4. Crystallization Techniques

4.1. Micro Reaction Chamber

The mode of crystallization of insulins upon neutralization was studied by microscopy using a micro reaction chamber (see Fig. 5). The technique enabled visual examination to determine whether the initial precipitations were amor-

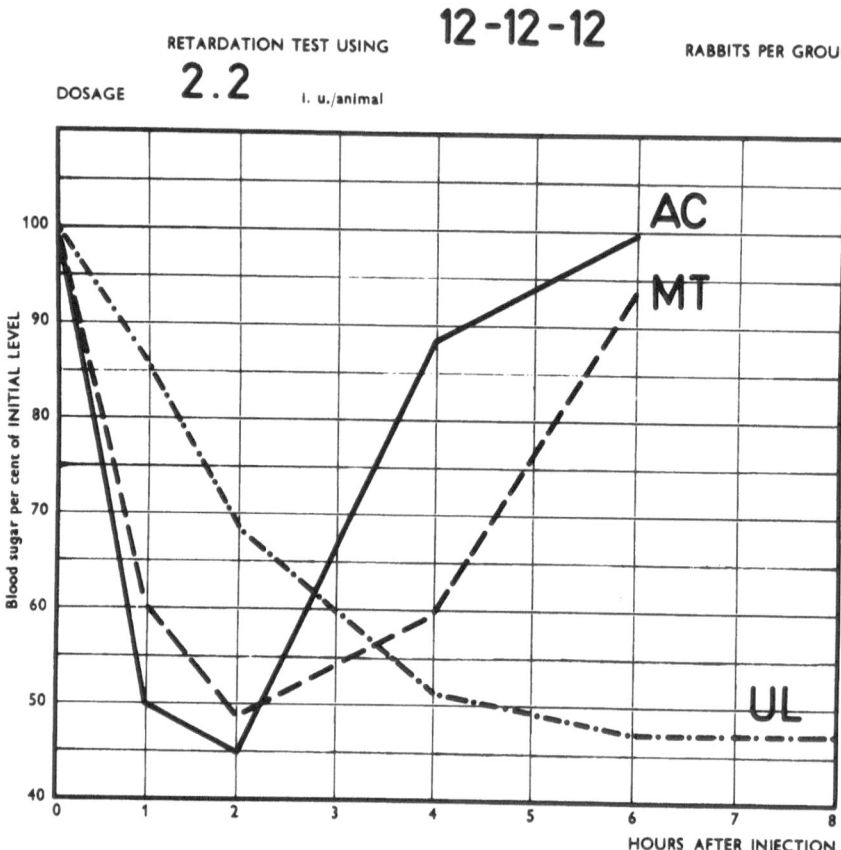

RETARDATION TEST USING **12-12-12** RABBITS PER GROUP

DOSAGE **2.2** i. u./animal

Figure 4. Glucose profile curves in the prolongation test in rabbits for a short-acting insulin formulation, AC (Actrapid, neutral solution), an intermediate-acting insulin formulation, MT (Monotard, neutral suspension of crystalline human insulin), and a long-acting formulation, UL (Ultralente, neutral suspension of bovine insulin crystals). An IP was defined using discriminant analysis on linear expressions of the glucose values and scaled to yield a value of 0 for AC and 100 for UL. MT scores an index of 42. (After Markussen *et al.*, 1987a, with permission.)

phous or crystalline. Soluble, prolonged-acting insulins instantly form 1–2 μm large crystals along the front, in vivid Brownian movement.

4.2. Hanging-Drop Crystallization

Another *in vitro* method of studying the crystallization of insulins is to make pH increase continuously in a hanging drop of 20 μl, by having a volatile acid–

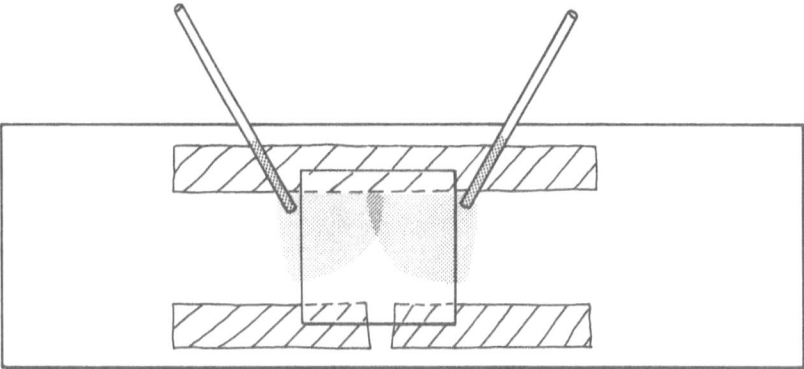

Figure 5. Micro reaction chamber for studying the formation of crystals under the microscope. The cover slip is lifted about 100 μm over the slide by tape and fixed by water-resistant tape. An acid insulin solution and a 1% NaHCO₃ solution are applied to each side by capillaries, and the solutions soak into the chamber by capillary attraction. Crystallization takes place at the front between the two solutions.

like acetic acid in the drop and a sodium hydroxide solution in the well (Markussen *et al.*, 1988b). This technique often yields large, well-shaped crystals, but correlations between prolonged-action and crystal forms remain to be established (cf. Fig. 6).

4.3. Crystallization In Vivo

In order to prove that the retarded absorption *in vivo* is due to crystal formation at the site of injection, rather than being an absorption phenomenon due to ion exchange or hydrophobic interaction, tissue sections were cut from the injection site in a pig 4 hr after injection of Arg^{B27},Thr^{B30}-NH$_2$ human insulin in a 240 μM acid solution. The sections were treated with guinea pig antiinsulin Ig followed by fluorescent labeled anti-guinea pig Ig, using an 0.1 M Tris, 1 mM Zn^{2+} pH 6.8 buffer for the obligatory washings. The sections were viewed using phase-contrast and fluorescent microscopy (see Fig. 7). The overlapping of crystalline structures in phase-contrast and enlightened spots in fluorescent light strongly verify the proposed mechanism of prolongation.

5. Results and Discussion

5.1. Substitutions that Enhance the Prolonged Action

A summary of the influence of the various substitutions is given in Table I.

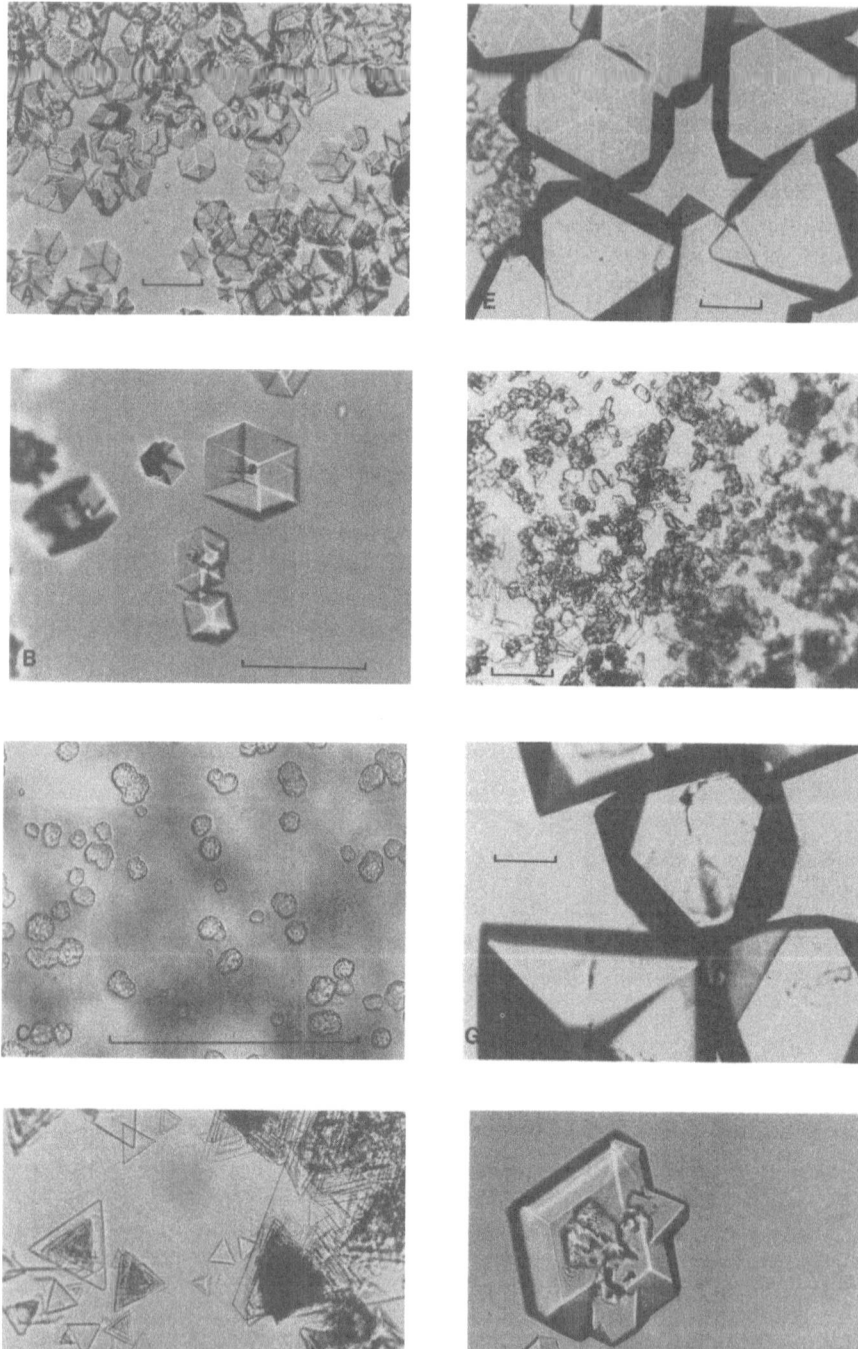

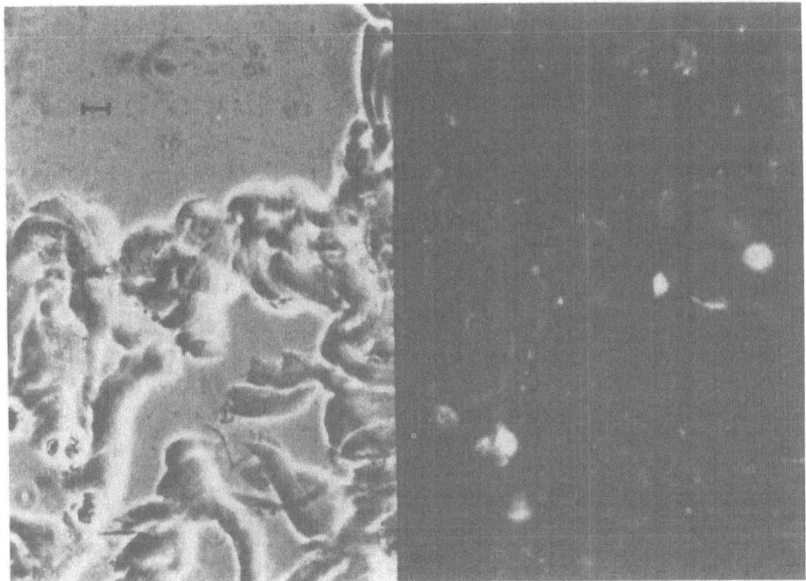

Figure 7. Tissue section taken from the site of injection in a pig 4 hr after injection of an acid solution of Arg^{B27},Thr^{B30}-NH_2 human insulin. The section is viewed in phase-contrast (left) and by fluorescent light (right). The length of the bar represents 10 μm. (Courtesy U. Ribel and C. Gotfredsen, Novo Research Institute.)

5.1.1. $Thr^{B30} \rightarrow Thr\text{-}NH_2$, $Lys\text{-}NH_2$, $Arg\text{-}NH_2$, $Lys\text{-}Lys\text{-}NH_2$, and $Arg\text{-}Arg\text{-}NH_2$

The blocking of the C-terminal of the B-chain by an amide group in the presence of excess Zn^{2+} (31 Zn^{2+}/hexamer of insulin) yielded a prolonged action with all insulins, ΔIP ranging from 21 to 43 (Markussen *et al.*, 1987a). The Arg^{B30}-NH_2 and Arg^{B30}-Arg^{B31}-NH_2 insulins were prolonged-acting with the native Zn^{2+} content, which is 3 Zn^{2+}/hexamer, with ΔIPs of 28 and 59, respectively. Substitutions in position B30 hardly influence the potency in the MBG assay; Lys^{B30}-NH_2 human insulin is 89% (100–77), with the slightly decreased value probably being due to the prolonged action. When the potency of analogs bearing multiple substitutions was resolved into contributions from

←

Figure 6. Crystals of insulins obtained by the hanging-drop method at slowly increasing pH in water. The four insulins at the left are short- or intermediate-acting, whereas the four at the right are prolonged-acting, when administered subcutaneously in an acid formulation containing 3 Zn^{2+}/hexamer. The structures deviate from human insulin as follows: A, Thr^{B30}-NH_2; B, Gly^{A21},Thr^{B30}-NH_2; D, Gln^{A17},Thr^{B30}-NH_2; C, Gln^{A17},Lys^{B30}-NH_2; E, Arg^{B27},Thr^{B30}-NH_2; F, Gly^{A21},Arg^{B27},Thr^{B30}-NH_2; G, Ser^{A21},Arg^{B27},Thr^{B30}-NH_2; H, Ser^{A21},Gln^{B21},Thr^{B30}-NH_2. The Zn^{2+} content of the insulins corresponds to 3 Zn^{2+}/hexamer. The length of the bars represents 100 μm.

Table I. Sites of Substitution in Human Insulin and Conserved Residues Important for the Transpeptidation Reaction

Site	Human insulin	Substitution	ΔCharge	ΔIP[e]	Stability	Potency factor[e] FFC	Potency factor[e] MBG
A17	Glu	Gln	+1	−2[a]		0.38	0.73
A21	Asn	Gly	0	−9[a]	+	0.67	0.78
A21	Asn	Ser	0	−11[a]	+	0.71	0.77
A21	Asn	Thr	0	−29	+	0.33	0.43
A21	Asn	Arg	+1	−31	+	0.20	0.75
A21	Asn	Asp	−1	+1[a]	+	0.73	0.88
B(−1)-B0	—	Lys-Lys	+2	−14[a]			
B13	Glu	Gln	+1	+8		0.34	0.72
B21	Glu	Gln	+1	+30		0.46	1.03
B27	Thr	Lys	+1	+48		0.99	0.81
B27	Thr	Arg	+1	+46		1.31	0.97
B30	Thr	Thr-NH$_2$	+1	+34		1.02	0.95
B30	Thr	Lys-NH$_2$	+2	+40		0.78	0.81
B30	Thr	Arg-NH$_2$	+2	+28			
B(30–31)	Thr-	Arg-Arg-NH$_2$	+3	+59			
B(30–31)	Thr-	Lys-Lys-NH$_2$	+3	+31		0.78	0.81
B22	Arg	None[b]					
B28	Pro	None[c]					
B29	Lys	None[d]					

[a]Not significantly different from 0.
[b]Arg^{B22} constrains the conditions of transpeptidation (Markussen, 1987).
[c]Pro^{B28} enables the introduction of a Lys or Arg in B27.
[d]Lys^{B29} makes up the principal site of reaction of the transpeptidations.
[e]Calculated by a multiple regression analysis from data on insulins featuring from one to four substitutions. The effect of Zn^{2+} concentration amounts to ΔIP = +2.6 for an increase of 1 Zn^{2+} per monomer of insulin.

each substitution, the potency factors of Thr^{B30}-NH$_2$ and Lys^{B30}-NH$_2$ were 1.02 and 0.78 in the FFC assay and 0.95 and 0.81 in the MBG assay, respectively (Markussen *et al.*, 1987b). The effect on the ΔIP of adding a charge by blocking the α-carboxyl group of the C-terminal appears to be greater than if the charge is added by substituting a basic amino acid without blocking the α-carboxyl group (Markussen *et al.*, 1988b).

Figure 8 shows a picture of the interface between hexamers of the 2 Zn^{2+} insulin crystal. Negative and positive protein-bound charges cancel each other in this area, and "there is a remarkably extended pattern of ionic contacts made possible by the changing of the conformations of the B-29 and B-30 side chains in the dimer molecules as well as of the initial A-chains" (Hodgkin *et al.*, 1983). Hence it is surprising that removing two negative charges from this locus increases crystallizability; that is, the packing of hexamers. It is even more surprising that the introduction of Arg-Arg-NH$_2$ in position B(30–31), which adds +6 charges, results in an extremely prolonged-acting insulin (cf. Table I).

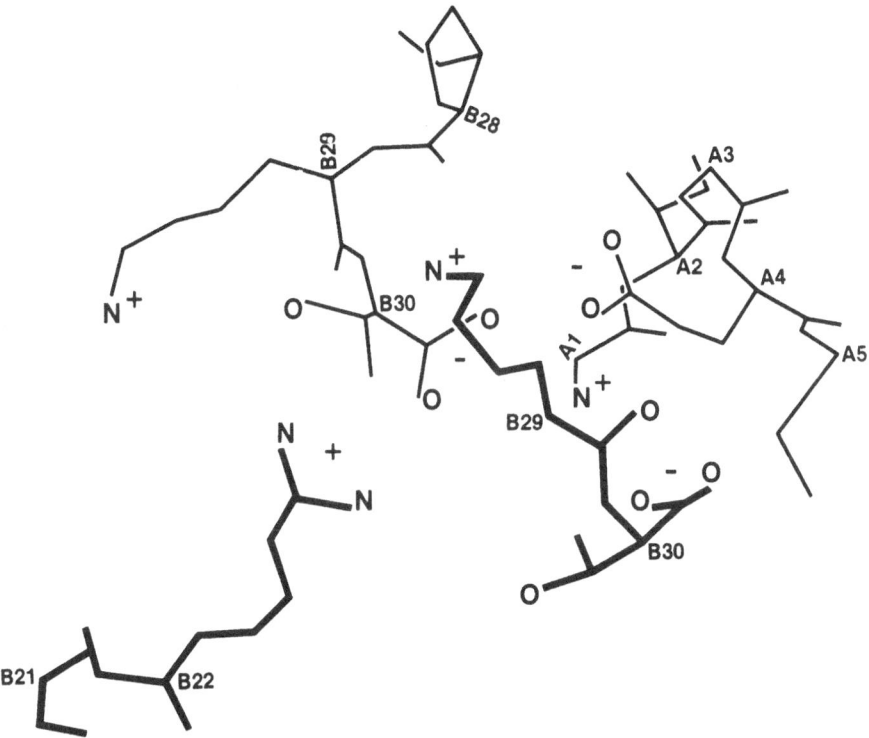

Figure 8. Molecular graphic depiction of the interface between two hexamers at the locus where each hexamer contributes a B30 residue. Thin lines represent monomer 1 of hexamer 1 and bold lines represent monomer 2 of hexamer 2, designated monomer 1 and 2 according to the Peking convention. (After Markussen *et al.*, 1987a, with permission.)

5.1.2. Thr^B27 → Lys and Arg

The most significant increase in IP by a single amino acid substitution was found with the Arg^{B27} and Lys^{B27} substitutions. The concomitant blocking of the C-terminal of the B-chain was a prerequisite for the prolonged action. An interesting feature of these two closely related amino acids is that whereas the Lys substitution does not alter the potency in the FFC test, the Arg substitution increases the potency by 31% (cf. Table I). The crystals of three Arg^{B27} insulins grown in water in a hanging drop are shown in Fig. 6. Whereas the Asn^{A21} and Ser^{A21} insulins form nonbirefringent octahedrons, the Gly^{A21} variety yields rhombic platelets.

Figure 9 presents a molecular graphic attempt to explain the role of the Arg^{B27} substitution for hexamer packing in the lattice. Whereas three of the six Arg^{B27} of a hexamer are beyond reach of another hexamer, the other three can be

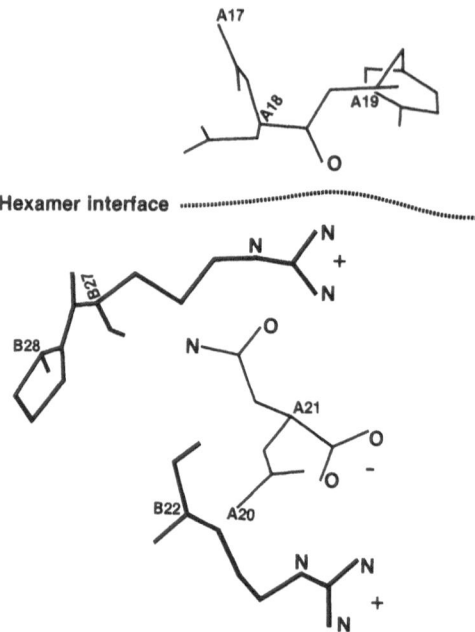

Figure 9. Molecular graphic depiction of a possible hydrogen-bond bridge between two hexamers, brought about by sandwiching the guanidino group of Arg[B27] between the β-carboxamide oxygen of Asn[A21] of molecule 2 of its own dimer and the backbone oxygen of Asn[A18] of molecule 2 of a second hexamer. (After Markussen *et al.*, 1987b, with permission.)

sandwiched by hydrogen bonds between the β-carboxamide oxygen of Asn[A21] of its own dimer and the backbone oxygen of Asn[A18] of a monomer 2 of another hexamer.

5.1.3. Glu[B13] → Gln

The six Glu[B13] residues of the insulin hexamer are located in the center of the hexamer (Blundell *et al.*, 1972). Substitution of Glu with a neutral amino acid should annihilate the electrostatic repulsion and thereby stabilize the hexamer. The Glu → Gln substitution was found to increase the IP significantly (see Table I). In the disappearance study by external γ-counting the half-life increased from 11 hr for Arg[B27],Thr[B30]-NH$_2$ human insulin to 35 hr for the analog featuring the Gln[B13] substitution in addition. Both preparations were 240 μM insulin and contained 3 Zn^{2+}/hexamer. The most prolonged-acting analog discovered so far is Gln[B13],Arg[B27],Thr[B30]-NH$_2$ human insulin. Unfortunately, the Gln[B13] substitution reduces the potency in the FFC assay.

5.1.4. $Glu^{B21} \rightarrow Gln$

The Gln^{B21},Thr^{B30}-NH_2 human insulin featured reduced potency in the FFC assay, whereas the MBG assay showed no change (see Table I). Glu^{B21} is located on the surface of the hexamer, but does not make contact with other hexamers. Nevertheless, Gln^{B21} increases IP markedly. The effects of the $Glu^{B21} \rightarrow Gln$ substitution resemble those of the $Glu^{B13} \rightarrow Gln$ substitution, despite the fact that the two residues are in opposite parts of the hexamer.

5.2. Substitutions without Significant Effect on Absorption

5.2.1. $Glu^{A17} \rightarrow Gln$

This substitution, albeit on the surface of the hexamer, failed to contribute to a prolonged absorption (see Table I). It may destabilize the hexamer, because the carboxyl groups of Glu^{A17} form intermolecular salt bridges to the amino group of N-terminal Phe^{B1} of the adjacent dimer in the hexamer. Furthermore, the potency becomes markedly reduced by the Gln^{A17} substitution (Table I).

5.2.2. B-Chain N-terminal Extension by $Lys^{B(-1)}$-Lys^{B0}

In contrast to the marked increase in IP by introduction of basic residues C-terminally in the B-chain, N-terminal addition of Lys-Lys decreased the IP, although not significantly (Table I). The results underline the fact that the net charge and isoelectric pH are far less important for crystallization and solubility than the actual position and type of substitution.

5.3. Substitutions that Enhance Stability

The exchange of Asn^{A21} with Gly, Ser, Thr, Asp, His, and Arg increased the stability in acid solution astonishingly. Thus, at pH 3 and at 37°C the rates of deamidation and of formation of monodesamido products decrease 25-fold and about 100-fold, respectively (Markussen *et al.*, 1988a). These results show that Asn^{A21} is the major site of by-product formation in acid solution, and that side reactions at the other carboxamide groups are markedly suppressed. Obviously, different mechanisms cover deamidations of internal and C-terminal asparagine residues (Leach and Lindley, 1953). One likely mechanism for the instability of C-terminal Asn is anchimeric catalysis by the protonated α-carboxyl group (see Fig. 10).

5.3.1. $Asn^{A21} \rightarrow Gly, Ser, and Thr$

The Gly and Ser substitutions caused small and similar decreases in potencies in both bioassays (see Table I). They also caused a small but statistically insignifi-

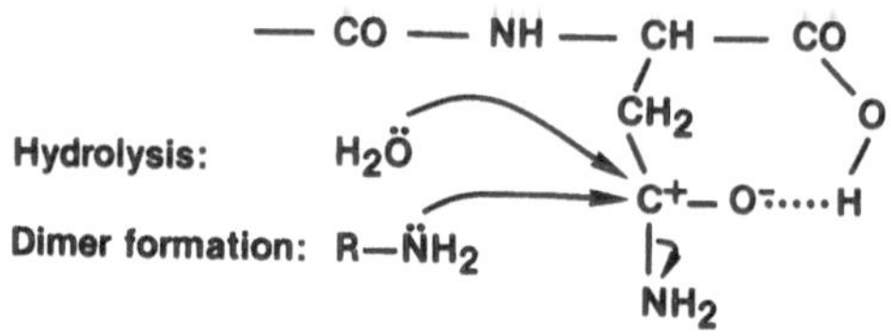

Figure 10. Proposed mechanism for the anchimeric catalysis of hydrolysis and aminolysis in a C-terminal asparagine by the α-carboxyl group. (After Markussen *et al.*, 1988b, with permission.)

cant drop in the IP. A difference was noted in the form two insulins crystallize from water in the hanging drop (Fig. 6). Whereas Gly^{A21},Arg^{B27},Thr^{B30}-NH$_2$ human insulin yields small rhombic platelets, Ser^{A21},Arg^{B27},Thr^{B30}-NH$_2$ human insulin crystallized as nonbirefringent, large octahedrons. Another difference was seen in the pharmacokinetic absorption studies when the concentration of Zn^{2+} was changed (see Section 5.5.2.). The Thr^{A21} substitution caused not only a marked decrease in potency, but also a significant drop in IP. Although the potency is thus somewhat sensitive to substitution in position A21, the complete conservation of Asn in all species remains a fact to think about.

5.3.2. Asn^{A21} → Arg and Asp

The attempt to combine an increase in charge and in stability by introducing Arg^{A21} failed, since the IP dropped markedly, as did the potency in the FFC assay (Table I). The similar effect observed by inserting basic residues in the B-chain N-terminal and A-chain C-terminal may reflect an increased solubility at pH 7 of such insulins, for example, as a result of hindrance of hexamer stacking by the charged groups.

The Asp^{A21} substitution was known from products of acid transformation of insulin (Sundby, 1962). It reduces potency only marginally (Brange *et al.*, 1987) and stabilizes insulin in acid solution (Markussen *et al.*, 1988a). Despite the fact that Asp^{A21} reduces the charge and renders the two examined insulins, Asp^{A21},Arg^{B27},Thr^{B30}-NH$_2$ and Asp^{A21},Arg^{B27},Lys^{B30}-NH$_2$ human insulins, negatively charged at pH 7.4, these were as prolonged-acting as the corresponding Asn^{A21} insulins (Markussen *et al.*, 1987b, 1988a).

5.4. Failure of Expression of Pro^{A21}

The attempt to produce a Pro^{A21} insulin failed, as the expression in yeast of the single-chain precursor was below the detection limit of the HPLC analysis (Markussen *et al.*, 1988a). In the crystal structure of insulin a hydrogen bond bridges the backbone nitrogen of A21 with the backbone carbonyl of B23 glycine. This hydrogen bond may guide the folding of the single-chain precursor prior to the formation of the disulfide bridges. Since a Pro residue lacks a hydrogen atom at its backbone nitrogen, this hydrogen bond cannot form, for which reason the folding might have failed.

5.5. Galenic Factors Influencing Absorption Kinetics

5.5.1. Concentration of Insulin

Insulin preparations for clinical use traditionally contained 40 U/ml, which corresponds to 240 μM. However, a change in concentration to 100 U/ml (600 μM) is being implemented worldwide. The concentration is of the utmost importance for the rate of nucleation and crystal growth (Schlichtkrull, 1958). Since the present attempt to achieve a prolonged action depends on crystallization *in situ*, it was not surprising that the concentration had an effect (see Fig. 11). For preparations containing 3 Zn^{2+}/hexamer of $Gly^{A21}, Arg^{B27}, Thr^{B30}$-$NH_2$ human insulin the half-life of the absorption process increases from 11 to about 25 hr by increasing the concentration of insulin from 240 to 600 μM. Note also that the approximation to first-order kinetics is best at the high concentration of insulin.

5.5.2. Concentration of Zinc Ions

The unit cell of most insulin crystals is a hexamer in which 2 Zn^{2+} aligned along the threefold axis coordinate to the 6 His^{B10} residues. The incorporation of

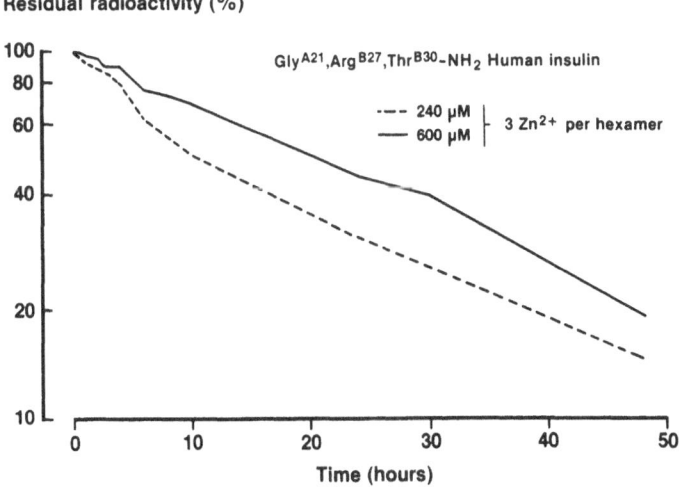

Figure 11. Absorption curves of $Gly^{A21}, Arg^{B27}, Thr^{B30}$-$NH_2$ human insulin in 240- and 600-μM formulations, containing 3 Zn^{2+}/hexamer. Isotonic agent 1.6% (w/v) glycerol; pH = 4.

Zn^{2+} In crystalline insulin suspensions is a well-known principle for achieving prolonged action (Schlichtkrull, 1958; Brange *et al.*, 1987). The influence of Zn^{2+} is best demonstrated at the low insulin concentration, where crystallization *in situ* may be incomplete (see Fig. 12). At the lowest Zn^{2+} concentration (1 Zn^{2+}/hexamer) the course of absorption is clearly biphasic, indicating that some insulin is being absorbed without having been in the crystalline state in the tissue.

The interplay between Zn^{2+} and the rate of absorption for two closely related insulins, Gly^{A21},Arg^{B27},Thr^{B30}-NH$_2$ and Ser^{A21},Arg^{B27},Thr^{B30}-NH$_2$ human insulin, is shown in Fig. 13. Whereas the absorption of the former depends only to an insignificant degree on the Zn^{2+} concentration in the formulation, the absorption of the latter is markedly controlled by Zn^{2+}. This difference in absorption undoubtedly reflects the difference in crystal forms (Fig. 6), with the one that crystallizes most rapidly and yields the smallest crystals being the one that is most independent of Zn^{2+}.

5.5.3. Buffer in the Formulation and Pain

Crystallization in the tissue can be delayed by using a buffer such as acetic acid/sodium acetate or citric acid/sodium citrate in the pH range from 3 to about 5 as isotonic agent. This will result in a biphasic absorption as part of the insulin escapes from the site of injection without having crystallized first (Markussen *et*

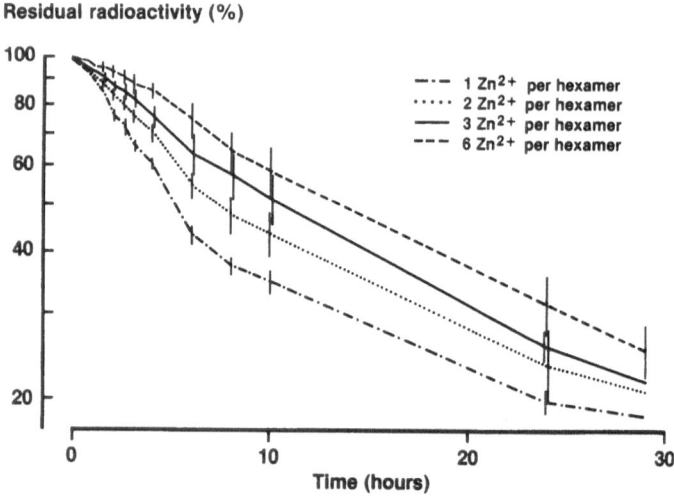

Figure 12. Disappearance of [^{125}I]Arg^{B27},Thr^{B30}-NH$_2$ human insulin in formulations containing 240 μM of the insulin and 1, 2, 3, and 6 Zn^{2+}/hexamer. Isotonic agent 1.6% (w/v) glycerol; pH = 4.5; dose = 1.2 nmoles/kg body wt. (After Markussen *et al.*, 1987b, with permission.)

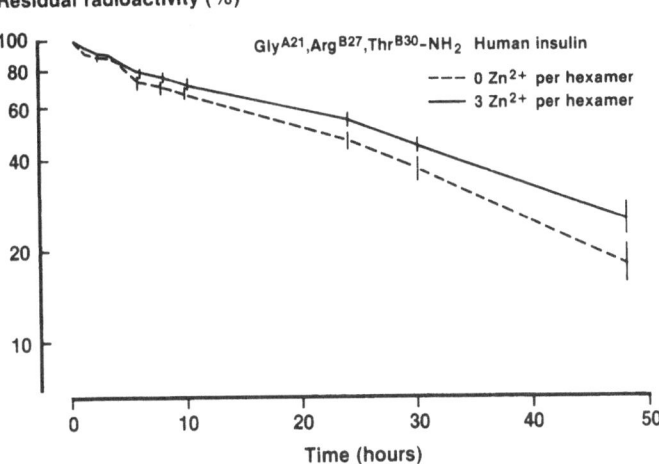

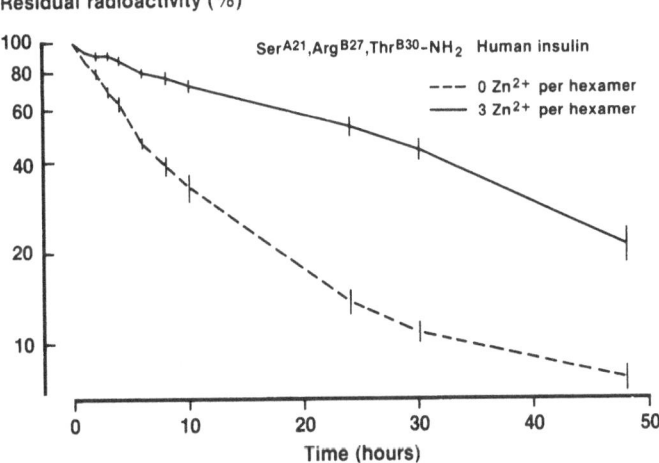

Figure 13. Absorption curves for two insulins: (A) $Gly^{A21}, Arg^{B27}, Thr^{B30}-NH_2$ and (B) $Ser^{A21}, Arg^{B27}, Thr^{B30}-NH_2$ human insulins in soluble formulations containing 600 μM insulin at two levels of Zn^{2+}/hexamer. Isotonic agent 1.6% (w/v) glycerol; pH = 4.

al., 1987b). Suspensions with biphasic absorption are used extensively in multiple daily injection regimes. However, the incorporation of an acid buffer in a drug for subcutaneous injection is excluded, since it causes a transient (10–20 sec) sting at the site of injection.

On the other hand, the injection of unbuffered insulin at pH 3 was found to

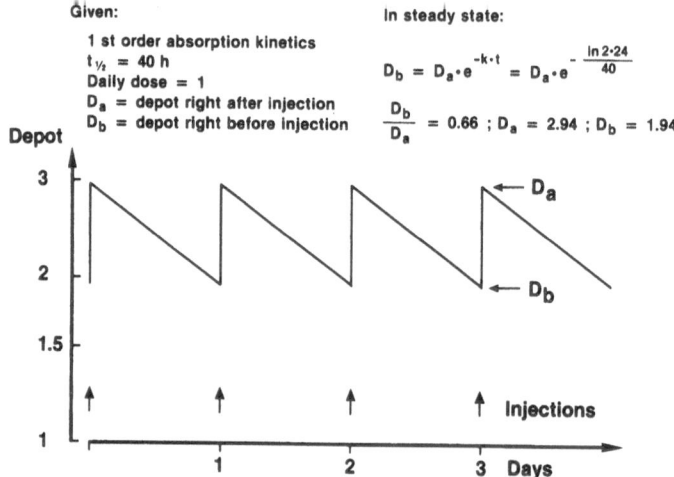

Given: In steady state:

1 st order absorption kinetics
$t_{1/2}$ = 40 h $D_b = D_a \cdot e^{-k \cdot t} = D_a \cdot e^{-\frac{\ln 2 \cdot 24}{40}}$
Daily dose = 1
D_a = depot right after injection
D_b = depot right before injection $\frac{D_b}{D_a}$ = 0.66 ; D_a = 2.94 ; D_b = 1.94

Figure 14. Calculation and graphic depiction of the subcutaneous depot of basal insulin in the steady state, given first-order absorption kinetics with a half-life of 40 hr and one daily injection. The depot varies ±20% around a mean of 2.44 times the size of the daily dose.

be painless apart from the prick of the needle; acid insulin formulations were used in the early years of insulin manufacture as a prerequisite for keeping foreign proteins in solution and contaminating proteases inactive.

6. Preclinical Studies and Considerations

6.1. Reproducibility of Absorption

The variation in absorption of crystalline insulin suspensions is large, both between and within patients (Binder, 1969; Lauritzen *et al.,* 1979, Lauritzen, 1985). The reproducibility of the absorption in pigs of $Gly^{A21}, Arg^{B27}, Thr^{B30}\text{-}NH_2$ human insulin in a 6 Zn^{2+}/hexamer solution at pH 3 and in a concentration of 600 μM was compared to that of the conventional suspensions. A comparison of the coefficient of variation of $t_{50\%}$ (i.e., the time when 50% of the insulin has disappeared from the injection site) is inadequate due to different absorption profiles. The insulin in soluble formulation is absorbed according to first-order kinetics, with a rate constant independent of dose. The absorption rate of the suspensions increases initially during the first hours after the injection; i.e., a time lag is encountered. Furthermore, the absorption rate in percent of dose decreases as the dose is increased. Consequently, the coefficient of variance of the absorption rates was used for the comparisons. During a 48-hr study this coefficient of variance for the soluble, prolonged-acting insulin was below that of the conventional suspensions at all time points, most markedly between 10 and 24 hr. Between 12 and 20 hr the coefficient of variance was about 0.27 for the soluble insulin and 1.0 for the insulin suspension, respectively. This property might be the most important advance of the soluble, prolonged-acting insulins in the clinical situation in which the primary goal is to normalize blood glucose.

The increased reproducibility in absorption could result from the combination of at least two factors. First, the injection of a solution ensures that the insulin dose is distributed in the total of the injected volume, whereas the crystals in a suspension might be trapped in the pores between the cells during the injection. Second, the low solubility of $Gly^{A21},Arg^{B27},Thr^{B30}-NH_2$ human insulin at pH 7 may make the dissolution the time-determining process, i.e., the process with the longest transit time, whereby the variations in physiological factors, such as blood flow and concentrations of Zn^{2+} complexing compounds, become of reduced significance to the course of absorption.

6.2. Calculation of Variations in Depots and Blood Levels of Basal Insulin in the Steady State

The half-life of the absorption in young pigs of 60–80 kg of Gly^{A21}, $Arg^{B27},Thr^{B30}-NH_2$ human insulin was found to be 25–30 hr. Experience from insulin suspensions is that absorption from the subcutis of the human thigh is somewhat slower. If we assume a half-life of 40 hr and set the daily dose arbitrarily to 1, the subcutaneous depot will vary in steady state from 1.94 right before the injection to 2.94 right after the injection (see calculation in Fig. 14). Given first-order absorption kinetics, the absorption is proportional to the depot. This means that blood levels of basal insulin will be proportional to the depots, provided that clearance also follows first-order kinetics, which is usually the case. Hence, the calculated variation in blood levels of basal insulin is ± 20% around the mean, a much smaller variance than is obtainable with the presently used suspensions featuring erratic absorption kinetics.

6.3. Potency and Units

The definition of the potency of a new insulin requires, of course, the establishment of a standard thereof. The human insulin unit equals 6 nmoles, and one definition of the unit of a new insulin could be to make it equal to 6 nmoles. If this new insulin, unit per unit, is as efficient in lowering blood glucose as an equivalently prolonged-acting preparation of ordinary insulin, the definition is safe from a clinical point of view. Since short-acting, monomeric insulins are under development too, in order to cope with the sharp rise in blood glucose following a meal (Brange *et al.*, 1988), the 6-nmole unit might in the long run result in the least confusion. The unit concept was introduced at a time when biological tests were necessary to establish the potency of products of highly variable purity. Today potency analysis by high-performance liquid chromatography is being implemented, but the unit has survived and may be difficult to remove from the labels of insulin vials because of its established use for 65 years.

6.4. Immunogenicity

The clinical use of a species of foreign insulin might expectedly give rise to antibodies against insulin. Since 94% of newly detected diabetic children developed insulin antibodies after 2 years of treatment with human insulin (Marshall *et al.*, 1988), the phenomenon is of a more quantitative than qualitative nature; will the daily use of a foreign insulin cause the formation of antibodies to an extent that it is clinically unacceptable? One unacceptable effect would be if the short-acting insulin that is administered shortly before the meals was bound to antibodies and then released later when the patient was fasting. However, it was concluded from a recent study that "insulin antibodies in insulin treated diabetics as a rule do not cause insulin resistance and neither deteriorate nor improve diabetic control" (Hübinger *et al.*, 1988). Immunization studies for 3 months in rabbits with soluble, prolonged-acting insulins, deviating from human insulin in three positions, have not resulted in the formation of high titer–high affinity antibodies. About 50% of the rabbits responded, and the titer peaked in the responders after 7 weeks of immunization. Only long-term clinical studies in human patients will answer whether the clinical benefits from a better glucose control can outweigh any drawbacks from potentially elevated antibody titers.

7. Summary

Soluble, long-acting insulins with increased stability and improved absorption reproducibility relative to existing insulin suspensions have been engineered. Clinical trials commenced in January 1988.

ACKNOWLEDGMENTS. The present work was brought about by participation of M. T. Hansen, F. Norris, K. Norris (genetics), I. Diers (fermentation), L. Drube, K. Müggler, H. O. Voigt, S. Weber (purification and synthesis), U. Damgaard, L. Snel, L. Thim (chemical analysis), A. R. Sørensen, E. Sørensen (bioassay), L. Andersen (immunogenicity), I. Jensen, A. Plum, U. Ribel, S. Jørgensen (pharmacokinetics), L. Langkjaer (formulations and stability), P. Hougaard, Aa. Vølund (statistical analysis), S. M. Lauersen, and R. K. Hjortkjaer (toxicology).

References

Binder, C., 1969, *Absorption of Injected Insulin*, Munksgaard, Copenhagen.

Blundell, T. L., Cutfield, J. F., Cutfield, S. M., Dodson, E. J., Dodson, G. G., Hodgkin, D. C., and Mercola, D. A., 1972, Three-dimensional atomic structure of insulin and its relationship to activity, *Diabetes* **21**(suppl. 2):492–505.

Bradbury, A. F., Finnie, M. D. A., and Smyth, D. G., 1982, Mechanism of C-terminal amide formation by pituitary enzymes, *Nature* **298**:686–688.

Brange, J., Skelbaek-Pedersen, B., Langkjaer, L., Damgaard, U., Ege, H., Havelund, S., Heding, L. G., Jørgensen, K. H., Lykkeberg, J., Markussen, J., Pingel, M., and Rasmussen, E., 1987, *Galenics of Insulin*, pp. 36–38, 54–57, Springer-Verlag, Berlin.

Brange, J., Ribel, U., Hansen, J. F., Dodson, G., Hansen, M. T., Havelund, S., Melberg, S. G., Norris, F., Norris, K., Snel, L., Sørensen, A. R., and Voigt, H. O., 1988, Monomeric insulins obtained by protein engineering and their medical implications, *Nature* **333**:679–682.

British Pharmacopoeia, 1980, Her Majesty's Stationery Office, London, A 142.

European Pharmacopoeia, 1984, V.2.2.3. Assay of insulin method C, Her Majesty's Stationery Office, London.

Geiger, T., and Clarke, S., 1987, Deamidation, isomerization, and racemization at asparaginyl and aspartyl residues in peptides, *J. Biol. Chem.* **262**:785–794.

Gliemann, J., and Gammeltoft, S., 1974, The biological activity and the binding affinity of modified insulins determined on isolated rat fat cells, *Diabetologia* **10**:105–113.

Helbig, H.-J., 1976, *Insulindimere aus der B-Komponente von Insulinpräparationen*, Rheinisch-Westfälischen Technischen Hochschule, Aachen.

Hodgkin, D. C., Dodson, E., Dodson, G., and Reynolds, C., 1983, Insulin, *Biochem. Soc. Trans.* **11**:411–417.

Hübinger, A., Becker, A., and Gries, F. A., 1988, Total insulin levels in Type 1 diabetic patients with insulin antibodies and their effect on insulin requirement and metabolic control, *Diabetes Res.* **7**:65–69.

Kobayashi, M., Ishibashi, O., Takata, Y., Haneda, M., Maegawa, H., Watanabe, N., and Shigeta, Y., 1985, Prolonged disappearance rate of a structurally abnormal mutant insulin from the circulation in humans, *J. Clin. Endocrinol. Metab.* **61**:1142–1145.

Lauritzen, T., 1985, *Pharmacokinetic and Clinical Aspects of Intensified Subcutaneous Insulin Therapy*, Laegeforeningens Forlag, Copenhagen.

Lauritzen, T., Faber, O. K., and Binder, C., 1979, Variations in ^{125}I-insulin absorption and blood glucose concentration, *Diabetologia* **17**:291–295.

Leach, S. J., and Lindley, H., 1953, The kinetics of hydrolysis of the amide group in proteins and peptides, *Trans. Faraday Soc.* **49**:915–925.

Markussen, J., 1982, U.S. patent 4,343,898.

Markussen, J., 1985, Comparative reduction/oxidation studies with single chain des-(B30) insulin and porcine proinsulin, *Int. J. Peptide Protein Res.* **25**:431–434.

Markussen, J., 1987, *Human Insulin by Tryptic Transpeptidations of Porcine Insulin and Biosynthetic Precursors*, MTP Press, Lancaster.

Markussen, J., Jørgensen, K. H., Sørensen, A. R., and Thim, L., 1985, Single chain des-(B30) insulin, *Int. J. Peptide Protein Res.* **26**:70–77.

Markussen, J., Hougaard, P., Ribel, U., Sørensen, A. R., and Sørensen, E., 1987a, Soluble, Prolonged-acting insulin derivatives. I. Degree of protraction and crystallizability of insulins substituted in the termini of the B-chain, *Protein Engineering* **1**(3):205–213.

Markussen, J., Diers, I., Engesgaard, A., Hansen, M. T., Hougaard, P., Langkjaer, L., Norris, K., Ribel, U., Sørensen, A. R., Sørensen, E., and Voigt, H. O., 1987b, Soluble prolonged-acting insulin derivatives. II. Degree of protraction and crystallizability of insulins substituted in positions A17, B8, B13, B27 and B30, *Protein Engineering* **1**(3):215–223.

Markussen, J., Damgaard, U., Diers, I., Fiil, N., Hansen, M. T., Larsen, P., Norris, F., Norris, K., Schou, O., Snel, L., Thim, L., and Voigt, H. O., 1987c, Biosynthesis of human insulin in yeast via single-chain precursors, in: *Peptides 1986* (D. Theodoropoulos, ed.), pp. 189–194, Walter de Gruyter, Berlin.

Markussen, J., Diers, I., Hougaard, P., Langkjaer, L., Norris, K., Snel, L., Sørensen, A. R., Sørensen, E., and Voigt, H. O., 1988a, Soluble, prolonged-acting insulin derivatives. III. Degree of protraction, crystallizability and chemical stability of insulins substituted in positions A21, B13, B23, B27 and B30, *Protein Engineering,* **2**(2):157–166.

Markussen, J., Hansen, M. T., Norris, K., and Sørensen, E., 1988b, Synthesis of insulins substituted in positions A17, B13, B27 and in the terminals of the B-chain, combining genetic engineering and tryptic transpeptidation in organic-aqueous medium, in: *Peptide Chemistry 1987* (T. Shiba, and S. Sakakibara, eds.), pp. 417–422, Protein Research Foundation, Osaka.

Marshall, M. O., Heding, L. G., Villumsen, J., Åkerblom, H. K., Baevre, H., Dahlquist, G., Kjaergaard, J-J., Knip, M., Lindgren, F., Ludvigsson, J., Persson, B., Rilva, A., Stenhammer, L., Strömberg, L., Søvik, O., Thalme, B., Vidnes, J., and Wefring, K., 1988, Development of insulin antibodies, metabolic control and B-cell function in newly diagnosed insulin dependent diabetic children treated with monocomponent human insulin or monocomponent porcine insulin, *Diabetes Res.* **9**:169–175.

Moody, A. J., Stan, M. A., and Stan, M., 1974, A simple free fat cell bioassay for insulin, *Horm. Metab. Res.* **6**:12–16.

Schlichtkrull, J., 1958, *Insulin Crystals,* Munksgaard, Copenhagen.

Schlichtkrull, J., Pingel, M., Heding, L. G., Brange, J., and Jørgensen, K. H., 1975, Insulin preparations with prolonged effect, in: *Handbook of Experimental Pharmacology,* Vol. XXXII/2 (A. Hasselblatt and F. v. Bruchhausen, eds.), pp. 729–777, Springer-Verlag, Berlin.

Sundby, F., 1962, Separation and characterization of acid-induced insulin transformation products by paper electrophoresis in 7 M urea, *J. Biol. Chem.* **237**:3406–3411.

Thim, L., Hansen, M. T., Norris, K., Hoegh, I., Boel, E., Forström, J., Ammerer, G., and Fiil, N. P., 1986, Secretion and processing of insulin precursors in yeast, *Proc. Natl. Acad. Sci. USA* **83**:6766–6770.

IX

DESIGN OF NOVEL ANTIVIRAL
AGENTS AND VACCINES

Studies on the Structure and Function of Ubiquitin

STANLEY T. CROOKE, CHRISTOPHER K. MIRABELLI,
DAVID J. ECKER, TAUSEEF R. BUTT,
SOBHANADITYA JONNALAGADDA, SCOTT DIXON,
LUCIANO MUELLER, FRANK BROWN, PAUL WEBER, and
BRETT P. MONIA

1. Introduction

Ubiquitin is a small (76 amino acid) protein found in all eukaryotes either free or covalently attached to proteins in the nucleus, cytosol, or plasma membrane (Wilkinson, 1986; Busch, 1984; Hershko, 1988). It is the most highly conserved protein yet discovered in eukaryotes, with only three conservative amino acid substitutions separating yeast ubiquitin from the human (Fig. 1). The protein is water soluble, extremely stable, and may be boiled without loss of activity. Ubiquitin is stable to proteases and has not been shown to be glycosylated (Ozkaynak *et al.*, 1984; Vierstra *et al.*, 1986; Schlesinger and Goldstein, 1975). It contains no cysteine and only one histidine at position 68. The crystal structure of human ubiquitin has been solved at 1.8 Å resolution and demonstrates that ubiquitin is a compact molecule with a hydrophobic core containing three and one-half turns of α helix and five strands of β sheet. At the carboxy terminus,

STANLEY T. CROOKE, CHRISTOPHER K. MIRABELLI, AND DAVID J. ECKER • Isis Pharmaceuticals, Carlsbad Research Center, Carlsbad, California 92008.   TAUSEEF R. BUTT, SOBHANADITYA JONNALAGADDA, SCOTT DIXON, LUCIANO MUELLER, FRANK BROWN, AND PAUL WEBER • SmithKline Beecham Laboratories, King of Prussia, Pennsylvania 19406.   BRETT P. MONIA • Department of Pharmacology, University of Pennsylvania, Philadelphia, Pennsylvania 19104; and Isis Pharmaceuticals, Carlsbad Research Center, Carlsbad, California 92008.

Ubiquitin Proteins

```
            1                                                                                              19
human:  met-gln-ile-phe-val-lys-thr-leu-thr-gyl-lys-thr-ile-thr-leu-glu-val-glu-pro
yeast:  ... ......... ........ ....... ............................. ..... ser
plant:  ... ......... ........ ....... ............................. ..... ser

            20              24              28 29                                        38
human:  ser-asp-thr-ile-glu-asn-val-lys-ala-lys-ile-gln-asp-lys-glu-gly-ile-pro-pro
yeast:  ........... asp ......... ser ......... ..... ......... ...
plant:  ........... asp ... ....... ala-gln ...... ........... ...

            39                                                                              57
human:  asp-gln-gln-arg-leu-ile-phe-ala-gly-lys-gln-leu-glu-asp-gly-arg-thr-leu-ser
yeast:  ... ............ .......... ... .................... ....
plant:  ... ............ .......... ... ........... ......... ....

            58                                                                              76
human:  asp-tyr-asn-ile-gln-lys-glu-ser-thr-leu-his-leu-val-leu-arg-leu-arg-gly-gly
yeast   ... ................ ....... ..... ..... ....... ... ...... ....
plant:  ... ................ ....... ..... ..... ....... ... ...... ....
```

Figure 1. Structure of human, yeast, and plant ubiquitins. Ubiquitin is conserved throughout the animal kingdom.

four amino acids protrude from the core of the molecule and have considerable freedom of motion (Vijay-Kumar *et al.*, 1987).

Ubiquitin genes are organized in two modes. Polygenes of lengths varying from 3 to more than 15 ubiquitin genes arranged in head-to-tail spacerless repeats and terminating in a ubiquitin in which the carboxy terminal glycine is blocked with an additional amino acid have been reported in a number of species (Baker and Board, 1987; Wilborg *et al.*, 1985; Finley *et al.*, 1987; Gausing and Barkardottir, 1986; Bond and Schlesinger, 1985; Dworkin-Rastl *et al.*, 1984; Izquierdo *et al.*, 1984; Ball *et al.*, 1987). The length of the polyubiquitin genes, the number of loci, and the amino acid employed to block the carboxy terminal glycine vary as a function of species. More recently, a group of genes in which a gene for ubiquitin is fused to a carboxy terminal extension protein of 52 or 76–80 amino acids in length has been identified (Lund *et al.*, 1985; Ozkaynak *et al.*, 1987; Monia *et al.*, 1989). The genes identified code for carboxy extension proteins that are similar to metal binding domains of proteins involved in gene regulation (Miller *et al.*, 1985; Rhodes and Klug, 1986).

Ubiquitin has been proposed to be involved in numerous cellular functions.

The best characterized activity is ATP-dependent proteolysis (Hershko, 1988; Butt *et al.*, 1989). However, some data support the notion that ubiquitin is involved in many other processes. In yeast, deletion of the polyubiquitin genes demonstrated that ubiquitin is an essential component of the stress response system (Finley *et al.*, 1987). Histone H_2A is stably ubiquitinated, and variations in the pattern of ubiquitination have been proposed to play a role in gene regulation (Rechsteiner, 1988). At least two membrane-localized receptors, platelet-derived growth factor and lymphocyte homing receptors, have been reported to be ubiquitinated (Rechsteiner, 1987; Busch, 1984), and some growth and differentiation factors have been shown to be stably ubiquitinated, suggesting a role for ubiquitin in receptor ligand interactions and signal transduction (Okabe *et al.*, 1986). Studies have suggested that at least 90% of the rapidly degraded proteins are degraded via the ubiquitin proteolytic pathway. A mutant mouse cell line that is defective in intracellular breakdown of abnormal or short-lived proteins due to a mutation in E_1, a ubiquitin pathway enzyme, arrests in G_2 during the cell cycle (Finley *et al.*, 1984; Ciechanover *et al.*, 1984). The discovery of stable ubiquitin–actin conjugates suggests that ubiquitin may be involved in thin filament assembly or function during muscle contraction. Furthermore, we have recently shown that overexpression of certain mutant ubiquitins in yeasts results in inhibition of cellular growth (Butt *et al.*, 1988). Ubiquitin has been found in the paired helical filaments, which are the principle constituents of the neurofibrillary tangles found in patients with Alzheimer's disease (Mori *et al.*, 1987). Moreover, recently interest has centered on the role of the ubiquitin carboxy extension proteins and the role that ubiquitin plays in their synthesis and regulation (Monia *et al.*, 1989). Figure 2 summarizes some of the activities that have been suggested for ubiquitin.

Current concepts about the ATP-dependent proteolytic processes in which ubiquitin participates are presented in Fig. 3 (Hershko and Ciechanover, 1986; Ciechanover, 1987). As only monoubiquitin is active and northern analyses demonstrate the existence of mRNA species corresponding to polyubiquitin, proteases that process multiple ubiquitins to active monoubiquitin have been proposed. We have recently identified and partially purified enzymes that selectively cleave either ubiquitin–ubiquitin junctions or ubiquitin–fusion protein junctions in which the nonubiquitin protein may be a variety of proteins (Jonnalagadda *et al.*, 1989).

Once processed, monoubiquitin may enter the ubiquitin-dependent proteolytic cycle. In the first step, in an ATP-dependent fashion, ubiquitin is conjugated to a thiol containing enzyme(s) (E_1) at the adenylate site. A second ubiquitin is subsequently added, and the first ubiquitin is transferred to the thiol site. When fully charged, E_1 can then transfer ubiquitin to one of a family of enzymes (E_2) via a sulfhydryl transfer reaction. E_2 then transfers ubiquitin to one of the E_3 enzymes responsible for transferring the ubiquitin to either the amino terminus or

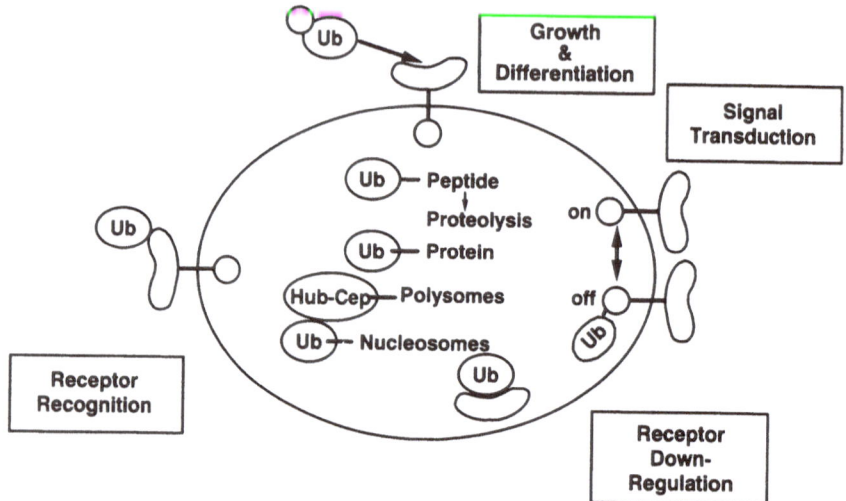

Figure 2. Possible functions of ubiquitin.

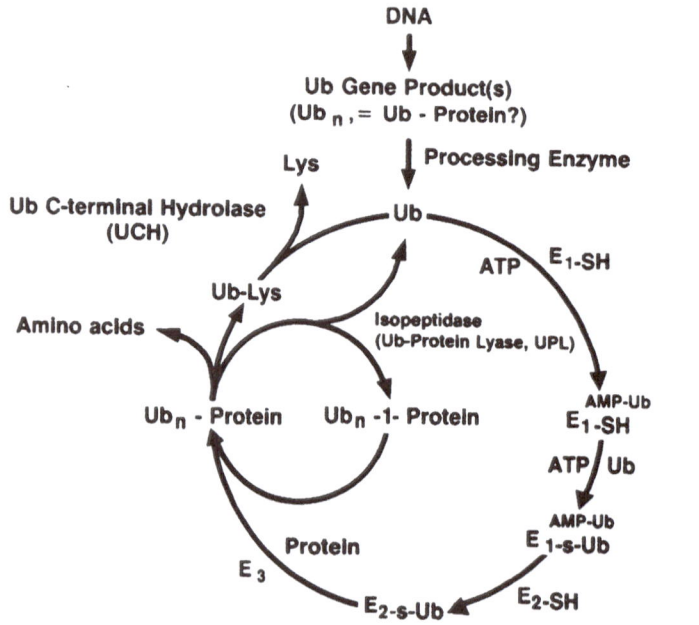

Figure 3. Ubiquitin-dependent proteolytic cycle.

the epsilon amino group of an internal lysine in the target protein. In fact, recent data suggest that proteolysis ensues primarily when proteins are internally ubiquitinated. Once ubiquitinated, the target protein may be degraded or a group of isopeptides may edit the ubiquitin off inappropriately targeted proteins. Ultimately, ubiquitin-lysine is released and processed to free ubiquitin by a ubiquitin C-terminal hydrolase.

The signals that determine the proteins that are degraded and the rate of degradation are not clearly defined. Based on studies employing a fusion gene of ubiquitin and a β-galactosidase gene with an amino terminal extension, Bachmair and colleagues (Bachmair *et al.*, 1986) proposed that the rate of processing and degradation of ubiquitinated proteins is governed by the amino terminus of the target protein. However, studies in Varshavsky's laboratory, our laboratory, and others have shown that the signals for proteolysis and processing are much more complex (Butt *et al.*, 1988).

Nevertheless, the proteolytic process provides a number of assays that are excellent tools for studying structure–activity relationships on ubiquitin, at least with regard to proteolysis. Figure 4 shows several steps in the ubiquitin-dependent proteolytic process that are employed as assays (Jonnalagadda *et al.*, 1988). The first two reactions are reversible, and can be employed to evaluate the first two steps in the proteolytic cycle. By performing experiments in the presence or absence of dithiotreitol, it is possible to evaluate the transfer of ubiquitin from the adenylate to thiol sites in E_1 [Eq. (3)]. The conjugation of ubiquitin to target

---

Equations

$$E1_{-SH} + ATP + Ub \rightleftharpoons E1_{-SH}^{-AMP-Ub} + PPi \qquad (1)$$

$$E1_{-SH}^{-AMP-Ub} + ATP + Ub \rightleftharpoons E1_{-S-Ub}^{-AMP-Ub} + PPi + AMP \qquad (2)$$

$$E1_{-S-Ub}^{-AMP-Ub} + E2\text{-}SH \text{ or (DTT)} \longrightarrow E1_{-SH}^{-AMP-Ub}$$
$$+ E2\text{-}S\text{-}Ub \text{ or (DTT-Ub)} \qquad (3)$$

$$E2\text{-}S\text{-}Ub + R\text{-}NH_2 \xrightarrow{E3} E2\text{-}SH + R\text{-}NH\text{-}Ub \qquad (4)$$

Figure 4. Ubiquitin-dependent proteolytic cycle: Equation 1: Ubiquitin is conjugated to the adenylate site on E_1. Equation 2: A second molecule of ubiquitin is added to the adenylate site and the first ubiquitin molecule is transferred to the thiol site. Equation 3: The ubiquitin at the thiol site is transferred either to E_2 or a thiol reagent such as DTT. Equation 4: The carboxy terminus of ubiquitin is covalently attached to either the amino terminus or an e-amino group of lysine on the target protein.

proteins by E_3 can be evaluated by gel electrophoresis and immunoblot analyses [Eq. (4)].

2. Synthesis and Expression of Ubiquitin and Mutant Genes in Yeast and Escherichia coli

To study the structure–activity relationships of ubiquitin, our strategy was to synthesize cassette-adapted ubiquitin genes and express them under a variety of controls in yeast and *E. coli*. Armed with these tools, we could then study the effects of overexpression of wild-type ubiquitin in a homologous system (yeast ubiquitin in yeast), the activities of mutant proteins purified from cells with no wild-type ubiquitin (*E. coli*), and the effects of mutant ubiquitin expression in yeast.

Figure 5 shows the cassette-adapted yeast ubiquitin gene (Ecker *et al.*,

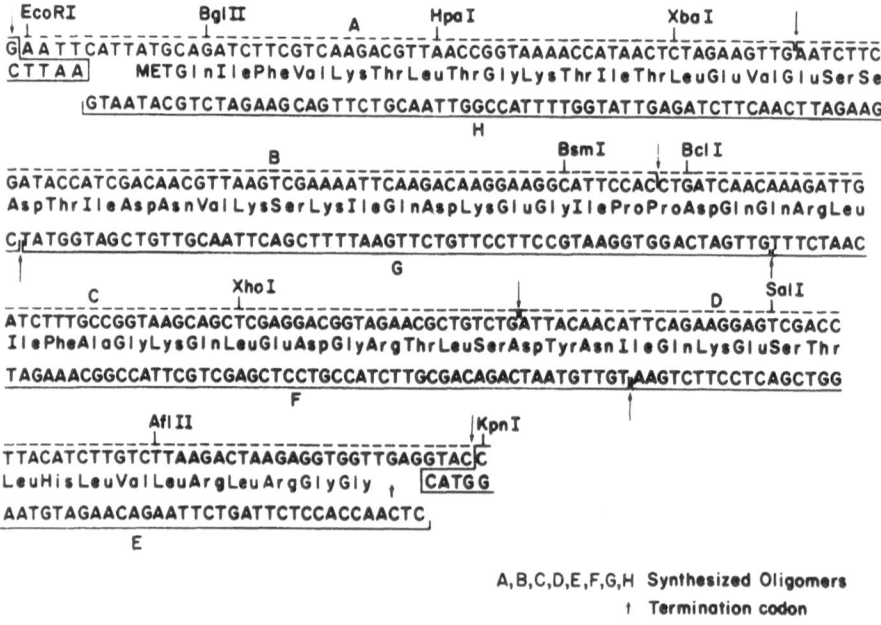

Figure 5. Assembly of the cassette-adapted synthetic yeast ubiquitin gene. Arrowheads indicate the ends of the synthetic oligonucleotide fragments. The annealed complementary oligomer pairs had nine-base cohesive overlaps with adjacent pairs. Details of the ligation scheme are given in the text. A-H, Synthesized oligomers; ↑, termination codon; |, unique restriction sites.

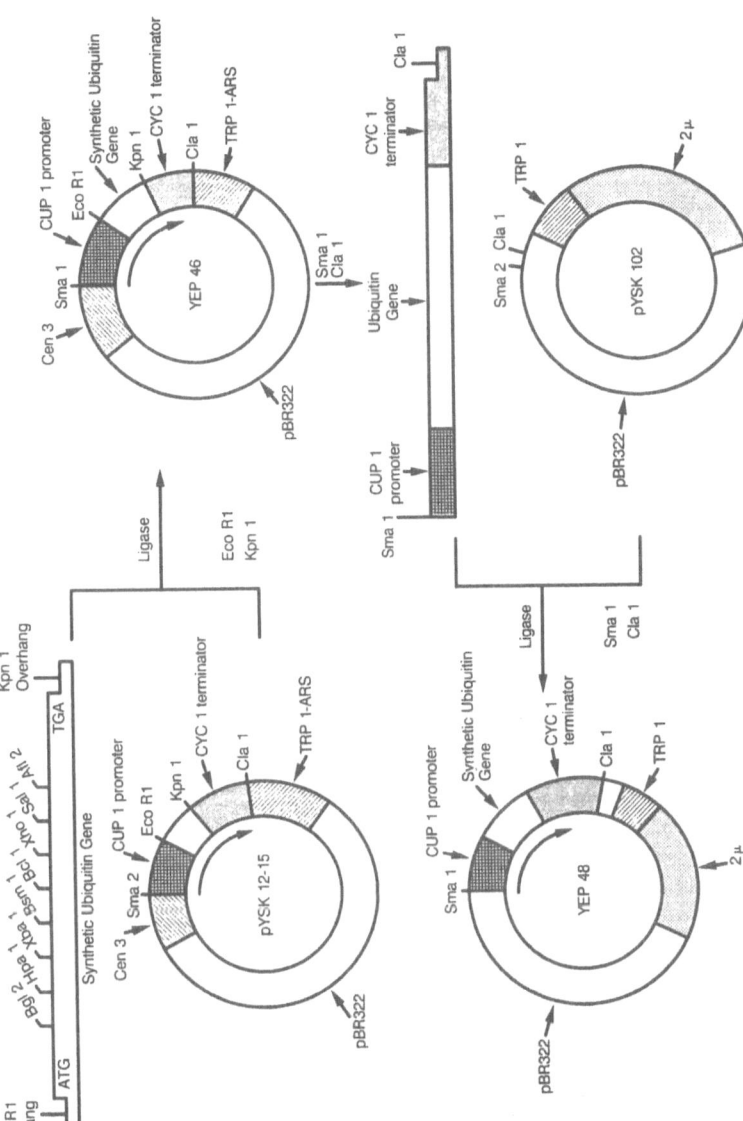

Figure 6. Yeast vectors for expression of the synthetic yeast ubiquitin gene under the *CUPI* (yeast metallothionein) promoter. Vectors containing the *Cen3* fragment (PYSK12-15, YEP46) are maintained in yeast at low copy number. Vectors with 2-μM sequences (PYSK102, YEP48) are maintained at 30–50 copies per cell.

1987a). The gene was assembled after synthesis of segments A–H containing the unique restriction endonuclease sites shown above the sequence and introduced by silent mutations. After synthesis the gene was cloned into a modified PUC 18 vector, amplified, recovered, and integrated into the yeast expression vectors under the control of the *CUP1* promoter, a copper-sensitive promoter (Fig. 6). When expressed in a strain of yeast with multiple copies of the *CUP1* locus (*CUP1ʳ*), ubiquitin was inducibly expressed (Fig. 7). When expressed in a strain

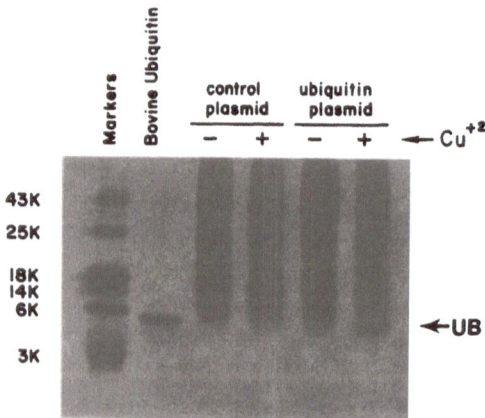

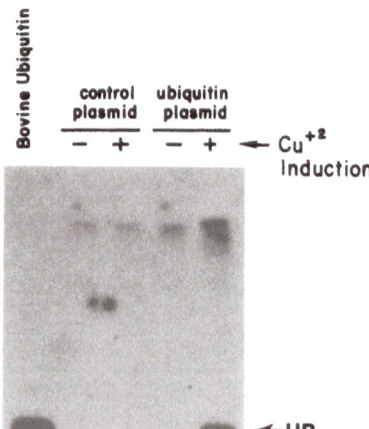

Figure 7. Copper-induced expression of the synthetic ubiquitin gene in CUPIʳ yeast. (Upper gel) *CUP1ʳ* yeast containing a control (PYSK102) or ubiquitin-expressing (YEP46) plasmid were incubated with or without 100 μM CuSO₄ for 2 hr. Total yeast proteins were prepared, run on an 18% SDS polyacrylamide gel, and stained with Coomassie blue. (Lower gel) Prepared identically to the above gel, except that after electrophoresis the gel was blotted onto nitrocellulose paper and probed with antiubiquitin antibodies.

in which the *CUP1* locus was deleted (*CUP1*), both the low and high copy number vectors constitutively produced ubiquitin (Fig. 8). Studies with the ubiquitin purified from these cells demonstrated that the ubiquitin derived from the synthetic gene was as active as wild-type yeast or animal ubiquitin. Furthermore, constitutive expression of high levels of yeast ubiquitin in yeast had no apparent deleterious effects.

In order to evaluate the activities of mutant ubiquitins, studies were performed in *E. coli*, which has no ubiquitin. The human ubiquitin gene was synthesized in a fashion analogous to the yeast gene (Fig. 9). It was then cloned into an *E. coli* expression vector under the control of the heat-inducible promoter, PL promoter (Fig. 10), and expressed in high quantity in *E. coli*. The ubiquitin expressed in *E. coli* was as active as yeast or bovine ubiquitin. Based on computer modeling (Fig. 11), several sites were selected for mutagenesis, expression, and purification. The activities of each mutant were determined (Table I). Figure 12 shows that each mutant protein was purified to homogenity. The

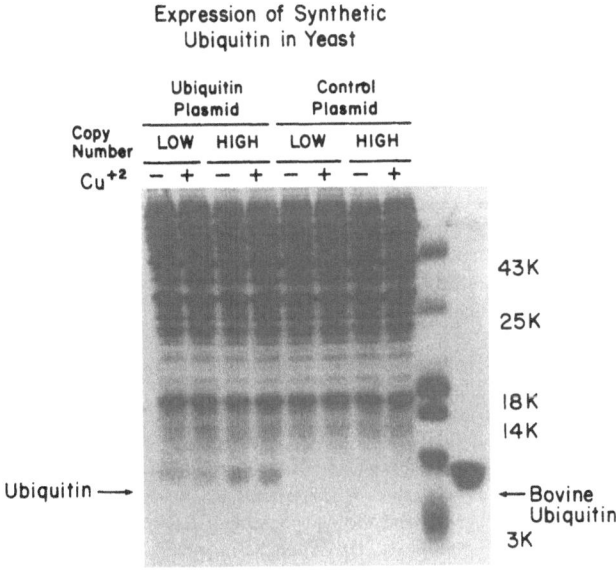

Figure 8. Constitutive expression of the synthetic ubiquitin gene in *CUP1ˢ* yeast. *CUP1ˢ* yeast containing high (PYSK102) and low (PYSK12-15) copy control plasmids and high (YEP46) and low (YEP44) copy ubiquitin expression plasmids were incubated with or without 25 μM CuSO₄ for 2 hr. Total yeast proteins were prepared, run on 18% SDS polyacrylamide gel, and stained with Coomassie blue.

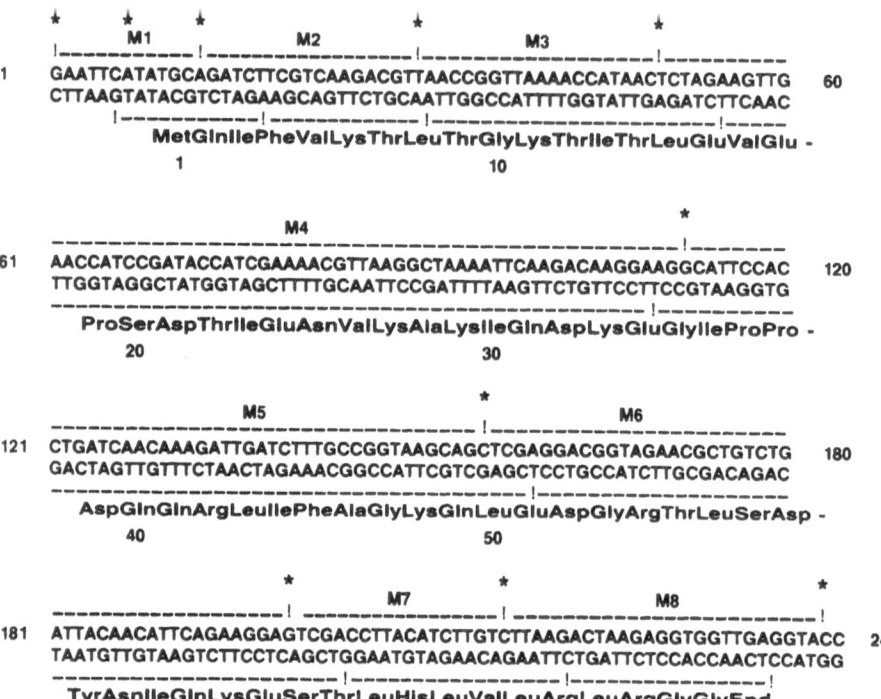

Figure 9. Synthetic modular human ubiquitin gene. The lengths of the mutagenesis modules (M1–M8) range from 12 to 64 base pairs; all regions of the synthetic gene are therefore accessible to site-specific or cassette mutagenesis, because a single stretch of synthetic double-stranded can bridge the gap between any adjacent restriction sites. In this study, modules M1, M4, M6, M7, and M8 were all replaced to produce the mutants listed in Table I.

Figure 10. Mutagenesis and module switching. (A) The modular mutagenesis strategy was used to generate a double mutant and two single mutations from one piece of synthetic double-stranded DNA. Synthetic DNA, encoding two specific mutations (Tyr-59→Phe, His-68→Lys), was ligated into the appropriately restricted wild-type ubiquitin gene, replacing modules M6 and M7 to generate a double mutant. To generate each single mutant, the double mutant gene was cut at a unique Sall site between the two modules and at the unique Aat2 site in the pUC vector. The resulting fragments were gel-purified and each was ligated with the complementary fragment from a similarly restricted plasmid containing the wild-type gene. (B) Cloning the synthetic ubiquitin gene in the *E. coli* expression vector pMG27N-S. The synthetic ubiquitin gene has an NdeI site that includes the initiation codon CA-T*ATG*. The ubiquitin gene was removed from the pUC plasmid by restriction with NdeI and PvuII. The gene-containing fragment was gel-purified and ligated into the *E. coli* expression vector pMG27N-S, which was restricted with HindIII, the overhang flush ending with DNA polymerase (Klenow fragment) and then cut with NdeI. Insertion of the ubiquitin gene at the NdeI site of the expression vector placed the reading frame downstream of the heat-inducible λ P_L promoter and adjacent to the λ cII gene ribosome binding site.

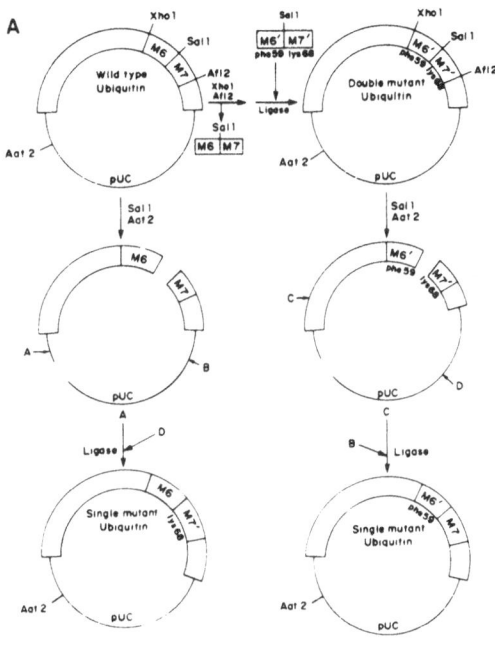

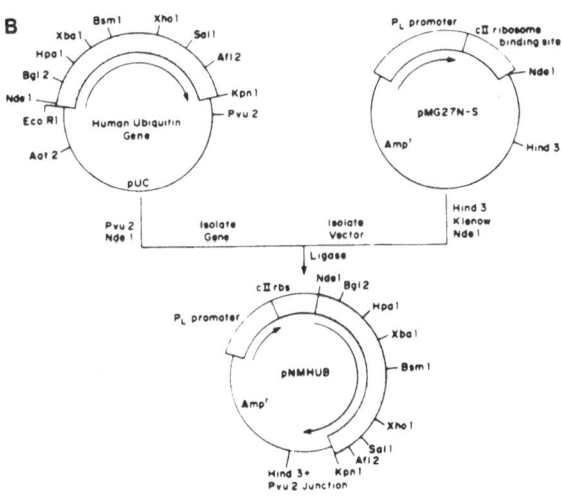

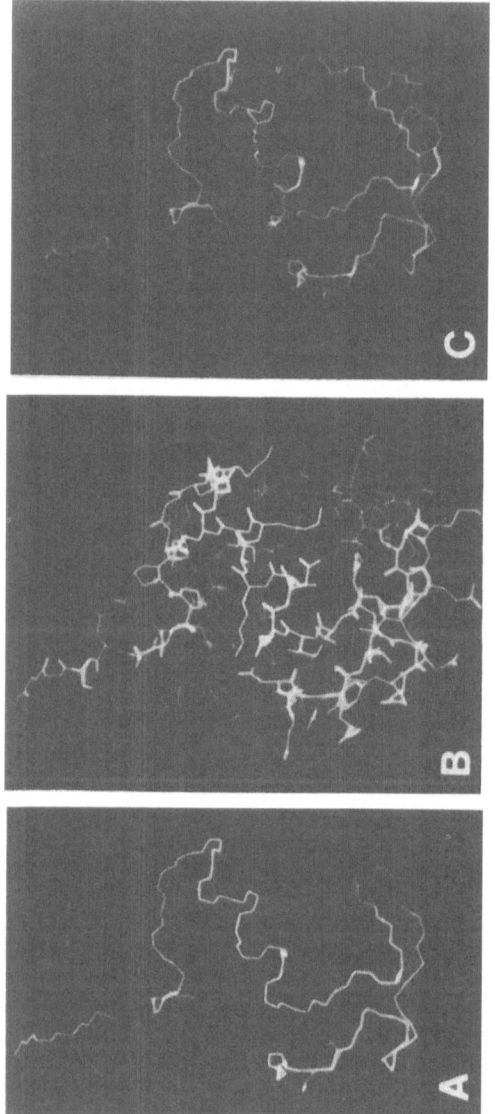

Figure 11. Structure of ubiquitin and location of the mutagenesis sites. The molecular graphics were generated from the 1.8-Å ubiquitin structure coordinates. (A) Ubiquitin backbone. (B) Ubiquitin with the mutagenized amino acid side chains. (C) Ubiquitin with the amino acid side chains.

Table I. Description and Location of Ubiquitin Mutations and Their Activities
in Supporting the *In Vitro* Degradation of ^{125}I-BSA by the Ubiquitin-Dependent Pathway[a]

Protein structure	Location of change	*In vitro* BSA degradation activity (% of bovine ubiquitin)
Animal ubiquitin (from cow)	—	100
Animal ubiquitin (expressed in *E. coli*)	—	100
pro 19 → ser, ala28 → ser, glu24 → asp (yeast ubiquitin, from *E. coli*)	Surface	100
pro19 → ser	Surface	100
leu67 → asn, leu69 → asn	Core	0
gly76 → ala	Tail	0–10
leu73 → Δ	Tail	0
leu73 → Δ, arg72 → ser	Tail	0
tyr59 → phe	Surface	70–100
his68 → lys	Surface	30
tyr59 → phe, his68 → lys	Surface	30

[a]The activities of mutant ubiquitins are reported relative to the activity of animal ubiquitin, isolated from bovine blood, in the degradation of substrate with 4 μg of ubiquitin or mutant ubiquitin per assay. Aliquots were removed at hourly intervals until the end of the experiment at 4 hr, and the rate of substrate degradation with time was measured.

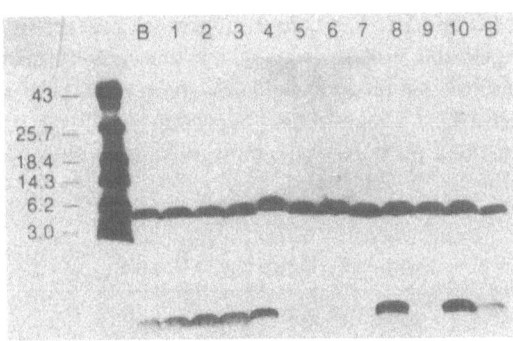

Figure 12. Purified site-specific ubiquitin mutant on an 18% SDS-polyacrylamide gel stained with Coomassie blue and an identical get blotted onto nitrocellulose paper and probed with antibodies raised to purified bovine ubiquitin. The number at the top of each lane corresponds with a ubiquitin mutation in the order given in Table I. B = bovine ubiquitin.

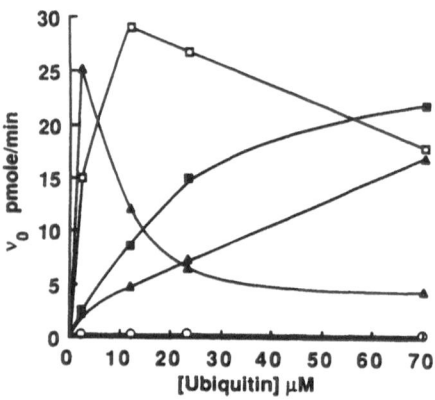

Figure 13. Ubiquitin and mutant ubiquitin concentration dependence of ATP:PP$_i$ exchange rates catalyzed by the ubiquitin-activating enzyme E$_1$. Each reaction contained purified E$_1$, 0.09 mg; Tris-HCl, 50 mM, pH 7.6; ATP, 1 mM PP$_i$, 100 μM, specific activity 6.6 10^4 dpm/nmole; MgCl$_2$, 10 mM; dithiothreitol, 0.1 mM; and varying ubiquitin concentrations. The reactions were stopped after 20 min and counts at ATP absorbed to charcoal were measured. Animal ubiquitin isolated from bovine blood, from expression in *E. coli,* yeast ubiquitin from expression in *E. coli,* and animal ubiquitin mutant Pro-19→Ser were identical in this assay and represented in the single curve (△). The experimental variation on each data point was approximately 5%–10%. Mutant in the tail, Gly-76→Ala, Leu-73→delete, Leu-73→delete, Arg-72→Ser, and Leu-67,69→Asn were completely inactive and shown in the single curve (o). □, Try-59→Phe; ■, His-68→Lys; ▲, double-mutant Tyr-59→Phe, His-68→Lys.

Western blot demonstrates that polyclonal antibodies to ubiquitin recognized each of the mutants except those with modifications in the carboxy tail.

The identity of each mutant and the effects of each mutation on the three-dimensional structure were determined by one- and two-dimensional nuclear magnetic resonance (NMR) analysis in which the resonance of each proton in the wild-type and mutant proteins was assigned (Ecker *et al.,* 1987a,b). Additionally, we studied the ability of each mutant to participate in the first step in the proteolytic cycle (Fig. 13). Based on these studies, we arrived at the conclusions shown in Tables II and III. The data demonstrated that the molecule is indeed a single-domain globular protein from which the carboxy terminal amino acids protrude. Moreover, interactions with E$_1$ are essential for initiating the proteolytic cycle, and these are absolutely dependent on the carboxy terminal amino acids. However, other portions of the molecule are required after charging of E$_1$

Table II. Conclusions. I. Structure of Ubiquitin

Carboxy terminus (residues 72–76) unhindered
His 68 → LYS induced subtle conformational changes in backbone
Leu 67 → ASN; Leu 69 → ASN may be less stable and is less soluble
TYR 59 → Phe induces minimal conformational effects; H-bonding not necessary for loop stability

Table III. Conclusions. II. Structure–Function Relationships

Animal, yeast, and plant ubiquitins are equivalent in conformation and function.

Mutations in animal vs. yeast ubiquitin do not represent "second site compensation."

The carboxy tail is essential for conjugation to E_1 and protein degradation.

The carboxy tail is the antigenically dominant epitope.

E_1 interacts with the carboxy tail and the globular domain of ubiquitin.

Surface mutations may hinder interactions of E_1 with the globular domain.

Interactions with E_1 required for proteolysis.

Selection pressure for proper folding.

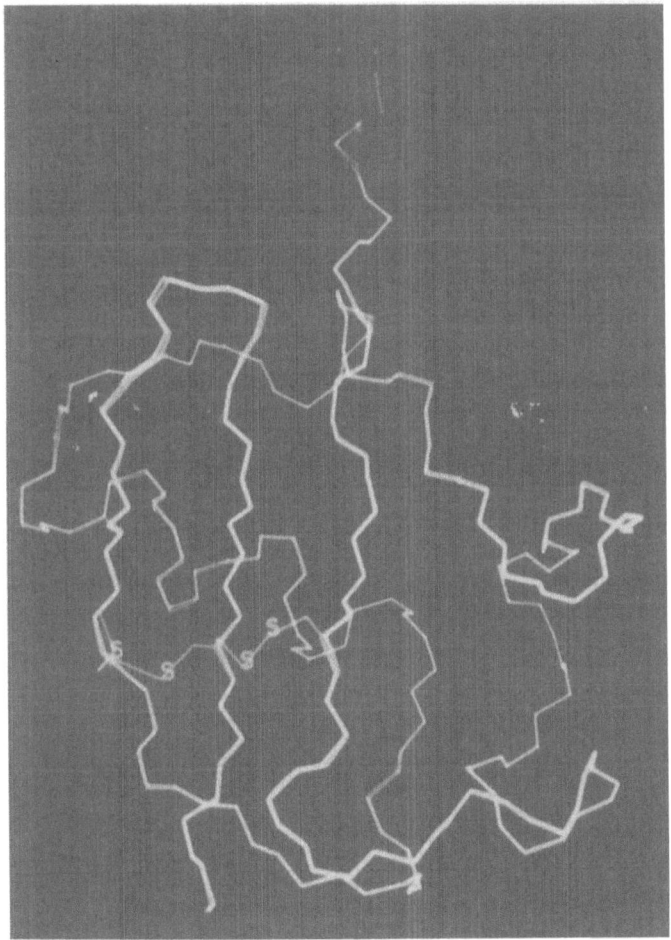

Figure 14. Placement of disulfide bonds in ubiquitin by molecular modeling. The phe4 cys + thr14→cys and phe 4→cys + thr66→cys; the crystal structure backbone trace is shown superimposed on the energy-minimized disulfide mutant structure.

with ubiquitin, and a conformational change in ubiquitin is probably required to propogate the sequence of events.

3. Disulfide Mutants

To prove that in order to propagate the proteolytic cycle, a conformational change in ubiquitin after being conjugated to the target protein is required, we introduced disulfide cross-links into ubiquitin (Ecker *et al.*, 1989). Our strategy was to engineer disulfides into ubiquitin that could be well tolerated by the native structure, but would prevent required conformational changes. We were aided by the lack of cysteines in ubiquitin and the previously defined crystal and NMR structure that allowed us to model proposed molecules with a high level of precision.

Figure 14 shows structures of two disulfide mutants, 4–14 and 4–66, projected by computer modeling techniques. The model predicts that both disulfides should be well tolerated and that both, particularly the 4–66 mutant, should hinder conformational mobility. Table IV shows the mutations made. Each mutant gene was expressed in *E. coli,* as previously described, and purified. Titrations were performed to determine free sulfhydryls; the results are shown in Table IV. The single-cysteine mutants displayed a single sulfhydryl per mole. The double mutants had not titratable sulfhydryls, suggesting that the hoped-for

Table IV. Positions of Mutations and Titratable Sulfydryls per Protein Monomer by Ellman's Reagent[a]

	Codon			
Wild-type ubiquitin	4	14	66	SH/monomer
	-TTC- PHE	-ACT- THR	-ACC- THR	0.0
PHE4 → CYS	-TGT- CYS	-ACT- THR	-ACC- THR	1.4
THR14 → CYS	-TTC- PHE	-TGT- CYS	-ACC- THR	0.8
THR66 → CYS	-TTC- PHE	-ACT- THR	-TGT- CYS	1.1
PHE4 → CYS, THR14 → CYS	-TGT- CYS	-TGT- CYS	-ACC- THR	0.0
PHE4 → CYS, THR66 → CYS	-TGT- CYS	-ACT- THR	-TGT- CYS	0.0
PHE4 → CYS, THR14 → CYS THR66 → CYS	-TGT- CYS	-TGT- CYS	-TGT- CYS	0.8

[a]The relative concentrations of ubiquitin monomers and dimers were determined by running the protein sample on nonreducing polyacrylamide gel electrophoresis followed by scanning a photographic negative of the gel.

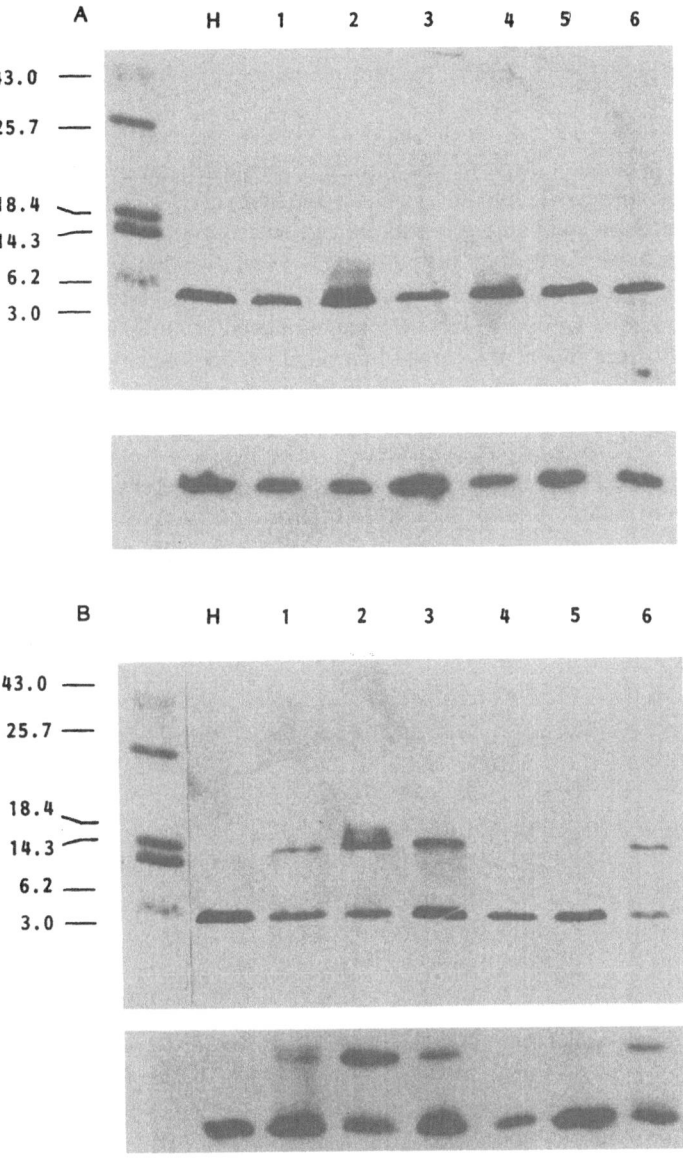

Figure 15. (A, top) Coomassie-stained 18% acrylamide gel containing SDS and 2-mercaptoethanol of purified human ubiquitin and the cysteine mutants. The lanes are identified as follows: H, human ubiquitin; 1, phe4→cys; 2, thr14→cys; 3, thr66→cys; 4, phe4→cys,thr14→cys; 5, phe4→cys,thr66→cys; 6, phe4→cys,thr14→cys,thr66→cys. (A, bottom) An identically prepared gel transferred to nitrocellulose and probed with antiubiquitin antibody and [125-l]iodinated protein A followed by autoradiography. (B) Identical to the gels shown above, except that 2-mercaptoethanol was deleted from the gel buffer.

intramolecular disulfides formed spontaneously. The triple mutants displayed the expected single sulfhydryl group per mole.

Figure 15 shows that all the mutants migrated as a single band when electrophoresed in the presence of 2-mercaptoethanol. In the absence of 2-mercaptoethanol, however, the single and triple mutants migrated as two bands, while the double mutants displayed only a single band, indicating that intermolecular disulfides formed in mutants with a free sulfhydryl, and that intramolecular disulfides form spontaneously and are favored.

Figure 16 shows that all the mutants except 4–66 and 4–14–66 were as active as wild-type ubiquitin in supporting proteolysis. Inasmuch as the 4–66 cysteine mutant was predicted to be well tolerated and to have no impact on the mobility or activity of the carboxy terminal peptide, our hypothesis suggested that all the mutants should be equally active in supporting conjugation to E_1, which was indeed the case. Thus, these studies confirm that the initial interactions with E_1 and conjugation to target proteins require the carboxy terminus, but that a conformational change in ubiquitin that is prevented by forming a disulfide at positions 4 and 66 is required to fully propagate the proteolytic cycle.

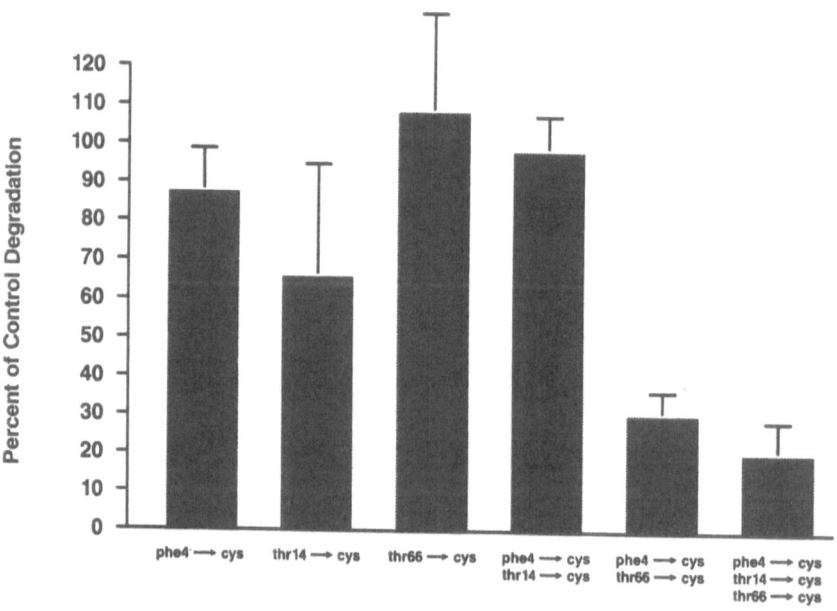

Figure 16. Activities of ubiquitin mutants in supporting the *in vitro* degradation of [125-1]BSA by the ubiquitin-dependent pathway. The activities of the mutants are reported relative to that of wild-type human ubiquitin, defined at 100%. The assay was performed as described previously. The error bars indicated the standard deviation of the five experiments.

Table V. Activity of Ubiquitin COOH-Terminal Peptides in Assays with Enzymes of ATP/Ubiquitin-Dependent Pathway

Ubiquitin COOH-terminal peptides	Structures of the COOH-terminal peptides	Final concentrations in the reaction	Hydrolysis of [3H] ATP with E_1 (%) [a]	Stimulation of 32[PPi]:ATP exchange with E_1 (activity compared to bovine ubiquitin) [b] (%)	Degradation of 125I-BSA with reticulocyte extracts [c] (%)
Tri-	Ac-Arg-Gly-Gly	5 mM	0	0.3	0
Tetra-	Ac-Leu-Arg-Gly-Gly	5 mM	0	2.9 ± 0.1	0.01
Penta-	Ac-Arg-Leu-Arg-Gly-Gly	5 mM	7.5 ± 0.5	8.6 ± 0.6	0.32
Hexa-	Ac-Leu-Arg-Leu-Arg-Gly-Gly	5 mM	13.5 ± 1.5	22.0 ± 0.5	0.22
Nonspecific penta-	Ac-Ser-gln-Asn-Tyr-Pro	5 mM	0	0	0.1
Bovine ubiquitin		2 µM	24.0 ± 3.0	100.0 ± 5.0	14.0 ± 0.5

[a] Estimation of the [3H]AMP released from the hydrolysis of [3H]ATP with ubiquitin-activating enzyme (E_1) in the presence of ubiquitin or ubiquitin COOH-terminal peptides. The experiment was performed three times; the values are given in this column.

[b] Analysis of the ubiquitin or ubiquitin COOH-terminal peptide-mediated 32PPi:ATP exchange with ubiquitin-activating enzyme (E_1). The experiment was performed two times; the values observed are represented in this column.

[c] Analysis of the ubiquitin or ubiquitin COOH-terminal peptide-mediated *in vitro* degradation of 125I-bovine serum albumin (BSA) with reticulocyte lysate. The lysate with 125I substrates such as soybean trypsin inhibitor or lysozyme gave similar results. The results were performed two times; the representative values for bovine serum albumin are given in this column.

4. Carboxy Terminal Peptides

To determine the degree to which the carboxy terminus is necessary and sufficient to interact at the adenylate site of E_1 and to assess whether a conformational change in ubiquitin is required for the transfer of ubiquitin from the adenylate to the thiol site on E_1, we tested a series of peptides homologous to the carboxy terminal hexapeptide in ubiquitin. We reasoned that the carboxy peptides should interact with the adenylate site on E_1, but that other portions of the ubiquitin structure are required to effect transfer to the thiol site, and thus the peptides should not be transferred. Table V shows the peptides that we tested and the results of several assays. Although they displayed dramatically lower poten-

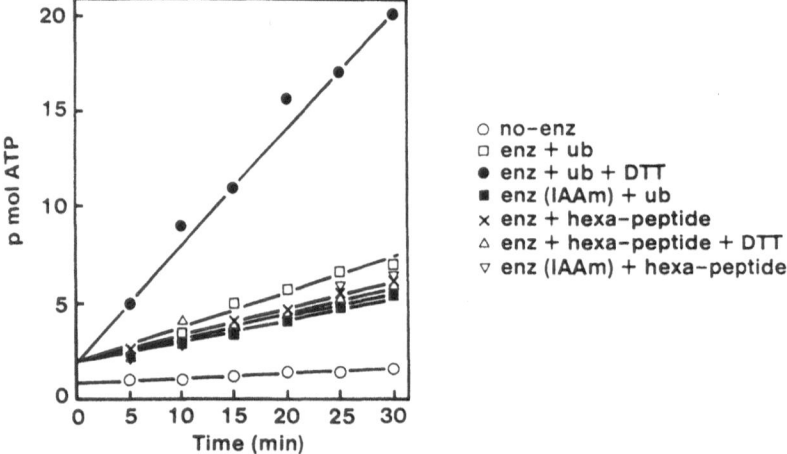

Figure 17. Analysis of ubiquitin and hexapeptide-mediated hydrolysis of [γ-^{32}P]ATP with ubiquitin-activating enzyme E_1 or E_1 pretreated with iodacetamide. E_1 was pretreated with iodacetamide according to a previously published procedure. E_1 or E_1 pretreated with iodacetamide was incubated with 2 μM ubiquitin or 5 mM ubiquitin COOH-terminal peptides in the absence or presence of DTT. Estimation of the [γ-^{32}P]ATP hydrolyzed was performed by measuring the ^{32}P$_i$ released during the reaction. Two moles of ^{32}P$_i$ are released for each mole of enzyme. The initial burst in the hydrolysis of [γ-^{32}P]ATP observed between no enzyme (o) and enzyme in the presence of substrates ($\square$, $\bullet$, $\blacksquare$, x, $\triangle$, $\triangledown$) is due to the rapid hydrolysis of the first molecule of ATP [Eq. (1)]. The enhanced rate of ATP hydrolyzed in the presence of E_1, ubiquitin, and DTT ($\bullet$) is due to the transfer of activated ubiquitin to DTT. This was not observed in the presence of E_1, hexapeptide, and DTT ($\triangle$), indicating the absence of an activated thiol-peptide on E_1. There is no significant increase in the rate observed with E_1 and hexapeptide (x), or E_1 pretreated with iodacetamide and hexapeptide ($\triangledown$) or ubiquitin ($\blacksquare$) in the presence of DTT ($\triangledown$) ($\blacksquare$), because the "thiol" site on E_1 is destroyed by treating the enzyme with iodoacetamide. The slight increase in the rate observed when E_1 is incubated with ubiquitin ($\square$) could be due to contaminating E_2s, DTT, or other proteins in the E_1 fraction that can accept the activated thiol-ubiquitin from E_1.

cies, the homologous pentapeptide and hexapeptide supported the first step in ubiquitin-dependent proteolytic cycle, as shown by both the ATP hydrolysis and the pyrophosphate exchange assays. However, neither supported degradation of the full proteolytic pathway. Random peptides were inactive. Figure 17 shows that indeed, although less potent, the homologous hexapeptide resulted in charging of E_1 at the adenylate site equivalent to ubiquitin in the absence of dithiothiotol, which serves as an acceptor of ubiquitin after transfer to the thiol site and thus enhances the rate of ATP hydrolysis.

From these studies, we can conclude that the carboxy terminal peptide of ubiquitin is necessary and sufficient to charge E_1 at the adenylate site, but that transfer to the thiol site requires the rest of the ubiquitin structure. Moreover, although the peptides tested did not behave as competitive antagonists to ubiquitin in the proteolytic pathway because of their low affinity for E_1, small molecule antagonists should be feasible.

5. Conclusions

Employing protein engineering, we have gained considerable insight into the structure and function of ubiquitin. Energy-minimized molecular models of ubiquitin prove to be quite accurate and allow us to be efficient in the choices of mutant proteins that were expressed and tested. Overexpression of the key mutant proteins in yeast and mammalian cells will provide additional insights into the potential cellular activities of ubiquitin and the impact of changes in the structure of ubiquitin and its performance.

References

Bachmair, A., Finley, D., and Varshavsky, A., 1986, *In vivo* half-life of a protein is a function of its amino-terminal residue, *Science* **234**:179–186.

Baker, R. T., and Board, P. G., 1987, The human ubiquitin gene family: Structure of a gene and pseudogenes from the UbB subfamily, *Nucleic Acid Res.* **15**:443–463.

Ball, E., Karlik, C. C., Beall, C. J., Saville, D. L., Sparrow, J. C., Bullard, B., and Fryberg, E. A., 1987, Arthrin, a myofibrillar protein of insect flight muscle, is an actin-ubiquitin conjugate, *Cell* **51**:221–228.

Bond, U., and Schlesinger, M. J., 1985, Ubiquitin is a heat shock protein in chicken embryo fibroblasts, *Mol. Cell. Biol.* **5**:949–956.

Busch, H., 1984, Ubiquitination of proteins, *Methods Enzymol.* **106**:238–262.

Butt, T. R., Khan, M. I., Marsh, J., Ecker, D. J., and Crooke, S. T., 1988, Ubiquitin-metallothionein fusion protein expression in yeast, *J. Biol. Chem.* **263**:16364–16371.

Butt, T. R., Jonnalagadda, S., Monia, B. P., Sternberg, E. J., Marsh, J. A., Stadel, J., Ecker, D. J., and Crooke, S. T., 1989, Ubiquitin fusion augments the yield of cloned gene products in Escherichia coli, *Proc. Natl. Acad. Sci. USA* **86**:2540–2544.

Ciechanover, A., 1987, Regulation of the ubiquitin-mediated proteolytic pathway: Role of the substrate α-NH₂ group and of transfer RNA, *J. Cell. Biochem.* **34**:81–100.

Ciechanover, A., Finley, D., and Varshavsky, A., 1984, Ubiquitin dependence of selective protein degradation demonstrated in the mammalian cell cycle mutant ts85, *Cell* **37**:57–66.

Dworkin-Rastl, E., Shrutkowski, A., and Dworkin, M. B., 1984, Multiple ubiquitin mRNAs during *Xenopus laevis* development contain tandem repeats of the 76 amino acid coding sequence, *Cell* **39**:321–325.

Ecker, D. J., Khan, M. I., Marsh, J., Butt, T. R., and Crooke, S. T., 1987a, Chemical synthesis and expression of a cassette adapted ubiquitin gene, *J. Biol. Chem.* **262**:3524–3527.

Ecker, D. J., Butt, T. R., Marsh, J., Sternberg, E. J., Margolis, N., Monia, B. P., Jonnalagadda, S., Khan, M. I., Weber, P. L., Mueller, L., and Crooke, S. T., 1987b, Gene synthesis, expression, structures, and functional activities of site-specific mutants of ubiquitin, *J. Biol. Chem.* **262**:14213–14221.

Ecker, D. J., Butt, T. R., Marsh, J., Sternberg, E. J., Dixon, J. S., Weber, P. L., and Crooke, S. T., 1989, Ubiquitin function studied by disulfide engineering, *J. Biol. Chem.* **264**:1887–1893.

Finley, D., Ciechanover, A., and Varshavsky, A., 1984, Thermolability of ubiquitin-activating enzyme from the mammalian cell cycle mutant ts85, *Cell* **37**:43–55.

Finley, D., Ozkaynak, O., and Varshavsky, A., 1987, The yeast polyubiquitin gene is essential for resistance to high temperatures, starvation, and other stresses, *Cell* **48**:1035–1046.

Gausing, K., and Barkardottir, R., 1986, Structure and expression of ubiquitin genes in higher plants, *Eur. J. Biochem.* **158**:57–62.

Hershko, A., 1988, Ubiquitin-mediated protein degradation, *J. Biol. Chem.* **263**:15237–15240.

Hershko, A., and Ciechanover, A., 1986, Transfer RNA is an essential component of the ubiquitin and ATP-proteolytic system, *Prog. Nucleic Acid Res. Mol. Biol.* **33**:19–56.

Izquierdo, M., Arribas, C., Galceran, J. Burke, J., and Cabrera, V. M., 1984, Characterization of a *Drosophila* repeat mapping at the early-ecdysone puff 63F and present in many eukaryotic genomes, *Biochim. Biophys. Acta* **783**:114–121.

Jonnalagadda, S., Ecker, D. J., Sternberg, E. J., Butt, T. R., and Crooke, S. T., 1988, Ubiquitin carboxyl-terminal peptides, *J. Biol. Chem.* **263**:5016–5019.

Jonnalagadda, S., Butt, T. R., Monia, B. P., Mirabelli, C. K., Gotlib, L., Ecker, D. J., and Crooke, S. T., 1989, Multiple (α-NH-ubiquitin) protein endoproteases in cells, *J. Biol. Chem.* **264**:10637–10642.

Lund, P. K., Moats-Staats, B. M., Simmons, J. G., Hoyt, E., D'Ercole, A. J., Martin, F., and Van Wyk, J. J., 1985, Nucleotide sequence analysis of a cDNA encoding human ubiquitin reveals that ubiquitin is synthesized as a precursor, *J. Biol. Chem.* **260**:7609–7613.

Miller, J., McLachlan, A. D., and Klug, A., 1985, Repetitive zinc-binding domains in the protein transcription factor IIIA from *Xenopus* oocytes, *EMBO J.* **4**:1609–1614.

Monia, B. P., Ecker, D. J., Jonnalagadda, S., Marsh, J., Gotlib, L., Butt, T. R., and Crooke, S. T., 1989, Gene synthesis, expression, and processing of human ubiquitin carboxyl extension proteins, *J. Biol. Chem.* **264**:4093–4103.

Mori, H., Kondo, J., and Ihara, Y., 1987, Ubiquitin is a component of paired helical filaments in Alzheimer's disease, *Science* **235**:1641–1644.

Okabe, T., Fujisawa, M., Mihara, A., Sato, S., Fujiyoshi, N., and Takaku, F., 1986, Amino-terminal amino acid sequence of a novel autocrine growth factor: Homology with ubiquitin, *J. Cell Biol.* **103**:442a.

Ozkaynak, E., Finley, D., and Varshavsky, A., 1984, The yeast ubiquitin gene: head-to-tail repeats encoding a polyubiquitin precursor protein, *Nature* **213**:663–666.

Ozkaynak, E., Finley, D., Solomon, M., and Varshavsky, A., 1987, The yeast ubiquitin genes: A family of natural gene fusions, *EMBO J.* **6**:1429–1439.

Rechsteiner, M., 1987, Ubiquitin-mediated pathways for intracellular proteolysis, *Annu. Rev. Cell Biol.* **3**:1–30.

Rechsteiner, M. (ed.), 1988, *Ubiquitin,* Plenum Press, New York.

Rhodes, D., and Klug, A., 1986, An underlying repeat in some transcriptional control sequences corresponding to half a double helical turn of DNA, *Cell* **46**:123–132.

Schlesinger, D. H., and Goldstein, G., 1975, Molecular conservation of 74 amino acid sequence of ubiquitin between cattle and man, *Nature* **255**:423–424.

Vierstra, R. D., Langan, S. M., and Schaller, G. E., 1986, Complete amino acid sequence of ubiquitin from the higher plant Avena sativa, *Biochemistry* **25**:3105–3108.

Vijay-Kumar, S., Bugg, C. E., and Cook, W. J., 1987, Structure of ubiquitin refined at 1.8 resolution, *J. Mol. Biol.* **194**:531–544.

Wilborg, O., Pedersen, M. S., Wind, A., Berglund, L. E., Marcker, K. A., and Vuust, J., 1985, The human ubiquitin multigene family: some genes contain multiple directly repeated ubiquitin coding sequences, *EMBO J.* **4**:755–759.

Wilkinson, K. D., and Mayer, A. N., 1986, Alcohol-induced conformational changes of ubiquitin, *Arch. Biochem. Biophys.* **250**:390–399.

Recognition at Membrane Surfaces
Influenza HA and Human HLA

DON C. WILEY

1. Introduction

The influenza virus hemagglutinin (HA) and the human histocompatibility antigens HLA-A2 and HLA-A28 are membrane glycoproteins whose three-dimensional structures have been determined by X-ray diffraction as part of an effort to understand their mechanisms of action.

Here we consider recent results and our current understanding of the following: How the influenza HA mediates binding of influenza virus to its cellular receptor; progress on how the influenza HA mediates viral entry into a cell by membrane fusion, and how histocompatibility antigens recognize processed foreign antigens and are themselves recognized by the receptors (TCR) on cytotoxic T lymphocytes (CTLs). Although our understanding of the first two points is incomplete, strategies for intervening to prevent virus infections can be suggested.

2. HA–Cell Receptor Interaction

Sialic acids are the only known components of the cell surface necessary for influenza virus binding. On human cells the predominant sialic acid is α-N-acetyl neuraminic acid (α-NeuAc), a sialic acid (Fig. 1) which is the only known component of the cell surface necessary for influenza virus binding; the virus

---

DON C. WILEY • Department of Biochemistry and Molecular Biology and Howard Hughes Medical Institute, Harvard University, Cambridge, Massachusetts 02138.

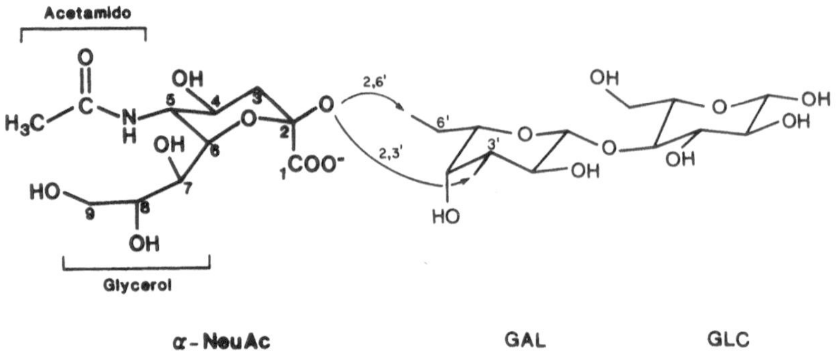

Figure 1. Sialyllactose: α-*N*-acetyl neuraminic acid (α-NeuAc), a sialic acid, linked α(2,6) or α(2,3) to lactose (galactose-β(1,4)-glucose), binds to the influenza HA. The α-anomers of sialic acids have their carboxylate axial to the pyranose ring at the 2 position. The acetamido and glycerol moieties of α-NeuAc are equatorial to the ring at the 5 and 6 positions, respectively.

does not bind to neuraminidase-treated cells (Paulson, 1985; Paulson *et al.*, 1979), and binding is restored by resialiation (Gottschalk, 1959) or addition of sialiated glycolipids (Bergelson *et al.*, 1982; Suzuki *et al.*, 1985). Oligosaccharides terminating in sialic acid are found on many glycoproteins and glycolipids on cellular membranes. Different strains of influenza virus can distinguish different linkages of sialic acid [α(2,6) and α(2,3), Fig. 1] to penultimate saccharides (Rogers and Paulson, 1982; Rogers *et al.*, 1983a).

The HA glycoprotein from influenza virus has a pocket composed of amino acid residues conserved throughout the sequence variation of the HA that is associated with recurrent epidemics and pandemics in the human population (Fig. 2A, B). Mutations at position 226 in this pocket (Fig. 2B) alter receptor specificity for the 1968 Hong Kong virus from a preference for α(2,6) linked α-NeuAc to α(2,3) linked α-NeuAc (Fig. 1) (Rogers *et al.*, 1983b). William Weis and colleagues (Weis *et al.*, 1988) have determined the three-dimensional structure of complexes of sialyllactose, an analog of the cell receptor, with the HA from both the wild-type and "α(2,3)" mutant virus by X-ray diffraction. In both cases, the difference Fourier electron density showed a flat central disk with in-plane protrusions interpreted as the ring substitutents of α-NeuAc and one out-of-plane protrusion extending toward the protein interpreted as the axial carboxylate substitutent at C2 (Fig. 1). No electron density was observed corresponding to the lactose, although chemical assays on crystals after the X-ray exposure confirmed its presence, suggesting that in the crystals the lactose is spatially disordered, an observation consistent with binding studies in solution (see Section 2.1).

On the basis of the two structures, crystallographic least squares refined to 3 Å resolution, an atomic model for the interaction of influenza virus with its

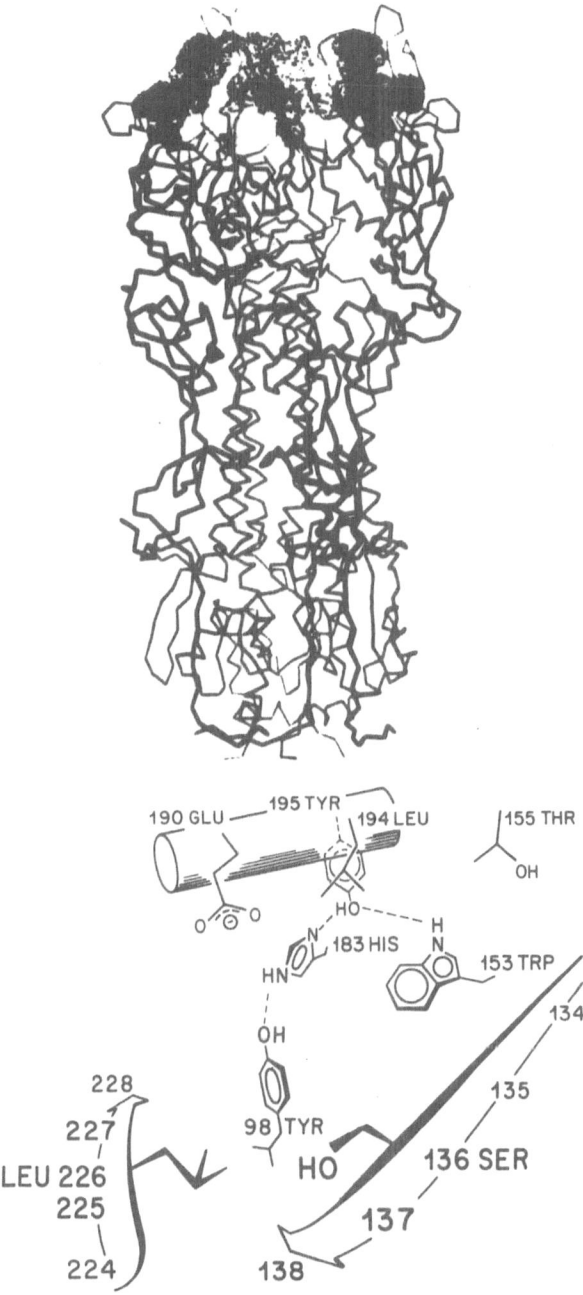

Figure 2. (A) The influenza virus hemagglutinin binding sites. Location of the three receptor binding sites on the influenza haemagglutinin. The HA is a trimer composed of three HA_1 chains and three HA_2 chains (only α carbons are shown). Each site is composed of conserved residues within a single HA_1 monomer; these residues are shown as a dot surface. (B) The receptor binding site showing hydrogen bonding among conserved residues in site.

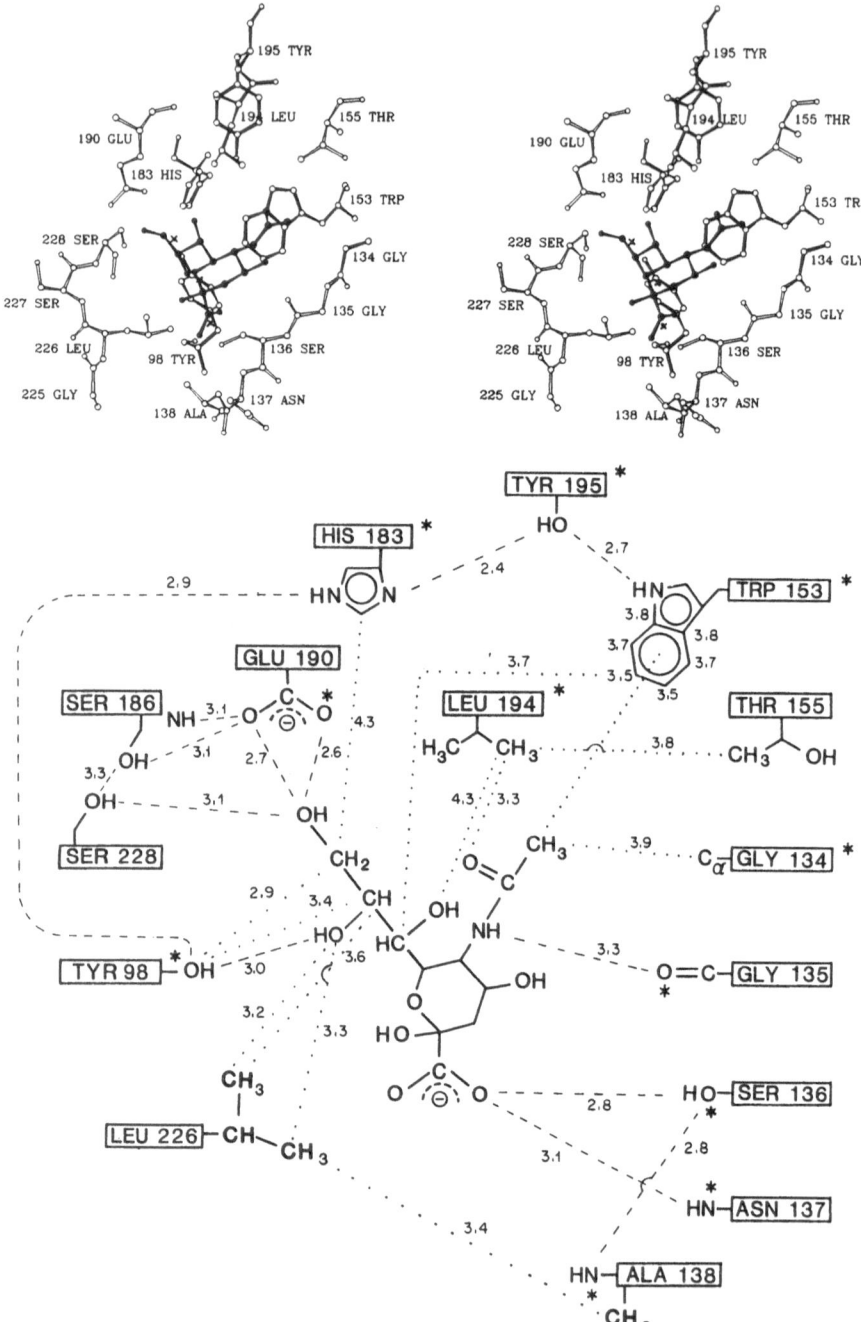

Figure 3. (A) Wild-type receptor site with sialic acid, stereo pair. Residues with solid bonds are conserved in all known HA sequences from natural isolates. Sialic acid is shown with solid spheres. Three water molecules displaced by sialic acid binding are indicated by X's. Diagram drawn with

receptor, sialic acid, was presented (Weis *et al.*, 1988) (Fig. 3A, B). Figure 3B shows potential interactions between the HA and sialic acid, including the observation that the *N*-acetyl methyl group appears to contact the center of the six-membered ring of the conserved tryptophan at position 153, which through a hydrogen bond network is mutually oriented by tyrosine 98, which contacts glycerol hydroxyl 8. This local stabilization among the binding site residues may be important, considering the large variation in amino acid sequence that occurs around the edges of the site (see Section 2.2). Figure 3A shows that the glycerol hydroxyls and the carboxylate oxygens appear to displace bound water molecules upon binding. Beth Wurzburg, in our laboratory, is constructing site-directed mutants to test the strength of some of the interactions seen in Fig. 3B.

2.1. Nuclear Magnetic Resonance and Solution Studies of Binding

Some evidence that the model based on crystal structures is applicable to HA–sialic acid binding in solution has resulted from the development of a nuclear magnetic resonance (NMR) binding assay (Sauter, Bednarski, Wurzburg, Whitesides, and Wiley, unpublished). Figure 4 shows a comparison of the 500 MHz NMR spectra of $\alpha(2,3)$ sialyllactose as a free ligand (bottom) and in the presence of the purified, soluble (Brand and Skehel, 1982) influenza virus HA glycoprotein. In the presence of HA one ligand resonance shifts upfield (the *N*-acetyl methyl resonance), and a number of references broaden (e.g., *N*-acetyl methyl protons, three axial and three equitorial protons), as expected for a small ligand interacting with a 200-kDa protein. (Experiments indicate that the broadening is an effect of the binding between HA and the sialic acid ligand, not an effect of viscosity.) Resonances from the lactose moiety (e.g., H_1' and H_1'') appear essentially unbroadened by the interaction, which is consistent with the observation that the lactose does not seem to interact strongly in the crystal. Titration of the line broadening and change in chemical shift of the *N*-acetyl methyl resonance (Fig. 4B) by altering the protein-to-ligand ratio permits the calculation of equilibrium dissociation constants (Kronis and Carver, 1982).

The dissociation constant for the HA from both wild-type and mutant HA measured this way is about 2 mM (Table I), which is weak. Since virus–cell

---

ARPLOT (Lesk and Hardman, 1982). (B) Potential interactions of α-NeuAc with wild-type (a) virus HA. Potential hydrogen bonds (dashed lines) and van der Waals contacts (dotted lines) in wild-type (D1112G) + $\alpha(2,6$ sialyllactose) (a). An asterisk (*) next to a box indicates that the residue is conserved in all known HA sequences from natural isolates; an asterisk next to an atom means that it is main chain or conserved in substitutions. (a) Coordinates from a refined model of D1112G + α-NeuAc. [The distances shown (in Å) are averages of the same distance in the three sites of the refined trimer models. Distances shown next to the atoms of the six-membered ring of Trp 153 are to the acetamido methyl group of NeuAc. Some atoms are shown with more hydrogen bonds than they can actually participate in, due to uncertainties in the refined coordinates (the coordinate error is approximately 0.3 Å).]

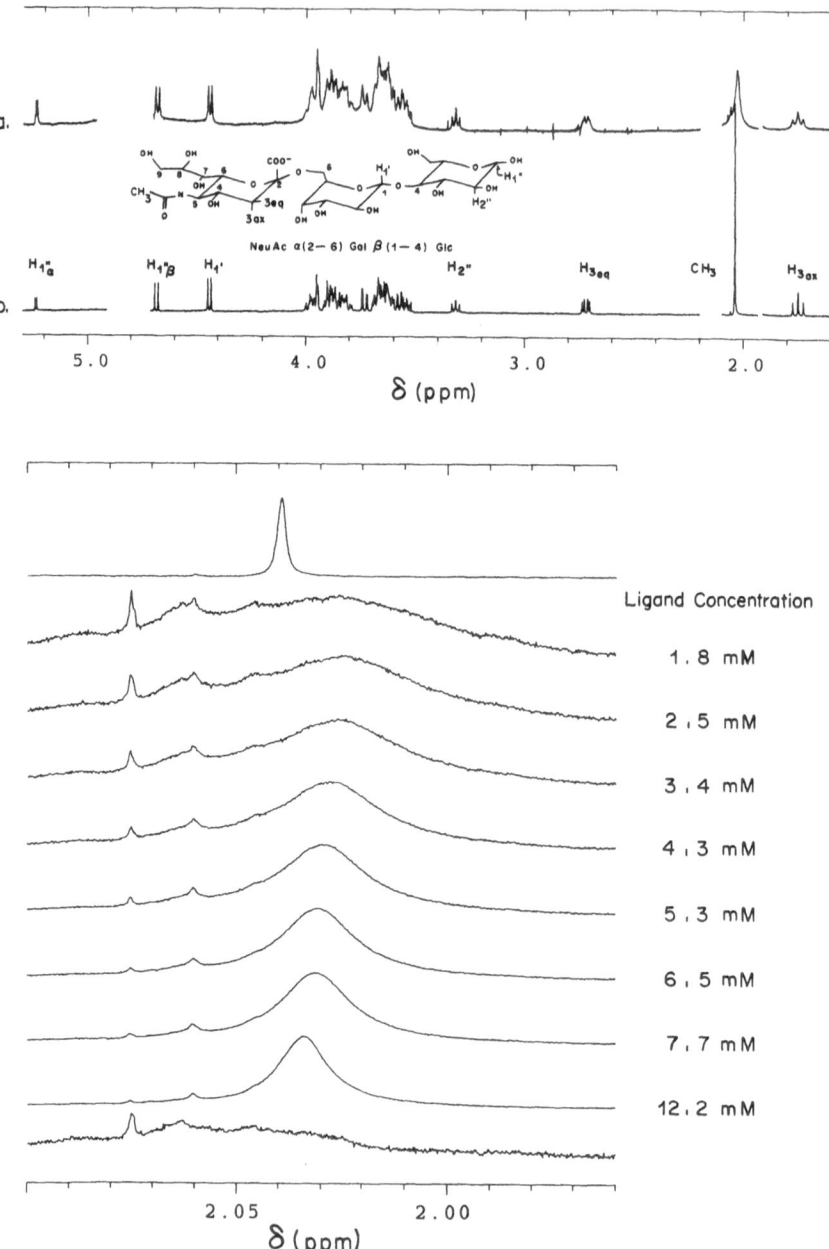

Figure 4. (A) 500-MHz proton NMR spectra of NeuAc α(2->6)Gal β(1->4)Glc in the presence (a) and absence (b) of X31 BHA. The three resolved resonances on the NeuAc moiety (H$_{3eq}$, H$_{3ax}$, and *N*-acetyl) exhibit considerable broadening in the presence of protein, and the *N*-acetyl displays a significant upfield chemical shift. Three resonances on the lactose moiety (H$_{1'}$, H$_{2''}$, and ;H$_{1\beta''}$) show a small amount of line broadening, and no discernible chemical shift. Samples were prepared in

Table I. Dissociation Constants Based on *N*-Acetyl Chemical Shifts
(pH 7.1 24°C)[a]

BHA variant	Ligand	K_D
X31 (Leu 226)	NeuAc $\alpha(2 \rightarrow 6)$Gal $\beta(1 \rightarrow 4)$Glc	2.1 ± 0.5 mM
	NeuAc $\alpha(2 \rightarrow 3)$Gal $\beta(1 \rightarrow 4)$Glc	3.2 ± 0.6 mM
	NeuAc $\alpha(2 \rightarrow)$ methyl	2.8 ± 0.3 mM
	NeuAc $\beta(2 \rightarrow)$ methyl	>20 mM
HAM (Gln 226)	NeuAc $\alpha(2 \rightarrow 6)$Gal $\beta(1 \rightarrow 4)$Glc	6 ± 1 mM
	NeuAc $\alpha(2 \rightarrow 3)$Gal $\beta(1 \rightarrow 4)$Glc	2.8 ± 0.5 mM
	NeuAc $\alpha(2 \rightarrow)$ methyl	4.7 ± 0.4 mM

[a]Hemagglutinin containing leucine at position 226 preferentially binds to sialosides containing an $\alpha(2 \rightarrow 6)$ glycosidic linkage, whereas the glutamine 226 variant exhibits specificity toward the $\alpha(2 \rightarrow 3)$ linkage. The trapped β-anomer of sialic acid binds to the protein at least 10-fold more weakly than the α-anomer.

interaction is essentially irreversible, it is likely that many HA–sialic acid interactions are maintained simultaneously by the hundreds of HAs and sialic acid molecules on the opposing membrane surfaces. Two other points are made by the NMR data in Table I: (1) Methyl sialosides (where a single methyl group replaces the lactose in sialyllactose) have approximately the same dissociation constants as sialyllactose, which is consistent with the crystallographic result indicating little or no interaction between the HA and the lactose moiety. (2) The α anomer (axial carboxylate) of sialosides binds with at least 50–100 times more affinity than the β anomer (equatorial carboxylate), which is consistent with the binding mode observed crystallographically, where the axial carboxylate appears to make a number of important contacts with the HA (Fig. 3B), which an equatorial carboxylate could not make. The magnitude of the totally bound upfield chemical shift of the *N*-acetyl methyl protons has been estimated as about 1.9 ppm (from experiments at different magnetic field strengths). Such a large upfield shift is consistent with the location of the methyl group over the face of an aromatic ring of a tryptophan as observed crystallographically; that position should result in a large upfield ring-current shift.

Similar correlations between the interaction observed in the crystals and the

deuterated phosphate-buffered saline, pH 7.1; and the chemical shift scales were aligned based on internal sodium trimethylsilylpropionate. (B) The line-broadening and chemical shift of the *N*-acetyl resonance provide a measure of the binding of sialyllactose to hemagglutinin. The spectrum at the top is NeuAc $\alpha(2\text{->}6)$Gal $\beta(1\text{->}4)$Glc. The amounts of line-broadening and upfield shift are proportional to the fraction of ligand bound to the protein. Therefore, at high concentrations of sialyllactose, when only a small fraction of it is bound to the protein, the *N*-acetyl resonance more closely resembles that of the free ligand.

effect of sialosides on the rate of binding assayed between virus and sparsely sialiated erythrocytes (Colman *et al.*, 1983; Richardson *et al.*, 1980) have been discussed previously (Weis *et al.*, 1988).

2.2. Antigenic Variation and Inhibitor Design

Figure 5 shows the top view of the HA trimer. There is a central hole in the molecule, and the three sialic acid binding sites are shown by arrows. The dot

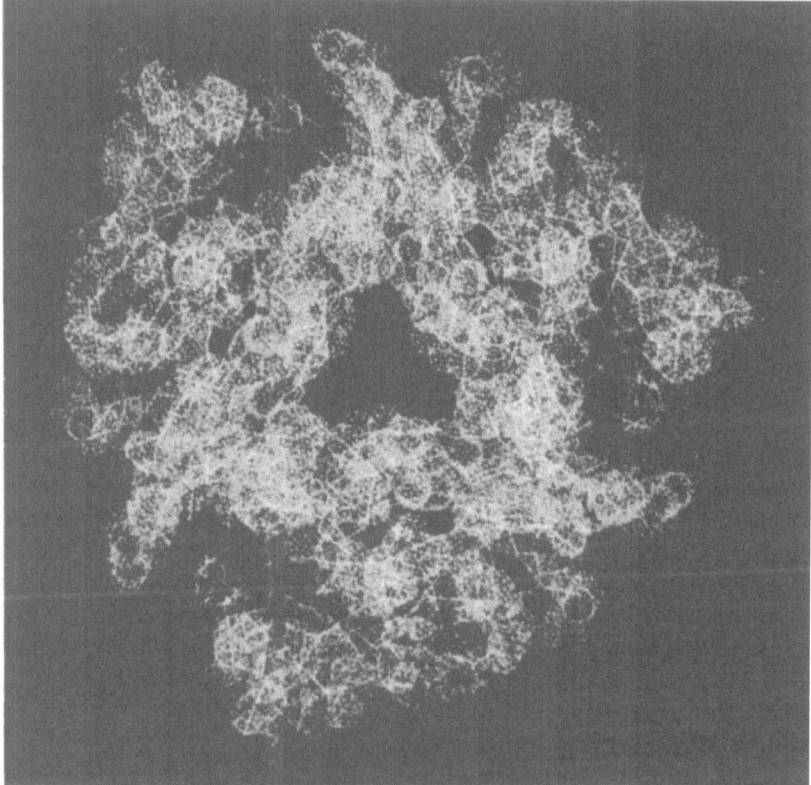

Figure 5. The location of the influenza virus haemagglutinin binding sites. Top view of the hemagglutinin trimer illustrating antigenic variation around three receptor-binding sites (arrows). Only the α carbons are shown. Residues covered in dot surface are those of A/Aichi/68 HA that have varied between 1968 and 1982 in natural isolates, and also those from antigenically distant viruses with single amino acid substitutions selected by growth in neutralizing monoclonal antibodies. The *N*-linked carbohydrate at Asn 165, which protects the protein surface from antibody pressure (Skehel *et al.*, 1984), is also shown as a dot surface. The bar represents 25 Å, the approximate size of an antibody "footprint" (Amit *et al.*, 1986; Colman *et al.*, 1987; Sheriff *et al.*, 1987). Antigenic sites surrounding receptor binding sites have been noted for influenza neuraminidase (Colman *et al.*, 1983), influenza HA (Wiley *et al.*, 1981), and rhinovirus (Rossmann *et al.*, 1986).

surface shows all those residues that have changed during antigenic variation since 1968 and the oligosaccharide at position 165 that covers part of the surface. It is clear that the sialic acid binding site is the only part of the top surface of the molecule that has been conserved. Since sialic acid essentially fills this site (Weis *et al.*, 1988), it seems likely that it is the only component of the cell surface receptor that is recognized by HA_1. No other region in this part of the HA appears to have been conserved through viral evolution for the binding function. An inhibitor designed to interact with the conserved residues in this site might block virus–cell binding and be effective against influenza virus of all subtypes (Weis *et al.*, 1988). Such an inhibitor would probably have to be presented as a polymer or on a microsurface to overcome the large effective affinity between the surfaces of the virus and the cell.

A corollary of the data in Fig. 5 is that an antibody directed at the conserved sialic acid binding site would unavoidably overlap [due to its approximately 20 × 30 Å footprint, (Colman, 1988), see 25-Å bar in Fig. 5] areas of the HA on the edge of the site that are variable, rendering it ineffective in the face of antigenic variation. Antigenic sites surrounding receptor or binding enzymatically active sites have been noted for influenza neuraminidase (Colman *et al.*, 1983), influenza HA (Wiley *et al.*, 1981), and rhinovirus (Rossmann *et al.*, 1986).

3. Virus Entry and Membrane Fusion

A number of membrane-containing animal viruses enter cells by fusing the viral membrane to a cellular membrane (for review, see White *et al.*, 1983). Influenza virus, like one class of those viruses, first binds to a cellular receptor and then is endocytosed. The low pH of the endosomal compartment triggers the viral HA molecule to mediate membrane fusion. A highly conserved, very hydrophobic, glycine-rich peptide (GLFGAIFGFI . . .) at the amino terminus of HA_2 (and some other viral fusion proteins) has been implicated in fusion (Skehel and Waterfield, 1975; Gething *et al.*, 1978; Scheid *et al.*, 1978; Richardson *et al.*, 1980). This "fusion peptide" is located in the interface between HA_2 subunits in the "stem" region of the HA about 100 Å from the distal tip of the molecule and 35 Å from the viral-membrane end of the molecule (Fig. 6). A large number of experiments indicate that at the pH optimum for virus fusion (around pH 5.0), the "fusion peptide" of the HA moves from its location in the pH 7.0 conformation of the HA to an exposed location, conferring new lipophilic properties on the protein (reviewed in Wiley and Skehel, 1987) (Fig. 6). The low-pH-induced conformational change appears to be irreversible, but not a general unfolding or denaturation, as properties such as sialic acid binding, antibody binding, and circular dichroism (CD) measurements of secondary structure content are preserved (Skehel *et al.*, 1982) (Fig. 6). Trypsin digestion of the low-pH-

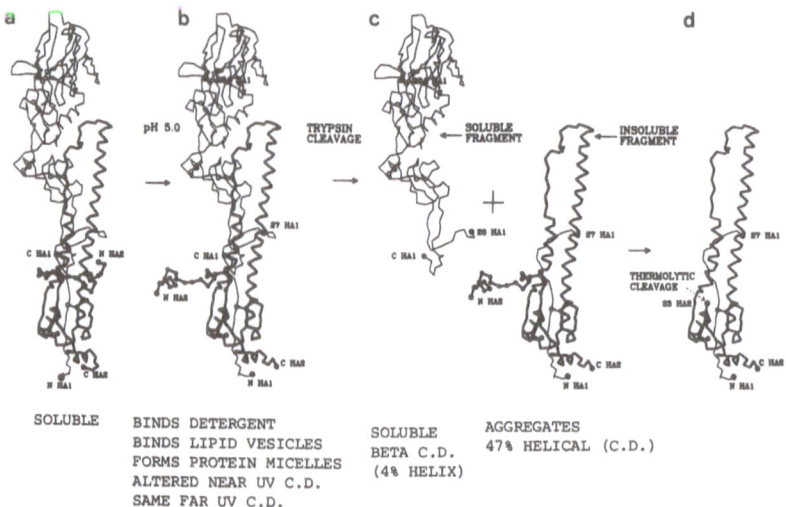

SOLUBLE   BINDS DETERGENT   SOLUBLE   AGGREGATES
               BINDS LIPID VESICLES   BETA C.D.   47% HELICAL (C.D.)
               FORMS PROTEIN MICELLES (4% HELIX)
               ALTERED NEAR UV C.D.
               SAME FAR UV C.D.

Figure 6. Schematic representation of the effect of incubating BHA at low pH. (a) pH 7.0 monomer structure. (b) After low pH, the hydrophobic N-terminal peptide of HA$_2$ is exposed (see text). (c) The low pH conformation of BHA can be cleaved *in vitro* by trypsin at HA$_1$ 27, which leads to release of the "top" globular domain from the stem, indicating a change in the tertiary interactions in the stem regions of the monomer. (d) The hydrophobic properties of the low pH form are low after thermolytic removal of residues 1–23 of HA$_2$.

induced conformation reveals new cleavage sites at the HA$_1$–HA$_2$ contact region near HA$_1$ residue 27 and in the HA$_1$–HA$_1$ contact region at residue 224 (Fig. 6), indicating structural alterations in those regions and providing a simple assay for the occurrence of the conformational change. Thermolytic digestion results in the "trypsin-cut" at HA$_1$ 27 and the removal of the "fusion peptide" residues 1–23 (Fig. 6) (or 1–35 under other conditions) (Ruigrok *et al.*, 1988). As thermolytic digestion results in a loss of the lipophilic properties of the low-pH-induced conformation, it seems likely that they resulted from the exposure of the "fusion peptide" (Ruigrok *et al.*, 1988).

Further information about the low-pH-induced conformation of the HA has been provided by the selection of viruses mutant in the pH optimum for membrane fusion (Daniels *et al.*, 1985) by growth in amantidine HCl, which raises the pH of endosomes. A list of 38 such mutants, and others isolated by other means, and their likely structural consequences is given elsewhere (Daniels *et al.*, 1985, 1987; Wiley and Skehel, 1987; Rott *et al.*, 1985; Doms *et al.*, 1986). Figure 7 shows their positions on the HA monomer. Three classes of mutation are evident: (1) (see triangles in Fig. 7) Amino acid substitutions on or near the N-terminal fusion peptide. All appear to destabilize the pH 7.0 location of the fusion peptide, requiring a smaller pH decrease to trigger its release. The struc-

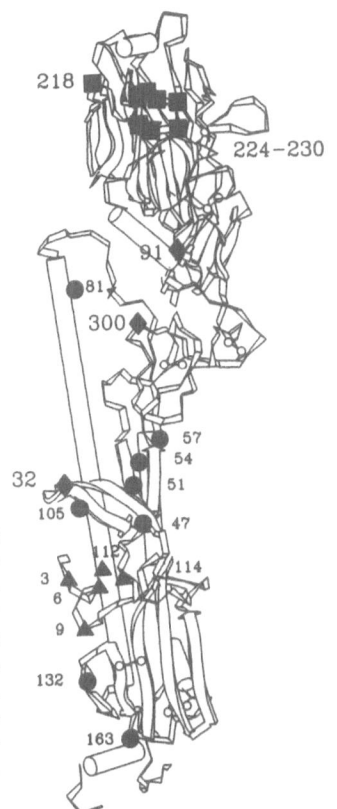

Figure 7. Schematic drawing of the HA monomer showing the location of single-amino-acid substitutions found in the membrane fusion mutants. ■ indicates a substitution located in the HA$_1$–HA$_1$ interface; ●, in the HA$_2$ trimer interface; ▲, in the vicinity of the N-terminal hydrophobic peptide of HA$_2$; and ♦, in the HA$_1$–HA$_2$ interface. Large number labels designate residues in HA$_1$. The N-terminal peptide of HA$_2$, marked N, is tucked into the interface between two HA$_2$ subunits. The amino acid numbers show the positions of single-amino-acid substitutions in membrane fusion mutants.

ture of one such mutant, Asp 112 to Gly has been determined by X-ray diffraction and results in the loss of about four intramolecular hydrogen bonds to the "fusion peptide" (Weis, Cusack, Brown, Daniels, Skehel, and Wiley, unpublished). (2) (see circles and diamonds in Fig. 7) Amino acid substitutions in the interface between HA subunits. All involve changes in the electrostatic charge of residues and appear to destabilize the pH 7.0 conformation of the interfaces in the "stem" region of the molecule, indicating that the conformational change probably takes place over the whole interface. Chemical cross-linking experiments indicate that the trimeric structure is preserved (Ruigrok *et al.*, 1988), although measurements of low protein concentrations argue for a weakening of the interface (Nestorowicz *et al.*, 1985; Doms and Helenius, 1986). (3) (see squares in Fig. 7) Amino acid substitutions and a deletion in the trimer interface between the globular HA$_1$ domains appear to indicate that the interface at the distal end of the structure changes or dissociates during the low-pH-induced conformational change. Evidence from monoclonal (Daniels *et al.*,

1983; Webster *et al.*, 1983; Jackson and Nestorowicz, 1985; Yewdell *et al.*, 1983) and antipeptide (White and Wilson, 1988) antibody binding to the low-pH-induced conformation also argues that the top globular HA_1 domains may dissociate from each other.

We have constructed site-directed mutations in the interface between top globular domains of HA_1 to see whether we could raise the pH optimum for fusion or prevent fusion entirely (Godley, Pfeifer, Shaw, Kaufmann, Wharton, Suchanek, Pabo, Skehel, and Wiley, unpublished). A mutant HA with serine 205 replaced by aspartic acid (located at the center of the HA_1–HA_1 interface) was expressed on Chinese hamster ovary (CHO) cells and compared to HA from wild-type virus expressed on CHO cells. While the wild-type HA undergoes the low-pH-induced conformational change near pH 5.2, the Asp 205 mutant changes conformation at a higher pH. The mutation designed to destabilize the HA_1–HA_1 interface by introducing an unpaired, charged residue appears to change the pH optimum for the conformational change. A disulfide bond was added to link the HA_1 chains around the trimer axis by introducing cysteines at 212 and 216. These positions were chosen from a short list of positions where a disulfide might be possible, which was produced by Carl Pabo and E. Suchanek (Johns Hopkins) using their protein engineering program PROTEUS. The mutant HA molecule, when expressed in CHO cells and precipitated by HA-specific antibody, had a molecular weight that was three times that of the monomer, and could be reconverted to the monomer molecular weight by reduction. This covalent trimer of HA does not undergo the low-pH-induced conformation change as monitored by trypsin digestion. Experiments with monoclonal antibodies that distinguish the pH7 from pH5 conformation and with membrane-fusion assays employing fluorescent liposomes and polykaryon formation all indicate that while wild-type HA expressed on CHO cells is fusion active, the disulfide mutant is not.

Our current understanding of membrane fusion is limited. At low pH, the "fusion peptide" appears to be uncovered, accompanied by other conformational changes throughout the molecule. Synthetic peptides with the same sequence as the N-terminus of HA_2 have been shown to promote fusion of small, highly curved lipid vesicles (Lear and Degrado, 1987; Wharton *et al.*, 1988), but not to fuse cell membranes. It is thus possible that membrane fusion is a simple consequence of unmasking this fusion-active peptide. A generally accepted view is that the low-pH conformational change unmasks the "fusion peptide," which then interacts with either the cellular or the viral membrane or both to bring the two membranes into apposition and possibly simultaneously to destabilize them. In the absence of further knowledge about the fusion mechanism, it is still possible to conceive of preventing the low-pH conformational change, as the disulfide mutant demonstrated, by stabilizing the pH 7.0 conformation of the molecule. This would be equivalent to the strategy used by Beddell and Goodford at Wellcome Ltd. (Beddell *et al.*, 1984); they designed small molecules that stabilized the oxy conformation of hemoglobin, effecting the oxy-to-deoxy conformational change required for the aggregation of hemoglobin found in sickle cell disease. A molecule that could stabilize the pH 7 fusion-inactive form of the HA might prevent expression of the membrane fusion activity and therefore block viral infection.

4. Human Histocompatibility Antigens

There are two classes of major histocompatibility antigen membrane glycoproteins: Class I antigens are composed of one membrane-anchored polypeptide chain with three domains, $\alpha1$, $\alpha2$, $\alpha3$, each 90 residues long, noncovalently associated with a soluble 99-residue IgG-like domain $\beta2$-microglobulin (Cresswell *et al.*, 1973; Grey *et al.*, 1973; Peterson *et al.*, 1972) (Fig. 8a). These antigens are found on almost all cell surfaces and are the target of

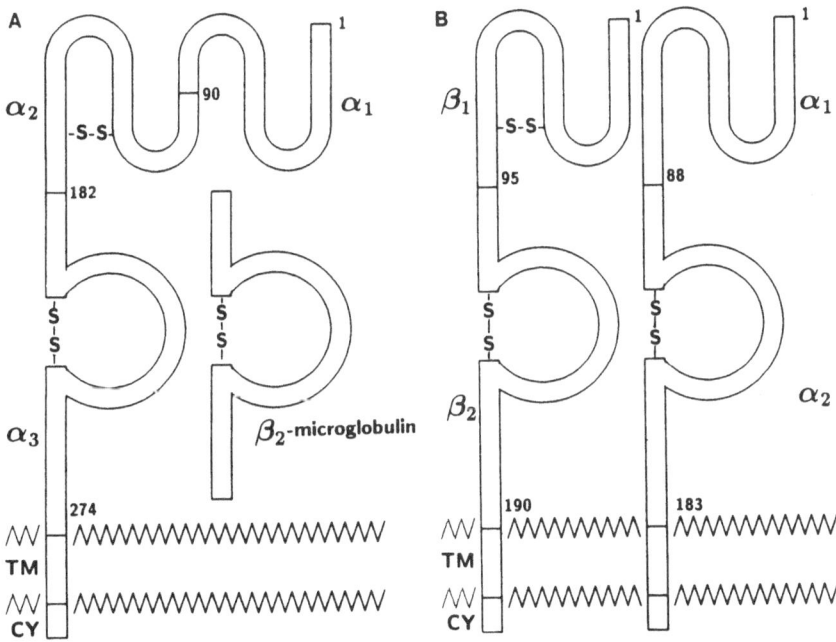

Figure 8. Class I (A) and class II (B) histocompatibility antigens appear to have similar domains distributed differently on polypeptide chains. Class I molecules have three domains, including one homologous to IgG constant regions (α_3) on one transmembrane polypeptide and a second IgG constant region homology on a separate chain (β_2M). Class II molecules have two transmembrane polypeptide chains, each with one IgG-like domain ($\alpha_2 + \beta_2$).

TCRs on CTLs. Similarly, class II antigens appear to have four structural domains, but they are distributed evenly on two transmembrane polypeptides (review in Kaufman *et al.*, 1984) (Fig. 8b); these antigens are found on the surface of specialized cells of the immune system, such as B cells and macrophages, and are normally the target of TCRs on T-helper lymphocytes (review in Schwartz, 1983).

Although these molecules were originally discovered as markers for immunological "self" and are the targets in the rejection of foreign tissue transplants (Bach and van Rood, 1976), their predominant function is to present processed foreign antigens (i.e., short peptides) for recognition by CTL and helper T cells (review in Schwartz, 1983; Townsend, 1987). Figure 9 shows schematically how foreign antigens, produced in this case by a viral infection, are synthesized in the cytoplasm, but end up on the cell surface as degraded peptides bound to histocompatibility antigens (Townsend, 1987). The T-cell receptors are hypervariable molecules closely homologous to immunoglobulin Fab fragments (Hedrick *et al.*, 1984; Yanagi *et al.*, 1984). CTLs recognize the complex of a processed foreign antigen and a specific histocompatibility antigen and kill the infected cell. A similar phenomenon takes place with class II antigens, but the foreign antigen originates from outside the cell. The foreign antigens appear to be degraded in intracellular vesicles (Ziegler and Unanue, 1982) and bound to a class II antigen. This complex is then recognized by the TCR of a helper T cell.

Histocompatibility antigens are highly polymorphic in the population, but a very limited number are expressed in an individual (e.g., individual humans

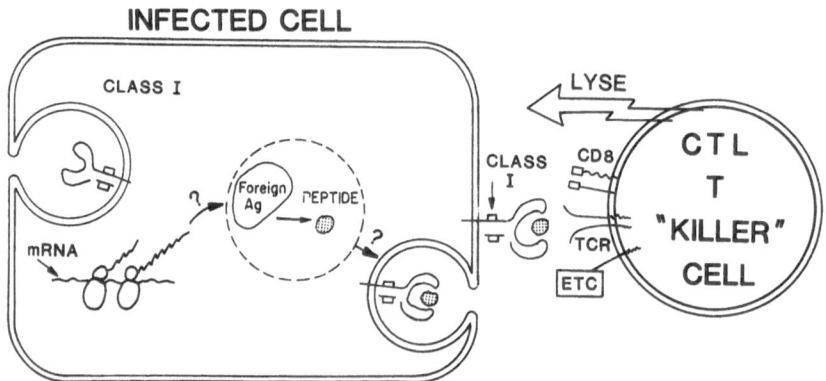

Figure 9. Cytotoxic-thymic Lymphocytes (CTL) recognize virally infected cells. The hypervariable T-cell receptor (TCR) on a CTL appears to bind to a complex antigen composed of the class I histocompatibility antigen and a foreign (viral) antigen. The viral antigen, synthesized in the cytoplasm of an infected cell, is broken down (processed) into antigenic peptides by an unknown pathway. The resulting peptide is "presented" on the cell surface complexed with a class I histocompatibility antigen.

express only six different class I polymorphs). An unusual aspect of this molecular recognition system is that since so many foreign antigen peptides must be recognized in each individual over a lifetime, each histocompatibility antigen must be able to bind thousands of different peptides. How this is accomplished is a major question.

4.1. The Structure of HLA-A2

We determined the three-dimensional structure of the human class I histocompatibility antigen, HLA-A2, to 3.5 Å resolution by X-ray diffraction in 1987 (Bjorkman *et al.*, 1987a) (Fig. 10). The structure was determined primarily through the efforts of Pamela Bjorkman, who, as a graduate student and postdoctoral fellow, crystallized the papain-soluble form of the molecule in collaboration with Jack Strominger (Bjorkman *et al.*, 1986), collected the high-resolution X-ray data on two crystal forms located isomorphous heavy atom derivatives, calculated electron density maps, and integrated and refined the structure. She was joined by Boudjema Samraoui and together they were able to completely trace the α3 and β2m domains, revealing two IgG-constant-like domains, as expected from sequence homology (Orr *et al.*, 1979a,b,c; Tragardh *et al.*, 1979; Peterson *et al.*, 1974; Smithies and Poulik, 1982), but which related to each other in a novel way (Bjorkman *et al.*, 1986). The electron density maps also showed all the secondary structural elements of α1 and α2, eight extended chains forming a β sheet and two long α-helical regions on top of the sheet (Fig. 10, Fig. 11), but did not allow unambiguous connection to be made among these elements. Mark Saper joined the group to carry out phase refinement by iterative phase averaging (Bricogne *et al.*, 1976) of two crystal forms of HLA-A2; 3.5 Å resolution electron density maps produced by this procedure were interpreted by Saper, Bjorkman, and Samraoui as an unambiguous fit to the amino acid sequence (Bjorkman *et al.*, 1987b). The α1 and α2 domains have a similar chain-fold and appear to form an "intramolecular dimer" by the edge-to-edge apposition of their four-strand β sheets, a dimerization motif observed previously in intramolecular interactions and therefore possible for α1 and β1 of class II antigens (Bjorkman *et al.*, 1987b).

The structure revealed a striking cleft between the α-helical regions of α1 and α2 (Fig. 10), the bottom of which was formed primarily by the first two β strands of each domain (Fig. 11). Furthermore, the cleft contained extraelectron density not accounted for by the HLA amino acid sequence, and thought possibly to be the image of a processed antigenic peptide or a mixture of peptides (Bjorkman *et al.*, 1987b). Potential sources for such a peptide (peptides) have been enumerated elsewhere (Bjorkman *et al.*, 1987a,b). We have argued that this cleft is the processed foreign antigen binding site on the basis of its size and position; because of the presence of the majority of polymorphic amino acid positions in

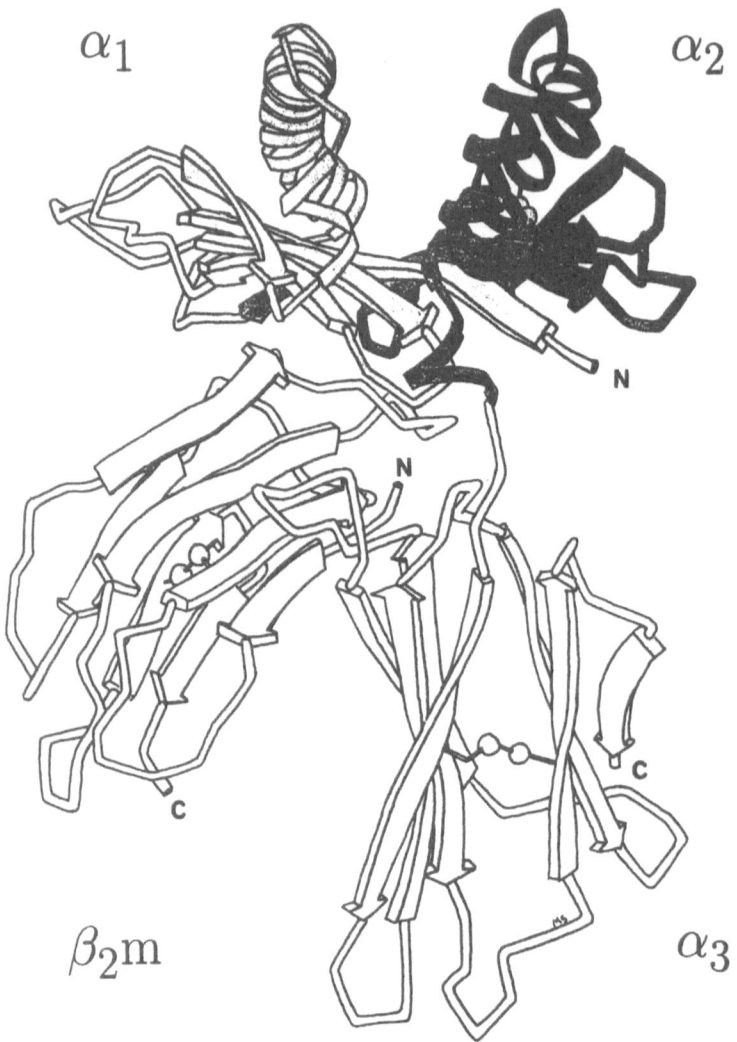

α_1

α_2

β_2m

α_3

Figure 10. Schematic representation of the structure of HLA-A2. The β-strands are shown as thick arrows in the amino-to-carboxy direction; α-helices are represented as helical ribbons. Connecting loops are depicted as thin lines. Disulfide bonds are indicated as two connected spheres. The molecule is shown with the membrane proximal immunoglobulin-like domains (α3,β₂m) at the bottom, and the polymorphic α1 and α2 domains at the top. The indicated C-terminus of α3 is the papain cleavage site. In the membrane-bound molecule, another 13 amino acids extend past the cleavage site toward the membrane, which we assume to be horizontal at the bottom of the figure. This orientation places the helices of the α1 and α2 domains, and the antigen recognition site located between them, on the top surface of the molecule. The domains α1 and α2 form a platform with a single eight-stranded β-pleated sheet (seen edge on) covered with α-helices.

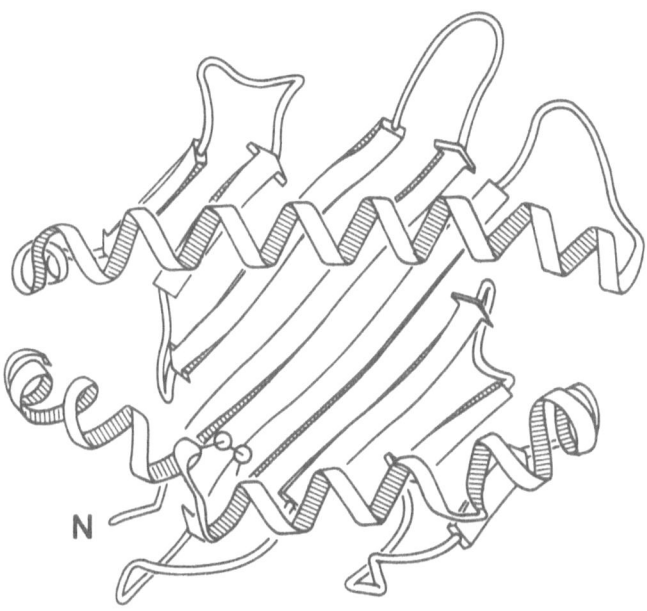

N

Figure 11. Schematic representation of the top surface of HLA-A2. The α1 and α2 domains are shown as viewed from the top of the molecule, showing the surface that is presumably contacted by a T-cell receptor (90° from Fig. 10). The N-terminus of α1 is indicated. Each domain consists of four antiparallel β-strands followed by a long helical region, and the domains pair to form a single eight-stranded β-sheet topped by α-helices. The disulfide bond connects residue 164 in the α2 long helix to residue 101 in the first β-strand of α2.

mouse and human in and around the cleft; and on the basis of the location of mutations that alter recognition by CTLs (Bjorkman *et al.*, 1987b). For example, Table II lists specific residues in mutant H-2K^b mice which result in histoincompatibility with their parental haplotype (review in Nathenson *et al.*, 1986) and residues in human class I proteins defining serological subtypes with altered reactivity for CTLs (Biddison *et al.*, 1980; review in Lopez de Castro *et al.*, 1985); underlined residues indicate that they face into the binding cleft or are located at the edge of the cleft pointing into solution (or toward a TCR). The table demonstrates that all but one of the mutants with altered recognition by T cells (see Table II caption) have altered residues in the cleft or on its edges, arguing that the cleft is the site of foreign antigen recognition and the site recognized by TCRs (Bjorkman *et al.*, 1987b). The cleft is about 25 Å long, 10 Å wide, and 10 Å deep (Bjorkman *et al.*, 1987a). At its widest point it appears that it could accommodate a helical peptide (Bjorkman *et al.*, 1987b).

HLA-A2 has been crystallographically least-squares-refined at 2.6 Å resolution by M. Saper (Saper, Bjorkman, Garrett, Samraoui, Bennett, Strominger,

Table II. Human and Murine Variants with Altered CTL
Reactivity[a]

Murine	
K^{bm1}	E152A, R155Y, L156Y
K^{bm8}	Y22F, M23I, E24S
K^{bm5,16}	Y116F
K^{bm6,7,9}	Y116F, C121R
K^{bm3}	D77S, K89A
K^{bm11}	D77S, T80N
K^{bm23}	R75H, D77S
K^{bm10}	T163A, V165M, K173E, N174L
K^{k}-TNP	A40V
R8.313(H-2K^{b})	L82F
Human	
HLA-A2.4	F9Y
HLA-A2.2Y	F9Y, 43, 95, 156
M7 (HLA-A2)	E43R, V95L, L156W
HLA-B27d	Y59H
HLA-B27F	D74Y, D77N, L81A
HLA-B27.2	D77N, T80I, L81A
HLA-B27.3	D77S, V152E
HLA-27.4	D77S, H114D, D116Y, V152E
SDM(Aw68)	I95V, M97R
KNE(HLA-A2.4)	Y99C
HLA-A3	G107W
HLA-Aw68	G107W
HLA-B7(CF)	Y116F
DK1(HLA-A2)	A149T, V152E, L156W
E1(HLA-A3)	E152V, L156Q

[a]Spontaneous mutants in the H-2K^{b} gene of mice (reviewed in Nathenson *et al.*, 1986) and human serological subtype variants (reviewed in Bjorkman *et al.*, 1987b) that exhibit altered T-cell recognition are listed with the altered amino acid positions. Underlined amino acids are believed to face into the class I binding cleft, dashed underlines face up away from the cleft, possibly toward the T-cell receptor (Bjorkman *et al.*, 1987b).

and Wiley, unpublished). Figure 12 shows a space-filling model of the HLA-A2 binding site. It is possible to imagine a T-cell receptor, which based on its sequence similarity may have a footprint like the approximately 20 × 30 Å footprint of Fabs, recognizing a peptide in the cleft and simultaneously those residues along the edge of the cleft in Figure 12. Evidence for that picture comes from a recent study of the recognition of single-amino-acid-substitution mutants of H-2K^{b} by single, cloned CTLs (Ajitkumar *et al.*, 1988). One CTL could distinguish amino acid substitutions at 80 and 82 near one end of the cleft and 162, 166, and 174 at the other end (Fig. 12). The most likely explanation is that the TCR of that cell contacted those regions simultaneously, covering the cleft in the process. The possibility that changes at those residues dislodged or reoriented a peptide in the cleft, leading to the observed perturbations in binding, appears

Figure 12. A van der Waals surface representation of HLA-A2 showing the location of a single amino acid substitution of the H-2K^b molecule that can be recognized by a single cloned CTL (Ajitkumar *et al.*, 1988). (The surface representation was produced on an Evans and Sutherland PS 390 using FRODO and CPK rendering software modified by M. A. Saper.)

unlikely, as the particular residues in consideration do not face into the site as potential ligands to a peptide might, but instead are on the edge of the site facing away from the cleft and potentially toward the TCR (Ajitkumar *et al.*, 1988). A similar argument has been made in the recognition of the class II antigen I-A^k (Mathis, personal communication). The observation that the third hypervariable region of the two TCR chains is by far the most hypervariable has led to the suggestion that those regions of the TCR may primarily contact the foreign peptide, while the other hypervariable regions primarily contact the less variable histocompatibility antigen (Davis and Bjorkman, 1988).

4.2. Peptide Binding

An α-helical conformation has been suggested for peptides bound to histocompatibility antigens (Watts *et al.*, 1985; Allen *et al.*, 1987), and there is evidence that some T-cell epitopes could form amphipathic α-helices (Delisi and

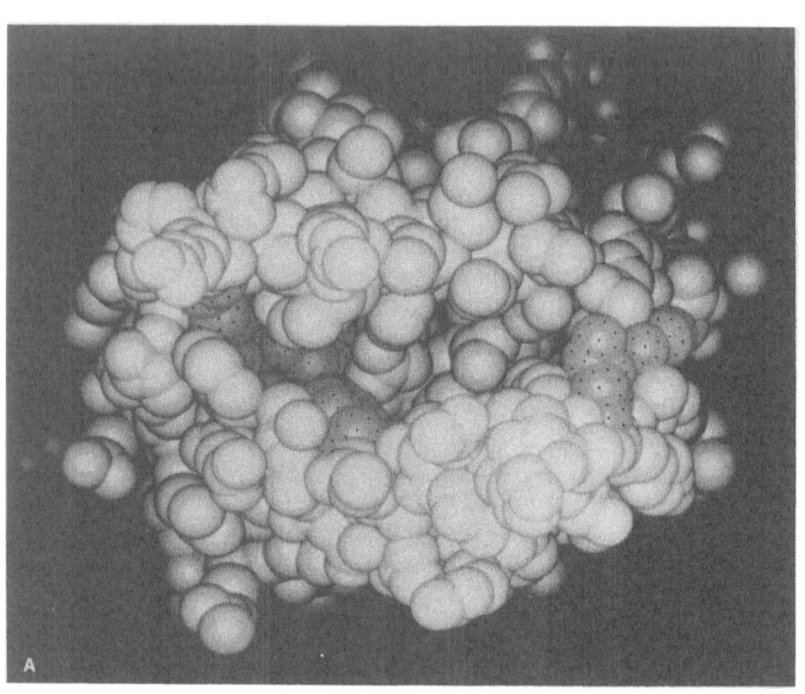

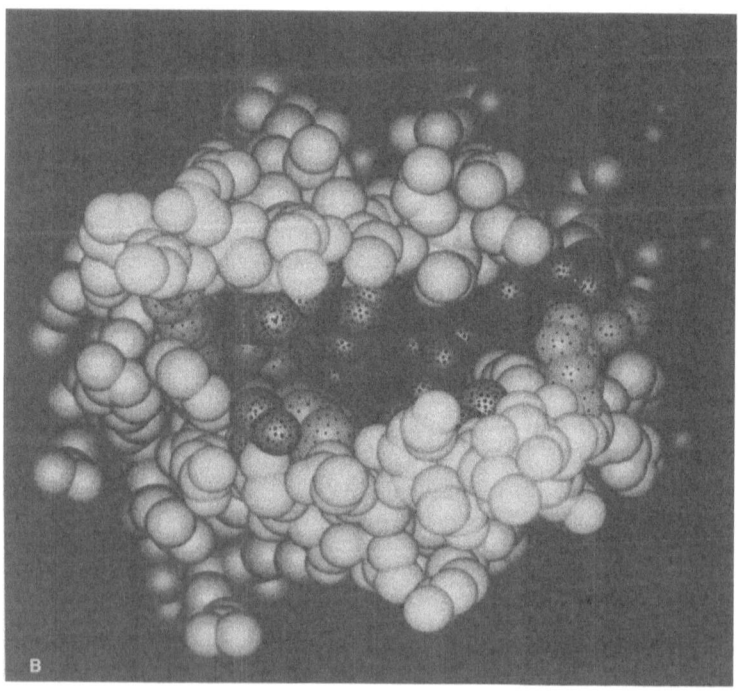

Berzofsky, 1985; Berkower *et al.*, 1986). However, other evidence argues equally convincingly against a helix and in favor of a more extended conformation (Sette *et al.*, 1987; Fox *et al.*, 1987).

The putative peptide binding site of class I histocompatibility antigens has a cluster of four tyrosines, 7, 59, 159, 171, at one end, which are conserved in all known class I sequences, and a conserved lysine (146), threonine (143), and tyrosine (84) at the other end (Fig. 13a). Most of the residues between the two ends and filling the bottom and sides of the site are the most polymorphic ones in human and mouse (Fig. 13b). The composition of the site includes many polar and charged residues, arguing against a single amphipathic helix as a universal motif. There are prominent ridges and grooves on the helical edges of the site which may provide crevices for peptide side chains, but no single peptide-helix-to-HLA-helix packing mode has yet been deduced. A number of pockets in the bottom and sides of the cleft suggest locations for peptide side chains. *A priori* it seems possible that polar atoms in the site may recognize main-chain atoms of a peptide, while some of the peptide side chains occupy the pockets. X-ray data are currently being collected to 2.2 Å resolution, in the hope that further refinement of the structure may help to define the mode of peptide binding.

4.3. The Structure of HLA-A28

The three-dimensional structure of HLA-A28, a second polymorphic class I molecule, has been determined by X-ray diffraction and crystallographically refined to 2.6 Å resolution (Garrett, Saper, Bjorkman, Strominger, and Wiley, unpublished). The structure was determined and refined ($R = 17\%$) by T. Garrett from crystals originally discovered by P. Bjorkman (Bjorkman *et al.*, 1986), using a starting model HLA-A2 coordinates that had been partially refined at 3.0 Å by M. Saper. HLA-A28 differs from HLA-A2 in 13 residues, 10 of which are polymorphic positions in human class I molecules. It may, therefore, offer an indication of the effect of polymorphic substitutions on the structure of class I antigens. All but three of the changes are in amino acids that face into the cleft, positions that are likely to be ligands to peptide antigens. One of those not in the cleft, 62, a change in charge from glycine in A2 and Arginine in A28, is on the long α-helix of the α1 domain at the end of the cleft in a position to be recog-

---

Figure 13. (A) Constant residue in the HLA-A2 foreign antigen binding site. Four tyrosines (59, 171, 7, 84) and methionine 5 on the left of the cleft and lysine 146, tyrosine 84, and threonine 143 on the right of the cleft are conserved in all amino acid sequences of class I molecules (Bjorkman *et al.*, 1987b). (B) Polymorphic residues in the HLA-A2 foreign antigen binding site. A number of the most polymorphic positions in human and mouse class I histocompatibility antigens are located in the center and along one edge of the prominent cleft in the HLA-A2 molecule. (Surfaces represented as in Fig. 12.)

nized directly by a T-cell receptor. Of those facing into the cleft there appear to be three main effects: (1) Four polar residues occupying one face of the long helix in the $\alpha 1$ domain have changed electrostatic charge (E63N, K66N, H70Q, H74D), substantially altering the character of the $\alpha 1$ side of the cleft. (2) A hydrophobic ridge is created in HLA-A28 by the protrusion of a tryptophan (156L in A2) across the bottom of the center of the cleft toward gln 70 and the alteration of the position of tyrosine 99 by the change of phenylalanine 9 to a tyrosine. (3) Four differences in the bottom of the site (V95I, R97M, H114R, Y116D) open a large pocket covered by R97 and filled by Y116 in HLA-A2. At one side of the pocket an aspartic acid (116) appears salt-linked to arginine 114, while on the other side two unneutralized aspartic acids, 74 and 77, line the edge of the pocket.

Overall, the effect of the 13 amino acid substitutions is to leave the structure largely unchanged. As expected, HLA-A28 has the same three-dimensional structure and large cleft as HLA-A2. However, the character of the cleft is changed dramatically in both electrostatic charge and in the composition and size of specific "holes" or pockets which may be binding sites for peptide side chains. The most obvious of such differences is the presence of a deep pocket with a negatively charged residue near the end. This pocket looks like an ideal location for a large, positively charged peptide side chain. The "extra electron density" in the cleft of the HLA-A28 structure includes a prominent feature that extends down into this pocket.

The existence in nature of an HLA polymorph HLA-Aw69 (Holmes and Parham, 1985) that is a hybrid of HLA-A2 and HLA-A28, with $\alpha 1$ identical to HLA-A28 and $\alpha 2$ identical to HLA-A2, offers us the possibility of seeing how the detailed pockets in the site can be affected by the two domains, once binding studies of peptide exclusive to the three sites A2, A28, and Aw69 can be accomplished.

4.4. Modeling Class II Histocompatibility Antigens

A hypothetical model of the foreign antigen binding site of class II histocompatibility antigens has been created (Brown *et al.*, 1988), based on the alignment of conserved features of the amino acid sequences and on the alignment of highly polymorphic positions in all of the known amino acid sequences of class I and class II histocompatibility antigens.

Acknowledgments. It is a pleasure to acknowledge my co-workers and collaborators in the work described herein. All of the work on influenza virus was done in close collaboration with Dr. John Skehel, National Institute of Medical Research, Mill Hill, London. My co-workers were (and are) Dr. William Weis, Nicholas Sauter, Dr. Mark Bednarski (in Professor George Whiteside's laboratory), Dr. Stephen Cusack, Jerry Brown, Dr. John Hanson, Dr. John Pfeifer, Lucy Godley, Dr. Stephen Wharton (Mill Hill), and Beth Wurzburg. Dr. Randy Kaufmann and Greg Shaw of Genetics Institute were instrumental in initiating the recombinant DNA work. The work on human histocompatibility antigens was done in collaboration with Professor Jack Strominger of Harvard University. My co-workers were Dr. Pamela Bjorkman, Dr. Mark Saper, Dr. Tom Garrett, Dr. Boudjema Samraoui, Dr. William Bennett, Jerry Brown, Dr. Theodore Jardetzky, and Dr. Joan Gorga (Professor Strominger's laboratory). It is a pleasure to acknowledge a collaboration with Dr. Stanley Nathenson and his co-workers (Albert Einstein Medical School) (see Ajitkumar *et al.*, 1988). I also thank David Stevens (Mill Hill) and Anastasia Haykov for excellent technical assistance, and Don Giard (MIT Cell Culture Facility), Drs. Dean Mann (NIH), D. Michael Strong, and James Woody (U.S. Naval Research Unit) for providing cells that made the histocompatibility antigen work possible. The work was supported by the NIH and NSF and by the Howard Hughes Medical Institute.

References

Ajitkumar, P., Geier, S. S., Kesari, K. V., Borriello, F., Nakagawa, M., Bluestone, J. A., Saper, M. A., Wiley, D. C., and Nathenson, S. G., 1988, Evidence that multiple residues on both the a-helices of the class I MHC molecule are simultaneously recognized by the T cell receptor, *Cell* **54**:47–56.

Allen, P. M., Matsueda, G. R., Evans, R. J., Dunbar, J. B., Jr., Marshall, G. R., and Unanue, E. R., 1987, Identification of the T-cell and Ia contact residues of a T-cell antigenic epitope, *Nature* **327**:713–715.

Amit, A. G., Mariuzza, R. A., Phillips, S. E. V., and Poljak, R. J., 1986, Three-dimensional structure of an antigen-antibody complex at 2.8 Å resolution, *Science* **233**:747–753.

Bach, F. H., and van Rood, J. J., 1976, The major histocompatibility complex—Genetics and biology, *N. Engl. J. Med.* **295**:806–813.

Beddell, C. R., Goodford, P. J., Kneen, G., White, R. D., Wilkinson, S., and Wootton, R., 1984, Substituted benzaldehydes designed to increase the oxygen affinity of human haemoglobin and inhibit the sickling of sickle erythrocytes, *Br. J. Pharm.* **82**:397–407.

Bergelson, L. D., Bukrinskaya, A. G. Prokazova, N. V., Shaposhnikova, G. I., Kocharov, S. L., Shevchenko, V. P., Kornilaeva, G. V., and Fomina-Ageeva, E. V., 1982, Role of gangliosides in reception of influenza virus, *Eur. J. Biochem.* **128**:467–474.

Berkower, I., Buckenmeyer, G. K., and Berzofsky, J. A., 1986, Molecular mapping of a histocompatibility-restricted immunodominant T cell epitope with synthetic and natural peptides: Implications for T cell antigenic structure, *J. Immun.* **136**:2498–2503.

Biddison, W. E., Ward, F. E., Shearer, G. M., and Shaw, S., 1980, The self determinants recognized by human virus-immune T cells can be distinguished from the serologically defined HLA antigens, *J. Immun.* **124**:548–552.

Bjorkman, P. J., Strominger, J. L., and Wiley, D. C., 1986, Crystallization and X-ray diffraction studies on the histocompatibility antigens HLA-A2 and HLA-A28 from human cell membranes, *J. Mol. Biol.* **186**:205–210.

Bjorkman, P. J., Saper, M. A., Samraoui, B., Bennett, W. S., Strominger, J. E., and Wiley, D. C.,

1987a, Structure of the human class I histocompatibility antigen, HLA-A2, *Nature* **329**:506–512a.

Bjorkman, P. J., Saper, M. A., Samraoui, B., Bennett, W. S., Strominger, J. E., and Wiley, D. C., 1987b, The foreign antigen binding site and T cell recognition regions of class I histocompatibility antigens, *Nature* **329**:512–518b.

Brand, C. M., and Skehel, J. J., 1972, Crystalline antigen from the influenza virus envelope, *Nature* **238**:145–147.

Bricogne, G., 1976, Methods and programs for direct-space exploitation of geometric redundancies, *Acta Crystallogr.* **A32**:832–847.

Brown, J. H., Jardetzky, T., Saper, M. A., Samraoui, B., Bjorkman, P. J., and Wiley, D. C., 1988, A hypothetical model of the foreign antigen binding site of class II histocompatibility molecules, *Nature* **332**:845–850.

Colman, P. M., 1988, Structure of antibody-antigen complexes: Implications for immune recognition, *Adv. Immunol.* **43**:99–132.

Colman, P. M., Varghese, J. N., and Laver, W. G., 1983, Structure of the influenza virus glycoprotein antigen neuraminidase at 29 Å resolution, *Nature* **303**:41–47.

Colman, P. M., Laver, W. G., Varghese, J. N., Baker, A. T., Tulloch, P. A., Air, G. M., and Webster, R. G., 1987, Three-dimensional structure of a complex of antibody with influenza virus neuraminidase, *Nature* **326**:358–363.

Cresswell, P., Turner, M. J., and Strominger, J. L., 1973, Papain-solubilized HL-A antigens from cultured human lymphocytes contain two peptide fragments, *Proc. Natl. Acad. Sci. USA* **70**:1603–1607.

Daniels, R. S., Douglas, A. R., Skehel, J. J., and Wiley, D. C., 1983, Analyses of the antigenicity of influenza haemagglutinin at the pH optimum for virus-mediated membrane fusion, *J. Gen. Virol.* **64**:1657–62.

Daniels, R. S., Downie, J. C., Hay, A. J., Knossow, M., Skehel, J. J., Wang, M. L., and Wiley, D. C., 1985, Fusion mutants of the influenza virus hemagglutinin glycoprotein, *Cell* **40**:401–439.

Daniels, R. S., Jeffries, S., Yates, P., Schild, G. C., Rogers, G. N., Paulson, J. C., Wharton, S. A., Douglas, A. R. Skehel, J. J., and Wiley, D. C., 1987, The receptor-binding and membrane-fusion properties of influenza virus variants selected using anti-haemagglutinin monoclonal antibodies, *EMBO J.* **6**:1459–1465.

Davis, M. M., and Bjorkman, P. J., 1988, T-cell antigen receptor genes and T-cell recognition, *Nature* **334**:395–402.

Delisi, C., and Berzofsky, J. A., 1985, T-cell antigenic sites tend to be amphipathic structures, *Proc. Natl. Acad. Sci. USA* **82**:7048–7052.

Doms, R. W., and Helenius, A., 1986, Quaternary structure of influenza virus hemagglutinin after acid treatment, *J. Virol.* **60**:833–839.

Doms, R. W., Gething, M. J., Henneberry, J., White, J., and Helenius, A. J., 1986, Variant influenza virus hemagglutinin that induces fusion at elevated pH, *Virology* **57**:603–13.

Fox, B. S., Chen, C., Fraga, E., French, C., Singh, B., and Schwartz, R. H., 1987, Functional distinct agretopic and epitopic sites, *J. Immun.* **139**:1578–1588.

Gething, M. J., White, J. M., and Waterfield, M. D., 1978, Purification of the fusion protein of Sendai virus: Analysis of the NH2-terminal sequence generated during precursor activation, *Proc. Natl. Acad. Sci. USA* **75**:2737–2740.

Gottschalk, A., 1959, *The Viruses*, Vol. 3, (F. M. Burnet and W. M. Stanley, eds), pp. 51–61, Academic Press, New York.

Grey, H. M., Kubo, R. T., Colon, S. M., Poulik, M. D., Cresswell, P., Springer, T., Turner, M., and Strominger, J. L., 1973, The small subunit of HL-A antigens is β2-microglobulin, *J. Exp. Med.* **138**:1608–1612.

Hedrick, S. M., Nielsen, E. A., Kavaler, J., Cohen, D., and Davis, M. M., 1984, Sequence relationships between putative T-cell receptor polypeptides and immunoglobulins, *Nature* **308:**153–158.

Holmes, N., and Parham, P., 1985, Exon shuffling in vivo can generate novel HLA class I molecules, *EMBO J.* **4:**2849–2854.

Jackson, D. C., and Nestorowicz, A., 1985, Antigenic determinants of influenza virus hemagglutinin, *Virology* **145:**72–83.

Kaufman, J. F., Auffray, C., Korman, A. J., Shackelford, D. A., and Strominger, J. C., 1984, The class II molecules of the human and murine major histocompatibility complex, *Cell* **36:**1–13.

Kronis, K. A., and Carver, J. P., 1982, Specificity of isolectins of wheat germ agglutinin for sialyloligosaccharides: A 360-MHz proton nuclear magnetic resonance binding study, *Biochemistry* **21:**3050–3057.

Lear, J. D., and Degrado, W. F., 1987, Membrane binding and conformational properties of peptides representing the NH_2 terminus of influenza HA-2, *J. Biol. Chem.* **262:**6500–6505.

Lesk, A. M., and Hardman, K. D., 1982, Computer-generated schematic diagrams of protein structures, *Science* **216:**539–540.

Lopez de Castro, J., Barbosa, J. A., Krangel, M. S., Biro, A., and Strominger, J. L., 1985, Structural analysis of the functional sites of Class I HLA antigens, *Immunol. Rev.* **85:**149–168.

Nathenson, S. G., Geliebler, J., Pfaffenbach, G. M., and Zeff, R. A., 1986, Murine major histocompatibility complex Class-I mutants: Molecular analysis and structure-function implications, *Annu. Rev. Immun.* **4:**471–502.

Nestorowicz, A., Laver, W. G., and Jackson, D. C., 1985, Antigenic determinants of influenza virus haemagglutinin, *J. Gen. Virol.* **66:**1687–1695.

Orr, H. T., Lancet, D., Robb, R. J., Lopez de Castro, J. A., and Strominger, J. L., 1979a, The heavy chain of human histocompatibility antigen HLA-B7 contains an immunoglobulin-like region, *Nature* **282:**266–270.

Orr, H. T., Lopez de Castro, J. A., Parham, P., Ploegh, H. L., and Strominger, J. L., 1979b, Comparison of amino acid sequences of two human histocompatibility antigens, HLA-A2 and HLA-B7: Location of putative alloantigenic sites, *Proc. Natl. Acad. Sci. USA* **76:**4395–4399.

Orr, H. T., Lopez de Castro, J. A., Lancet, D., and Strominger, J. L., 1979c, Complete amino acid sequence of a papain-solubilized human histocompatibility antigen, HLA-B7. 2. Sequence determination and search for homologies, *Biochemistry* **18:**5711–5719.

Paulson, J. C., 1985, Interactions of animal viruses with cell surface receptors, in: *The Receptors,* Vol. 2 (P. M. Conn, ed.), pp. 131–219, Academic Press, Orlando, FL.

Paulson, J. C., Sadler, J. E., and Hill, R. L., 1979, Restoration of specific Myxovirus receptors to Asialoerythrocytes by incorporation of sialic acid with pure sialyltransferases, *J. Biol. Chem,* **254:**2120–2124.

Peterson, P. A., Cunningham, B. A., Berggard, I., and Edelman, G. M., 1972, β_2-Microglobulin— A free immunoglobulin domain, *Proc. Natl. Acad. Sci. USA* **69:**1697–1701.

Peterson, P. A., Rask, L., and Lindblom, J. B., 1974, Highly purified papain-solubilized HL-A antigens contain β_2-Microglobulin, *Proc. Natl. Acad. Sci., USA* **71:**34–39.

Pritchett, T. J., 1987, Ph.D. thesis, Univ. of Calif., Los Angeles.

Richardson, C. D., Scheid, A., and Choppin, P. W., 1980, Specific inhibition of Paramyxovirus and Myxovirus replication by oligopeptides with amino acid sequences similar to those at the N-termini of the F_1 and HA_2 viral polypeptides, *Virology* **105:**205–222.

Rogers, G. N., and Paulson, J. C., 1982, Survey of receptor specificities of human and animal influenza viruses, *1982 Fed. Proc.* **41:**5880.

Rogers, G. N., and Paulson, J. C., 1983, Receptor determinants of human and animal influenza virus isolates: Differences in receptor specificity of the H3 hemagglutinin based on species of origin, *Virology* **127:**361–373.

Rogers, G. N., Pritchett, T., and Paulson, J. C., 1983a, Correlation of receptor specificity and glycoprotein inhibitor sensitivity for influenza H3 hemagglutinins, *1983 Fed. Proc.* **42**:2181.

Rogers, G. N., Paulson, J. C., Daniels, R. S., Skehel, J. J., Wilson, I. A., and Wiley, D. C., 1983b, Single amino acid substitutions in influenza haemagglutinin change receptor binding specificity, *Nature* **304**:76–79.

Rossmann, M. G., Arnold, E., Erickson, J. W., Frankenberger, E. A., Griffith, J. P., Hecht, H-J., Johnson, J. E., Kamer, G., Luo, M., Mosser, A. G., Rueckert, R. R., Sherry, B., Vriend, G., 1986, Structure of a human common cold virus and functional relationship to other picornaviruses, *Nature* **317**:145–153.

Rott, R., Orlich, M., Klenk, H.-D., Wang, M. L., Skehel, J. J., and Wiley, D. C., 1985, Studies on the adaptation of influenza viruses to MDCK cells, *EMBO J.* **3**:3329–3332.

Ruigrok, R. W. H., Aitken, A., Calder, L. J., Martin, S. R. Skehel, J. J., Wharton, S. A., Weis, W., and Wiley, D. C., 1988, Studies on the structure of the influenza virus haemagglutinin at the pH of membrane fusion, *J. Gen. Virol.* **69**:2785–2795.

Saito, H., Kranz, D. M., Takagaki, Y., Hayday, A. C., Eisen, H. N., and Tonegawa, S., 1984, A third rearranged and expressed gene in a clone of cytotoxic T lymphocytes, *Nature* **312**:36–40.

Sauter, N. K., Bednarski, M. D., Wurzburg, B. A., Hanson, J. E., Whitesides, G. M., Skehel, J. J., and Wiley, D. C., 1989, Hemagglutinins from two influenza virus variants bind to sialic acid derivatives with millimolar dissociation constants: a 500 MHz proton nuclear magnetic resonance study, *Biochemistry* **28**:8388–8396.

Scheid, A., Graves, M., Silver, S., and Choppin, P. W., 1978, in: *Negative Strand Viruses and the Host Cell,* (B. W. J. Mahy and R. D. Barry, eds.), pp. 181–193, Academic Press, London.

Schwartz, R. H., 1983, The role of gene products of the major histocompatibility complex in T cell activation and cellular interactions, in: *Fundamental Immunology* (W. E. Paul, ed.), pp. 379–399, Raven Press, New York.

Sette, A., Buus, S., Colon, S., Smith, J. A., Miles, C., and Grey, H. M., 1987, Structural characteristics of an antigen required for its interaction with Ia and recognition by T cells, *Nature* **328**:395–399.

Sheriff, S., Silverton, E. W., Padlan, E. A., Cohen, G. H., Smith-Gill, S. J., Finzel, B. C., and Davies, D. R., 1987, Three-dimensional structure of an antibody–antigen complex, *Proc. Natl. Acad. Sci. USA* **84**:8075–8079.

Skehel, J. J., and Waterfield, M. D., 1975, Studies on the primary structure of the influenza virus haemagglutinin, *Proc. Natl. Acad. Sci. USA* **72**:93–97.

Skehel, J. J., Bayley, P. M., Brown, E. B., Martin, S. R., Waterfield, M. D., White, J. M., Wilson, I. A., and Wiley, D. C., 1982, Changes in the conformation of influenza virus hemagglutinin at the pH optimum of virus-mediated fusion, *Proc. Natl. Acad. Sci. USA* **79**:968–972.

Skehel, J. J., Stevens, D. J., Daniels, R. S., Douglas, A. R., Knossow, M., Wilson, I. A., and Wiley, D. C., 1984, A carbohydrate side chain on hemagglutinins of Hong Kong influenza viruses inhibits recognition by a monoclonal antibody, *Proc. Natl. Acad. Sci. USA* **81**:1779–1783.

Smithies, O., and Poulik, M. D., 1982, Initiation of protein synthesis at an unusual position in an immunoglobulin gene? *Science* **175**:187–189.

Suzuki, U., Matsunaga, M., and Masumoto, M., 1985, N-Acetylneuraminyllactosylceramide, $G_{M3-NeuAc}$, a new influenza A virus receptor which mediates the adsorption-fusion process of viral infection, *J. Biol. Chem.* **260**:1362–1365.

Townsend, A., 1987, Recognition of influenza virus proteins by cytotoxic T lymphocytes, *Immunol. Res.* **6**:80–100.

Tragardh, L., Rask, L., Wiman, K., Fohlman, J., and Peterson, P. A., 1979, Amino acid sequence of an immunoglobulin-like HLA antigen heavy chain domain, *Proc. Natl. Acad. Sci. USA* **76**:5839–5842.

Watts, T. H., Gariepy, J., Schoonik, G. K., and McConnell, H. M., 1985, T-cell activation by peptide antigen: Effect of peptide sequence and method of antigen personation, *Proc. Natl. Acad. Sci. USA* **82**:5480–5484.

Webster, R. G., Brown, L. E., and Jackson, D. C., 1983, Changes in the antigenicity of the hemagglutinin molecule of H3 influenza virus at acidic pH, *Virology* **126**:587–599.

Weis, W., Brown, J. A., Cusack, S., Paulson, J. C., Skehel, J. J., and Wiley, D. C., 1988, Structure of the influenza virus haemagglutinin complexed with its receptor, sialic acid, *Nature* **333**:426–431.

Wharton, S. A., Martin, S. R., Ruigrok, R. W. H., Skehel, J. J., and Wiley, D. C., 1988, Membrane fusion by peptide analogues of influenza virus haemagglutinin, *J. Gen. Virol.* **69**:1847–1857.

White, J., and Wilson, I., 1988, Anti-peptide antibodies detect steps in a protein conformational change: Low-pH activation of the influenza virus hemagglutinin, *J. Cell. Biol.* **105**:2887–2896.

White, J., Kielian, M., and Helenius, A., 1983, Membrane fusion proteins of enveloped animal viruses, *Q. Rev. Biophys.* **16**:151–195.

Wiley, D. C., and Skehel, J. J., 1987, The structure and function of the hemagglutinin membrane glycoprotein of influenza virus, *Annu. Rev. Biochem.* **56**:365–94.

Wiley, D. C., Wilson, I. A., and Skehel, J. J., 1981, Structural identification of the antibody-binding sites of Hong Kong influenza haemagglutinin and their involvement in antigenic variation, *Nature* **289**:373–378.

Yanagi, Y., Yoshikai, Y., Leggett, K., Clark, S. P., Aleksander, I., and Mak, T. W., 1984, A human T cell-specific cDNA clone encodes a protein having extensive homology to immunoglobulin chains, *Nature* **308**:145–149.

Yewdell, J. W., Gerhard, W., and Bachi, T., 1983, Monoclonal anti-hemagglutinin antibodies detect irreversible antigenic alterations that coincide with the acid activation of influenza virus A/PR/834-mediated hemolysis, *J. Virol.* **48**:239–248.

Ziegler, H. K., and Unanue, E. R., 1982, Decrease in macrophage antigen catabolism caused by ammonia and chloroquin is associated with inhibition of antigen presentation to T cells, *Proc. Natl. Acad. Sci. USA* **79**:175–178.

Index